# State-of-the-Art Sensors Technology in Spain 2017

## Volume 2

Special Issue Editor

**Gonzalo Pajares Martinsanz**

MDPI • Basel • Beijing • Wuhan • Barcelona • Belgrade

**MDPI**

Special Issue Editor
Gonzalo Pajares Martinsanz
Department Software Engineering and Artificial Intelligence
University Complutense of Madrid
Spain

*Editorial Office*
MDPI
St. Alban-Anlage 66
Basel, Switzerland

This edition is a reprint of the Special Issue published online in the open access journal *Sensors* (ISSN 1424-8220) from 2017–2018 (available at: http://www.mdpi.com/journal/sensors/special_issues/SA_sensors_technology_spain_2017).

For citation purposes, cite each article independently as indicated on the article page online and as indicated below:

Lastname, F.M.; Lastname, F.M. Article title. *Journal Name* **Year**, *Article number, page range.*

**First Edition 2018**

Volume 2
ISBN 978-3-03842-959-3 (Pbk)
ISBN 978-3-03842-960-9 (PDF)

Volume 1–2
ISBN 978-3-03842-961-6 (Pbk)
ISBN 978-3-03842-962-3 (PDF)

# Table of Contents

# About the Special Issue Editor

**Gonzalo Pajares Martinsanz** received the Ph.D. degree in Physics from Distance University from Spain in 1995, discussing a thesis on stereovision. Since 1988 he worked at Indra in critical real-time software development. He also was working at Indra Space and INTA in advanced image processing for Remote Sensing. He joined the University Complutense of Madrid in 1995 on the Faculty of Informatics (Computer Science) at the Department of Software Engineering and Artificial intelligence. His current research interests include computer and machine visual perception, artificial intelligence, decision making, robotics and simulation with a lot of publications, including several books, on these topics. He is director of the ISCAR Research Group. He is Associated Editor in the indexed journal of Remote Sensing and serves as a member of the Editorial Board in the following indexed journals: Sensors, EURASIP Journal of Image and Video processing, Pattern Analysis and Applications. He is the Editor-in-Chief of Journal of Imaging.

# Preface to "State-of-the-Art Sensors Technology in Spain 2017"

Since 2009 six special issues has been published on sensors and technologies in Spain, where researchers present their successful progress. Forty high quality works have been collected and reproduced on this book, demonstrating significant achievements.

They are self-contained works addressing different sensor-based technologies, procedures and applications on several areas with a wide range of devices and useful developments. Readers will also find an excellent source of resources when necessary in the development of his research, teaching or industrial activity.

Although the book is focused on sensors and technologies in Spain, it describes worldwide advanced developments and references on the covered topics useful on the contexts addressed. Some works have been or come from international collaborations.

Our society is demanding new technologies for data acquisition, processing and transmission for immediate actions or knowledge, with important contributions to the welfare or actuations when required.

The international scientific and industrial communities worldwide are also indirect beneficiary. Indeed, the book provides insights and solutions for the different problems covered. Also, it lays the foundation for future advances toward new challenges. In this regard, new sensors contribute to the solution of existing problems, but also conversely, where the need to resolve certain problems demands the development of new technologies or procedures. This book is in this way.

We are grateful to all the persons involved in the edition of this book. Without the invaluable contribution of authors together with the excellent help of reviewers, this book would not have seen the light. More than 200 authors have contributed to this book.

Thanks to Sensors journal and the team of people involved in the edition and production of this book for the support and encouragement.

**Gonzalo Pajares Martinsanz**
*Special Issue Editor*

*sensors*

MDPI

*Article*

# Smart CEI Moncloa: An IoT-based Platform for People Flow and Environmental Monitoring on a Smart University Campus

Manuel Alvarez-Campana * , Gregorio López , Enrique Vázquez, Víctor A. Villagrá

Departamento de Ingeniería de Sistemas Telemáticos, Universidad Politécnica de Madrid, Avenida Complutense 30, 28040 Madrid, Spain; gregorio.lopez.lopez@upm.es (G.L.); enrique@dit.upm.es (E.V.); villagra@dit.upm.es (V.A.V.); berrocal@dit.upm.es (J.B.)
* Correspondence: mac@dit.upm.es; Tel.: +34-915-495-700

Received: 8 November 2017; Accepted: 6 December 2017; Published: 8 December 2017

**Abstract:** Internet of Things platforms for Smart Cities are technologically complex and deploying them at large scale involves high costs and risks. Therefore, pilot schemes that allow validating proof of concepts, experimenting with different technologies and services, and fine-tuning them before migrating them to actual scenarios, are especially important in this context. The IoT platform deployed across the engineering schools of the Universidad Politécnica de Madrid in the Moncloa Campus of International Excellence represents a good example of a test bench for experimentation with Smart City services. This paper presents the main features of this platform, putting special emphasis on the technological challenges faced and on the solutions adopted, as well as on the functionality, services and potential that the platform offers.

**Keywords:** Arduino; environmental monitoring; IoT; MQTT; people flow monitoring; Raspberry Pi; sensor networks; smart cities; visual analytics; Wi-Fi

## 1. Introduction

In the context of a globally accelerated urbanism process, the sustainable development of cities has become one of the major challenges of society nowadays. As a result, during the last years many initiatives and projects have been launched under the new paradigm of the so-called Smart Cities. The main goal of Smart Cities is indeed to make cities a better place to live now and in the long run, place the citizens at the center. In order to achieve this goal, Smart Cities rely on the Internet of Things (IoT) paradigm, so Information and Communication Technology (ICT) is intensively applied to all the areas related to citizens' welfare, such as transport and mobility, healthcare, energy or the environment [1,2]. This intensive use of ICT encompasses from the massive deployment of all types of sensors and actuators and Machine-to-Machine (M2M) communications infrastructures, to the processing of the huge amount of gathered data to provide value-added services and applications, as Figure 1 illustrates.

In order to promote multidisciplinary research, development and innovation activities related to Smart Cities, in 2013 Universidad Politécnica de Madrid (UPM) launched the UPM City of the Future initiative [3]. One of the main projects within this initiative was the design, development and deployment of an IoT-based platform which allowed for the experimentation and evaluation of Smart City services in the Moncloa Campus of International Excellence (CEI Moncloa) [4].

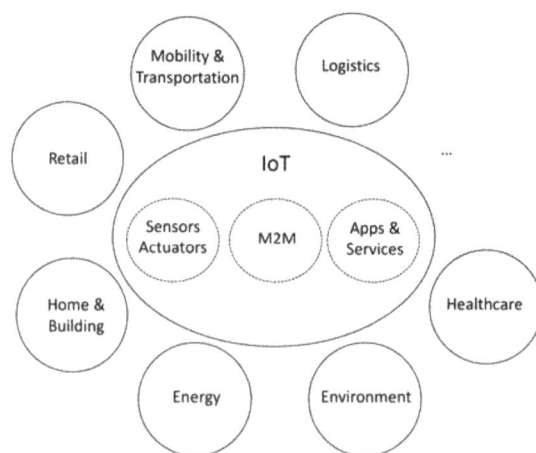

**Figure 1.** Overview of the core technologies and main application areas covered under the Smart City paradigm.

Figure 2 shows a map of the CEI Moncloa. This campus presents some specific features that make it especially appropriate as a test bench for Smart City services. Unlike other university campuses which are in the outskirts of cities, CEI Moncloa is integrated in the metropolitan area of Madrid, the capital city of Spain. The campus covers an area of 5.5 km² and comprises 144 buildings, including schools, research centers, student residences, and sport premises. The daily flow in the campus goes up to 120,000 people and the daily road traffic accounts for thousands of vehicles, most of which do not stay in the campus but just cross it. In addition, the campus has a significant public transport infrastructure, including two underground lines and 13 bus lines.

Being an eminently university area, the industrial and commercial activity on the campus is very low. As a result, this allows performing experiments that would be very difficult to put into practice in other areas of the city without causing major inconveniences to the citizens. Thus, it is possible to set up, test and fine tune smart parking or smart lighting pilot schemes before deploying them in operational environments. Likewise, the fact that it is not a residential neighborhood facilitates the deployment of the required infrastructure during the nights or the weekends.

The main goal of this paper is to present the IoT-based platform deployed in the CEI Moncloa campus, putting special emphasis on the main technological challenges that were faced and on the solutions that were adopted, as well as on the functionality, services and potential that the platform offers.

The paper is structured as follows: Section 2 first presents an overview of the system architecture and then describes the design and implementation details of each of the subsystems which compose the platform, comparing them with related work available in the state of the art when appropriate. Section 3 goes through some of the most relevant information provided as figures or graphs by the platform in order to facilitate visual analytics. Section 4 presents three selected use cases which illustrate the potential of the platform. Section 5 discusses some key issues of IoT platforms for Smart Cities and how the IoT-based platform Smart CEI Moncloa contributes to them. Finally, Section 6 summarizes the main contributions of the paper and draws conclusions.

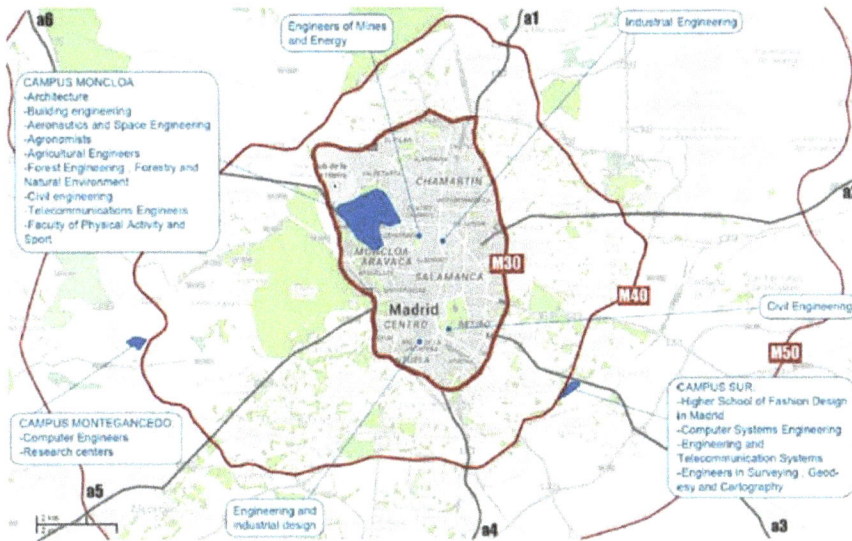

**Figure 2.** Location of Moncloa CEI Campus in Madrid.

## 2. Architecture and Subsystem Design

As it has been mentioned in Section 1, the main goal of the IoT-based platform Smart CEI Moncloa is to facilitate the experimentation in the area of Smart Cities, allowing us to carry out studies and prove concepts which can be later put into practice in actual cities with guarantees of success. Therefore, the initial idea was to deploy a platform which incorporated the ICT infrastructure required to provide a set of basic services, along with a sensor network which allowed demonstrating the capacity and scalability of the adopted solution. Thus, the platform needed to meet the following main requirements: (1) it should be low cost and scalable; and (2) it should be based on standardized and open solutions which enable incorporating new services to the platform seamlessly.

As a result, the platform Smart CEI Moncloa is based on an open architecture, aligned with the most recent standards within the scope of the Future Internet and Web engineering, which facilitates adding new functionalities and services by either UPM research groups or other companies or interested entities. Figure 3 shows an overview of the system architecture of the Smart CEI Moncloa, highlighting the sensing layer, the networking and data communications layer, and the applications or service layer, as defined by the IEEE P2413 [5,6] and other related work available in the literature [7]. This three layers are also aligned with the three domains defined in the ETSI M2M reference architecture (later transferred to OneM2M [8]), namely the device domain, the network domain, and the application domain [9].

The sensing layer is composed by a set of sensors (and eventually also actuators) that are geographically distributed across the campus. These devices are responsible for measuring the parameters of interest to the services that are delivered or the experiments that are carried out on top of the platform. Figure 3 shows some examples of sensor networks related to services that can potentially be tested and delivered in Smart CEI Moncloa, such as building energy monitoring, environmental monitoring, traffic and people flow monitoring, or control of city services (e.g., parking, lighting watering).

The networking and data communications layer represents the core IP-based communications infrastructure which enables the secure data exchange between the sensing layer and the application layer. This layer makes the most out of the wireless and wired communications infrastructure available in the campus, as it will be described in Section 2.3.

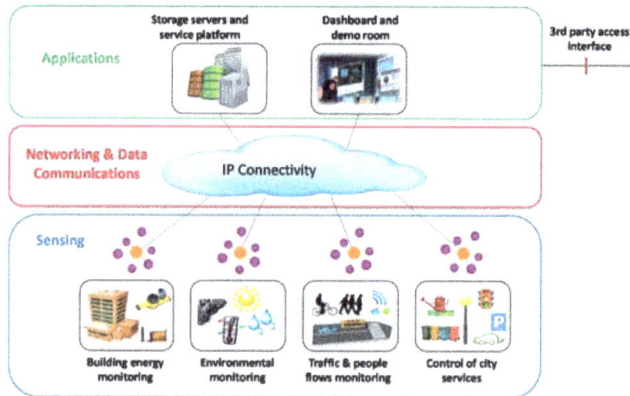

**Figure 3.** Overview of the system architecture and main layers of the Smart CEI Moncloa.

The application layer is where the service logic and algorithms reside. It is composed of the storage servers, the service platform, and the dashboard. The storage servers are responsible for the massive storage of the data generated by the sensors and for providing access to such data to the applications that require so. They provide Big Data capabilities for handling high volumes of data, as well as Open Data format support for third party applications. The service platform is responsible for executing the applications based on SOA (Service-Oriented Architecture). The dashboard is accessible via web, which allows ubiquitous access to the control and visualization data of the platform. In addition, there is a demo room located in the Telecommunications Engineering School and equipped with big screens which allows for showing the platform capabilities to students, researchers or other interested parties. Regarding the services, Smart CEI Moncloa includes a set of initial services with the idea of progressively including new ones through its life cycle. The initial pilot services are:

- People flow monitoring, which allows counting people and associated applications, such as movement pattern analysis, places with higher transit of people, stay time in relevant places, etc.
- Environmental monitoring, which allows analyzing several environmental parameters, such as temperature, humidity, light, noise level or air composition, both indoors and outdoors.

As Figure 3 shows, each of these pilot services have an associated sensor network. Next, these sensor networks are described in detail, together with the rest of subsystems of the platform.

### 2.1. Sensing Layer

#### 2.1.1. People Flow Monitoring Sensor Network

People flow monitoring represents a topic of especial interest with a wide range of applications, which spans from crowd monitoring in events such as demonstrations or concerts, to user monitoring in public transport infrastructures such as undergrounds or airports, or client monitoring in the retail sector.

There are different ways to approach this issue. For instance, it can be performed based on video analysis [10]. Although this is a quite appropriate approach to detect crowds, it is not so appropriate for a more finely-tuned detection. In addition, this kind of solutions required the deployment of video-cameras and the processing of the recorded video on the fly. The ubiquity of mobile phones invites to consider solutions based on them. Thus, there are proposals in the state of the art that use the Global Positioning System (GPS) receiver of the phone for this purpose [11]. However, this kind of solutions present problems in indoor environments. Passive monitoring of Wi-Fi devices represents a very low cost solution for people flow monitoring that does work both indoors and outdoors [12–18].

This approach is based on monitoring the Wi-Fi radio frames, which include the Medium Access Control (MAC) addresses that univocally identify each device worldwide. Due to the fact that nowadays almost everybody carries a smart phone that uses Wi-Fi, it allows obtaining a reasonable estimation of the number of people in the surroundings of the listening device. It is worthwhile to mention that the Wi-Fi devices do not only transmit when they are connected to an Access Point (AP), but also when they are not, sporadically sending probe frames in order to find out if there are known APs (e.g., home) in the vicinity.

The fact that the MAC address is unique allows for establishing time correlations to compute parameters of interest, such as the average stay time of a smart phone in a given place. By using different listening devices, it is also possible to establish spatial correlations to determine, for instance, the path followed by the smartphones in a mall or underground network. Nevertheless, the Wi-Fi tracking also has drawbacks, such as that it does not allow counting or tracking people who do not use smartphones or who turn the Wi-Fi interface off. Likewise, it may also count as smart phones other devices, which may distort the measurements.

Another drawback of the Wi-Fi tracking has to do with the privacy implications of gathering MAC addresses, since they may eventually allow identifying the owners of the devices. In order to avoid privacy concerns, the companies that use this technology usually go for trade-off solutions such as announcing that Wi-Fi signals are being monitored so that people who do not want to be tracked can turn the Wi-Fi interface off. It is also common to find applications that allow requesting not to be tracked through a web form (opt-out). Some other applications go further by applying a hash function to the MAC address. Nevertheless, this anonymity technique has issues since, although the relation with the MAC address gets blurred, what happen actually is that one ID (the MAC address) is replaced by other (the hash code).

As a result, during the last years several proposals have appeared to protect Wi-Fi communications by means of MAC address anonymization. In the beginning, these proposals appeared as apps for smart phones, but during the last years the smart phone manufacturers themselves have started including these techniques in the latest versions of their OS (Operating Systems) (e.g., iOS, Android, Windows).

Such MAC address anonymization techniques aim to avoid using the actual MAC address until the device gets connected to the Wi-Fi network (i.e., they use a fake MAC address in the probe frames). The solutions depend on the manufacturer and OS [19]. Thus, in the case of iOS, the solution involves sending locally administered MAC addresses in the probe frames, randomly selecting the 3 less significant bytes of the MAC address. In the case of Android, some manufacturers have decided to use random MAC addresses in the probe frames from the MAC address ranges assigned by the IEEE to them.

In the context of Smart CEI Moncloa, privacy issues have been considered carefully. Thus, the developed Wi-Fi tracking sensors apply a hash function to the MAC addresses so that they are never stored nor processed. In addition, the data are carried securely up to the platform servers where they are handled in an aggregate manner, instead of individually.

Furthermore, the software of the developed sensors has been modified in order to avoid that the aforementioned MAC anonymization mechanisms affect the obtained measurements. Thus, the Wi-Fi frames with locally administered MAC addresses or including special MAC address ranges are discarded, so these devices are not taken into account. Anyway, it is worthwhile to mention that this is not actually such a big deal in the case of the Smart CEI Moncloa platform, since most of the users are connected to the Eduroam free Wi-Fi access, so their smart phones ends up using their actual MAC address. A similar approach is used to mitigate the effect of the MAC address anonymization techniques in malls and other public infrastructures where free Wi-Fi access is offered if service conditions are accepted.

As Figure 4 shows, the Wi-Fi sensors have been developed based on a Raspberry Pi board [20] equipped with a USB Wi-Fi dongle configured in monitor mode. The cost of this solution is in the order of tens of euros, which represents a remarkable cost cut compared to other options. One of the basic software components is the wifimon program, which processes the headers of the IEEE 802.11 frames detected by the dongle on the fly. This program, inspired in the airmon-ng of the aircrack-ng suite [21], is developed in C and interacts with the Wi-Fi interface driver. By default, the program periodically scans all the Wi-Fi channels, both in the 2.4 GHz and in the 5 GHz bands, generating a report every 5 s including the detected devices, as well as a more detailed report on the activity of the detected devices every 15 min. Nevertheless, the channels to be monitored, the measurement time for each channel or the periodicity of the reports, can be modified through a configuration file.

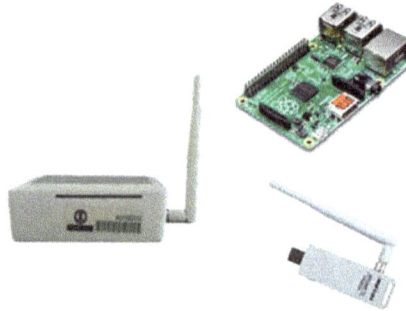

**Figure 4.** Listening device used in the people flow monitoring sensor network.

It should be noted that the channel scan implies that not all the channels are monitored at a time, but they are periodically sampled, so it may happen that a frame transmitted in a channel that is not being listened at a given a moment of time is not detected. This issue is not actually relevant since the objective of the application is to detect the presence of devices during a significant period of time and in this case the probability of being detected is very high. Anyhow, the developed software allows using several Wi-Fi dongles in parallel, although this option was finally discarded for cost reasons and taking into account the good results obtained using only one Wi-Fi dongle.

The Smart CEI Moncloa platform currently comprises 52 listening devices deployed across the 13 engineering schools, as shown in Figure 5.

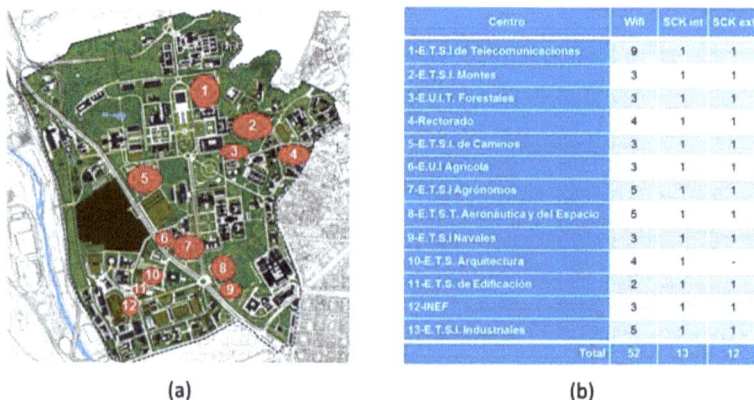

| Centro | Wifi | SCK int | SCK ext |
|---|---|---|---|
| 1-E.T.S.I de Telecomunicaciones | 9 | 1 | 1 |
| 2-E.T.S.I. Montes | 3 | 1 | 1 |
| 3-E.U.I.T. Forestales | 3 | 1 | 1 |
| 4-Rectorado | 4 | 1 | 1 |
| 5-E.T.S.I. de Caminos | 3 | 1 | 1 |
| 6-E.U.I Agrícola | 3 | 1 | 1 |
| 7-E.T.S.I Agrónomos | 5 | 1 | 1 |
| 8-E.T.S.T. Aeronáutica y del Espacio | 5 | 1 | 1 |
| 9-E.T.S.I Navales | 3 | 1 | 1 |
| 10-E.T.S. Arquitectura | 4 | 1 | - |
| 11-E.T.S. de Edificación | 2 | 1 | 1 |
| 12-INEF | 3 | 1 | 1 |
| 13-E.T.S.I. Industriales | 5 | 1 | 1 |
| Total | 52 | 13 | 12 |

(a)            (b)

**Figure 5.** Summary of the sensors deployed in the Smart CEI Moncloa (November 2017): (**a**) Location of the schools within the Campus; (**b**) number and type of sensors per site.

### 2.1.2. Environmental Monitoring Sensor Network

If global warming and climate change were a matter of opinion at some point, it is not the case anymore. Nowadays, they represent a reality without a doubt whatsoever. As a result, environmental monitoring has become an IoT application of capital importance for sustainable development and to improve quality of life, especially in the cities [22–27].

This is why it was selected as one of the initial pilot services in the Smart CEI Moncloa. The sensor network for this service is based on the Smart Citizen Kit (SCK) [28], shown in Figure 6. This device meets the main two requirements pointed out at the beginning of this section. First, it is a low cost solution (few hundreds euros) compared to expensive air quality monitoring stations (thousands of euros). Second, it is based on Arduino [29], and both the hardware and software are open source. As Figure 6 shows, a sensor shield is incorporated on top of the Arduino mother board including an array of sensors that allow measuring temperature, humidity, light, noise, CO and $NO_2$. In addition, the device can be used both indoors and outdoors, if protected with the appropriate box (Figure 6b).

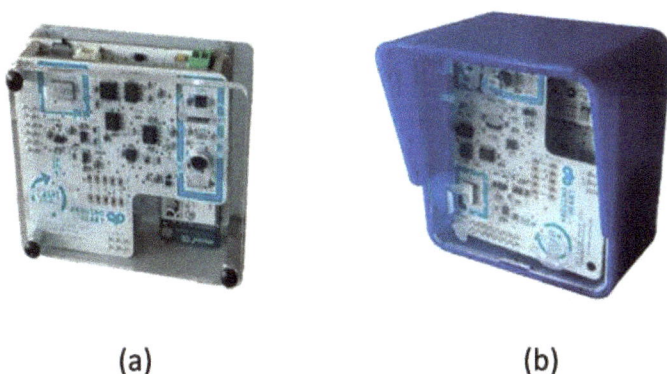

**(a)**                                                              **(b)**

**Figure 6.** Environmental monitoring device: (**a**) Indoors; (**b**) Outdoors.

Being open source, the SCK allowed being finely tuned to the specific requirements of the Smart CEI Moncloa. From the software point of view, the original firmware assumes the connection to the smartcitizen.me platform [28], where the raw data from the sensors (typically electrical resistance) are processed. Such a platform fosters the involvement of citizens by allowing them to register their own devices and offers public visualization of the location of the registered nodes together with their measurements (including both real time and historical data). In the case of the Smart CEI Moncloa, the data were required to be stored in the storage servers to facilitate studies and the development of applications especially focused on the campus. Nevertheless, this was solved by developing a specific piece of software at the backend that works as gateway, since this solution offers a great trade-off between compatibility and software development effort.

In addition, the firmware of the devices was modified in order to allow automatic restarting (e.g., if the communication with the servers fails) and remote restarting upon request from the dashboard. These changes improve dramatically the operation and maintenance of the platform, since they avoid having to physically access these devices (which is typically difficult) just to hard-reset them.

The CO and $NO_2$ measurements are taken by means of the MiCS-4514 MicroElectroMechanical Sensor (MEMS) [30]. This is a sensor typically used in automotive applications (e.g., to measure emissions from automobile exhausts). Hence, it is optimized for high concentration of gases, but it has attracted the interest of the IoT research community due to its very low cost. In order to obtain the concentration from the measured resistance ($R_S$), the graphs shown in Figure 7 are provided. As it can be seen, these graphs represent the normalized resistance versus the concentration. However, the datasheet only provides the maximum and minimum value for $R_0$ (sensing resistance in air), needed to compute the normalized resistance, whose actual value depends on specific ambient conditions (notably temperature and humidity).

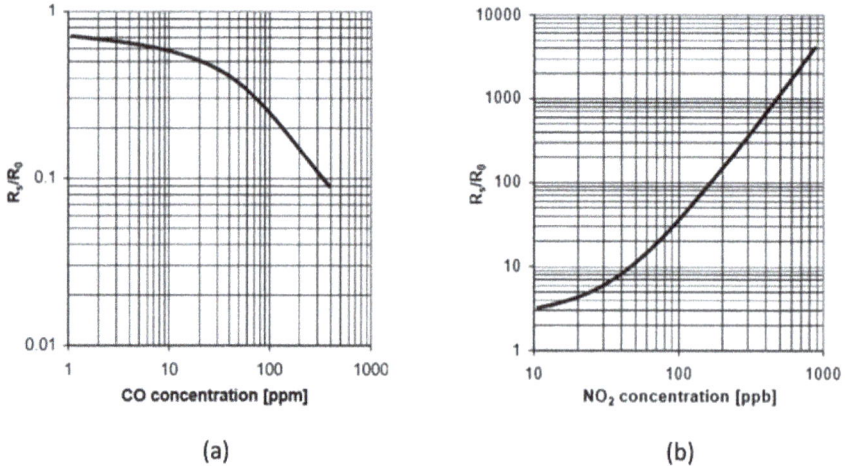

**Figure 7.** MiCS-4514 calibration curves: (**a**) CO concentration; (**b**) $NO_2$ concentration [30].

Several approaches have been explored within the context of the Smart CEI Moncloa to tackle this issue. First, the maximum and minimum $R_S$ were considered as $R_0$ for the CO and $NO_2$ sensors respectively. However, the analyses of the gathered data revealed that the sensors somehow get contaminated with time, as Figure 8 illustrates, so it was decided to automatically update it over periods of 30 days. Although the obtained calibration is not very precise, the measurements taken by these sensors are still valuable to detect remarkable variations or trends, as it will be illustrated in Sections 3 and 4.

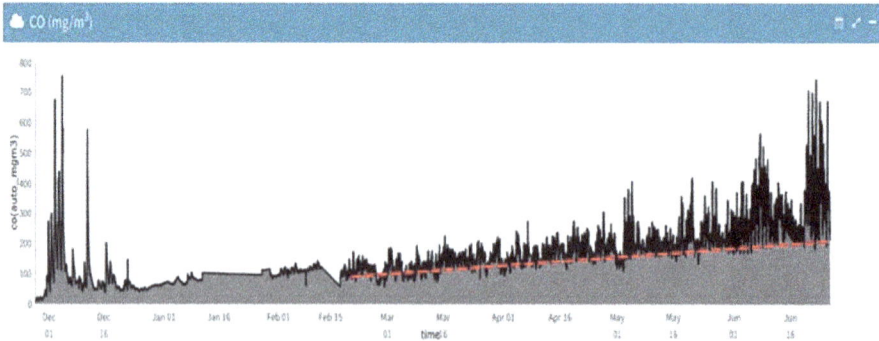

**Figure 8.** Evolution of CO concentration $(mg/m^3)$ from December 2015 to June 2016.

As Figure 5 shows, 25 SCKs have been installed throughout the campus, 12 of them being outdoors and 13 being indoors. The main goal of the outdoor sensors is to enable the realization of environmental studies focused on the campus. Hence, they have allowed verifying significant differences in temperature and humidity across the campus, which may be caused by the Manzanares river (see Figure 2). Likewise, they have also allowed validating that there are areas with higher levels of noise and pollution, which is related to higher road traffic.

The indoor sensors have been installed in the libraries, which are one of the common places that present higher activity in the campus. In this case, the sensors are being used to monitor temperature, humidity, noise level, and air quality. In addition, this information can be combined with the information coming from the people flow monitoring service to come up with value-added applications for the university community, such as the one presented in Section 4.3.

## 2.2. Networking and Data Communications Layer

M2M communications infrastructures represent the backbone of any IoT platform, enabling the secure massive data exchange between the sensors and actuators and the information systems. M2M communications architectures can be classified into two main categories: monolithic and hierarchical. Both approaches present complementary pros and cons [31]. Hierarchical architectures are more flexible, versatile and scalable than monolithic ones, but it is also more complex to manage and secure them, since they comprise several network segments (and so several intermediate nodes) and combine different communications technologies, thus the attack surface being larger.

As it has just been mentioned, on the one side, the hierarchical architectures involve heterogeneous infrastructures which combine short range and moderate data rate technologies (e.g., WPAN technologies, such as Bluetooth or Zigbee, or WLAN technologies, such as Wi-Fi) with long range high data rate technologies (both wireless, such as cellular technologies, and wired, based on twisted pair or fiber optics). On the other side, monolithic architectures generally rely on a single technology to connect the sensors and actuators and the information systems. In this case, it is quite common that the communications infrastructure is owned by a telecom operator, which is responsible for its management and maintenance, at the expense of increasing the operational costs of the IoT platform. In this context, it is worthwhile to mention the increase popularity of LPWAN (Low Power Wide Area Networks) technologies, such as SigFox, LoRa, NB-IoT or LTE-M [32]. These technologies provide wide coverage, low consumption and low data rate, so they are especially suitable for scenarios with medium to high density of fixed sensors/actuators, as well as for applications which cover wide areas and require long-life batteries.

At application layer, apart from the option of Web Services on top of Hypertext Transfer Protocol/File Transfer Protocol (HTTP/FTP), two protocols stand out, namely: Message Queue Telemetry Transport (MQTT) and Constrained Application Protocol (CoAP) [33]. Table 1 summarizes their main features.

**Table 1.** Comparison of MQTT and CoAP.

| Feature | MQTT | CoAP |
| --- | --- | --- |
| Standard | ISO/IEC PRF 20922 [34] | RFC 7252 [35] |
| Communication paradigm | Publish-subscribe | Request-response |
| Number of messages | 16 | 4 |
| Header Size | 2 bytes | 4 bytes |
| Transport layer protocol | TCP | UDP |
| RESTful | No | Yes |

Figure 9 shows an overview of the communications architecture and protocol stack of Smart CEI Moncloa. It can be seen that the platform presents a hybrid communications architecture, which makes the most out of the communications infrastructure available at the UPM, thus minimizing the deployment costs.

The people flow monitoring sensors (on the right hand side of Figure 9) are directly connected to the information system, located at the Telecommunications Engineering School, via the Ethernet network of the UPM. The communications are protected end-to-end by the use of Transport Layer Security (TLS) on top of Transport Control Protocol/Internet Protocol (TCP/IP). The measurements are periodically sent using MQTT. As it can be seen in Table 1, this application protocol has been selected due to the following reasons: (1) it provides reliability, since it works on top of TCP; (2) it is very lightweight; (3) the versatility and flexibility provided by the publish-subscribe mechanism itself and by the hierarchical MQTT topics. To be more precise, the publish/subscribe mechanism allows the sensors not only to send measurements periodically, but also to receive management commands (e.g., to reboot them or to perform a remote firmware update).

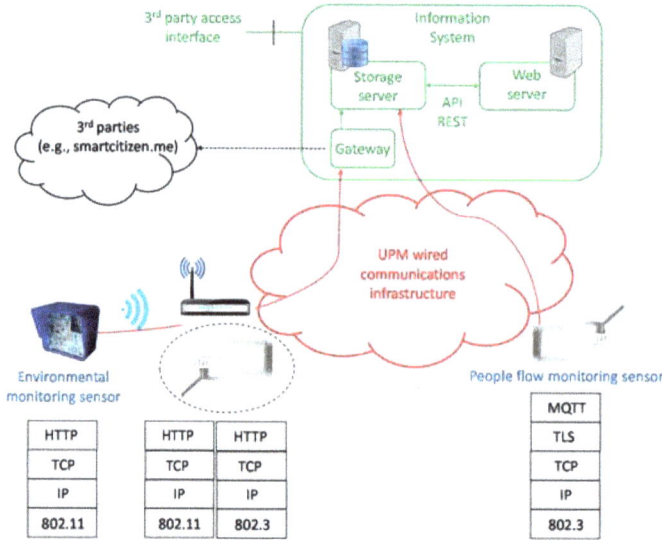

**Figure 9.** Overview of the communications architecture and protocol stack of the Smart CEI Moncloa platform.

The implemented solution consists on the deployment of a Mosquitto MQTT broker [36] at the information system, to whom the MQTT clients installed in the people flow monitoring sensors get connected. Such clients have being developed based on the Mosquitto C library and can be easily configured using a text file. This configuration file basically sets the events which are enabled in the sensor for both publish and subscribe. In the latter case, a script can be invoked upon reception of the event (e.g., to reboot the sensor). Thus, new functionality can be easily added to the sensors.

At the information system, there are also MQTT clients that subscribe to the events of the sensors in order to gather the data to be processed. Such clients are based on the same solution explained in previous paragraph, so the aforementioned configuration files can be used to establish what to do with each event.

Taking advantage of the hierarchical structure of the MQTT topics, all the publish events follow the structure *SERVICE/ID/EVENT(/TIMESTAMP)*. This structure allows easy storage the data coming from the sensors in a UNIX folder sorting them by service, sensor (ID), type of event and timestamp, which refers to the period when the measurement was taken. As it is explained in detail in Section 2.3, this solution has proved to be very efficient, avoiding the overhead associated to the use of databases and, at the same time, facilitating the processing of time series by using the timestamp.

In the case of the people flow monitoring sensors, the publish events start by WIFI, followed by the MAC address of its Ethernet interface, which is used as unique ID. These sensors generate different types of events. For instance, when the sensor starts working, the event *WIFI/<mac address>/START* is sent including a timestamp, what is useful for operation and maintenance purposes. Another example is the event *WIFI/<mac address>/SUM*, which includes a message with several counters related to the devices detected during the last 5 s, such as how many of them are active, how many have appeared, how many have gone, or how many are machines (devices that are always present). It is worthwhile to highlight also the event *WIFI/<mac address>/STA/TIMESTAMP* which includes a summary of the activity of the detected devices during the last 15 min. Recall that all the time intervals (e.g., 5 s, 15 min) can be configured via the configuration file.

The format of the content published under the different topics is Comma Separated Value (CSV). This is one of the most widely used data formats in IoT environments. It is especially appropriate when the structure of the sent data is fixed. In this kind of situations, it outperforms other options, such as Extensible Markup Language (XML) or JavaScript Object Notation (JSON), in terms of overhead, since CSV includes the header that explains the meaning of each field of the subsequent lines only at the beginning of the file.

As it can be observed in the left hand side of Figure 9, in the case of the environmental monitoring sensors, the communications solution was given, since they are only equipped with a Wi-Fi interface. These results in a hierarchical architecture where a given Wi-Fi network of the UPM is used to get access to the Ethernet infrastructure. It should be noted that the measurements are not encrypted end-to-end although, in the Wi-Fi link, WPA2, which used to be a very secure protocol until the publication of the vulnerabilities that have been recently found [37], together with MAC filtering is used. Additional security may be provided at application layer by means of the so-called API keys. Although in this case interoperability prevails over security, there could be scenarios where it would be of capital importance to secure this sensing infrastructure (see Section 2.1.2).

The environmental monitoring sensors assume the use of an API REST for time synchronization and for sending the measurements to the smartcitizen.me platform. Therefore, it was necessary to modify the firmware to redirect it to an API REST deployed in the information system of the Smart CEI Moncloa platform. In order to homogenize the system architecture and to make the most out of the flexibility of MQTT, such an API REST server also works as HTTP/MQTT gateway (see Figure 9). It could also be possible to install Mosquitto in the environmental monitoring sensors (e.g., [38]). However, as it was mentioned in Section 2.1.2, the adopted solution allows sharing the gathered data seamlessly with minimum changes in the sensors.

The events coming from the environmental monitoring sensors have the format *SCK/<mac address>/EVENT*. Thus, the event that is used to send the environmental parameters is *SCK/<mac addres>/RAW/<timestamp>*. These events include a CSV file with all the raw data measured by the sensors of the SCK. At the server, the MQTT client that processes these events store them in the subfolder *SCK/<mac address>/RAW*.

*2.3. Application Layer*

IoT applications generate unprecedented volumes of data. Therefore, data mining is key to extract valuable information from them [39], either by means of Artificial Intelligence (AI) and Machine Learning (ML) algorithms in search of patterns that allow automatically optimize processes and solve problems [40] or by means of visual representations that support data interpretation and decision making (the so-called visual analytics) [41].

As it has already been explained in Section 2.2, one of the key features of the Smart CEI Moncloa is the use of MQTT, which allows easily structuring the measurements sent by the sensors by means of folders and files at the storage server. Alternatively, such measurements could have been stored in some of the databases commonly used in IoT applications. However, as it has already been mentioned too, the measurements in the Smart CEI Moncloa represent time series and it is well-known that regular databases do not work properly when dealing with them. In order to solve this, the so-called time series databases, whose storage and search mechanisms are especially optimized for this kind of data, having being proposed in the literature [42].

Although this kind of databases could have been deployed in the storage server of Smart CEI Moncloa, for the sake of simplicity and performance, it was finally preferred taking advantage of the hierarchical structure of the MQTT topics, which include a timestamp. Such timestamps follow the YYYYMMDDhhmmss format, which dramatically facilitates the search for files by date. This feature, together with an own developed C library for handling and processing CSV files, allows developing simple scripts for time series search with very low response time. This is observed in the dashboard of the platform, where the time series queries are served very quickly, without penalizing the on-line usability of the platform, as it would happen if relational databases were in place.

In addition, the platform is processing the stored measurements in background mode and generating files with more elaborated data, which help reducing even more the query response time. Thus, in the case of the people flow monitoring sensors, for instance, daily CSV files are generated including a digest of the activity of all the devices detected throughout the day. The analysis of these files allows computing the average stay time and discriminating between machines, passers-by and stayers based on it. Another example of the benefits that this background on-the-fly processing brings, has to do with the analysis of the people flow in a whole building. In this case, the files associated to the measurements of different sensors of the same building are compared in order to avoid counting the same mobile/person several times (e.g., due to overlapping of Wi-Fi cells). The resulting file allows analyzing the people flow at building level, identifying the total stay time or the frequency of the visited places.

Finally, the dashboard of the platform is a web application developed under the Responsive Web Design (RWD) approach in order to ensure usability in all types of devices (e.g., smart phones, tablets, laptops) [43]. From the client side, the application is based on the Bootstrap framework [44], which combines Hypertext Markup Language (HTML), Cascading Style Sheets (CSS), and JavaScript. For the data representation, the well-known JavaScript libraries D3.js [45] and NVD3.js [46] are used, together with the library nvd3_graphs.js, which was specifically developed for the Smart CEI Moncloa platform. Furthermore, the JavaScript library leaflet.js [47] is used for the maps. From the server side, the application is based on Node.js [48] and Express [49]. Node.js is a well-known server platform based on JavaScript which allows developing scalable web applications (due to its event-oriented asynchronous behavior) in a fast and simple way (since clients and server are develop in the same programming language). The Express web framework makes the configuration of the Node.js server simpler.

## 3. Overview of the Provided Data

As it has just been explained, with the aim of demonstrating the functionality and potential of the deployed platform, a web-based dashboard has been developed. This application offers different functionalities depending on the user profile (namely, guess, partner, and administrator). These user profiles can visualize different graphs, maps, and tables, which are based on the data gathered by the sensor networks and facilitate performing visual analytics. In addition, the application also allows administrators to manage the deployed sensors remotely.

Figure 10 shows the welcome page of the dashboard [43]. This page provides an overview of the activity in the campus. On the left hand side, it can be seen a heat map that shows a real-time estimation (updated every 5 s) of the density of persons in the different premises of the campus (based on the Wi-Fi devices detected by the people flow monitoring sensor network). On the right hand side, the graph on the top shows the time evolution of the number of detected devices during the last 7 days; whereas the graph on the button shows a bar-chart that compares the number of devices detected during the last 15 min in each school. As it can be observed, the devices are classified into three categories (namely, machines, passers-by, and stayers) based on the average stay time and some heuristics. This allows differentiating between devices belonging to people who come to the school for work or studying, those that belong to people who are just passing by (e.g., couriers), and those that are actually fixed devices (e.g., printers).

In addition, the registered users can get access to custom graphs tailored to the different schools or even to individual sensors. For instance, Figure 11 shows the view associated to the sensor installed in the library of the Civil Engineering School. In the time evolution graph (left hand side of Figure 11), the opening hours can be observed together with two occupation peaks (in the morning and in the afternoon) with an off-peak period in between (due to the lunch time). This per-sensor view also includes a histogram of the stay time (in minutes) of the Wi-Fi devices in the surroundings of the sensor (right hand side of Figure 11).

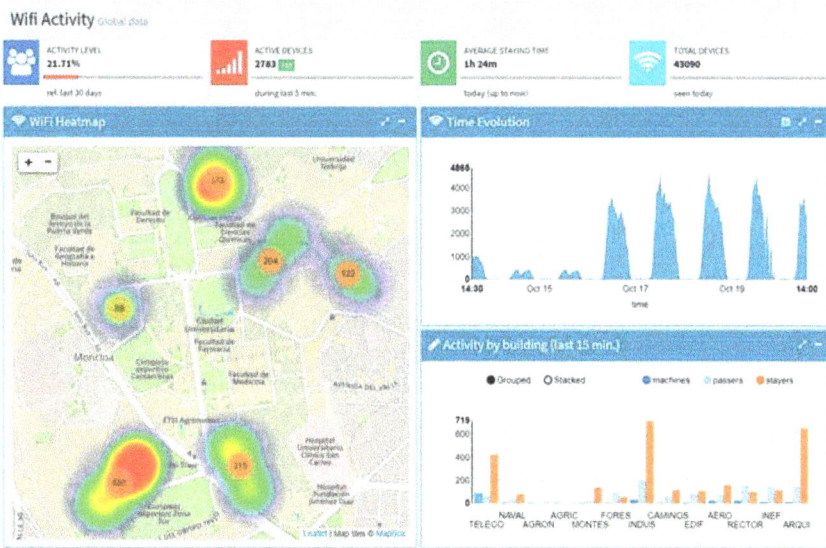

**Figure 10.** Welcome page of the Smart CEI Moncloa dashboard.

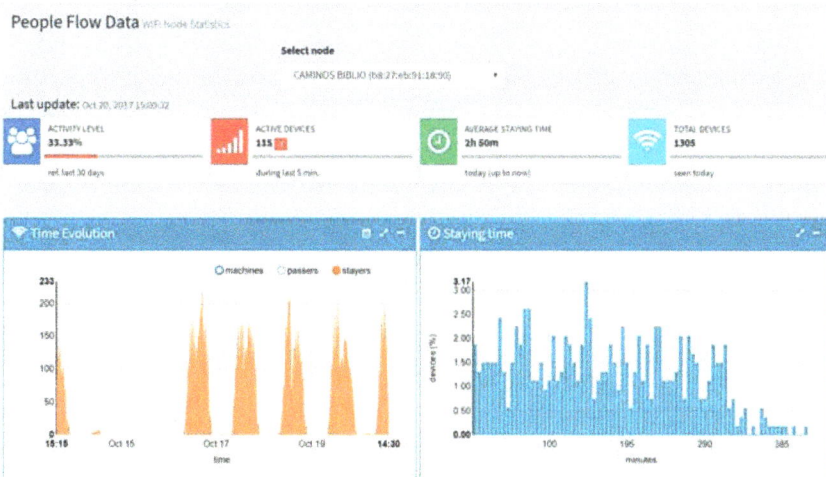

**Figure 11.** View associated to the people flow monitoring sensor installed in the library of the Civil Engineering School.

Figure 12 shows a snapshot of the activity in the Industrial Engineering School. In the first row of graphs, going from the left to the right, first it can be seen a heat map that illustrates the distribution of people in the different premises of the school (entrance area, classrooms, library, canteen, parking, and sport premises). Next, it can be seen a graph that shows the evolution of the number of detected devices during the last 7 days. As it has already been pointed out, these devices are classified as machines, passers-by, and stayers. The histogram of the stay time is also provided as last graph of the first row. In the second row of graphs, it can be seen two bar-charts that allow comparing the number of devices detected in each of the monitored premises. It should be kept in mind that in the building analysis the data from all the installed sensors are correlated, avoiding counting the same devices multiple times (e.g., due to Wi-Fi cell overlap).

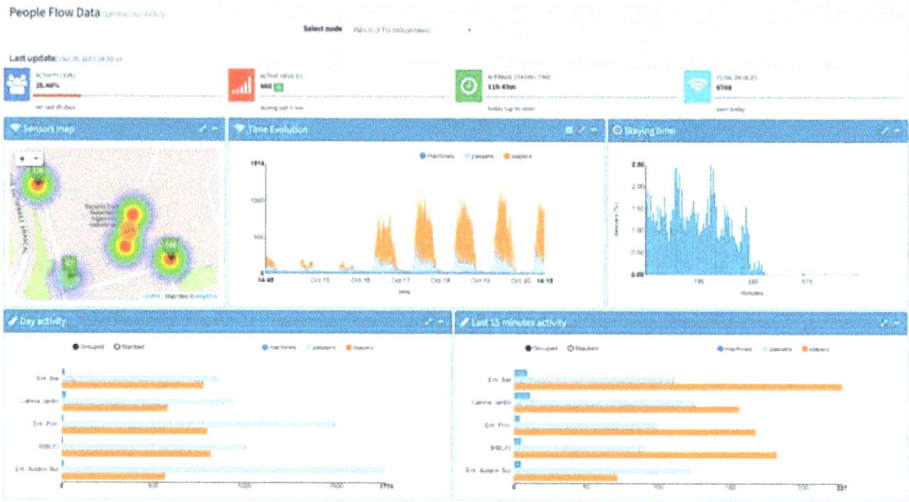

**Figure 12.** View associated to the people flow monitoring service for the Industrial Engineering School.

Both the per-sensor and per-school views allow visualizing either real-time information or historic data. Thus, on the top of Figure 13, it can be seen the distribution of the people flow monitoring sensors at the Telecommunications Engineering School and at the Industrial Engineering School respectively; whereas on the bottom of Figure 13, it can be seen a snapshot of the occupation in the different premises of each school at a given moment of time.

(a)

(b)

**Figure 13.** *Cont.*

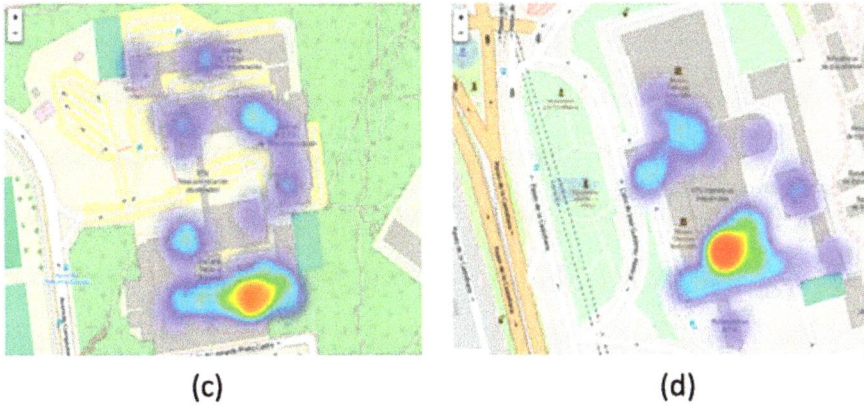

**Figure 13.** (a,b) Distribution of the people flow monitoring sensors at the Telecommunications Engineering School and at the Industrial Engineering School respectively; (c,d) Snapshot of the occupation in the different premises of the Telecommunications Engineering School and the Industrial Engineering School respectively.

Figure 14, on the other hand, illustrates how the historical occupation data can be consulted, in this case for the Telecommunication Engineering School.

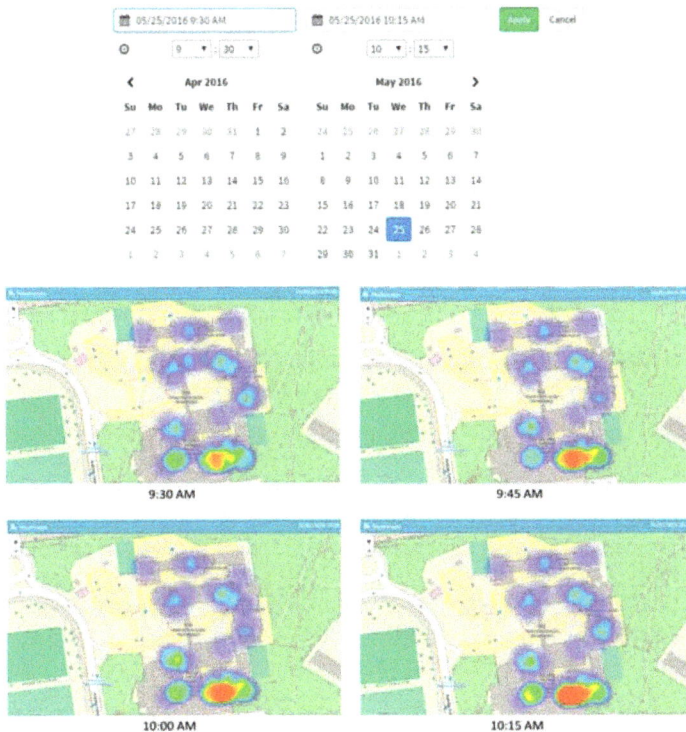

**Figure 14.** Historical occupation data for the Telecommunications Engineering School. It allows selecting the day and the time frame.

To conclude this overview of the data provided by the dashboard, Figure 15 shows the view of the environmental monitoring service accessible to everybody. On the left hand side of Figure 15, it can be seen a real-time summary of the monitored parameters. On the right hand side of Figure 15, it can be seen the weekly graphs for each parameter. These graphs allow observing a clear correlation between the monitored parameters depending on whether it is day or night. It should be noted that no big difference is observed between weekdays and weekend because the libraries of the schools are open during the weekend. The registered users can also get access to the data gathered from specific sensors by selecting them.

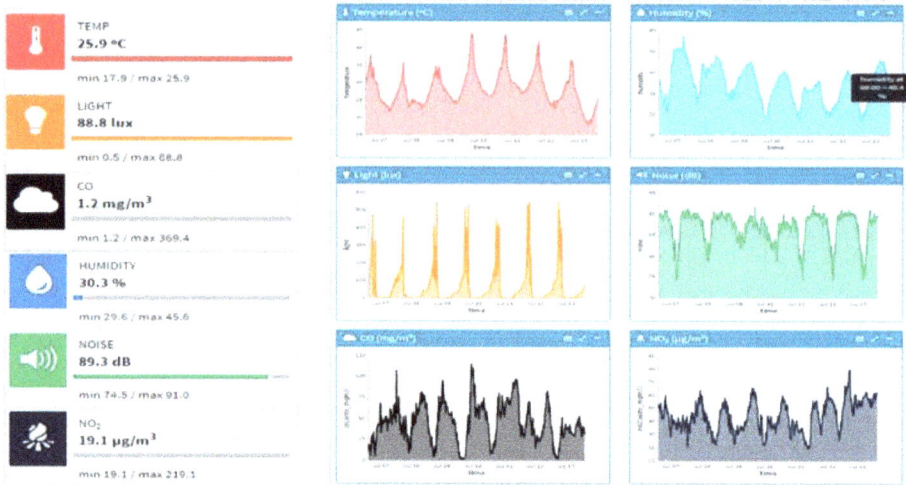

**Figure 15.** Overview of the environmental monitoring data.

## 4. Use Cases

The information currently provided by the Smart CEI Moncloa platform can be used and combined to come up with a wide range of value-added services for Smart Cities. This section goes through just a few examples with the aim of illustrating such a potential and fostering new ideas.

*4.1. Use Case 1: Smart Emergency Management Application*

Public safety is one of the main application areas of the people flow monitoring service. The heat maps, for instance, allow identifying the areas where there are higher concentrations of people. Thus, a potential application may focus on these areas and compare the number of detected devices with a given threshold in order to detect and avoid overcapacity. This service may also be so useful for searching for people inside buildings in case of disasters such as fire or earthquakes. This was indeed the idea that a team of UPM students presented at Hack for Good 2016 and that was awarded by Telefonica [50].

The application on Smart Emergency Management was able to determine where people were located based on the data collected by the people flow monitoring sensor network of Smart CEI Moncloa. To probe the effectiveness of the solution, the students took a data set collected during December 2015 at the Telecommunication Engineering School. During the Hackaton sessions, they developed an application that processed the data in order to estimate the location of Wi-Fi devices. Figure 16 shows the developed web-based graphical user interface, which allows tracking the distribution of the devices within the building along the time.

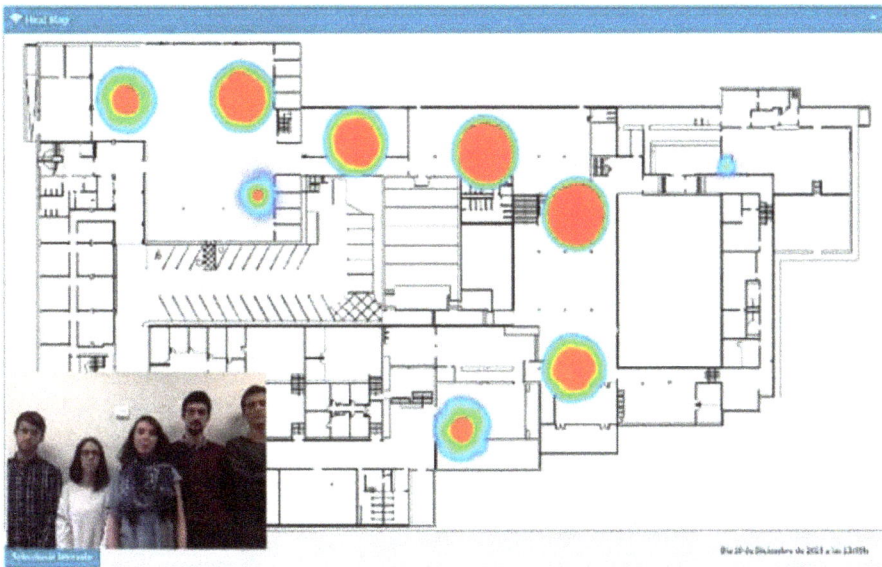

**Figure 16.** Smart Emergency Management web-based application.

*4.2. Use Case 2: Traffic Restriction Application*

As it has already been mentioned, environmental monitoring is of capital importance especially in the cities, where pollution is reaching levels that jeopardize citizens' health. As a result, several important cities around the world, such as Stockholm, London, Paris or Santiago de Chile, have been applying road traffic restrictions in specific areas, such as downtown, for quite a few years. These measures are also being applied in Madrid for a few months. These measures range from forbidding the use of public parking or reducing the speed limit in the main highways surrounding the city to allowing only the entrance to the city of vehicles with odd or even license plate. The severity of the applied restrictions depends typically on the $NO_2$ concentration, which is measured by a few expensive air quality monitoring stations.

Figure 17 compares the $NO_2$ concentration measured by one of those monitoring stations with the $NO_2$ concentration measured by the outdoors environmental sensor of the Industrial Engineering School, which is in front of the municipal monitoring station. It should be noted that on October 24 and October 25 the Madrid city council imposed parking and speed limit restrictions due to the pollution levels. Although, as it is explained in Section 2.1.2, the CO and $NO_2$ measurements from the environmental sensors used in the Smart CEI Moncloa platform are not precisely calibrated, Figure 17 shows that they are valid for identifying sharp increases and, in general, trends.

Therefore, this kind of low-cost sensors can be combined with more expensive air quality monitoring stations to increase the granularity of the measurements, enabling more specific or focused restrictions which can be taken for less severe situations (e.g., forbid the road traffic in a specific area instead of in the whole city center and update the citizen with these restrictions dynamically via an app, which may even communicate or be embedded in the GPS navigator, thus trying to act on the traffic not only based on the traffic conditions but also on environmental variables). In this kind of hypothetical applications, security issues would need to be addressed carefully, since an attacker who got the control of a remarkable amount of low-cost sensors could provoke chaotic situations. In addition, the measurements taken by the expensive air quality monitoring stations can be used to dynamically calibrate the low-cost sensors.

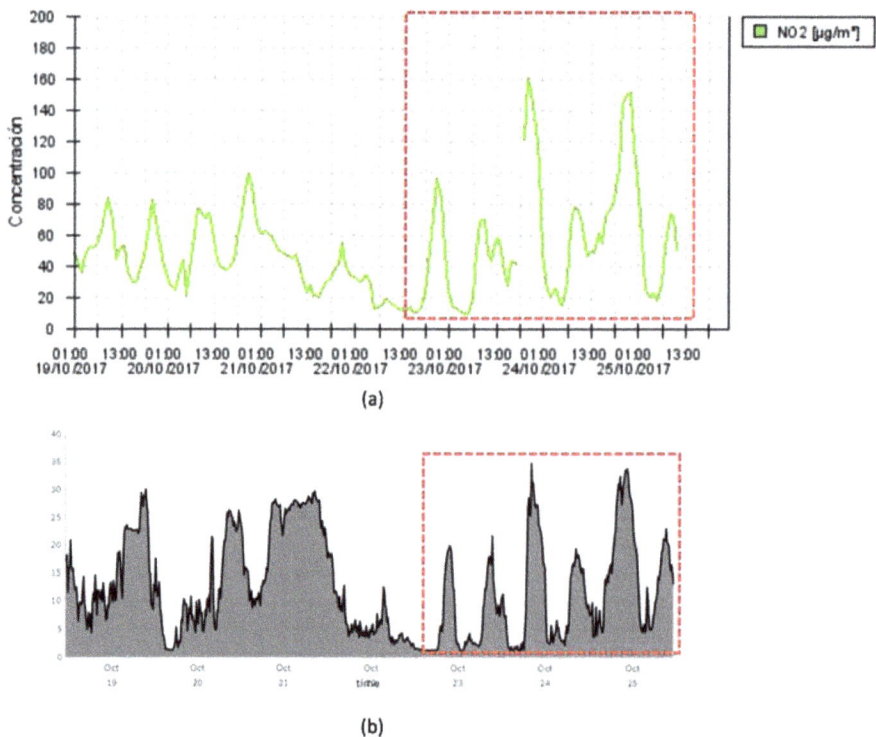

Figure 17. (a) $NO_2$ concentration measured by a municipal air quality monitoring station [51]; (b) $NO_2$ concentration measured by the outdoor environmental sensor of the Industrial Engineering School, which is in front of the municipal monitoring station.

## 4.3. Use Case 3: Library Application

This use case represents an example of how the information from the people flow monitoring sensors and the environmental sensors can be combined resulting in value added applications, notably to the university community in this specific case. Figure 18 shows the library application developed by a UPM student [52]. This application allows selecting one of the libraries of the different schools and provides very useful information, such as the percentage of occupation and free seats (which is an estimation based on the Wi-Fi detected devices) and the temperature, levels of noise, and CO and $NO_2$ concentration (which are taken from the indoors environmental sensors). The latter information is definitely interesting to student due to the effect that these parameters have on the concentration and so on the effectiveness of the study time. This information can be combined with the distance between the users' location and the libraries to enable trade-off selections.

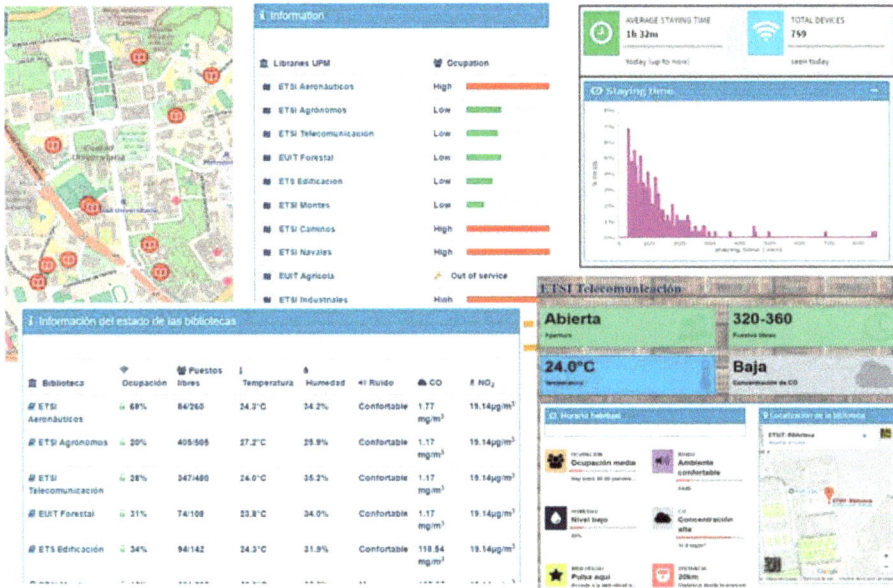

**Figure 18.** Web-based application to aid library selection.

## 5. Discussion

IoT-based platforms for Smart Cities involve a huge amount of sensors and actuators and a wide variety of technological features and decisions, ranging from the hardware and firmware of such sensors and actuators to the plethora of backend technologies and potential services, and the myriad of communications technologies and protocols available to connect both ends. As a result, this kind of platforms are technologically complex and putting them into practice at large scale implies huge costs and risks. Therefore, pilot schemes that allow validating proof of concepts, experimenting with different technologies and services, and fine-tuning them before deploying them massively, are especially important in this context.

The Smart CEI Moncloa presents a set of features that make it a great test bench for Smart City services. It covers a fairly large area (5.5 km²) integrated into the metropolitan area of Madrid, which ensures a high traffic of people and vehicles. It involves a remarkable number of sensors (77) spread across the different schools that comprise the UPM. These sensors rely on two of the most widely used hardware platforms for IoT, notably Raspberry Pi and Arduino. The platform also uses especially relevant protocols for IoT such as MQTT and provides a human-friendly web interface based on RWD that allows consulting the available information from any kind of device (e.g., smartphone, tablet, laptop). Furthermore, the two initial pilot services (namely, people flow monitoring and environmental monitoring) are also especially relevant for Smart City scenarios, as it has been illustrated throughout the paper, and new services can be easily included.

In addition to all this, the fact that the platform is deployed in a university campus makes that the provided value goes beyond the services and applications themselves, fostering the learning and training of students, and the research and innovation from the whole university community. As a matter of fact, Section 4 presents several examples of value-added applications developed by UPM students in different contexts (e.g., hackathons, final thesis, etc.). Moreover, the Moncloa CEI includes all kind of engineering schools along with humanities schools, which make it a remarkable multidisciplinary environment, something that is so important for Smart City applications.

The lessons learned from the design, development, and deployment of the Smart CEI Moncloa can be of special interest for undertaking similar projects in the future. At the sensing layer, the paper illustrates one of the main challenges when dealing with a huge amount of low-cost sensors: keeping them properly calibrated [53]. Typically, automatic procedures to try to dynamically update and fix the calibration of the sensors will be required. In Section 2.1.2, a proposal to dynamically calibrate the data based on the gathered measurements is explained. Moreover, as it is pointed out in Section 4, if this kind of sensors are deployed together with a few high-quality sensors that drop more reliable measurements, they can be dynamically calibrated based on them. Anyway, in the case of the sensors for CO and $NO_2$ concentration used in the Smart CEI Moncloa, it has been proved that, despite they are not accurately calibrated, they are still valid for detecting sharp variations and trends. As it is also mentioned in Section 4, if this kind of low cost sensor infrastructures were used for providing Smart City services, security mechanisms, such as the ones described in Section 2.2, would be necessary.

At the networking and data communications layer, taking into account the volume of devices and the associated investments that this kind of systems involve, an important takeaway is that, despite there are communications technologies that are especially appropriate for Smart City applications (such as the LPWAN technologies [32]) and so must be considered, the available communications infrastructure must be used whenever possible, in order to reduce deployment costs. Thus, if a city council offers free Wi-Fi all over the city, the AP being located, for instance, in the streetlights, and they want to start offering a smart parking service, the first approach they should consider if it is possible to send the data from the parking sensors via Wi-Fi, guaranteeing the required level of reliability and without compromising the QoS (Quality of Service) of the Wi-Fi users (which seems quite improbable, since the parking sensors can be optimized so that they only transmit a small packet when a car parks or leaves). Likewise, if a city council wanted to offer an energy management service in their public buildings which are equipped with wired and wireless infrastructures, they should first consider using such infrastructure to carry the measurements and commands back and forth.

Taking into account the wide variety of devices this kind of systems involve, it is also very common that different communications technologies and protocols co-exist within the same platform. The use of gateways or middleware is key to solve this kind of issues [54–57]. Middleware platforms are also valuable to provide common policies to access the available data, thus solving security and privacy issues [58,59]. Furthermore, it is also worthwhile to remark upon related middleware platforms focused on enabling ad hoc proximity-based Device-to-Device (D2D) communications to support so-called Mobile Social Networks in Proximity (MSNP) applications [60].

At the application layer, there are also several data formats available whose use depends on the specific features of the target application. If there are different types of devices providing heterogeneous unstructured data, JSON may be the most appropriate solution, taking into account its great flexibility and that it can be easily integrated with NoSQL databases for on-line Big Data, such as MongoDB [61]. However, if the data are heterogeneous but they follow always a fixed structure, as in the case of the Smart CEI Moncloa platform, CSV represents a great solution, since it is the most lightweight data format and meets the data requirements. The paper also presents a simple and scalable approach to store and handle time series data.

Finally, the paper also reveals that in order to obtain the maximum value from Smart City services and applications, they need to be approached holistically, since there are so many possible synergies between them. Section 4, for instance, presents the example of how combining the information from the people flow monitoring sensor network with the information from the environmental sensor network can allow developing an added-value application to aid library selection. But there are many other examples. For instance, a smart lighting service would not be actually smart if it did not take into account information related to light or to the presence of people, if available. Likewise, a smart watering service would not be actually smart if it did not take into account information related to the humidity of the ground or even the weather forecast, if available.

## 6. Conclusions

This paper presents the Smart CEI Moncloa: an IoT-based platform deployed in the Moncloa Campus of International Excellence, located in the Madrid metropolitan area, and whose main goal is to enable experimenting with Smart City services and fine-tuning them before eventually deploying them at large scale.

The platform currently offers two pilot services, namely people flow monitoring based on Wi-Fi tracking and environmental monitoring, including the following parameters: temperature, humidity, light, noise, CO concentration, and $NO_2$ concentration. The paper describes the main challenges that were found and the design and implementation decisions that were made at each of the main layers of the platform, namely the sensing layer, the networking and data communications layer, and the application layer. In addition, the paper also presents the main functionalities that the platform currently offers, mostly related to performing visual analytics through a web-based dashboard which guarantees ubiquitous access, and discusses its potential by means of a few selected use cases.

As for future work, energy monitoring is the next service that is planned to be included in the platform due to its great relevance for sustainable development and to the great synergies with the already delivered services. Thus, the people flow monitoring information and the environmental information can be considered to smartly control the university Heating, Ventilation, and Air Conditioning (HVAC) systems, e.g., switching off (or properly adjusting the working point of) the heating or air conditioning when nobody is in a room or switching on the heating a couple of hours before people start arriving at the university after a very cold night or weekend. This kind of control would enable saving energy and improving comfort levels at the same time. Furthermore, the Distributed Energy Resources (DER) deployed throughout the campus (e.g., distributed generation based on solar photovoltaics and wind turbines, or electric vehicles charging spots) may be also taken into account in order to perform consumption and generation matching at campus level.

In addition, it is also planned to encourage the university community to apply ML and AI algorithms to the available data to come up with novel applications that bring value to the university itself and foster the creation of cutting-edge technological spin-offs focused on Smart Cities.

**Acknowledgments:** This work has been partly funded by the Spanish Ministry of Economy and Competitiveness through the DHARMA (Dynamic Heterogeneous Threats Risk Management and Assessment) project (TIN2014-59023-C2-2-R) and by the UPM City of the Future initiative. The authors would like to thank all the students that have participated in the development and deployment of the Smart CEI Moncloa: Pablo Arias, Marta Martín, Carlos Moreno, Ana Martín, Carlos Gómez and Almudena Fernández.

**Author Contributions:** Manuel Álvarez-Campana led the design, development and deployment of the Smart CEI Moncloa platform, which has been used by all the authors for educational and research purposes. Manuel Álvarez-Campana and Gregorio López led the paper preparation, which includes contributions from all the authors.

**Conflicts of Interest:** The authors declare no conflict of interest.

## Abbreviations

| | |
|---|---|
| AI | Artificial Intelligence |
| AP | Access Point |
| API | Application Programming Interface |
| MAC | Medium Access Control |
| CEI | Campus of International Excellence |
| CoAP | Constrained Application Protocol |
| CSV | Comma Separated Value |
| D2D | Device-to-Device |
| DER | Distributed Energy Resources |
| ETSI | European Telecommunications Standards Institute |
| FTP | File Transport Protocol |

| GPS | Global Positioning System |
|-----|---------------------------|
| HTTP | Hypertext Transfer Protocol |
| HVAC | Heating, Ventilation, and Air Conditioning |
| ICT | Information and Communications Technologies |
| IEEE | Institute of Electrical and Electronics Engineers |
| IP | Internet Protocol |
| IoT | Internet of Things |
| JSON | JavaScript Object Notation |
| LPWAN | Low Power Wide Area Network |
| LTE | Long Term Evolution |
| M2M | Machine-to-Machine |
| MEMS | Microelectromechanical Sensor |
| ML | Machine Learning |
| MQTT | Message Queuing Telemetry Transport |
| MSNP | Mobile Social Networks in Proximity |
| NB-IoT | Narrowband-IoT |
| OS | Operating System |
| QoS | Quality of Service |
| RWD | Responsive Web Design |
| SCK | Smart Citizen Kit |
| TCP | Transport Control Protocol |
| TLS | Transport Layer Security |
| UDP | User Datagram Protocol |
| UPM | Universidad Politécnica de Madrid |
| WLAN | Wireless Local Area Network |
| WPAN | Wireless Personal Area Network |
| XML | eXtensible Markup Language |

## References

1. Al-Fuqaha, A.; Guizani, M.; Mohammadi, M.; Aledhari, M.; Ayyash, M. Internet of Things: A Survey on Enabling Technologies, Protocols, and Applications. *IEEE Commun. Surv. Tutor.* **2015**, *17*, 2347–2376. [CrossRef]
2. Zanella, A.; Bui, N.; Castellani, A.; Vangelista, L.; Zorzi, M. Internet of Things for Smart Cities. *IEEE Internet Things J.* **2014**, *1*, 22–32. [CrossRef]
3. UPM City of the Future Initiative. Available online: http://blogs.upm.es/cityofthefuture-upm/en/initiative/ (accessed on 29 October 2017).
4. Moncloa Campus of International Excellence. Available online: http://www.campusmoncloa.es/en/cei-campus-of-international-excellence.php (accessed on 29 October 2017).
5. Minerva, R.; Biru, A.; Rotondi, D. Towards a definition of the IoT. *IEEE Internet Initiative.* 2015. Available online: https://iot.ieee.org/images/files/pdf/IEEE_IoT_Towards_Definition_Internet_of_Things_Revision1_27MAY15.pdf (accessed on 29 October 2017).
6. IEEE P2413. Available online: http://grouper.ieee.org/groups/2413/ (accessed on 29 October 2017).
7. Xu, L.D.; He, W.; Li, S. Internet of Things in Industries: A Survey. *IEEE Trans. Ind. Inform.* **2014**, *10*, 2233–2243. [CrossRef]
8. OneM2M. Available online: http://www.onem2m.org/ (accessed on 29 October 2017).
9. Dohler, M.; Boswarthick, D.; Alonso-Zárate, J. Machine-to-machine in smart cities & smart grids. In Proceedings of the IEEE Global Communications Conference Tutorial, Anaheim, CA, USA, 3–7 December 2012.
10. Li, T.; Chang, H.; Wang, M.; Ni, B.; Hong, R.; Yan, S. Crowded Scene Analysis: A Survey. *IEEE Trans. Circ. Syst. Video Technol.* **2015**, *25*, 367–386. [CrossRef]
11. El Mallah, J.; Carrino, F.; Khaled, O.A.; Mugellini, E. Crowd Monitoring. In *Proceedings of the Third International Conference on Distributed, Ambient, and Pervasive Interactions, Los Angeles, CA, USA, 2–7 August 2015*; Streitz, N., Markopoulos, P., Eds.; Springer: New York, NY, USA, 2015; Volume 9189, pp. 496–505. [CrossRef]

12. Fernández-Ares, A.; Mora, A.M.; Arenas, M.G.; García-Sanchez, P.; Romero, G.; Rivas, V.; Castillo, P.A.; Merelo, J.J. Studying real traffic and mobility scenarios for a Smart City using a new monitoring and tracking system. *Future Gener. Comput. Syst.* **2017**, *76*, 163–179. [CrossRef]

13. Alessandrini, A.; Gioia, C.; Sermi, F.; Sofos, I.; Tarchi, D.; Vespe, M. WiFi positioning and Big Data to monitor flows of people on a wide scale. In Proceedings of the 2017 European Navigation Conference (ENC), Lausanne, Switzerland, 9–16 May 2017; pp. 322–328. [CrossRef]

14. Zhou, M.; Sui, K.; Ma, M.; Zhao, Y.; Pei, D.; Moscibroda, T. MobiCamp: A Campus-wide Testbed for Studying Mobile Physical Activities. In Proceedings of the 3rd International Workshop on Physical Analytics (WPA 2016), Singapore, 26 June 2016; ACM: New York, NY, USA, 2016; pp. 1–6. [CrossRef]

15. Suciu, G.; Vochin, M.; Diaconu, C.; Suciu, V.; Butca, C. Convergence of software defined radio: WiFi, ibeacon and epaper. In Proceedings of the 2016 15th RoEduNet Conference: Networking in Education and Research, Bucharest, Romania, 7–9 September 2016; pp. 1–5. [CrossRef]

16. Wi-Fi Tracking at London Underground. Available online: http://www.gizmodo.co.uk/2017/09/london-underground-wifi-tracking-heres-everything-we-learned-from-tfls-official-report/ (accessed on 29 October 2017).

17. Wi-Fi Tracking in British Museums. Available online: http://www.gizmodo.co.uk/2017/04/exclusive-heres-what-museums-learn-by-tracking-your-phone/ (accessed on 29 October 2017).

18. Wi-Fi Tracking at Helsinki Airport. Available online: http://www.telegraph.co.uk/travel/news/Big-Brother-airport-installs-worlds-first-real-time-passenger-tracking-system/ (accessed on 29 October 2017).

19. Martin, J.; Mayberry, T.; Donahue, C.; Foppe, L.; Brown, L.; Riggins, C.; Rye, E.C.; Brown, D. A Study of MAC Address Randomization in Mobile Devices and When it Fails. *arXiv* **2017**. [CrossRef]

20. Raspberry Pi. Available online: https://www.raspberrypi.org/ (accessed on 29 October 2017).

21. Aircrack-ng. Available online: https://www.aircrack-ng.org/ (accessed on 29 October 2017).

22. Kotsev, A.; Schade, S.; Craglia, M.; Gerboles, M.; Spinelle, L.; Signorini, M. Next Generation Air Quality Platform: Openness and Interoperability for the Internet of Things. *Sensors* **2016**, *16*, 403. [CrossRef] [PubMed]

23. Shah, J.; Mishra, B. IoT enabled environmental monitoring system for smart cities. In Proceedings of the 2016 International Conference on Internet of Things and Applications (IOTA), Pune, India, 22–24 January 2016; pp. 383–388. [CrossRef]

24. Brienza, S.; Galli, A.; Anastasi, G.; Bruschi, P. A Low-Cost Sensing System for Cooperative Air Quality Monitoring in Urban Areas. *Sensors* **2015**, *15*, 12242–12259. [CrossRef] [PubMed]

25. MIT Tata Center Air Pollution Monitoring Using Low-Cost Sensors. Available online: https://tatacenter.mit.edu/portfolio/air-pollution-sensors/ (accessed on 29 October 2017).

26. Ibrahim, M.; Elgamri, A.; Babiker, S.; Mohamed, A. Internet of things based smart environmental monitoring using the Raspberry-Pi computer. In Proceedings of the 2015 Fifth International Conference on Digital Information Processing and Communications (ICDIPC), Sierre, Switzerland, 7–9 October 2015; pp. 159–164. [CrossRef]

27. Kelly, S.D.T.; Suryadevara, N.K.; Mukhopadhyay, S.C. Towards the Implementation of IoT for Environmental Condition Monitoring in Homes. *IEEE Sens. J.* **2013**, *13*, 3846–3853. [CrossRef]

28. Smart Citizen. Available online: https://smartcitizen.me/ (accessed on 29 October 2017).

29. Arduino. Available online: https://www.arduino.cc/ (accessed on 29 October 2017).

30. MiCS-4514 Datasheet. Available online: http://files.manylabs.org/datasheets/MICS-4514.pdf (accessed on 29 October 2017).

31. Lopez, G.; Moreno, J.I.; Amarís, H.; Salazar, F. Paving the road toward Smart Grids through large-scale advanced metering infrastructures. *Electr. Power Syst. Res.* **2015**, *120*, 194–205. [CrossRef]

32. Raza, U.; Kulkarni, P.; Sooriyabandara, M. Low Power Wide Area Networks: An Overview. *IEEE Commun. Surv. Tutor.* **2017**, *19*, 855–873. [CrossRef]

33. Karagiannis, V.; Chatzimisios, P.; Vázquez-Gallego, F.; Alonso-Zarate, J. A Survey on Application Layer Protocols for the Internet of Things. *Trans. IoT Cloud Comput.* **2015**, *1*, 11–17.

34. International Organization for Standardization. *ISO/IEC 20922:2016 Information Technology—Message Queuing Telemetry Transport (MQTT) v3.1.1*; International Organization for Standardization: Geneva, Switzerland, 2016.

35. Shelby, Z.; Hartke, K.; Bormann, C. The Constrained Application Protocol (CoAP). RFC 7252, 2014. Available online: https://tools.ietf.org/html/rfc7252 (accessed on 29 October 2017).

36. Mosquitto. Available online: https://mosquitto.org/ (accessed on 29 October 2017).

37. Vanhoef, M.; Piessens, F. Key Reinstallation Attacks: Forcing Nonce Reuse in WPA2. In Proceedings of the ACM Conference on Computer and Communications Security, Dallas, TX, USA, 30 October–3 November 2017; Available online: https://www.krackattacks.com/ (accessed on 29 October 2017).

38. Arduino Client for MQTT. Available online: https://pubsubclient.knolleary.net/ (accessed on 29 October 2017).

39. Chen, F.; Deng, P.; Wan, J.; Zhang, D.; Vasilakos, A.V.; Rong, X. Data Mining for the Internet of Things: Literature Review and Challenges. *Int. J. Distrib. Sen. Netw.* **2015**. [CrossRef]

40. Mahdavinejad, M.S.; Rezvan, M.; Barekatain, M.; Adibi, P.; Barnaghi, P.; Sheth, A.P. Machine learning for Internet of Things data analysis: A survey. *Digit. Commun. Netw.* **2017**. [CrossRef]

41. Fekete, J.D. Visual Analytics Infrastructures: From Data Management to Exploration. *Computer* **2013**, *46*, 22–29. [CrossRef]

42. Jensen, S.K.; Pedersen, T.B.; Thomsen, C. Time Series Management Systems: A Survey. *IEEE Trans. Knowl. Data Eng.* **2017**, *29*, 2581–2600. [CrossRef]

43. Smart CEI Moncloa Dashboard. Available online: http://ceiboard.dit.upm.es/dashboard/ (accessed on 29 October 2017).

44. Bootstrap. Available online: http://getbootstrap.com/ (accessed on 29 October 2017).

45. D3.js (Data-Driven Documents). Available online: https://d3js.org/ (accessed on 29 October 2017).

46. nvd3.js (Reusable charts for d3.js). Available online: http://nvd3.org/examples/index.html (accessed on 29 October 2017).

47. leaflet.js. Available online: http://leafletjs.com/ (accessed on 29 October 2017).

48. JavaScript Node.js. Available online: https://nodejs.org/en/ (accessed on 29 October 2017).

49. Node Express.js. Available online: http://expressjs.com/es/ (accessed on 29 October 2017).

50. Smart Emergency Management Application. Available online: http://cityofthefuture-upm.com/upm-students-awarded-by-telefonica-for-smart-city-emergency-management/ (accessed on 29 October 2017).

51. Madrid City Council Air Quality Monitoring System. Available online: http://www.mambiente.munimadrid.es/sica/ (accessed on 29 October 2017).

52. Library Application. Available online: http://ceiboard.dit.upm.es/smartcampus/biblio (accessed on 29 October 2017).

53. Castell, N.; Dauge, F.R.; Schneider, P.; Vogt, M.; Lerner, U.; Fishbain, B.; Broday, D.; Bartonova, A. Can commercial low-cost sensor platforms contribute to air quality monitoring and exposure estimates? *Environ. Int.* **2017**, *99*, 293–302. [CrossRef] [PubMed]

54. Atzori, L.; Iera, A.; Morabito, G. The Internet of Things: A survey. *Comput. Netw.* **2010**, *54*, 2787–2805. [CrossRef]

55. Razzaque, M.A.; Milojevic-Jevric, M.; Palade, A.; Clarke, S. Middleware for Internet of Things: A Survey. *IEEE Internet Things J.* **2016**, *3*, 70–95. [CrossRef]

56. Thread. Available online: http://threadgroup.org/ (accessed on 4 December 2017).

57. OpenHAB. Available online: http://www.openhab.org/ (accessed on 4 December 2017).

58. Sicari, S.; Rizzardi, A.; Miorandi, D.; Coen-Porisini, A. Dynamic Policies in Internet of Things: Enforcement and Synchronization. *IEEE Internet Things J.* **2017**. [CrossRef]

59. Sicari, S.; Rizzardi, A.; Miorandi, D.; Capiello, C.; Coen-Porisini, A. A secure and quality-aware prototypical architecture for the Internet of Things. *Inf. Syst.* **2016**, *58*, 43–55. [CrossRef]

60. Wang, Y.; Wei, L.; Jin, Q.; Ma, J. Alljoyn Based Direct Proximity Service Development: Overview and Prototype. In Proceedings of the 2014 IEEE 17th International Conference on Computational Science and Engineering, Chengdu, China, 19–21 December 2014; pp. 634–641. [CrossRef]

61. MongoDB. Available online: https://www.mongodb.com/ (accessed on 29 October 2017).

![sensors logo] *sensors*     MDPI

*Article*

# A Portable Dynamic Laser Speckle System for Sensing Long-Term Changes Caused by Treatments in Painting Conservation

**Alberto J. Pérez [1], Rolando J. González-Peña [2,\*], Roberto Braga Jr. [3], Ángel Perles [1,4], Eva Pérez–Marín [5] and Fernando J. García-Diego [6]**

[1]   Dep. d'Informàtica de Sistemes i Computadors (DISCA), Universitat Politècnica de València, 46022 Valencia, Spain; aperez@disca.upv.es (A.J.P.); aperles@disca.upv.es (A.P.)

[2]   Dep. Fisiología. Unidad de Biofísica y Física Médica. Facultad de Medicina y Odontología, Universitat de València, 46010 Valencia, Spain

[3]   Dep. Engenharia (DEG), Universidade Federal de Lavras (UFLA), 3037 Lavras, Brazil; robbraga@deg.ufla.br

[4]   ITACA Institute, Camino de Vera, s/n, 46022 Valencia, Spain

[5]   Dep. de Conservación y Restauración de Bienes Culturales, Universitat Politècnica de València, 46022 Valencia, Spain; evpema@crbc.upv.es

[6]   Dep. de Física Aplicada, Universitat Politècnica de València, 46022 Valencia, Spain; fjgarcid@upv.es

\*   Correspondence: rolando.j.gonzalez@uv.es; Tel.: +34-963983776

Received: 15 November 2017; Accepted: 9 January 2018; Published: 11 January 2018

**Abstract:** Dynamic laser speckle (DLS) is used as a reliable sensor of activity for all types of materials. Traditional applications are based on high-rate captures (usually greater than 10 frames-per-second, fps). Even for drying processes in conservation treatments, where there is a high level of activity in the first moments after the application and slower activity after some minutes or hours, the process is based on the acquisition of images at a time rate that is the same in moments of high and low activity. In this work, we present an alternative approach to track the drying process of protective layers and other painting conservation processes that take a long time to reduce their levels of activity. We illuminate, using three different wavelength lasers, a temporary protector (cyclododecane) and a varnish, and monitor them using a low fps rate during long-term drying. The results are compared to the traditional method. This work also presents a monitoring method that uses portable equipment. The results present the feasibility of using the portable device and show the improved sensitivity of the dynamic laser speckle when sensing the long-term process for drying cyclododecane and varnish in conservation.

**Keywords:** dynamic speckle; activity; temporal history speckle pattern; varnish; cyclododecane

## 1. Introduction

Digital holography [1,2], a well-known speckle interferometric technique, is generally used to measure paints as they dry, however, the technique demands complex experimental configurations that limit its use [3,4]. Other digital speckle pattern interferometry techniques, such as shearography, are applied for art objects, particularly canvas and panel paintings [5,6], but require a similar experimental holographical configuration. In art restoration, treatments usually change the visual aspect of the art objects. As in the case of the application of varnish in paintings, changes may appear in the brightness and light saturation [7–9].

The processes that provide fixation and consolidation in painting conservation use numerous substances, among which is cyclododecane (which forms bright artefacts on the surface, thus changing the final aspect of the object) [10–12]. One way to control the impact of art restoration is by managing the drying times: by setting the best manipulation times and so reducing surface artefacts. The timing of the evaporation of the solvents present in inks can be measured using thermogravimetric techniques [13], as well as by weighting the canvas during drying [14] (with some obvious limitations for in situ restoration).

Dynamic laser speckle (DLS), or biospeckle laser (BSL), is a non-destructive technique that monitors biological and non-biological activity. This technique can be applied on various materials with differing behaviors. Complex fluid applications include the motility of frozen bovine semen [15], blood flow [16,17], bacterial chemotaxis [18], as well as in the reaction of MEL-RC08 line cancer cells to the application of the drug Colcemid™ (Gibco BRL, Grand Island, NY, USA) [19]. In fluids, the approaches must be biased to accommodate the considerable movement produced by light scatterers within colloid or biological tissues.

The dynamics of the processes are critical in DLS, and adjustments are required in the setup, as well as in the choice of the best image processing method. In colloids, which is the case of paintings, the literature discusses DLS and paint drying [20–23], where the level of activity of the scatterers greatly varies from the moment the ink is applied, with high levels of volatilization until the end of the drying process. Thus, a method of monitoring the process without compromising the observation of the phenomenon deserves to be evaluated and tested.

The literature describes the use of commercial equipment to evaluate drying ink [24] or paint, and is presented as an alternative for use outside of optical laboratories, where the many external influences present a challenge [25] when dealing with slow paint drying processes.

Therefore, the challenge is to monitor the speckle phenomenon in situ where external influences create a barrier to many applications, and evaluate the robustness of the portable equipment [25,26]. The adjustment of the speckle analysis must also be considered, particularly for processing events that occur over a long period of time. Therefore, the installed equipment must capture the information, as well analyzing the biased information.

This work aims to test the portable equipment for monitoring the activity of treatments during painting conservation in situ by means of a modified image processing biased for long-term measurement. Two types of chemicals were tested during drying, and they were evaluated using three laser wavelengths. The portable equipment was compared to a DLS laboratorial setup, as well as weight monitoring methods.

## 2. Design and Control of Portable System

### 2.1. Structural Design

Figure 1 shows the elements of the portable system for dynamic speckle patterns.

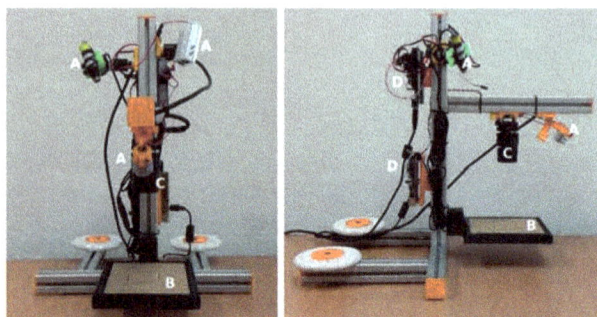

**Figure 1.** Portable experimental setup for the dynamic laser speckle. A: lasers; B: removable platform; C: camera; D: laser controller.

It can control up to four diode-lasers, enabling rear and frontal illumination using prisms and a removable platform. The images are captured with a CCD. All the elements are assembled using aluminum T-slots. This structure is sturdy enough to capture stable images. In addition, cushion pads are glued in the base to absorb vibrations. Camera height and laser orientation can be adjusted. The B platform (Figure 1) is removable for the placement of translucent material. This enables a rear illumination setup by means of prisms and laser reorientation. As portable equipment, it can be taken to the place of the conservation and the orientation of the camera and lasers can be adjusted to monitor the horizontal and vertical disposition of the paintings.

## 2.2. Electronic Design

The portable system control, as shown in Figure 2, was designed and developed at the Universitat Politècnica de València, in Spain.

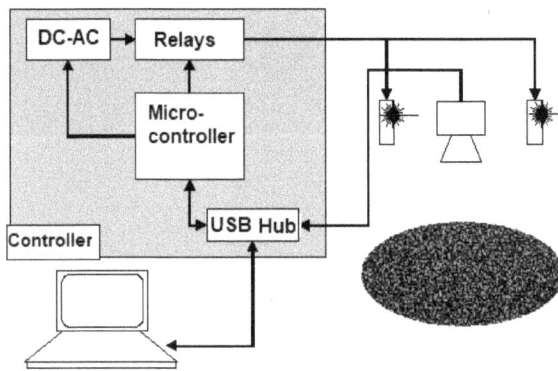

**Figure 2.** Diagram of portable system controller with the gray box representing the device control system connected to the computer, lasers, and camera.

For the experiments, three lasers were installed (infrared: 808 nm, 50 mW; red: 650 nm, 20 mW; green: 432 nm, 20 mW; from Prophotonix Limited, Salem, NH, USA). These lasers can be turned on/off using relays controlled by a microcontroller (Arduino, BCMI US LLC, Boston, MA, USA) connected to a laptop through a USB port. The camera is a USB 3.0 camera (See3Cam 2304 × 1296 pixels, 2.2 μm, no IR filter, e-con Systems Ltd, Chennai, India), also connected to a laptop. Software was developed to control image grabbing and laser synchronization. The lasers employed were not heavy-duty components; thus they can be turned on lasers on at a preselected time before capture. This reduces laser deterioration, and allow stabilization before each new capture. The software enables the periodical capture of groups of images at a specific frame rate. The capture surface can be illuminated with different lasers during the experiment. Figure 3 shows a portion of the captured speckle pattern generated on the surface by the red laser.

## 3. Materials and Methods

### 3.1. Specimen

The surface of an acrylic painted canvas was protected with two solutions:

- commercial acrylic resin varnish (Titan retouching varnish, INDUSTRIAS TITAN S.A., Barcelona, Spain, diluted in aliphatic and aromatic hydrocarbons).
- cyclododecane: saturated and chemically stable cyclic hydrocarbon ($C_{12}H_{24}$). This chemical is applied in conservation treatments on a variety of surfaces as a temporary protector. It was used as a solution of 5 g in 10 mL of ligroin (an aliphatic hydrocarbon mixture).

The process was performed by a professional painting conservator using a brush to apply homogeneous layers. Layers of a thickness of approximately 100 µm were applied considering the density of the products and the difference in weight of the canvas before/after application. The experimental conditions were 22–23 °C and 51–53% of humidity.

**Figure 3.** A portion of the captured speckle pattern created by the red laser diode used by the portable system (shows a real area of 4 × 2 cm).

### 3.2. Experimental Setup Employed for Non-Portable Dynamic Laser Speckle

The experimental setup for the non-portable system consisted of a linearly polarized He-Ne laser beam (633 nm, 35 mW, Research Electro-Optics, Inc., Boulder, CO, USA). The beam size was expanded using only a microscope objective with a 10× magnification in order have a round illuminated area with 100 mm of diameter covering a 40 mm square area.

The images were acquired from a TV zoom lens with a focal length of 50 mm, numerical aperture of f/11 (speckle size was 13.57 µm), connected to an AVT Marlin F-145B CCD camera (pixel size of 4.65 µm, Allied Vision Technologies, Stadtroda, Germany) [26].

### 3.3. Protocol for Speckle Pattern Acquisition

First experiment: validation of the portable system. A collection of 64 images (8 bits, 640 × 480 pixels, and an exposure time of 1/125) were acquired using a traditional experimental system (at the beginning of each minute, during 16 min at a rate of 10 frames per second). The portable system was configured in the same way: sets of 64 images, 640 × 480, 10 fps, during 16 min. In both cases, canvases of 60 × 60 mm were painted using the cyclododecane and varnish.

Second experiment: a comparison of back and forward scattering using the portable equipment was made. The portable device has the flexibility to produce dynamic laser speckle using back and forward scattering approaches (related to the reflection and transmission of light on and through the sample respectively). The transmission can only be used when the sample allows the light to pass through the sample to the camera. A layer of cyclododecane was applied to a glass surface that was illuminated using back and forward scattering, and using the same image time rate and processing that was presented in the first experiment.

Third experiment: time rate changes and DLS with different lasers were compared to the weight monitoring method. A uniform layer was applied on the canvases. One of the canvases was placed in the dynamic laser speckle capture system and another was placed on a scale to be weighed.

The canvas was weighed using a 1 mg precision scale (GEM50 Smart Weight, Better Basics Ltd., Chestnut Ridge, NY, USA) every five minutes initially—and then every 10 min—and finally every 30 min.

The canvas was illuminated every minute by the three lasers (infrared, red, and green) alternatively, and 20 images were captured at 10 fps (frames per second). Before each image acquisition, the corresponding laser was connected for five seconds before starting the capture of images to ensure light stability. Light stability was checked for every laser using an optical power meter (OPM 842-PE, Newport Ltd., Irvine, CA, USA).

Five sets of images were created after the capture process for each laser:

- Set A: Capture of 20 images at 10 fps every minute (for fast dynamics).
- Sets B, C, D, E: Capture of {30, 15, 6, 3} images at {1 image every 1 min, 1 image every 2 min, 1 image every 5 min, 1 image every 10 min} every 30 min (for slow dynamics).

This enables measuring the drying process in the same experiment using two methods. In all cases, image quality was tested to avoid speckle grains with unneeded information about the phenomena. Therefore, the setup was biased to avoid speckle with a blurred appearance and saturated areas, and to avoid inhomogeneity in accordance with the proposed quality test protocol [27,28].

*3.4. Methodology to Process Dynamic Speckle Images*

The dynamics of the speckle variation were monitored using second-order statistics [29], building the time history speckle pattern (THSP) matrices and co-occurrence matrix (COM) using a selection of random points in the prime image to create the THSP [30]. From the COM, we obtained the absolute values of differences (AVD) method [31], expressed in Equation (1):

$$AVD = \sum_{ij} COM_{ij}|i - j| \tag{1}$$

where the COM is the co-occurrence matrix related to the THSP, and the $i$ and $j$ variables represent the $i$ line and the $j$ column of each point of the COM matrix:

$$COM = [N_{ij}] \tag{2}$$

The entries are the number of occurrences (N) of a certain intensity value $i$ that is immediately followed by an intensity value $j$.

In Figure 4, it is possible to see the THSP created—instead of using random points in the prime image in the selected ROI.

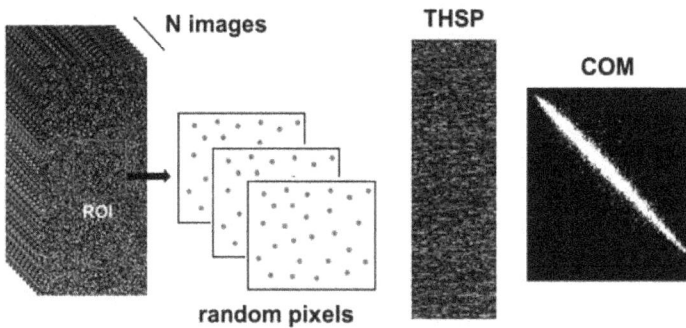

**Figure 4.** Scheme of the THSP and COM construction using random pixels.

Because the pixel set to compute AVD is randomly chosen, an AVD index was computed averaging the AVD values from ten pixels sets. The error bars appearing in the figures represent the standard deviations of those values.

The AVD index is used as a measure of light scattering that can be generally associated with activity. Activity is attributed to the numerous phenomena present during the drying processes. Thus, in this context, activity should be read as solvent evaporation and other curing reactions [32] that occur during the drying process, including the adjustment of the surface.

## 4. Results

### 4.1. Validation of the Portable Device to Monitor Drying Processes in Painting Conservation

Figure 5 shows the AVD index during the varnish and cyclododecane drying process using the traditional experimental configuration and the proposed portable system. The comparison between the traditional method and the portable system was made using a characteristic drying curve, expressed by an exponential [33,34]. The fitting of the data to an exponential behavior was tested using the $R^2$ index. The portable system had a lower fitting using the common method to monitor the short-term process, but still had a high $R^2$ value, which revealed its reliability.

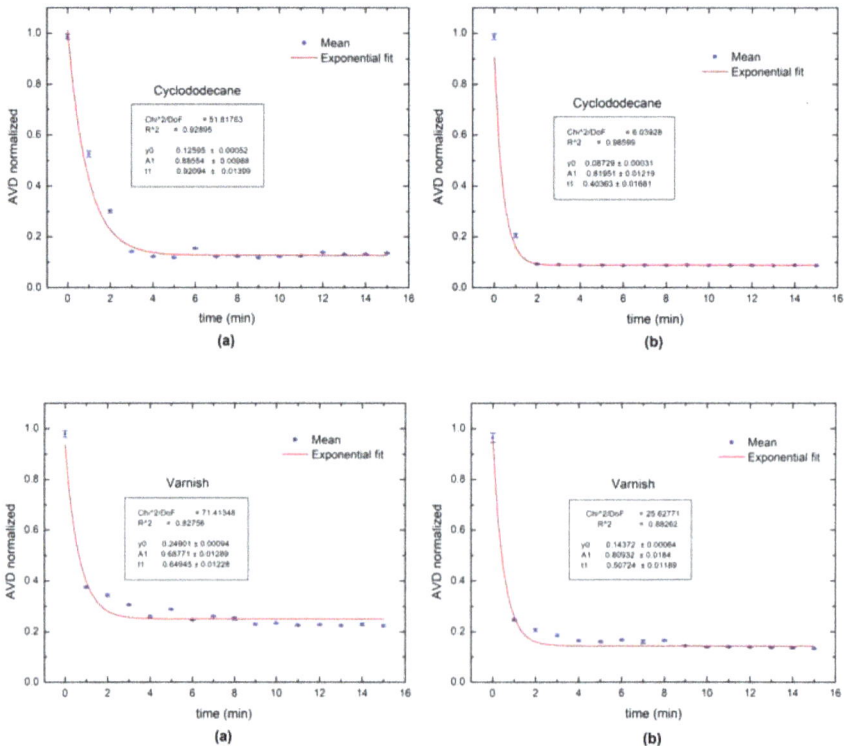

**Figure 5.** Dynamic laser speckle activity levels during the paint drying process for each painting treatments: (**a**) portable system and (**b**) lab system.

### 4.2. Comparison of Back and Forward Scattering Using the Portable Equipment

Figure 6 shows the AVD index with an exponential tendency curve when using a red laser. Two experimental configurations were used to acquire the data: back and forward scattering. In Figure 6a, the drying process of cyclododecane was sensed from 1 to approximately 0.1 in normalized values, while in Figure 6b the drying process was sensed from 1 to approximately 0.3 in normalized values.

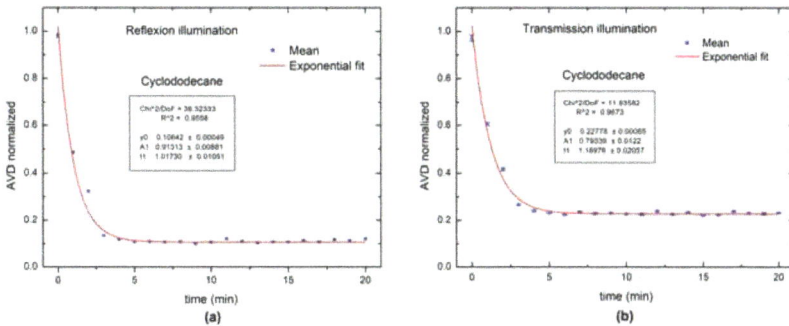

**Figure 6.** Dynamic laser speckle activity levels during paint drying process for cyclododecane: (**a**) back scattering (reflection illumination); (**b**) forward scattering (transmission illumination).

*4.3. Third Experiment: Changes in Time Rate Using Different Lasers Compared to Weight*

The results of the proposed method to monitor long-term activities in the drying process can be seen in Figure 7, where the drying times for cyclododecane (Figure 7a) and varnish (Figure 7b) are shown. In a time-rate of 1 frame per minute, 1 frame each 2 min, 1 frame each 5 min, and 1 frame each 10 min, the drying process was monitored during 16 h. The AVD index was based on 30, 15, 6 and 3 images respectively. In Figure 7a, we can divide the curve in two parts, the first from zero to 3 h, and the second part from 8 to 16 h. The zone between 3 and 8 h can be considered as transient. In the first part, the curve behaves similarly to the drying process presented in traditional measurements of dynamic laser speckle using higher time rates (10 fps, for example, in the case of Figure 6). It is comparable with the loss of mass monitored by the scale shown in Figure 8.

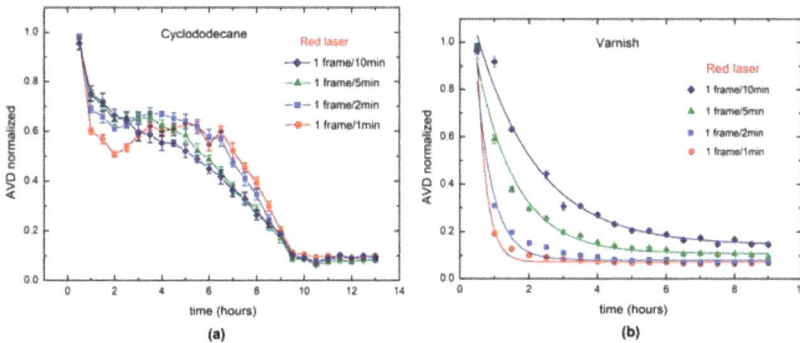

**Figure 7.** Dynamic laser speckle activity levels during the paint drying process for each painting treatment: (**a**) cyclododecane and (**b**) varnish.

However, the similarity was greater for the higher time rates, such as 1 frame per minute (1 frame/1 min) and 0.5 frame per minute (1 frame/2 min). While in the second part of the curve, the behavior is completely different from the traditional curve, shown in Figure 6, with an ability to sense small changes in the cyclododecane drying surface (where the traditional method presented a flat curve). For varnish, the behavior shown is similar to the traditional methods of drying, and this can be validated using a scale to weigh the mass loss.

Figure 8 shows the weighing process during the drying process using a scale. The speed of the drying process for the two treatments (varnish and cyclododecane) is shown. The cyclododecane shows a slower drying curve.

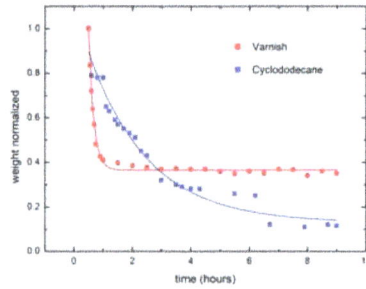

**Figure 8.** Weights of drying paint treatments over time.

The DLS analysis with different wavelength lasers for the varnish and cyclododecane is presented in Figure 9. Normalized AVD indexes were obtained. The varnish was more sensitive to the wavelengths than the cyclododecane. An fps rate of 10 was used to acquire the images in all cases. For fast dynamics, the cyclododecane was the first colloid to dry, in contrast with the weight monitoring observation (Figure 8).

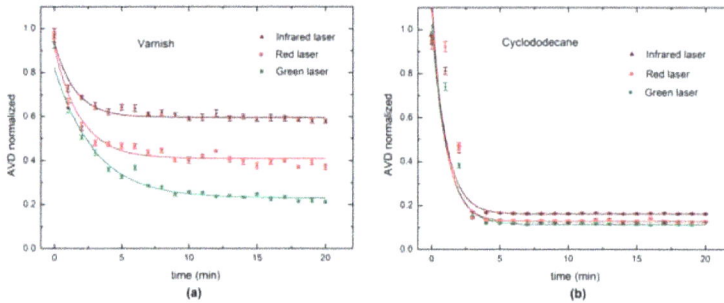

**Figure 9.** Dynamic laser speckle for fast drying dynamics for colloids using three wavelengths for each painting treatment: (**a**) varnish and (**b**) cyclododecane.

Figure 10 shows the DLS and weighing values for the three wavelength lasers. Varnish and cyclododecane were monitored during 20 min, and analysed using fast dynamics, i.e., 10 fps. The fast dynamics presented the best ability to follow the fastest drying process produced by the varnish, but the DLS could not sense the cyclododecane that presented the slowest drying dynamic.

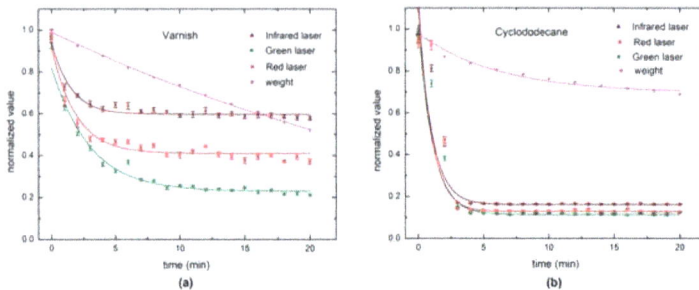

**Figure 10.** Dynamic laser speckle index for fast dynamics (10 fps) and the weighing for each painting treatment: (**a**) varnish and (**b**) cyclododecane.

The slow (long-term) dynamic is presented in Figure 11 for varnish drying, and the slow dynamic is also compared for weighing and the three lasers (wavelengths). Four time-rates were used to evaluate the long-term process response. The slower the time-rate, the greater is the difference in the weighing process. In this case, the acquisition of 1 frame each 10 min provided the best ability to sense changes in long-term monitoring. Otherwise, weighing output did not have the ability to follow the drying process after one hour.

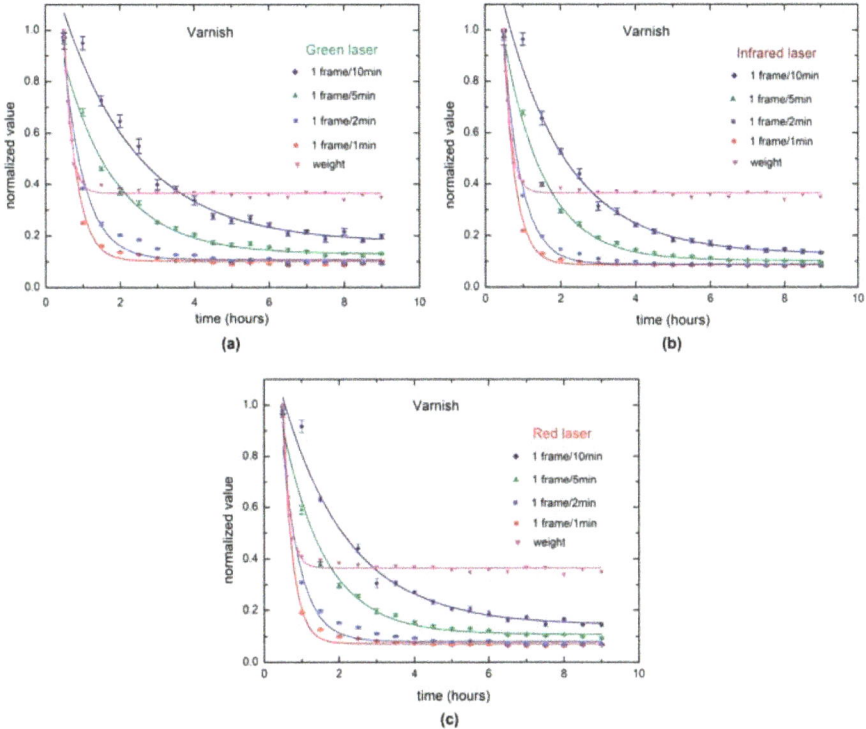

**Figure 11.** Dynamic laser speckle in low dynamics for varnish drying using the three wavelengths: (a) green laser (b) IR laser (c) red laser, as observed in four different time-rates and compared to the weighing.

The data was adjusted (Equation (3)), and the value of the variable 'y'; 'y$_0$'; t$_1$'; and 'A$_1$'are shown in Table 1.

$$y = y_0 + A_1 e^{\left(\frac{-x}{t_1}\right)} \tag{3}$$

**Table 1.** Exponential fit for drying varnish with low dynamics using three wavelengths.

| | Green Laser | | | | Infrared Laser | | | | Red Laser | | | |
|---|---|---|---|---|---|---|---|---|---|---|---|---|
| | y$_0$ | A$_1$ | t$_1$ | % | y$_0$ | A$_1$ | t$_1$ | % | y$_0$ | A$_1$ | t$_1$ | % |
| 1 image/1 min | 0.10 | 0.84 | 0.35 | 91.5 | 0.09 | 0.83 | 0.31 | 94.0 | 0.07 | 0.85 | 0.53 | 89.6 |
| 1 image/2 min | 0.11 | 0.79 | 0.55 | 93.6 | 0.09 | 0.85 | 0.47 | 97.4 | 0.08 | 0.75 | 0.96 | 91.6 |
| 1 image/5 min | 0.13 | 0.75 | 1.41 | 98.4 | 0.10 | 0.86 | 1.07 | 98.9 | 0.11 | 0.80 | 2.26 | 97.7 |
| 1 image/10 min | 0.17 | 0.90 | 2.04 | 98.2 | 0.13 | 0.97 | 1.67 | 98.7 | 0.14 | 0.89 | 3.50 | 98.7 |

Table 1 shows the parameters of the fitting curves for low dynamics and the expected characteristic curve of the drying varnish is clearly seen.

In Figure 12, the result of a long-term monitoring of cyclododecane is presented in comparison to the weighing and to three lasers with different wavelengths. The ability of each time-rate to follow the process was also observed. The cyclododecane dries in the external layer first, and then the drying process continues to the inner layers. In this case, the weighing process could follow the slow drying during the hours before stabilization (which also happened with the DLS outputs using low time-rates).

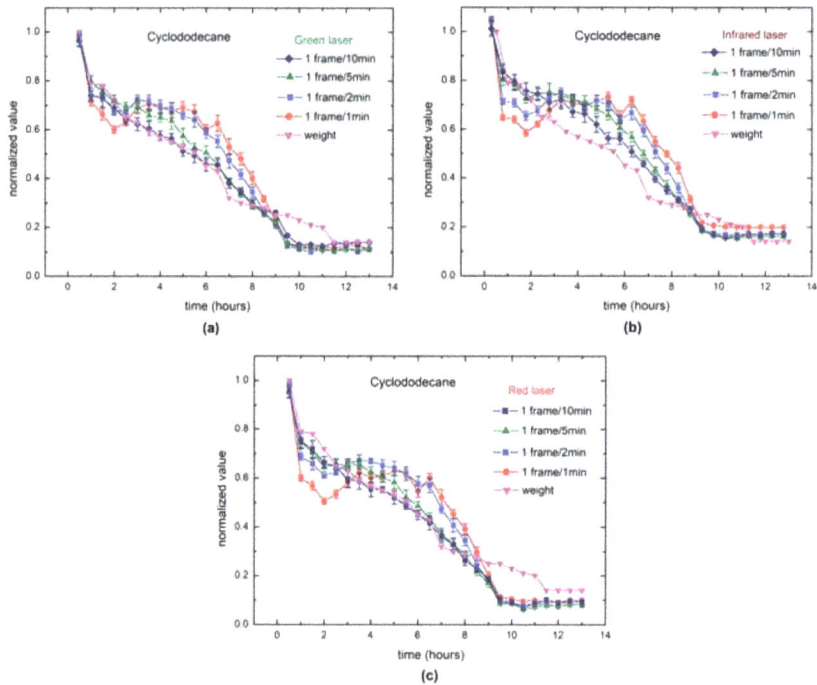

**Figure 12.** Dynamic laser speckle for drying cyclododecane with low dynamics using three wavelengths: (**a**) green laser (**b**) IR laser (**c**) red laser, observed in four different time-rates and compared to the weighed values.

## 5. Discussion

### 5.1. Validation of the Portable Device for Monitor Drying Processes during Painting Conservation

Beyond the ability of the proposed device to follow the drying process, it is relevant to highlight its ability to sense the process more smoothly than the traditional setup because of the different cameras used. In the portable device, the size of the pixel in the CCD camera was 2.2 μm, while the size of the pixel in the optical laboratory was 4.65 μm. The smoothness of the exponential curve from the proposed device enables us to follow the drying process for a longer time than when using the camera in the optical laboratory. In short, pixel size matters, and can improve the sensitivity of our sensor [26].

### 5.2. Comparison of Back and Forward Scattering Using the Portable Equipment

The ability of the back scattering to sense the drying process of cyclododecane for a wider range (from 1 to approximately 0.1 in normalized AVD index values) than forward scattering (from 1 to approximately 0.3 in normalized AVD index values) can be attributed to the lesser sensitivity of the transmission in the dynamic laser speckle outcomes [35]. This leads us to adopt the back scattering reflection as prior when possible.

*5.3. Third Experiment: Changes in Time Rate Using Different Lasers Relative to Weight*

The long-term monitoring by means of DLS revealed a better ability to follow the drying process than the weighing process. This means that over the long-term, DLS when monitored is a better sensor than the traditional DLS with high time-rate (10 fps); and also better than the weighing process that could be considered as the 'gold standard'. The monitoring of drying paint using the traditional fps is usually restricted to the first minutes [20,23,36].

The different dynamics of drying presented by varnish and cyclododecane were better followed by the long-term methodology proposed, where for the cyclododecane, the presence of two phases of drying can explain the relation of the time-rate with fast and slow drying, each linked to the surface and inner layers of the sample.

Varying the laser wavelength can be worthwhile in fast dynamics, such it happens in biological tissues, particularly in arterial pulse [37], where a lower wavelength is more sensitive to small changes. For the varnish, the lower the wavelength, the longer the process can be sensed (Figure 11). The composition of colloid and fast drying at the surface may explain the ability of the lower wavelengths. In long-term dynamics, varying the wavelength does not show a difference in sensitiveness, but it can be useful to make the image capture adjustments in function of the color of the studied surface.

The portable equipment produced reliable results, thus offering the facility to be used in situ. The use of the IR laser (offering independence from external light) makes the equipment more robust. Commercial equipment using DLS for drying ink also uses IR [24], but is limited to fast drying dynamics and this restricts its use in long-term monitoring.

## 6. Conclusions

Portable equipment for monitoring the activity of painting treatments during the restoration of paint in situ presented reliable outcomes that were comparable with the equipment used in an optical laboratory. In addition, the portable equipment is highly configurable (different wavelength lasers, back and forward scattering illumination, automatic image capture, and laser control). With these features, the system is especially interesting when used for painting conservation (where painting treatments should be monitored in place). In other areas of knowledge, the portable equipment can be used to monitor activity in the analysis of seeds, sperm motility, fruit maturation, as well as in food evaluation tasks completed outside optical laboratories.

Modified image processing biased to a long-term measurement presented better results than the traditional method driven by fast dynamics data acquisition and analysis. The proposed long-term methodology sensed different drying dynamics. The testing of laser wavelengths proved that the most accurate measurements can be obtained in fast dynamics, and improvements can generally be obtained regarding the sample and light interaction. Weight monitoring proved to be less sensitive to long-term changes in colloids during a slow drying process, as is the case of varnish. The weight monitoring setup may not be possible in real painting conservation work, or even if it is possible, it could be more difficult to implement that the proposed DLS method.

**Acknowledgments:** This work was partially funded by Generalitat Valenciana project AICO/2016/058 and by the Plan Nacional de I+D, Comisión Interministerial de Ciencia y Tecnología (FEDER-CICYT) under the project HAR2013-47895-C2-1-P and project HAR2017-85557-P.

**Author Contributions:** Alberto J. Pérez, Rolando J. González-Peña, and Roberto Braga Jr. conceived, designed, and performed the experiments; and analyzed the data and wrote the paper. Ángel Perles analyzed the data and wrote the paper. Eva Pérez-Marín and Fernando J. Garcia-Diego contributed reagents/materials and wrote the paper.

**Conflicts of Interest:** The authors declare no conflict of interest.

## References

1. Schnars, U.; Falldorf, C.; Watson, J.; Jüptner, W. *Digital Holography and Wavefront Sensing*; Springer: Berlin/Heidelberg, Germany, 2015.
2. Kim, M.K. Principles and techniques of digital holographic microscopy. *SPIE Rev.* **2010**, *1*, 018005. [CrossRef]
3. Yokota, M.; Kawakami, T.; Kimoto, Y.; Yamaguchi, I. Drying process in a solvent-based paint analyzed by phase-shifting digital holography and an estimation of time for tack free. *Appl. Opt.* **2011**, *50*, 5834–5841. [CrossRef] [PubMed]
4. Yamaguchi, I.; Ida, T.; Yokota, M.; Kobayashi, K. Monitoring of Paint Drying Process by Phase-Shifting Digital Holography. In *Adaptive Optics: Analysis and Methods/Computational Optical Sensing and Imaging/Information Photonics/Signal Recovery and Synthesis Topical Meetings on CD-ROM, Paper DWC2*; Optical Society of America: Washington, DC, USA, 2007. [CrossRef]
5. Krzemień, L.; Łukomski, M.; Kijowska, A.; Mierzejewska, B. Combining digital speckle pattern interferometry with shearography in a new instrument to characterize surface delamination in museum artefacts. *J. Cult. Herit.* **2015**, *16*, 544–550. [CrossRef]
6. Lasyk, L.; Lukomski, M.; Bratasz, L. Simple digital speckle pattern interferometer (DSPI) for investigation of art objects. *Opt. Appl.* **2011**, *41*, 688–700.
7. Trumpy, G.; Conover, D.; Simonot, L.; Thoury, M.; Picollo, M.; Delaney, J.K. Experimental study on merits of virtual cleaning of paintings with aged Varnish. *Opt. Express* **2015**, *23*, 33836–33848. [CrossRef] [PubMed]
8. Reifsnyder, J.M. A note on a traditional technique of Varnish application for paintings on panel. *Stud. Conserv.* **1996**, *41*, 120–122. [CrossRef]
9. Caley, T. Aspects of Varnishes and the Cleaning of Oil Paintings Before 1700. *Stud. Conserv.* **1990**, *35*, 70–72. [CrossRef]
10. Brückle, I.; Thornton, J.; Nichols, K.; Strickler, G. Cyclododecane: Technical Note on Some Uses in Paper and Objects Conservation. *J. Am. Inst. Conserv.* **1999**, *38*, 162–175. [CrossRef]
11. Rowe, S.; Rozeik, C. The uses of cyclododecane in conservation. *Stud. Conserv.* **2008**, *53*, 17–31. [CrossRef]
12. Maish, J.P.; Risser, E. A case study in the use of cyclododecane and latex rubber in the molding of marble. *J. Am. Inst. Conserv.* **2002**, *41*, 127–137. [CrossRef]
13. Lenk, R.S.; Fellers, J.F.; White, J.L. Comparative Study of Polyamides from Bisacid A2. *Polym. J.* **1977**, *9*, 9–17. [CrossRef]
14. Koleske, J.V. *Paint and Coating Testing Manual*, 15th ed.; ASTM International: West Conshohocken, PA, USA, 2012.
15. Carvalho, P.H.A.; Barreto, J.B.; Braga, R.A., Jr.; Rabelo, G.F. Motility parameters assessment of bovine frozen semen by biospeckle laser (BSL) system. *Biosyst. Eng.* **2009**, *102*, 31–35. [CrossRef]
16. Richards, L.M.; Kazmi, S.M.S.; Davis, J.L.; Olin, K.E.; Dunn, A.K. Low-cost laser speckle contrast imaging of blood flow using a webcam. *Biomed. Opt. Express* **2013**, *4*, 2269–2283. [CrossRef] [PubMed]
17. Ganilova, Y.A.; Ulyanov, S.S. A study of blood flow in microvessels using biospeckle dynamics. *Biophysics* **2006**, *51*, 299–304. [CrossRef]
18. Murialdo, S.E.; Sendra, G.H.; Passoni, L.I.; Arizaga, R.; Gonzalez, J.F.; Rabal, H.; Trivi, M. Analysis of bacterial chemotactic response using dynamic laser speckle. *J. Biomed. Opt.* **2009**, *14*, 064015. [CrossRef] [PubMed]
19. González-Peña, R.J.; Braga, J.; Roberto, A.; Cibrián, R.M.; Salvador-Palmer, R.; Gil-Benso, R.; Miguel, T.S. Monitoring of the action of drugs in melanoma cells by dynamic laser speckle. *J. Biomed. Opt.* **2014**, *19*, 057008. [CrossRef] [PubMed]
20. Arizaga, R.; Grumel, E.E.; Cap, N.; Trivi, M.; Amalvy, J.I.; Yepes, B.; Ricaurte, G. Following the drying of spray paints using space and time contrast of dynamic speckle. *JCT Res.* **2006**, *3*, 295–299. [CrossRef]
21. Faccia, P.A.; Pardini, O.R.; Amalvy, J.I.; Cap, N.; Grumel, E.E.; Arizaga, R.; Trivi, M. Differentiation of the drying time of paints by dynamic speckle interferometry. *Prog. Org. Coat.* **2009**, *64*, 350–355. [CrossRef]
22. Mavilio, A.; Fernández, M.; Trivi, M.; Rabal, H.; Arizaga, R. Characterization of a paint drying process through granulometric analysis of speckle dynamic patterns. *Signal Process.* **2010**, *90*, 1623–1630. [CrossRef]
23. Budini, N.; Mulone, C.; Balducci, N.; Vincitorio, F.M.; López, A.J.; Ramil, A. Characterization of drying paint coatings by dynamic speckle and holographic interferometry measurements. *Appl. Opt.* **2016**, *55*, 4706–4712. [CrossRef] [PubMed]

24. Brunel, L.; Brun, A.; Snabre, P. Microstructure movements study by dynamic speckle analysis. *Int. Soc. Opt. Photonics* **2006**, *6341*, 634127. [CrossRef]

25. Braga, R.A. Challenges to apply the biospeckle laser technique in the field. *Chem. Eng. Trans.* **2017**, *58*, 577–582. [CrossRef]

26. Braga, R.A.; González-Peña, R.J. Accuracy in dynamic laser speckle: Optimum size of speckles for temporal and frequency analyses. *Opt. Eng.* **2016**, *55*, 121702. [CrossRef]

27. Moreira, J.; Cardoso, R.R.; Braga, R.A. Quality test protocol to dynamic laser speckle analysis. *Opt. Lasers Eng.* **2014**, *61*, 8–13. [CrossRef]

28. Ansari, M.Z.; Nirala, A.K. Biospeckle numerical assessment followed by speckle quality tests. *Opt. Int. J. Light Electron Opt.* **2016**, *127*, 5825–5833. [CrossRef]

29. Rabal, H.J.; Braga, R.A., Jr. *Dynamic Laser Speckle and Applications*; CRC Press: Broken Sound Parkway NW, FL, USA, 2008; ISBN 978-1-4200-6016-4.

30. Oulamara, A.; Tribillon, G.; Duvernoy, J. Biological Activity Measurement on Botanical Specimen Surfaces Using a Temporal Decorrelation Effect of Laser Speckle. *J. Mod. Opt.* **1989**, *36*, 165–179. [CrossRef]

31. Braga, R.A.; Nobre, C.M.B.; Costa, A.G.; Sáfadi, T.; da Costa, F.M. Evaluation of activity through dynamic laser speckle using the absolute value of the differences. *Opt. Commun.* **2011**, *284*, 646–650. [CrossRef]

32. Narita, T.; Beauvais, C.; Hébraud, P.; Lequeux, F. Dynamics of concentrated colloidal suspensions during drying–aging, rejuvenation and overaging. *Eur. Phys. J.* **2004**, *14*, 287–292. [CrossRef] [PubMed]

33. Puspasari, I.; Talib, M.Z.M.; Daub, W.R.W.; Tasirin, S.M. Characteristic Drying Curve of Oil Palm Fibers. *Int. J. Adv. Sci. Eng. Inf. Technol.* **2014**, *4*, 20–24. [CrossRef]

34. Bellagha, S.; Amami, E.; Farhat, A.; Kechaou, N. Drying kinetics and characteristic drying curve of lightly salted sardine (Sardinella aurita). *Dry. Tech.* **2020**, *20*, 1527–1538. [CrossRef]

35. Trivi, M. Dynamic Speckle: Origin and Features. In *Dynamic Laser Speckle and Applications*; CRC Press: Broken Sound Parkway NW, FL, USA, 2008.

36. Kooij, H.M.; Fokink, R.; Gucht, J.; Sprackel, J. Quantitative imaging of heterogeneous dynamics in drying and aging paints. *Sci. Rep.* **2016**, *6*, 34383. [CrossRef] [PubMed]

37. Vaz, P.; Pereira, T.; Figueiras, E.; Correia, C.; Humeau-Heurtier, A.; Cardoso, J. Which wavelength is the best for arterial pulse waveform extraction using laser speckle imaging? *Biomed. Signal Proc. Control* **2016**, *25*, 188–195. [CrossRef]

*sensors*

MDPI

*Article*

# Improving Odometric Accuracy for an Autonomous Electric Cart

**Jonay Toledo \*, Jose D. Piñeiro, Rafael Arnay, Daniel Acosta and Leopoldo Acosta**

Computer Science and System Department, Universidad de La Laguna, 38200 Santa Cruz de Tenerife, Spain; jpineiro@ull.es (J.D.P.); rarnayde@ull.es (R.A.); daniel.leo.ah@gmail.com (D.A.); lacosta@ull.es (L.A.)
* Correspondence: jttoledo@ull.es; Tel.: +34-922-8286

Received: 13 December 2017; Accepted: 10 January 2018; Published: 12 January 2018

**Abstract:** In this paper, a study of the odometric system for the autonomous cart Verdino, which is an electric vehicle based on a golf cart, is presented. A mathematical model of the odometric system is derived from cart movement equations, and is used to compute the vehicle position and orientation. The inputs of the system are the odometry encoders, and the model uses the wheels diameter and distance between wheels as parameters. With this model, a least square minimization is made in order to get the nominal best parameters. This model is updated, including a real time wheel diameter measurement improving the accuracy of the results. A neural network model is used in order to learn the odometric model from data. Tests are made using this neural network in several configurations and the results are compared to the mathematical model, showing that the neural network can outperform the first proposed model.

**Keywords:** autonomous vehicles; odometry; neural networks; Robotics

## 1. Introduction

Odometry is one of the basic localization systems in any autonomous vehicle [1,2]. It is based on the use of data from on-board sensors in order to estimate changes in position and orientation from the vehicle itself, and is subsequently used in many autonomous systems to estimate their position relative to a starting location, by integrating sensors measurements. However, this method is sensitive to errors due to the integration over time and the final position is usually not very accurate. Usually, the odometry output of a robot is very poor, it is only valid for a few meters, and needs others sensors to obtain a good localization system. Any small increment in odometry accuracy can improve the whole localization system a lot.

Usually, odometry is used in combination with positioning systems like GPS, lasers, radio frequency markers, natural or artificial beacons, and others [3]. When mechanical odometry is not available, visual odometry can be used [4] to estimate vehicle position from changes between images. A complete sensorial system for an autonomous vehicle is based on multiple sensors combined to get position and orientation [5,6]. Some algorithms used for this purpose are Kalman filters [7], particle filters based on Montecarlo simulation [8], etc. Sensors excluding odometry usually need external information obtained from the environment, so in many situations these sensors simply do not work correctly. For example, GPS loses coverage when the vehicle does not have a full sky vision [9]. As another examples, if we introduce beacons, we need to structure the whole environment, or a laser needs available features to recognize in the environment, and these features should be in the range of action of the laser (typically between 10–20 m) [10].

The main advantage of odometry is that all localization information comes from the robot itself. Odometry information is always available and usually it is the only localization information when other sensors are not able to provide data, so a good odometry based localization system is always necessary and it is usually the first step to localization [11,12], obstacle detection [13], and navigation [14].

In wheeled robots, odometry is based on the movement of each wheel. A rotation sensor (rotation optical encoder) is attached to each drive wheel of the robot, and, knowing the wheel diameter, it is possible to approximate linear displacement of each wheel. Using each wheel traslation and the separation between wheels, position, and orientation of the robot (pose) is calculated. All of the calculation is based on optical encoder information, which obtains, in real time, the rotation angle of each wheel. The sensor and all of the parameters can be affected by errors, so final pose based on odometry usually is very noisy. For more details, see Section 2.

The main disadvantage of odometry is incremental error; odometry starts in a known pose, and this pose is updated with small increments using the integration of information acquired from sensors. Errors grow very fast due to the integration of sensor data, so a continuous calibration system is crucial.

In this paper, an intelligent odometry system for an autonomous vehicle is presented. The VERDINO project, Figure 1, is being developed at La Laguna University Robotics Group (GRULL) in Tenerife (Spain). The project is aimed to navigate in ITER (Technological Institute for Renewable Energies) facilities in Tenerife.

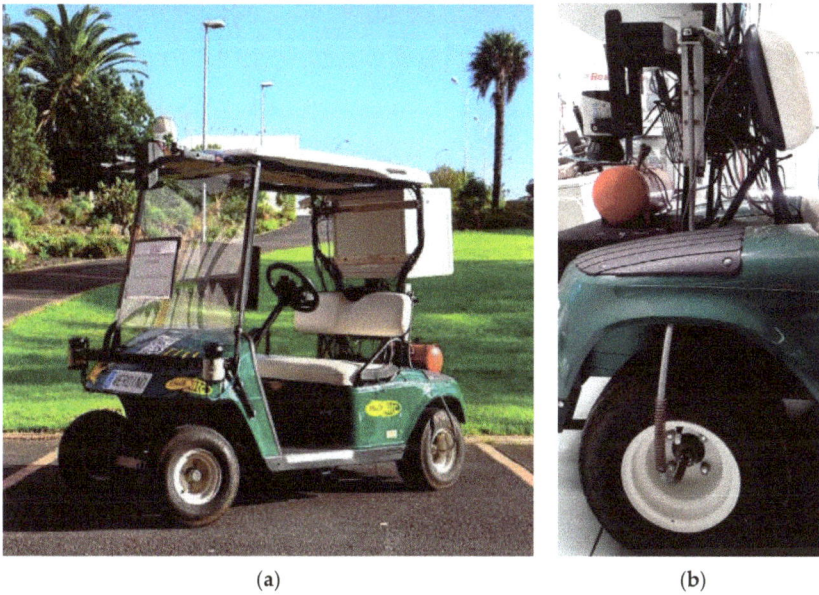

(a)                                           (b)

**Figure 1.** (a) Verdino autonomous golf cart; and, (b) odometric sensor based on a flexible link from the center of the wheel to the encoder placed in the vehicle.

The project consists of an electric vehicle, a standard EZGO golf cart, electronically and mechanically adapted for computer control, aimed for people transportation within a bioclimatic housing development. In the bioclimatic resort, no combustion cars are allowed, so all of the transport is made based on electric vehicles. The objective of the project is to build an on-demand autonomous transportation system, so the inhabitants of the village can use it as the internal transportation system between the houses and the central parking. The users have a request button in each house, so the car will go in an autonomous way to their entrance. The user rides on the car and selects its destination between available places, so the golf cart can drive to that destination, leave the passenger, and come back to the central station. The autonomous cart works like an internal taxi inside the bioclimatic area.

This vehicle features the following systems:

- Optic encoder based odometry system.
- Centimeter GPS system.
- Multiple LIDAR units.
- Stereo camera.

Based on this hardware infrastructure, the following features and capabilities have been developed:

- Vision algorithms for non-structured road edge detection and obstacle detection.
- Stereoscopic vision algorithms for object detection (in particular, pedestrians and animals).
- LIDAR map based localization.
- Obstacle detection and avoidance module.
- Navigation module with local and global objectives.
- Telemetry system to detect possible failures from the base station.
- User interface to set destination.

The vehicle is able to navigate autonomously in unstructured environments using the sensors that were located in the prototype, but there are situations in which no sensor besides odometry can provide information, so improving the odometry system will improve the full system performance. The localization subsystem of the prototype is based on odometry, the rest of sensors are included to improve accuracy when their information is available.

Other sensors, like gyroscopes, can be used in order to complement odometric information and improve accuracy. As examples, in [15], the information of the odometric system is fused with gyro information according to a set of rules; In [16], the rate angle is measurement using a gyroscope and the displacement using only one encoder; In [17], the authors intend to correct odometry errors when the robot is traversing a bump, trying to reduce non-systematic errors and getting robot angle in the bump based only on gyro information. In [18], an Unscented Kalman filter is used to fuse information from odometry and gyroscope, improving final accuracy.

The sensor used to detect front wheels angle is an encoder coupled to the steering wheel, this encoder gives enough information to drive autonomously the cart, but the accuracy of the angle is low. Some problems like the home position of steering angle and non-linearity between steering wheel and steering angle reduce its applicability to improving odometric accuracy. With this sensor, the model for obtaining a position based on speed and steering angle is worse than a badly calibrated odometry, so the authors have discarded it in order to improve odometry.

The main contribution of this paper is the application of a sensor set, including real time wheel diameter measurement to improve the accuracy of the odometric system for a golf cart, or similar Ackerman steering vehicles. The information generated by the sensors is processed using a neural network to include effects, which a static model does not represent. Previous works are based on differential robots with solid wheels that do not include this sensorial system. The application of neural networks to odometry in previous works is based on the detection of errors, and making parameter corrections in a mathematical model. In this paper, a continuous learning system based on neural networks is presented, able to adapt the model to changes in the system. This paper is organized as follows: in Section 2, equations that describe cart odometry are obtained. In Section 3, the model previously obtained is validated with real data. In Section 4, the odometric system is updated to include a wheel radius sensor, and Section 5 presents a neural network based odometry with some variations designed to improve its accuracy.

## 2. Odometric System

The odometric system is based on encoders coupled to rear wheels, as shown in Figure 1. Each encoder provides 1024 pulses per revolution and each revolution of the wheel generates a revolution of the encoder (1:1 coupling). Wheel rotation is transferred to encoders through a flexible mechanical transmission system that goes from the center of each wheel to the encoder placed on the side of the vehicle (see Figure 1). Encoder output is connected to an ad hoc electronics that samples the encoders signal every 0.5 ms. The electronics is designed to measure and integrate the encoder signals and the output is transmitted to the on-board computer every integration period of 20 ms. The integration is made in the microcontroller installed in the ad hoc electronics, based on Euler integration, collecting encoder increments for the integration time.

Kinematic state of the cart (pose) is described by its position (x, y) with respect to a fixed reference system and its orientation θ (angle between X axis of reference system and the cart longitudinal position).

When the prototype is turning, a circular trajectory is followed. The integration time is small enough to consider the trajectory curvature as constant. In Figure 2, initial (x$_i$, y$_i$) and final (x$_f$, y$_f$) position after and integration time are shown. Rear wheels displacements are obtained from the encoders and wheel size.

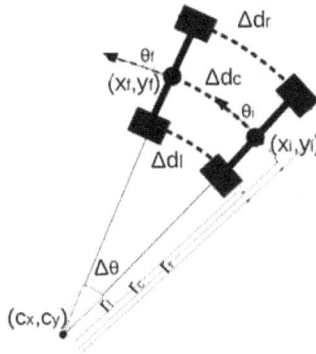

**Figure 2.** Odometric model.

The arc length of right (Δ$_{dr}$) and left (Δ$_{dl}$) rear wheels can be calculated based on encoder measurement (Δ$_{cr}$, Δ$_{cl}$), wheel radius (R$_r$, R$_l$), and encoder resolution (Enc$_r$, Enc$_l$).

$$\Delta d_r = \frac{2\pi R_r \Delta c_r}{Enc_r}$$
$$\Delta d_l = \frac{2\pi R_l \Delta c_l}{Enc_l}$$

(1)

Curvature radius for each wheel and center is calculated based on wheel distance $d_w$.

$$r_r = \frac{\Delta d_r d_w}{\Delta d_r - \Delta d_l}$$
$$r_l = \frac{\Delta l_r d_w}{\Delta d_r - \Delta d_l}$$
$$r_c = \frac{d_w}{2} \frac{\Delta d_r + \Delta d_l}{\Delta d_r - \Delta d_l}$$

(2)

With these assumptions, angle Δ$_θ$ and position (Δ$_x$, Δ$_y$) increments are:

$$\Delta_\theta = \frac{\Delta d_r - \Delta d_l}{d_w}$$
$$\Delta x = r_c(\cos(\theta)\sin(\Delta_\theta) - \sin(\theta)(1 - \cos(\Delta_\theta)))$$
$$\Delta y = r_c(\sin(\theta)\sin(\Delta_\theta) + \cos(\theta)(1 - \cos(\Delta_\theta)))$$

(3)

and in the last step, position and orientation are updated with the last computed increments:

$$\theta_{i+1} = \Delta_\theta + \theta_i$$
$$x_{i+1} = \Delta x + x_i$$
$$y_{i+1} = \Delta y + y_i$$

(4)

The odometry model only depends on three free parameters, wheel radiuses ($R_r$, $R_l$) and wheel separation $d_w$. According to this model, the only step to tune the odometric system is to measure these parameters as accurately as possible.

## 3. Odometric Validation

In order to validate equations presented in Section 2, some tests are made using our platform Verdino. A differential GPS is used as a ground truth, and data from odometry synchronized with GPS positioning is obtained. The GPS is a centimetric DGPS (a JAVAD GNSS Triumph-14, JAVAD GNSS Inc., Rock Avenue, San Jose, CA 95131, USA) with a horizontal precision below 1 cm and a vertical precision of around 1.5 cm using Differential GPS (DGPS) at a 5 Hz frequency. DGPS gives real time information about position accuracy; this information is calculated based on the number of available satellites, and position optimization results. In order to get the ground truth, the data is captured with good weather conditions with clear sky. The data is analyzed and if some error is detected, then this dataset is discarded. The data set used in this study has an error less than 1.5 cm in the entire traveled path.

The calibration of mobile robots odometry is a well-studied problem. In [19], a study of odometry error sources is presented, classifying errors in systematic due to errors in parameters and non-systematic due to wheel slippage and similar errors. A method to correct systematic odometry errors in mobile robots, based on a known indoor circuit is presented in [20]. The final position, and the ideal final position are compared, and the parameters are adjusted offline to minimize odometry errors. This method works well in controlled situations, with indoor circuits with smooth surfaces and curves, where nonsystematic errors are minimized. However, in outdoors, external environments, where non-systematic errors can appear, the circuit is not even or smooth, it is more convenient a continuous pose ground truth calibration as presented below.

Figure 3 shows an image of the Computer Science parking at University of La Laguna where tests were made. The red line shows the actual path of the vehicle based on GPS sensor. The parking has a slope of 10 degrees and some curves; these features allow for us to reproduce in the tests many possible situations present in a real road.

**Figure 3.** Parking of Computer Science in University of La Laguna, in red the actual path traveled in the experiments.

The first tests were made using the model presented in Section 2. A least square optimization of wheel radiuses ($R_r$, $R_l$) and wheel distance $d_w$ is made in order to get the best static parameters in order to minimize global error. The function to minimize is shown in Equation (5).

$$\min_{R_l, R_r, d_w} \|f(R_l, R_r, d_w)\|^2 = \min_{R_l, R_r, d_w} \left( \begin{array}{c} x_{(i+N)} - xGPS_{(i+N)} \\ y_{(i+N)} - yGPS_{(i+N)} \end{array} \right)^2 \tag{5}$$

In order to get data for minimization, the position is started at current position ($x_i$, $y_i$), and evolved $N$ steps using data from encoders, according to Equation (4). Odometric and GPS data are synchronized, so error is obtained when comparing the position after evolving the odometry pose ($x_{i+N}$, $y_{i+N}$) with the real position N steps away from the gps data ($xGPS_{i+N}$, $yGPS_{i+N}$). This error is used to tune the free optimization parameters ($R_r$, $R_l$, $d_w$). This process is made for all of the points in the data set, getting the parameter values which minimize the global error.

The number $N$ of iterations is fixed to 250, this distance is 5.5 m in average with a maximum distance of 10.3 m which is a medium distance in order to calculate the error. In 250 iterations, the distance traveled by the cart is not too small so the error between odometric position and GPS position is appreciable and not so big that the influence in each parameter is blurred due to the long path traveled and compensating errors at the final position. In Table 1, the measured and calculated parameters are shown. The results from the optimization process are quite similar and just fix small imprecisions in the odometry building process; however, the difference in the final trajectories with these parameters is clear.

**Table 1.** Parameter estimation.

| Parameters | $R_r$ | $R_l$ | $d_w$ |
|---|---|---|---|
| Estimated | 0.2332 m | 0.2295 m | 0.978 m |
| Measured | 0.215 m | 0.215 m | 0.97 m |

Results from measurement and estimated parameters are shown in Figure 4, where red plot represents ground truth. The starting point is marked by an O and the final point with a X. The green solid plot in left represents position evolution using the model described in Section 2 and the measured parameters from Table 1. Results are reasonable in straight path sections, but when Verdino takes a curve, the error increases quickly. This result is not a bad result, odometry is only valid for a few meters and outputs like this are expected in this kind of scenarios, due to parameter change dependence an error accumulated in the integration phase. As expected, odometry gives good positions when the displacement is small, but error increases rapidly. Odometry based on optimized parameters is shown Figure 4 right. Behavior is better and the final position of ground truth and odometry is close but still not as good as desired. The error between ground truth and odometric pose based on measurement parameters is $2.3204 \times 10^7$ m. This error is calculated as the sum of all the differences between odometric position and ground truth, so it includes the accumulative error in the whole path. In the case of optimized parameters the error is $1.1444 \times 10^7$ m using the same method, as expected the error is reduced significantly. In the parameter measurement experiment, the difference between ground truth and odometric pose after 1 s evolution is in average 0.2431 m, and 1.234 m after 5 s. For optimized parameters, this error for 1 s is 0.1976 m and for 5 s is 0.9305 m.

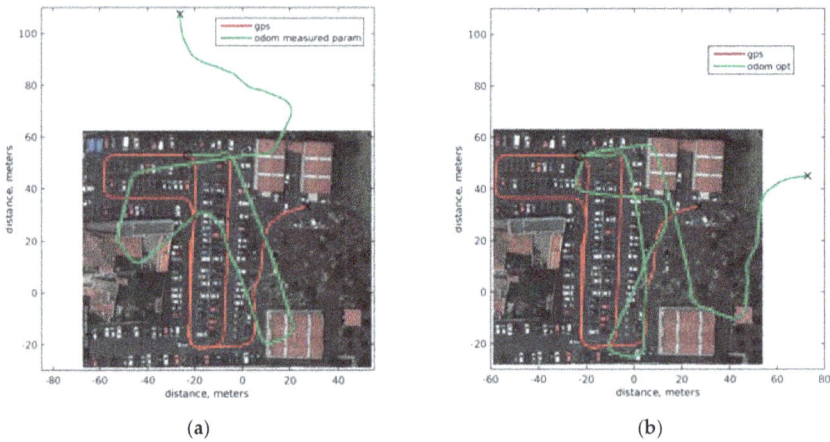

**Figure 4.** Optimization model results, solid red ground truth based in GPS. (**a**) solid green measurement parameters; (**b**) solid green optimization results.

This test shows that the model presented in Section 2 is not good enough to represent all of the effects that are present in cart odometry. Wheels are not rigid solids, so their radiuses change according to road conditions and the forces exerted. Changes in wheel diameters that cannot be described in the model are: changes in wheel diameters due to change in inclination during curves, loss of pressure of wheels, or changes in distribution of weight in the cart. All of these effects cannot be reflected in the model and can reduce the accuracy of results.

## 4. Wheel Radius Model

In order to improve accuracy results, a new sensor is included in the system to measure in real time wheel radius Figure 5. The range sensor is based on a Sharp short range optical sensor (2D120X measurement between 4~30 cm) placed close to wheel shaft and pointing to ground at a distance of 8 cm. The range sensor gives a measurement each 40 ms. The sensor is installed under the vehicle, so it is not affected by sun light. It includes an output capacitor to reduce electronic noise. Sensor output is shown in Figure 6, the output oscillation is due to vibrations and road surface rugosity. Using this sensor, a synchronized measurement of the radiuses can be made in real time. The sensor installed in Verdino is shown in Figure 5 and the output data representing wheel distance to the floor in Figure 6. Sensor is connected to an ad hoc electronics that gets the floor distance to sensor and calculates the wheel diameter. The sensor has millimetric resolution and is placed close to the ground in order to improve accuracy.

**Figure 5.** Range sensor used to get a real time radius measurement used to improve odometry accuracy.

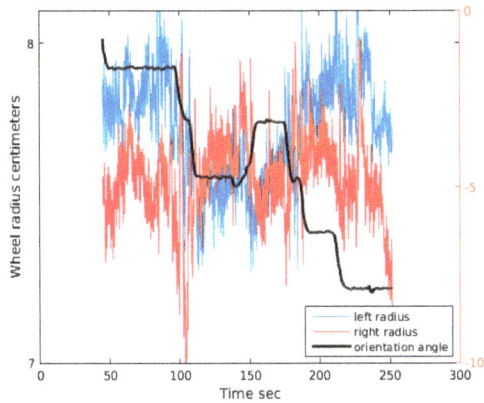

**Figure 6.** Wheel radius for left (blue) and right (red) wheels. In black is the orientation angle of the cart in radians.

As seen in Figure 6, the wheel radius is far from being a constant and depends on the current cart maneuver. For example, when the cart is turning left, approximately at 100 s. in the chart, the right radius increases and the left radius decreases. This small change in wheel sizes is not included in the previous model, so integrating this information in the model, a similar optimization process like that described in Section 3 can be carried out, but now including two new parameters to account for the wheel radius changes.

Figure 7 shows in solid green the optimization including the real time wheel radius calculation. Results are better with radius as expected, so information of wheel radiuses increases accuracy in final position calculation. As seen in chart, the main difference between the two techniques is in the turns, straight paths look quite similar, but turns are better represented including wheel radius changes. The error for odometric based optimization accumulated for all points is $1.1444 \times 10^7$ m, and when wheel radius changes are considered the error is $4.8561 \times 10^6$ m, giving a 56% reduction. The average difference between position in ground truth and the radios optimization after 1 s of evolution is 0.1972 m and after 5 s is 0.9147 m.

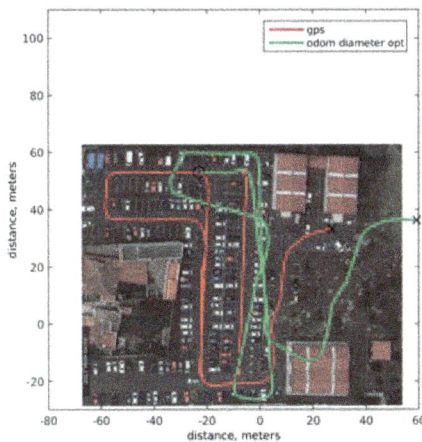

**Figure 7.** Optimization model including wheel radius change sensor; solid red ground truth based GPS, green, optimization result including radius.

It is clear that including a sensor to measure wheel radius improves the odometric localization, but this sensor has not yet enough accuracy to be the only navigation sensor. In order to improve results, new techniques are tested and compared with model based optimizations.

## 5. Neural Network Model

Another way to face of the odometry problem is to avoid calculating a mathematical model from cart dynamic equations, but instead learning a model based on data. This kind of models are very flexible and can take into account other factors which affect odometry and are not considered in a static mathematical model like that presented in Section 2. Some effects that can be reflected by a learning model are speed, turn direction, changes in spin rate, wheel slippage, etc.

Neural Networks are learning algorithms inspired by biological neurons [21,22]. Networks can be used for learning nonlinear complex mathematical functions from data. The function to learn is represented by a set of artificial neurons that receive an input and compute an output as weighted product sums. The weights are computed on a training phase where network output is compared to desired output. Function gradient changes are used to tune weight from various iterations until error reach a minimum or maximum number of iterations are reached. Neural networks are a general interpolator, so with careful training a good generalization can be obtained.

Other authors have used neural network in order to improve odometry accuracy, as in [23] where a neural networks is used to correct the errors generated in simulation by a mathematical fixed odometric model. In [24], neural networks are used to estimate the odometry error of a mobile robot.

In this paper, a feed-forward neural network is trained using data from odometry and GPS data is used as a ground truth. The network is shown in Figure 8. It has two inputs, the incremental counts between current and previous iterations of left and right encoders from the odometry system. The activation function is a sigmoid; the training set is composed of 7100 encoder data points with their correspondence ground truth based on GPS; the validation set is composed of 3000 points from the same path, but these points are not used to train the network. The tests are made with other data set of 11,500 points in the same scenario. The data of the GPS is interpolated to get the current GPS position of the cart when the odometry sensor sends encoder data, so odometry and GPS data are based on the same clock. The initial training is made off line, so all of the data is available in order to get interpolated positions.

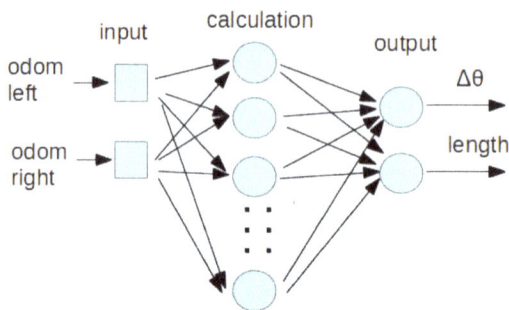

**Figure 8.** Feed-forward neural network schematic for odometry.

The network outputs angle change ($\Delta_\theta$) and length traveled by the cart due to change in odometry sensors. Therefore, the trained neural network works like a modeling function that gets odometry as input and gives displacement and angle as output.

The network is trained using a Levenberg-Marquadt supervised algorithm with a mean squared error. It has two neurons as inputs, fifty hidden neurons and two outputs. The training dataset is obtained from the global dataset, getting the angle increment and displacement due to each odometric input. This data is used to train the network without any other information.

Using neural network output, Equation (6) is applied in order to calculate pose ($X$, $Y$, $\theta$). Neural model gives the incremental angle and the displacement. In order to calculate the final pose, all of the incremental outputs of the network are integrated.

$$\theta_{i+1} = \Delta_\theta + \theta_i$$
$$X_{i+1} = X_i + length \cdot \cos(\theta_{i+1})$$
$$Y_{i+1} = Y_i + length \cdot \sin(\theta_{i+1})$$
$$\text{(6)}$$

Figure 9 shows in solid green results of applying neural model as odometry calculation. Final position is worse than the model based optimization, but during the path, neural computation is closer to ground truth than the model optimization. The accumulative error for neural computation is $1.0361 \times 10^7$: slightly smaller than the one obtained with the model based optimization, which is $1.1444 \times 10^7$. The average difference between ground truth and neural network position after 1 s is 0.1843 m, and after 5 s is 0.8192 m. This results shows that the learning based model can give better results than many mathematical models, even when only uses input and output snapshot, and not including any historical or state information.

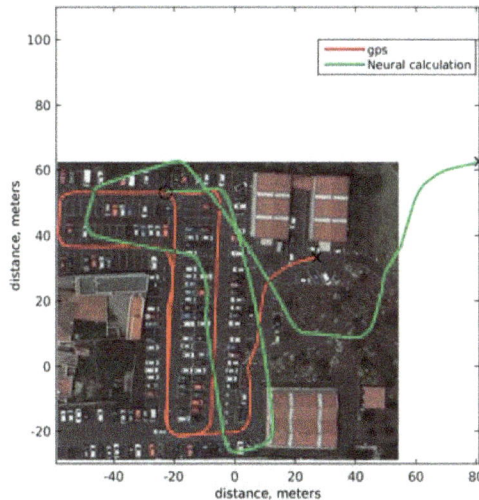

**Figure 9.** Odometric model based on neural network. Solid red ground truth GPS route. Solid green neural network odometric model path.

In order to increase the accuracy in pose calculation the previous neural network is modified by including some historical information about the path traveled by the cart. The new network has four inputs: odometry increments from left and right wheels from current and previous measurements. This information gives the network more information to generate the odometry function, including speed information and acceleration. In Figure 10, a path generated by the neural network with historical information is shown. The route at the final position is quite similar, but during the trace, the position is closer to ground truth. The accumulative error for neural historic is $9.8488 \times 10^6$, which gives the best results for all of the experiments presented. The average difference between ground truth and neural model including history after 1 s is 0.1839 m and after 5 s is 0.8179 m.

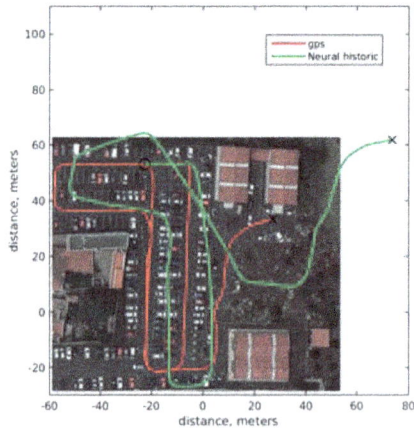

**Figure 10.** Neural model trains including history information. Solid red ground truth GPS route. Solid green historical neural calculation route.

A neural network model would increase its accuracy, as more information is available. So, in the last test, a neural network with six inputs is studied. The inputs are encoders from left and right wheels from present and previous measurements and real time wheel diameter obtained from wheel diameter sensor. The network is trained with data output from GPS ground truth. This network has all the information available for optimizing the model. The number of hidden and output neurons of the network is the same than in the previous tests. In Figure 11 the test made with this network is shown in solid green. The path followed based only in odometry information follows pretty good the ground truth. The accumulative error for neural network, including historical information and diameter is $1.4801 \times 10^6$, the smallest error from all the tests. The average difference between ground truth and neural odom including diameter and historic information after 1 s is 0.1739 m, and after 5 s is 0.8079 m.

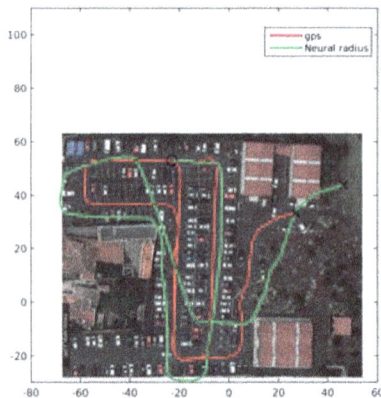

**Figure 11.** Neural odometric model test, including history information and wheel radius. Solid red ground truth GPS route. Solid green neural calculation route.

These tests show that neural networks are a very powerful replacement for mathematical models, getting better accuracy and adjusting the behavior of a real system when they have enough information. The neural network can use the relation between inputs and outputs that the mathematical model does not include, getting a very accurate output.

One of the problems of a real odometric system is that model parameters change over time. These changes can be classified as fast or slow changes. Fast changes can be generated by a difference in the weight of the passengers, so suddenly the model changes. Slow changes are due to changes in the characteristics of the environment, like pressure wheel loss, change in temperature, or wheel tear.

A static mathematical model and a learned model cannot face this problem, so the accuracy of the model will decrease along time. To solve this difficulty, a continuous training schema is used in order to get real time actualization of the model. When the cart is navigating, the localization subsystem is used as a source to tune the neural odometric system.

The navigation data (encoders and wheel radius) is included in a historical buffer to re-train the network using the last information in order to maintain an accurate model of the real system. The main problem is to obtain good localization data to train the network. The localization subsystem of the prototype is based on the fusion of multiple sensors and the localization in a map that is built for the prototype [8]. Accuracy in the localization of the cart depends on multiple factors, and localization quality can change depending on sensor information. Monte Carlo Localization (MCL) is a popular technique used to estimate the pose of a mobile robot, using a map to find the actual pose, which allows for the fusion of heterogeneous sensor data. The localization system of the prototype is based on an Adaptive MCL algorithm, which combines data from wheel odometry, an inertial measurement unit, a global positioning system and laser scanning. A particle weighting model that integrates GPS measurements is applied, which increases performance as compared with a particle generation approach. The output of the algorithm is the estimated position and a covariance matrix that measures the localization uncertainty.

The neural network is only retrained to include in the model possible changes in the system when accuracy of the ground truth is bigger (covariance less than 0.5). This schema allows forgetting a tuned model able to face fast and slow model changes, getting an accurate, real time, adaptive model even when the training data is not always accurate. Each new measurement with covariance less than 0.5 are applied to the neural model once, so the computational cost of retraining the network is very small.

In Figure 12, a test made with the prototype 20 days after previous optimization is shown. In solid green, the outputs of the static mathematical model and in solid red GPS based ground truth. Model parameters have changed due to change in temperature, wheel pressure, etc. In Table 2 the measurement parameters, the initial estimated parameters and the current parameters are shown, the change in parameters is small, but due to odometry integration, the pose changes considerably. However, neural network estimation based odometry with on line tuning, in solid blue, is able to adapt to new conditions and get a more accurate result.

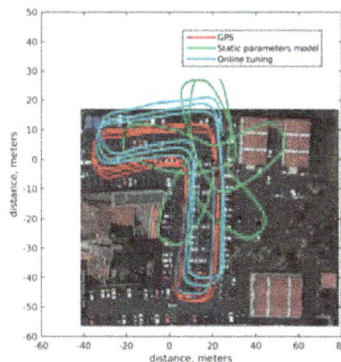

**Figure 12.** Test after 20 days of tuning odometry. Solid red, GPS based ground truth. green static parameter model. Blue on line tuned neural network model.

**Table 2.** Parameter estimation after 15 days.

| Parameters | $R_r$ | $R_l$ | $d_w$ |
|---|---|---|---|
| Measured | 0.215 m | 0.215 m | 0.97 m |
| Initial estimation | 0.2332 m | 0.2295 m | 0.978 m |
| Current estimation | 0.2105 m | 0.2095m | 0.971 m |

## 6. Conclusions

The odometric system of an autonomous vehicle is one of the main sensors for position and orientation estimation in robots and autonomous vehicles. However, its accuracy tends to be small in large distances. In short movements, position estimations are quite precise, but the errors increase quickly and divergences to the ground truth arise when the traveled distance increases. This paper is centered in the odometric system of the autonomous cart Verdino and the solutions applied in order to increase its accuracy.

As first attempt to calculate the position and orientation of Verdino using only odometry, a mathematical model based on the movement equations of the prototype is defined. The main problem of this approach is that many effects that occur on the systems, like wheel diameter or wheel separation changes, are not reflected in the model. To test this model results, a least squares optimization of model parameters are made in order to get the best parameter set. In a second step, and with the objective of improving accuracy, a new sensor is included for measuring wheel radius in real time. This new model increases accuracy, but this is still not enough to be a viable autonomous navigation system based only on odometry.

The parameters set for this model is obtained by an optimization process, so the cart pose based on odometry is close to the best that can be obtained based on this mathematical model and this sensor set. The odometry is well calibrated, so it is necessary another approach to improve the accuracy of the pose. Usually, the odometry output of a robot is very poor, and it needs others sensors to obtain a good localization system. Any small increment in odometry accuracy can improve the whole localization system a lot.

To improve odometry quality, a neural network model is used to represent cart motion. This model has the advantage that learns the input-output relations from data, and can include some effects that a simple mathematical model does not. This neural network model is tested in three variants. In the first a simple input output relation is learned, where the neural network only takes one odometry data point and gives angle and displacement variations. The position accuracy is improved when some historical information is presented, so the model can consider variables like speed or acceleration. To get this historical information, the input of the model consists on the last and the previous odometric information. The best results are obtained when the neural network not only gets odometry information, but real time wheel radiuses, giving the most accurate results. Neural network is continuously trained based on localization information as ground truth when this localization is good keeping updated the model over time.

In Table 3, a results summary is presented. This table is sorted based on the accuracy of the method, and, as expected, the worst behavior occurs when the parameters are measured directly from the cart. In the case where radius information is not available, better results are obtained with neural networks than with mathematical models, and when the network has historical state inputs, the results are even better. When diameter information is available, results clearly improve for both approaches, model, and neural networks, but the neural network outperform the simple model results.

In conclusion, neural network based optimization can be a very powerful system to improve the accuracy of the odometric system, due to its considerations of other effects not included in a mathematical model and its capacity to keep the model update over time. Even small improvements in localization for an odometry system, can improve the whole localization system, and in adverse circumstances, it may be the only available sensor.

**Table 3.** Tests results.

| Test | Accumulative Error | Position Error after 1 s | Position Error after 5 s |
|---|---|---|---|
| Measured parameters model | $2.3204 \times 10^7$ m | 0.2431 m | 1.234 m |
| Optimized parameters model | $1.1444 \times 10^7$ m | 0.1976 m | 0.9305 m |
| Neural Model | $1.0361 \times 10^7$ m | 0.1843 m | 0.8192 m |
| Historical Neural Model | $9.8488 \times 10^6$ m | 0.1839 m | 0.8179 m |
| Optimized real time wheel diameter | $4.8561 \times 10^6$ m | 0.1972 m | 0.9147 m |
| Historical wheel radius Neural Model | $1.4801 \times 10^6$ m | 0.1739 m | 0.8079 m |

**Acknowledgments:** The authors gratefully acknowledge the contribution of the Spanish Ministry of Science and Technology under SAGENIA project DPI2013-46897-C2-1-R and the project MAIH (2016TUR13) mobility and smart accessibility in Hotels, granted by CajaCanarias Foundation.

**Author Contributions:** Jonay Toledo had designed the odometry hardware, Jose D. Piñeiro obtained the mathematical model, Rafael Arnay developed the localization subsystem used to get the ground truth, Daniel Acosta had obtained the dataset for the experiments, Leopoldo Acosta is the project coordinator and he designed the experiments.

**Conflicts of Interest:** The authors declare no conflict of interest.

## References

1. Geiger, A.; Lenz, P.; Urtasun, R. Are we ready for autonomous driving? The KITTI vision benchmark suite. In Proceedings of the 2012 IEEE Conference on Computer Vision and Pattern Recognition (CVPR), Providence, RI, USA, 16–21 June 2012; pp. 3354–3361.
2. Sanchez, J.; Denis, F.; Checchin, P.; Dupont, F.; Trassoudaine, L. Global Registration of 3D LiDAR Point Clouds Based on Scene Features: Application to Structured Environments. *Remote Sens.* **2017**, *9*, 1014. [CrossRef]
3. Espinosa, F.; Santos, C.; Marrón-Romera, M.; Pizarro, D.; Valdés, F.; Dongil, J. Odometry and laser scanner fusion based on a discrete extended Kalman filter for robotic platooning guidance. *Sensors* **2011**, *11*, 8339–8357. [CrossRef] [PubMed]
4. Scaramuzza, D.; Fraundorfer, F. Visual Odometry Part I: The First 30 Years and Fundamentals. *IEEE Robot. Autom. Mag.* **2011**, *18*, 80–92. [CrossRef]
5. Montemerlo, M.; Becker, J.; Bhat, S.; Dahlkamp, H.; Dolgov, D.; Ettinger, S.; Haehnel, D.; Hilden, T.; Hoffmann, G.; Huhnke, B.; et al. Junior: The stanford entry in the urban challenge. *J. Field Robot.* **2008**, *25*, 569–597. [CrossRef]
6. Vivacqua, R.; Vassallo, R.; Martins, F. A Low Cost Sensors Approach for Accurate Vehicle Localization and Autonomous Driving Application. *Sensors* **2017**, *17*, 2359. [CrossRef] [PubMed]
7. Bento, L.C.; Nunes, U.; Moita, F.; Surrecio, A. Sensor fusion for precise autonomous vehicle navigation in outdoor semi-structured environments. In Proceedings of the 2005 IEEE Intelligent Transportation Systems, Vienna, Austria, 13–16 September 2005; pp. 245–250.
8. Ström, D.P.; Hernandez-Aceituno, J.; Morell, A.; Toledo, J.; Hamilton, A.; Acosta, L. MCL with Sensor Fusion Based on a Weighing Mechanism versus a Particle Generation Approachable. In Proceedings of the 2013 16th International IEEE Conference on Intelligent Transportation Systems (ITSC), The Hague, The Netherlands, 6–9 October 2013; pp. 166–171.
9. Wei, L.; Cappelle, C.; Ruichek, Y. Unscented information filter based multi-sensor data fusion using stereo camera, laser range finder and GPS receiver for vehicle localization. In Proceedings of the 2011 14th International IEEE Conference on Intelligent Transportation Systems (ITSC), Washington, DC, USA, 5–7 October 2011; pp. 1923–1928.
10. Moreira, M.A.G.; Machado, H.N.; Mendonca, C.F.C.; Pereira, G.A.S. Mobile robot outdoor localization using planar beacons and visual improved odometry. In Proceedings of the 2007 IEEE/RSJ International Conference on Intelligent Robots and Systems, San Diego, CA, USA, 29 October–2 November 2007; pp. 2468–2473.
11. Arnay, R.; Acosta, L.; Sigut, M.; Toledo, J. Ant colony optimisation algorithm for detection and tracking of non-structured roads. *Inst. Eng. Technol.* **2008**, *44*, 725–727. [CrossRef]

12. Hernández-Aceituno, J.; Arnay, R.; Toledo, J.; Acosta, L. Using Kinect on an Autonomous Vehicle for Outdoors Obstacle Detection. *IEEE Sens. J.* **2016**, *16*, 3603–3610. [CrossRef]
13. Morales, N.; Toledo, J.T.; Acosta, L.; Arnay, R. Real-time adaptive obstacle detection based on an image database. *Comput. Vis. Image Underst.* **2011**, *115*, 1273–1287. [CrossRef]
14. Morales, N.; Toledo, J.; Acosta, L. Generating automatic road network definition files for unstructured areas using a multiclass support vector machine. *Inf. Sci.* **2016**, *329*, 105–124. [CrossRef]
15. Maeyama, S.; Ishikawa, N.; Yuta, S. Rule based filtering and fusion of odometry and gyroscope for a fail safe dead reckoning system of a mobile robot. In Proceedings of the 1996 IEEE/SICE/RSJ International Conference on Multisensor Fusion and Integration for Intelligent Systems (Cat. No. 96TH8242), Washington, DC, USA, 8–11 December 1996; pp. 541–548.
16. Chen, D.; Bai, F.; Wu, L. Kinematics control of wheeled robot based on angular rate sensors. In Proceedings of the 2008 IEEE Conference on Robotics Automation and Mechatronics, Chengdu, China, 21–24 September 2008; pp. 598–602.
17. Borenstein, J.; Feng, L. Gyrodometry: A new method for combining data from gyros and odometry in mobile robots. In Proceedings of the IEEE International Conference on Robotics and Automation, Minneapolis, MN, USA, 22–28 April 1996; pp. 423–428.
18. Houshangi, N.; Azizi, F. Mobile Robot Position Determination Using Data Integration of Odometry and Gyroscope. In Proceedings of the 2006 World Automation Congress, Budapest, Hungary, 24–26 July 2006; pp. 1–8.
19. Borenstein, J.; Feng, L. Measurement and correction of systematic odometry errors in mobile robots. *IEEE Trans. Robot. Autom.* **1996**, *12*, 869–880. [CrossRef]
20. Borenstein, J.; Feng, L. Correction of systematic odometry errors in mobile robots. In Proceedings of the 1995 IEEE/RSJ International Conference on Intelligent Robots and Systems, Human Robot Interaction and Cooperative Robots, Pittsburgh, PA, USA, 5–9 August 1995; Volume 3, pp. 569–574.
21. Kim, K.K.K.; Patrón, E.R.; Braatz, R.D. Universal approximation with error bounds for dynamic artificial neural network models: A tutorial and some new results. In Proceedings of the 2011 IEEE International Symposium on Computer-Aided Control System Design (CACSD), Denver, CO, USA, 28–30 September 2011; pp. 834–839.
22. Han, G.; Fu, W.; Wang, W.; Wu, Z. The Lateral Tracking Control for the Intelligent Vehicle Based on Adaptive PID Neural Network. *Sensors* **2017**, *17*, 1244. [CrossRef] [PubMed]
23. Petrović, N.P.I.; Ivanjko, E. Neural Network Based Correction of Odometry Errors in Mobile Robots. In Proceedings of the FIRA Congress, Seoul, Korea, 26–29 May 2002; pp. 91–96.
24. Xu, H.; Collins, J.J. Estimating the Odometry Error of a Mobile Robot by Neural Networks. In Proceedings of the 2009 International Conference on Machine Learning and Applications, Miami Beach, FL, USA, 13–15 December 2009; pp. 378–385.

*sensors*

MDPI

Article

# New Control Paradigms for Resources Saving: An Approach for Mobile Robots Navigation

Rafael Socas * , Raquel Dormido and Sebastián Dormido

Departamento de Informática y Automática, Universidad Nacional de Educación a Distancia, Juan del Rosal 16, Madrid 28040, Spain; raquel@dia.uned.es (R.D.); sdormido@dia.uned.es (S.D.)
* Correspondence: rsocas@telefonica.net; Tel.: +34-629-578-386

Received: 16 December 2017; Accepted: 11 January 2018; Published: 18 January 2018

**Abstract:** In this work, an event-based control scheme is presented. The proposed system has been developed to solve control problems appearing in the field of Networked Control Systems (NCS). Several models and methodologies have been proposed to measure different resources consumptions. The use of bandwidth, computational load and energy resources have been investigated. This analysis shows how the parameters of the system impacts on the resources efficiency. Moreover, the proposed system has been compared with its equivalent discrete-time solution. In the experiments, an application of NCS for mobile robots navigation has been set up and its resource usage efficiency has been analysed.

**Keywords:** Event-based control; Networked Control Systems (NCS); resources efficiency; mobile robots; robot navigation

---

## 1. Introduction

In recent years, Networked Control Systems (NCS) have been gaining importance in the control community [1]. NCS are distributed architectures composed of controllers, sensors that can obtain information from the environment, actuators for acting on them, and a communication network that connects all the elements to achieve a common goal. Therefore, NCS is a field which includes different disciplines such as control theory, communications, software engineering and computer science. Typical applications where these control systems are being used are: space or terrestrial explorations, factory automation, remote diagnostics and troubleshooting, hazardous environments, experimental facilities, mobile robots, multi-vehicles networks, aircraft, manufacturing plant monitoring, nursing homes or hospitals, tele-robotics, tele-operation, etc.

The elements of an NCS system are called the agents. These elements use a communication network to exchange the information between them. Depending on the application where the NCS is used, this network can be deployed using wireline or wireless technology. These communication networks use digital technology to transmit the information which has constraints in delays and limited bandwidth. The information packaging and the constraints due to the limited resources of the network produce undesirable effects such as packet losses, variable delays and signal quantization issues among others. These effects may disturb the stability and performance of the system [2].

Therefore, reducing the traffic in the network is a critical aspect. If the number of packets is decreased can be guaranteed a predictable bandwidth, and at the same time, the analysis of the delays of the network is simplified [3]. As a conclusion, an important issue in the design of these control systems is to implement protocols for transmitting the sensor signals, the state of the system and the control information in a more effective way.

Some researchers have investigated the timing issues in NCS [4]. In traditional approaches, the controllers are used under the assumption of perfect communication, and then, the Maximum

Allowable Transfer Interval (MATI) between two subsequent message transmissions that ensures closed loop stability under a network protocol is determined. Try Once Discard (TOD) and Round Robin (RR) are protocols implemented based on this philosophy. On the other hand, the MATI protocol is often deployed in a centralized way, therefore, it is not practical for systems of large-scale.

Other proposals have achieved an important reduction of network resources usage without a significant loss of performance. Two approaches have been raised: Model-Based Networked Control Systems (MBNCS) and Event-Based Control (EBC). The basics of the MBNCS have been developed in [5,6], and they have been considered in networks of coupled systems in [7] using periodic communication. Another approach to deal with this problem has been built on an event-based feedback scheme in NCS [8–11]. In event-based control systems, the agent information is broadcast only when some measures of the state error cross a specified level (the event threshold). This control scheme is decentralized in the sense that an agent can broadcast its state using the local information and in an asynchronous way.

In an EBC system, the impact of noise in the sensors increases the number of events and therefore a degradation of the performance. If the disturbance is known or can be modeled, the controller can be properly set to reduce its effects. However, in most applications it is difficult either to estimate the noise level or to have a reliable model. In these cases, it becomes a hard work to tune the controller in a proper way. In [12,13] these problems have been investigated where a new control scheme has been proposed which are dynamically adjusted depending on the conditions of the environment. In both proposals, the algorithms work with an estimation of the noise previously calculated. In [14], the estimation of the noise and the tune of the controller is made in real time.

The EBC strategies have been widely used to control dynamical processes while decreasing considerably the number of packets that the sensors have to send to the controller over the network. In [15] the events based on state errors have been investigated. In [16–18], similar proposals have been analysed to apply in networked interconnected systems. In these works, the use of a zero-order hold (ZOH) in the controller is a common feature. In [19], an event-triggering in networked systems with probabilistic sensor and actuator fault has been investigated to reduce the computation load. An overview on sampled-data-based event-triggered control and filtering for networked systems has been presented in [20]. In this research, a deep investigation of the sampled data-based event-triggered scheme has been made. In general, the event-based control architectures can be a good solution for the systems with limited resources and they could be a more efficient control scheme than the classical ones [21,22].

In this paper, a new event-based control architecture based on a simple event-based control scheme for NCS environments is presented. Making use of the control strategy implemented a full analysis of different resource consumption is carried out. The use of bandwidth, computational load and energy resources are analysed. Several methods and methodologies to measure the efficiency in the resources consumption are also proposed. The main contributions of this work are the models development of resource consumption and their parametrization. Finally, the ideas presented in this work are applied to an NCS mobile robots system to solve the navigation problem.

The paper is organized as follows. Section 2 presents an overview of the event-based control. In Section 3, the principles of sampling criteria are described. In Section 4, the proposed control strategy is presented. Section 5 shows the resource usage in the proposed system. Section 6 presents the experimental results. Finally, the conclusions and future work are discussed in Section 7.

## 2. Event-Based Control Overview

The event-based control has motivated the interest of the control community in the last few years, multiple control architectures and new applications have been propossed based on these ideas. In [23], an event-driven sampling method called the area-triggered method has been proposed. In this scheme, sensor data are sent only when the integral of the differences between the current sensor value and the last transmitted one is greater than a given threshold. The proposed system reduces the data transmission rate and also improves the estimation performance in comparison with the conventional time-driven technique. In [24], a greenhouse climate is controlled by an event-based control system. The system is based on a network of wireless sensors to control the low frequency dynamics of the environment. In this case, the control actions are calculated by considering the events that produce the external disturbances. The proposed system increases the actuators life and allows cost savings by minimizing the wear while maintaining a good performance. In [25], an event-based sampling according to a constant energy of sampling error is investigated. The defined criterion is suitable for applications where the energy of the sampling error should be bounded (e.g., in greenhouse climate monitoring and control or in building automation). Finally, in [26], a fault isolation filter to apply on discrete-time networked control systems based on a particular form of the Kalman filter is proposed. The scheme makes an efficient use of the resources with a good estimation of failures and its effect on the performance. The sampled-data-based event-triggered control schemes is another emerging event-based control technique. The reliable control design for networked control system under event-triggered scheme is investigated in [19]. The key idea of this work is that only the newly sampled sensor measurements that violate specified triggering condition will be transmitted to the controller. The main advantage of this approach is that the proposed event-triggered scheme only needs a supervision of the system state in discrete instants and there is no need to retrofit the existing system. Finally, in [20], an overview and a deep investigation on sampled-data-based event-triggered control and filtering for networked systems has been done. Compared with some existing event-triggered and self-triggered schemes, a sampled-data-based event-triggered scheme can ensure a positive minimum inter-event time and make it possible to jointly design suitable feedback controllers and event-triggered threshold parameters.

In event-based control systems, information is exchanged between the elements (controller, sensors and actuators) depending on the state of the system [27]. When the system variables exceed a certain level an event is generated in the system and the control actions are executed. This means that the activity of the controller and the use of resources to communicate the different elements are restricted to the time intervals in which a control action must inevitably be taken to guarantee the system specifications.

In Figure 1, the basic scheme of an event-based control strategy is presented [28,29].

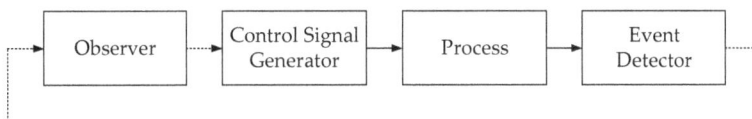

**Figure 1.** Basic scheme of an event-based control system. Solid arrows represents continuous signal and dashed arrows represents event-based signals.

The control scheme is composed of an event detector, an observer, and a control signal generator. The event detector generates an output signal when an event occurs, it happens when the error signal crosses a threshold. When an events occurs, the observer is updated and it passes the information to the control signal generator. With this information, the control signal generator generates the input signal to control the process. An important aspect of this strategy is that the observer and the control signal generator works in open loop between events.

This control architecture combines feedback and feedforward strategies. When an event is generated there is a feedback action. On the other hand, the feedforward actions happen when the actuators are driven by the control signal generator in open loop between events.

## 3. Event-Based Sampling Schemes

Different sampling criteria have been proposed in the event-based control schemes [27]. In the event-based sampling methods, the system acts only when the variables of the plant are in transient state. In the steady state the system does not act and some resources can be saved. On the other hand, in discrete time schemes the sampling is executed periodically (with the period $T = 1/f_s$ where $f_s$ is the sampling frequency) and it does not depend on the state of the system (Figure 2a).

Send-on-delta and integral criterion are the most widely used techniques in event-based control schemes. In the following sections, these methods will be defined and their effectiveness will be discussed.

*3.1. Send-On-Delta*

The send-on-delta sampling algorithm is the most natural signal-dependent strategy; in the literature, it is also known as level-crossing or deadbands sampling. In the send-on-delta technique, the sensors do not broadcast a new message if the signal remains within a certain level of confidence $\bar{e}_S$ (resolution) (Figure 2b). The sampling criterion is defined as

$$|y(t) - y(t_k)| \geq \bar{e}_S \tag{1}$$

The ratio of events that the send-on-delta algorithm generates can not be calculated in a general way, but its average value $N_S$ may be estimated by the following expression [30,31]:

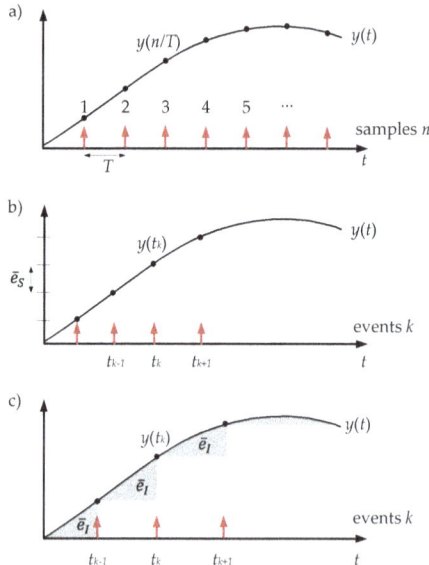

**Figure 2.** Sampling strategies: (**a**) periodic sampling; (**b**) send-on-delta; and (**c**) integral criterion.

$$N_S = \frac{1}{\overline{\Delta t}} \tag{2}$$

where $\overline{\Delta t}$ is the mean period between events considering the analysis interval $(t_0, t_n)$

In the send-on-delta algorithm, an event occurs when

$$|y(t_k) - y(t_{k-1})| = \bar{e}_S \quad , \quad k = 1, 2, ..., n \tag{3}$$

and the interval time between the events $k - 1$ and $k$ is expressed as

$$\Delta t_k = t_k - t_{k-1} = \frac{\bar{e}_S}{|\dot{y}(t_k)|} \tag{4}$$

where $\overline{|\dot{y}(t_k)|}$ is defined as

$$\overline{|\dot{y}(t_k)|} = \frac{1}{\Delta t_k} \int_{t_k}^{t_{k+1}} |\dot{y}(t)| dt \tag{5}$$

at this point, the mean period between events $\overline{\Delta t}$ can be defined as

$$\overline{\Delta t} = \frac{\sum_{k=1}^{n} \Delta t_k}{n} \tag{6}$$

in [31], the following relationship has been proven

$$\overline{\Delta t} = \frac{\bar{e}_S}{|\dot{y}(t)|} \tag{7}$$

where $\overline{|\dot{y}(t)|}$ was defined as

$$\overline{|\dot{y}(t)|} = \frac{1}{t_n - t_0} \int_{t_0}^{t_n} |\dot{y}(t)| dt \tag{8}$$

taking Equations (2) and (7) into account, the mean rate of events $N_S$ can be written as

$$N_S = \frac{\overline{|\dot{y}(t)|}}{\bar{e}_S} \tag{9}$$

In this case, the ratio of events $N_S$ in the system depends on two parameters:

- the resolution $\bar{e}_S$ of the sampling (the event threshold); and
- the mean of the absolute value of the first time-derivative $\overline{|\dot{y}(t)|}$ during the analysis interval.

As presented in Equation (9), the message rate in the send-on-delta strategy is a trade off between the resolution $\bar{e}_S$ and the average slope of the signal $y(t)$.

### 3.2. Integral Criterion

There are some reasons to apply the event-based integral criterion in control systems. This method has a high efficiency for sampling burst signals. Likewise, this technique is a good solution in applications where a critical problem of sampling process is the accuracy of approximation of a continuous-time signal by a sequence of discrete-time samples. In this algorithm, the sampling criterion (Figure 2c) is defined by Equation (10), where $\bar{e}_I$ is the resolution of the method:

$$\int_{t_k}^{t} |y(t) - y(t_k)| dt \geq \bar{e}_I \tag{10}$$

then, the mean rate of events $N_I$ can be estimated as follows [32]:

The system generates an event when the following condition is satisfied

$$\int_{t_k}^{t_{k+1}} |y(t) - y(t_k)| dt = \bar{e}_I \tag{11}$$

taking into account the time interval between the events $k$ and $k + 1$ ($\Delta t_k = t_{k+1} - t_k$), in [32], it has been demonstrated that the mean period between events considering the interval $(t_0, t_n)$ is

$$\overline{\Delta t} = \frac{\sum_{k=1}^{n} \Delta t_k}{n} = \frac{\sqrt{2\bar{e}_I}}{\sqrt{|\dot{y}(t)|}} \tag{12}$$

where $\overline{\sqrt{|\dot{y}(t)|}}$ is defined by

$$\overline{\sqrt{|\dot{y}(t)|}} = \frac{1}{t_n - t_0} \int_{t_0}^{t_n} \sqrt{|\dot{y}(t)|} dt \tag{13}$$

Finally, the mean rate of events $N_I$ based on the integral criterion strategy can be expressed as:

$$N_I = \frac{1}{\Delta t} = \frac{\overline{\sqrt{|\dot{y}(t)|}}}{\sqrt{2\tilde{e}_I}} \tag{14}$$

Taking Equation (14) into account, the mean rate of events depends on the mean of the square root of the signal derivative absolute value and the resolution $\tilde{e}_I$ used in the algorithm

*3.3. Effectiveness of Event-Based Sampling*

To study how effective the presented sampling algorithms are, they will be compared with periodic sampling strategies. In the send-on-delta algorithm, the mean rate of events $N_S$ was estimated in Equation (9). If the periodic sampling algorithm is considered, the sampling period $T$ can be obtained by Equation (15) where the accuracy of the algorithm is $\tilde{e}_P$

$$T = \frac{\tilde{e}_P}{|\dot{y}(t)|_{max}} \tag{15}$$

Now, the ratio of samples $N_P$ can be calculated as

$$N_P = \frac{1}{T} = \frac{|\dot{y}(t)|_{max}}{\tilde{e}_P} \tag{16}$$

If the send-on-delta is compared with the periodic sampling considering the same resolution for both methods ($\tilde{e}_S = \tilde{e}_P$), the following equation is obtained

$$\frac{N_S}{N_P} = \frac{|\dot{y}(t)|}{|\dot{y}(t)|_{max}} \tag{17}$$

In this case, $N_S = N_P$ if $y(t)$ is a linear signal, for the rest of the continuous-time signal $N_S < N_P$ is fulfilled. The last expression implies that the send-on-delta is more efficient than the periodic sampling algorithm.

Taking into account the integral criterion, the number of samples in the periodic sampling strategy is given by (see [32]):

$$N_P = \frac{\sqrt{2\tilde{e}_P}}{\sqrt{|\dot{y}(t)|_{max}}} \tag{18}$$

If both methods (event-based and periodic) have the same accuracy ($\tilde{e}_I = \tilde{e}_P$), the following equation is obtained from Equations (14) and (18):

$$\frac{N_I}{N_P} = \frac{\overline{\sqrt{|\dot{y}(t)|}}}{\sqrt{|\dot{y}(t)|_{max}}} \tag{19}$$

which shows that for the integral criterion the event-based strategy is more efficient than the periodic sampling $N_I < N_P$ in a general way.

It can be concluded that the event-based sampling algorithms are more efficient than the periodic strategies. It means that the event-based solution generates less events than the periodic sampling scheme.

## 4. NCS Control Architectures

In this section, two NCS schemes are discussed. First, the classical discrete-time architecture is presented, and then the event-based solution proposed in this work is investigated in detail. In these NCS, the agents (the controller and the remote node) are connected by a wireless network.

The basic scheme of a discrete-time NCS is presented in Figure 3a. The signals $u[n]$ (control signal) and $y[n]$ (sensor signal) are sampled with the frequency $f_s$. The signal $u[n]$ is sent to remote node over the communication channel $Ch_1(t)$. In the remote node, this information is used to act over the actuators. The signal $y[n]$ is sent to the controller over the $Ch_2(t)$. Finally, the controller calculates $u[n]$ considering the reference signal $w[n]$ and $y[n]$. Therefore, this architecture exchanges information over the communication channels every period of time defined by $T = 1/f_s$. When the plant is in a steady state, it is not necessary to interchange information between the elements of the system because the plant does not need new control actions. However, in this scheme, the communication channels $Ch_1(t)$ and $Ch_2(t)$, the controller, the actuators and the sensors are busy every period of time $T$.

The proposed event-based NCS is presented in Figure 3b. As mentioned above, the proposed system follows the basic principles of an event-based control system, being the evaluation of the consumption of resources the main objective of this work. The system is composed of a controller, an event generator (EG) and a memory block (M). In the EG the signals, $y[n]$ and $w[n]$ are compared. If the difference between these signals crosses the event threshold $\bar{e}$, the system generates an event $k$ and the signal $e[n_k]$ (error signal) is sent to the controller. Different methods can be applied to produce events, in [21], a review of these methodologies is presented. In a general way, the event threshold $\bar{e}$ is defined as a constant value. This parameter can be defined as a function of the noise in the sensors or as a function of other relevant variables of the process to generate the events in a more accuracy way. In [12,14], these ideas have been explored. Therefore, the event threshold has to be set as a trade off between the accuracy of the system and the number of events. The communication channel $Ch_2(t)$ is used to send the signal $e[n_k]$ to the controller. In this case, the RF channel is occupied only when the EG generates events. On the other hand, when the signals $w[n]$ and $y[n]$ are very similar, the plant/process is in a steady state. In this case, no events occur in the system and the channel $Ch_2(t)$ is free. When the signal $u[n_k]$ arrives in the remote node it is stored in the memory M. Therefore, every time the system generates an event, the memory is updated with a new value. In the periods of time between events, the information stored in the memory is used to control the remote node. Besides, when an event is generated in the system, the controller receives the signal $e[n_k]$ and it calculates the signal $u[n_k]$. Afterwards, this control signal is sent to the remote node via $Ch_1(t)$. As result of this, the communication resources $Ch_1(t)$ and $Ch_2(t)$ are used only when the system is generating events.

In general, the communication channels $Ch_1(t)$ and $Ch_2(t)$ use Industrial, Scientific and Medical (ISM) bands which are not exclusive and many agents may be using them at the same time. When other devices are using the same channel, the interferences, the packet dropouts and an excessive delay in the network could affect the performance of the control system. To avoid these effects, a free channel in the radio link between the remote node and the controller has to be selected. For example, if the 2.4 GHz band is used, there are around 126 available channels so there is a high probability of finding interference-free channels in the wireless network.

As a conclusion, the proposed event-based solution has three main advantages:

1. The RF channels are busy only when the system generates events, see Figure 3c.
2. The controller does not need to compute control signals when the plant is in the steady state.
3. The system is protected from Zeno phenomenons. In the proposed control scheme the events are generated in the system only at certain instants of time (Figure 3c). This behaviour sets a minimum time interval between events defined by $1/f_s$. This is a typical mechanism to avoid the Zeno phenomenons in the event-based control systems [33].

In a practical way, to compare the responses of both NCS architectures (discrete-time and event-based), some conditions must be imposed:

- The clocks of the systems are synchronized (Figure 3c). This implies that the events and the samples in the system are generated at the same instant of time.
- The accuracy of both systems has to be the same. In this case, the event threshold has to be set up considering the sampling frequency of the discrete-time system and the sampling criterion in the event-based one.

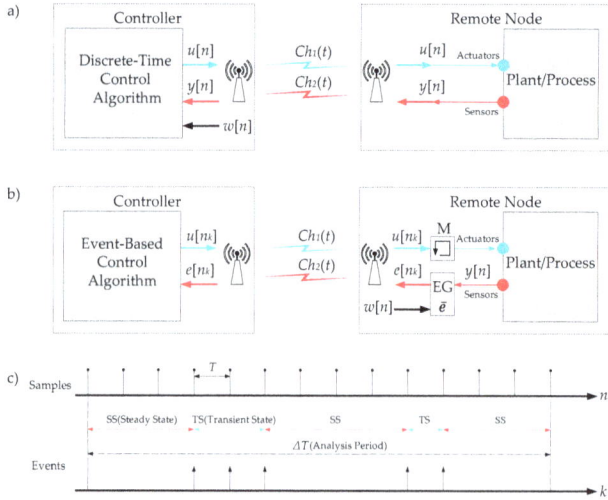

**Figure 3.** NCS control schemes: (**a**) discrete-time; (**b**) event-based; and (**c**) time diagrams. The samples represent the activity of the discrete-time system; for the event-based solution, the activity is represented by the events.

## 5. Resources Usage

To measure the performance of a control system, the Integral Absolute Error $IAE$ is applied. This criterion is widely used in continuous-time and discrete-time control systems. The $IAE$ calculates the difference between the output of the system ($y(t)$ for continuous-time or $y[n]$ for discrete-time) and the reference signal ($w(t)$ or $w[n]$) by the following equations:

$$IAE = \int_{t_0}^{t} |w(t) - y(t)| dt$$

$$IAE = \sum_{n_0}^{n} |w[n] - y[n]|$$

(20)

Therefore, the analysed system has a good performance if the $IAE$ is small.

Another way to evaluate the performance of a control system is by the Integral Absolute Error compared to Periodic loop $IAEP$. This criterion is mainly used in event-based control systems [34,35]. This indicator compares the output of the event-based system $y_{eve}[n]$ with the output of its equivalent discrete-time system $y[n]$, as presented in Equation (21).

$$IAEP = \sum_{n_0}^{n} |y_{eve}[n] - y[n]|$$

(21)

Although the $IAEP$ is a good criterion to measure the performance of an event-based control system, in practice, many researchers [36–40] use the ratio of events $N_N$. This parameter calculates the activity of an event-based control system versus an equivalent discrete-time control system, the ration of events is given by:

$$N_N = \frac{N_{eve}}{N_{per}}$$

(22)

where $N_{eve}$ is the number of events in the analysed system and $N_{per}$ the number of samples in the equivalent discrete-time system. In this case, the number of events can be defined as $N_{eve} = N_S \Delta T$ for the send-on-delta algorithm and $N_{eve} = N_I \Delta T$ for the integral criterion, where $\Delta T$ is the analysis period.

In the same way, the number of samples is defined as $N_{per} = N_P \Delta T$ (see Section 3). In this context, if $N_N < 1$ the event-based system has less activity than the equivalent discrete-time system and consequently, it uses less resources. As was previously analysed, the event-based sampling criteria generate less events than the periodic sampling techniques using the same resolution in both methods. It means that $N_P \geq N_{EVE}$ (where $N_{EVE} = N_S$ if the send-on-delta is used or $N_{EVE} = N_I$ if the integral criterion is used (see Equations (17) and (19)). If the analysis period $\Delta T$ is taken into account, it can be written $N_{eve} = N_{EVE} \Delta T$ and $N_{per} = N_P \Delta T$, then the ratio of events is given as:

$$N_N = \frac{N_{eve}}{N_{per}} = \frac{N_{EVE}}{N_P} \tag{23}$$

where $0 \leq N_N \leq 1$.

Using the event efficiency $\eta_N$, this can be expressed as:

$$\eta_N = (1 - N_N)100 \tag{24}$$

In a general way, if the resolutions in the sampling methods are the same, the activity of the event-based control systems is more efficient than the discrete-time solutions ($0\% \geq \eta_N \geq 100\%$).

In this work, to analyse the resources usage efficiency, the ratio $N_R$ is defined as follows

$$N_R = \frac{(R)_{eve}}{(R)_{dis}} \tag{25}$$

where $R$ indicates which resource is analysed, $(R)_{eve}$ is the usage of the resource $R$ in the event-based control system and $(R)_{dis}$ denotes the usage of the same resource in the equivalent discrete-time control system. In the same way, the efficiency in the usage of the resource $R$ can be expressed as

$$\eta_R = (1 - N_R)100 \tag{26}$$

In the following sections, the usage efficiency of the resources of the control system such as bandwidth, computational load and energy are investigated.

### 5.1. Bandwidth Usage

The model depicted in Figure 4a has been used to analyse the bandwidth usage.

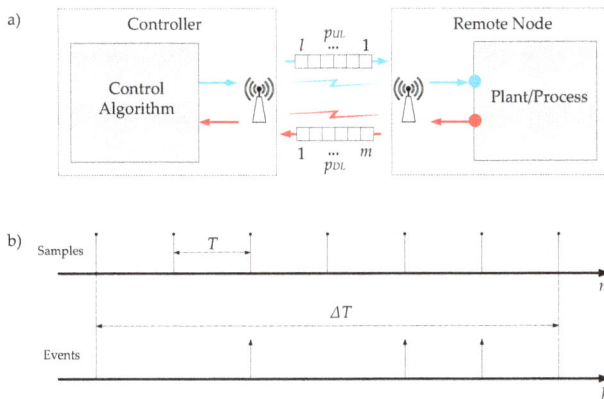

**Figure 4.** Bandwidth utilization model: (**a**) control architecture; and (**b**) time diagrams for discrete-time and event-based schemes.

The information in the uplink direction is $p_{UL} = l$ bits and in the downlink is $p_{DL} = m$ bits. The analysis period $\Delta T$ is considered. In this interval time, the discrete-time system generates $N_{per}$ samples and the event-based system $N_{eve}$ events. Then, the bandwidth usage, defined as the number of bits transmitted per second, can be written for the discrete-time system by

$$(BW_{UL})_{dis} = \frac{lN_{per}}{\Delta T} \; ; \; (BW_{DL})_{dis} = \frac{mN_{per}}{\Delta T} \tag{27}$$

and for the event-based solution is given by

$$(BW_{UL})_{eve} = \frac{lN_{eve}}{\Delta T} \; ; \; (BW_{DL})_{eve} = \frac{mN_{eve}}{\Delta T} \tag{28}$$

where $BW_{UL}$ and $BW_{DL}$ represent the bandwidth usage in the uplink and in the downlink direction, respectively.

The bandwidth usage ratio for the uplink $N_{BW_{UL}}$ and for the downlink $N_{BW_{DL}}$ can be expressed by the following equations

$$N_{BW_{UL}} = \frac{(BW_{UL})_{eve}}{(BW_{UL})_{dis}} = \frac{N_{eve}}{N_{per}} = N_N \tag{29}$$

$$N_{BW_{DL}} = \frac{(BW_{DL})_{eve}}{(BW_{DL})_{dis}} = \frac{N_{eve}}{N_{per}} = N_N \tag{30}$$

considering $N_{BW_{UL}} = N_{BW_{DL}}$, the bandwidth usage ratio for both directions $N_{BW}$ can be defined as

$$N_{BW} = \frac{N_{eve}}{N_{per}} = N_N \tag{31}$$

and the bandwidth usage efficiency is given by

$$\eta_{BW} = \eta_N \tag{32}$$

As Equation (32) shows, the bandwidth usage efficiency is the same as the event efficiency.

### 5.2. Computational Load Reduction

To obtain a model of the computational load in the analysed control systems, the scheme depicted in Figure 5 has been used.

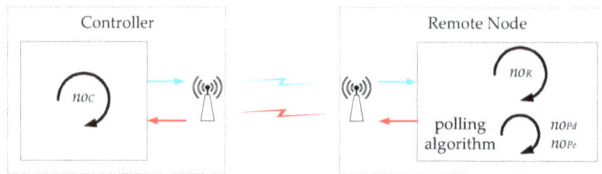

**Figure 5.** Computational resources model.

The control algorithm is composed of two elements: the algorithm in the controller and the algorithm in the remote node. In the remote node, the algorithm is divided in two blocks, the control algorithm and the polling algorithm. The polling block is used to get the measures from the sensors and in the event-based solution it also generates the events. In the controller, $no_C$ operations are executed each time the algorithm is run. On the other hand, in the remote node, $no_R$ operations are executed in the control algorithm either $no_{Pd}$ when the discrete-time solution is used or $no_{Pe}$ when the event-based technique is selected.

Taking into account the previous assumptions, the computational load in the discrete-time system $(CL)_{dis}$ and in the event-based control system $(CL)_{eve}$ considering the analysis period $\Delta T$ can be expressed by

$$(CL)_{dis} = \frac{(no_C + no_R + no_{Pd})N_{per}}{\Delta T} \tag{33}$$

$$(CL)_{eve} = \frac{(no_C + no_R)N_{eve} + no_{Pe}N_{per}}{\Delta T} \tag{34}$$

then, the computational load ratio $N_{CL}$ is given by

$$N_{CL} = \frac{(CL)_{eve}}{(CL)_{dis}} = \frac{(no_C + no_R)N_N + no_{Pe}}{no_C + no_R + no_{Pd}} \tag{35}$$

and the computational load efficiency can be written as

$$\eta_{CL} = (1 - N_{CL})100 \tag{36}$$

In general, if $N_N$ increases, the efficiency $\eta_{CL}$ decreases. If the computational load of the polling algorithms is smaller than the control algorithms ($no_{Pd} << (no_C + no_R)$ and $no_{Pe} << (no_C + no_R)$), $N_{CL} \approx N_N$ and $\eta_{CL} \approx \eta_N$. On the other hand, when the load of the polling algorithms increases, the computational load ratio ($N_{CL} \to 1$) and the event-based control system does not have efficiency in computational load $\eta_{CL} \to 0\%$. As a conclusion, the computational load efficiency of the system has a high dependence on the computational load of the polling algorithms.

From a practical point of view, to evaluate the computational burden of the system a simple procedure has to be followed. First, it is necessary to distinguish which part of the algorithm is executed periodically and which one is executed eventually when the events occur in the system. Then, the parameters $no_C$, $no_R$, $no_{Pd}$ and $no_{Pe}$ can be obtained doing a high level analysis of the control algorithm. In Section 6.3.2 and in Appendix A, this procedure has been applied in the application used as practical example in this work.

### 5.3. Energy Consumption

In this section, the energy efficiency of the presented event-based control system is investigated. In Figure 6, the simplified energy model of the system is depicted.

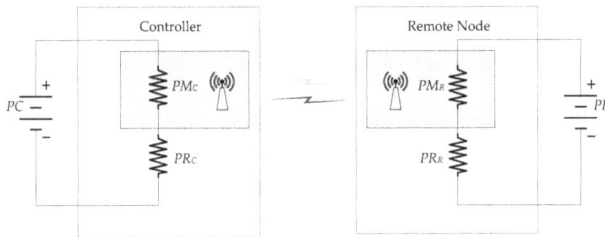

**Figure 6.** Energy model for the controller and for the remote node.

The electric power in the controller can be obtained adding up the power that the modem needs to transmit and receive the information $PM_C$ and the power in the rest of the system of the controller $PR_C$. When the discrete-time system is considered, the power of the controller is given by

$$PC_{dis} = PM_C N_{per} + PM_C \left(\frac{1 - \beta}{\beta}\right) \tag{37}$$

where $\beta$ is the power ratio defined by

$$\beta = \frac{PM_C}{PM_C + PR_C} \tag{38}$$

Using the same argument, the power of the controller when the event-based architecture is used can be written as

$$PC_{dis} = PM_C N_{eve} + PM_C \left( \frac{1-\beta}{\beta} \right) \tag{39}$$

Taking into account the analysis period $\Delta T$, the energy usage in the discrete-time controller $(EC)_{dis}$ and for the event-based controller $(EC)_{eve}$ are given by the following equations

$$(EC)_{dis} = \left( PM_C N_{per} + PM_C \left( \frac{1-\beta}{\beta} \right) \right) \Delta T \tag{40}$$

$$(EC)_{eve} = \left( PM_C N_{eve} + PM_C \left( \frac{1-\beta}{\beta} \right) \right) \Delta T \tag{41}$$

and the energy ratio in the controller $N_{EC}$ can be expressed as

$$N_{EC} = \frac{(EC)_{eve}}{(EC)_{dis}} = \frac{N_{eve}\beta + (1-\beta)}{N_{per}\beta + (1-\beta)} \tag{42}$$

and the energy efficiency in the controller can be written as

$$\eta_{EC} = (1 - N_{EC})100 \tag{43}$$

Using the same reasoning in the remote node, the energy usage in the discrete-time implementation $(ER)_{dis}$ and in the event-based solution $(ER)_{eve}$ can be written as

$$(ER)_{dis} = \left( PM_R N_{per} + PM_R \left( \frac{1-\gamma}{\gamma} \right) \right) \Delta T \tag{44}$$

$$(ER)_{eve} = \left( PM_C N_{eve} + PM_C \left( \frac{1-\gamma}{\gamma} \right) \right) \Delta T \tag{45}$$

where $\gamma = \frac{PM_R}{PM_R + PR_R}$. Then, the energy ratio in the remote system is defined by

$$N_{ER} = \frac{(ER)_{eve}}{(ER)_{dis}} = \frac{N_{eve}\gamma + (1-\gamma)}{N_{per}\gamma + (1-\gamma)} \tag{46}$$

and the energy efficiency in the remote node can be written as

$$\eta_{ER} = (1 - N_{ER})100 \tag{47}$$

If the power of the modem in the controller is high ($\beta \to 1$), the efficiency in energy usage in the proposed event-based system is the event efficiency $N_{EC} \to N_N$ and $\eta_{EC} \to \eta_N$. On the other hand, when the power of this device is low ($\beta \to 0$), the system is not energy efficient, $N_{EC} \to 1$ and $\eta_{EC} \to 0\%$. Similar conclusions can be obtained in the remote node considering the parameter $\gamma$. In conclusion, the power of the modems determines the energy efficiency of the proposed event-based control scheme.

## 6. Experimental Results

To check the ideas presented in this work, a test laboratory to investigate wireless control systems has been developed (Figure 7). In this platform, the controller has been implemented in a laptop and the remote nodes are the mobile robots. For this purpose, mOway mobile robots [41] have been used. Using this platform, an analysis of the behaviour of the event-based control schemes presented in this paper is carried out as well as a comparison with their equivalent discrete-time implementation.

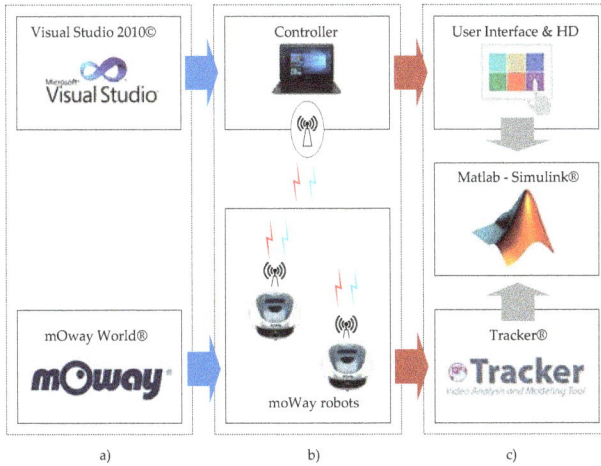

**Figure 7.** Laboratory for the experiments: (**a**) development module; (**b**) mobile robots environment; and (**c**) analysing tools.

The control algorithms have been programmed in C++ for the controller and in the mOway World environment for the robots (Figure 7a). A radio link interface is used to communicate the controller and the robots (Figure 7b). Finally, other applications such as the Tracker, the Matlab/Simulink and some scripts in the controller can be used to analyse the experimental results (Figure 7c).

The structure and components of the robots are depicted in Figure 8. The robots have four infra-red obstacle sensors with a maximum range of 3 cm and a sensor to measure the battery level (Figure 8a). The wireless communication system (robot RF interface (Figure 8b), and PC RF interface (Figure 8c)) works in the worldwide ISM frequency band at 2.400–2.4835 GHz and uses GFSK modulation. In the system 126 channels can be configured, the rate for each channel is 2 Mbps. The angular speeds of the wheels can be varied from 0 to 10.9 rad/s and their geometrical parameters $L = 6.6$ cm (distance between wheels) and $r = 1.6$ cm (radius of the wheels).

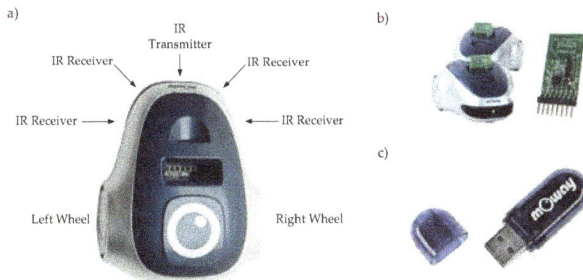

**Figure 8.** Robot platform: (**a**) components and sensor distribution; (**b**) robot wireless interface; and (**c**) PC wireless interface.

To analyse the event-based control architecture proposed in this work and its resources usage efficiency, some experiments have been set up. In these experiments robot navigation algorithms will be checked in the laboratory. In the following sections, the proposed navigation algorithms, their implementation in the system and the experimental results will be analysed in detail.

*6.1. Navigation Algorithms*

In navigation applications for mobile robots, the algorithms such as Go To Goal (GTG), Obstacle Avoidance (OA) and Wall Following (WF) are widely used [42–46]. In these applications, it is also critical to define a precise positioning mechanism to guarantee the convergence of the navigation algorithm [47,48].

In this work, the behaviour and the resource usage of the OA and WF algorithms are investigated in the proposed event-based control architecture. To define the navigation algorithms, the position and the orientation of the obstacle sensors have to be taken into account (Figure 9a). The variables that contain the measurements of the sensors are defined by two indexes as follows: $ll$ lateral left, $fl$ front left, $fr$ front right and $lr$ lateral right. The parameter $bl$ stores the battery level. In the actuators, the linear speeds of the wheels are $sl$ speed left and $sr$ speed right. Finally, the parameter $rn$ (robot number) identifies the robot in the platform.

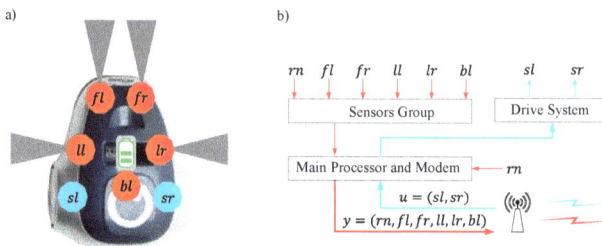

**Figure 9.** Sensors and actuators of the mOway robot: (**a**) position and orientation of the IR obstacle sensors ($fl, fr, ll, lr$), the battery sensor ($bl$), and the wheel actuators ($sl, sr$); and (**b**) flow diagram for the communication between sensors, actuators and the RF interface.

The sensor information $y = (rn, ll, fl, fr, lr, bl)$ and the control signals $u = (sl, sr)$ are sent and received by the RF interface, as shown in Figure 9b.

The architecture of the algorithm in the robots is depicted in Figure 10.

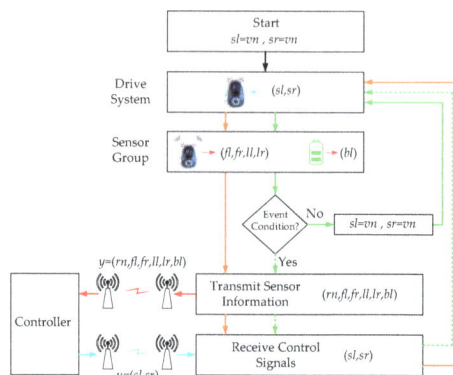

**Figure 10.** Robot control algorithm. The orange arrows represent the discrete-time solution and the green arrows the event-based implementation. The blocks with continuous arrows are executed each period of time $T$ and the blocks with dotted arrows when an event is generated in the system.

The orange arrows represent the discrete-time implementation. The green ones are used for the event-based solution. The discrete-time algorithm works as follows:

1.  The navigation speed $v_n$ is assigned to $sl$ and $sr$.
2.  The speeds $sl$ and $sr$ are applied to the actuators.
3.  The sensors group gets the information from the sensors.
4.  The sensor information $y$ is sent to the controller.
5.  The control signals $u$ are received from the controller.
6.  Finally, the speeds are applied to the robot wheels.

In the event-based solution, Steps 4 and 5 are executed only if the event condition is fulfilled. Furthermore, depending on which algorithm is executing (OA or WF) the event condition will be different. In the following subsections, these aspects and the controller algorithm will be defined in detail.

6.1.1. Obstacles Avoidance Algorithm

In the event-based implementation of the OA algorithm, the event condition is given by

$$\begin{aligned}
\text{if } ((fl > \bar{e}_{OA}) \text{ OR } (fr > \bar{e}_{OA}) \text{ OR } ... \\
...(ll > \bar{e}_{OA}) \text{ OR } (lr > \bar{e}_{OA})) \\
\{event = true\} \\
\text{else } \{event = false\}
\end{aligned} \tag{48}$$

where $\bar{e}_{OA}$ is the event threshold. As shown in Equation (48), if any of the obstacle sensors $(fl, fr, ll, lr)$ exceeds the value of the event threshold $(\bar{e}_{OA})$, an event is generated in the system and the robot sends the sensor information to the controller. The value of the event threshold determines the accuracy of the algorithm and the number of the events that the robot generates. If the threshold value is high, the number of events decreases at the same time as the accuracy of the algorithm does. On the other hand, there will be an inverse behaviour when the value of the event threshold decreases.

The algorithm in the controller is shown in Figure 11 and in Equation (49). In this case, the same algorithm is use for the two implementations (discrete-time and event-based).

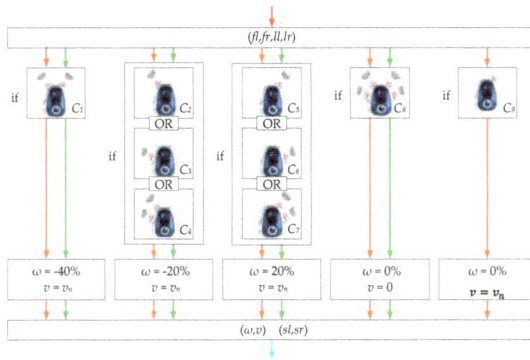

**Figure 11.** Obstacles avoidance algorithm in the controller. The blocks with orange arrows are executed in the discrete-time implementation and those with green arrows in the event-based one.

$$\begin{aligned}
\text{if } (C_1) \ \{\omega = -40\%, v = v_n\} \\
\text{else if } (C_2 \text{ OR } C_3 \text{ OR } C_4) \ \{\omega = -20\%, v = v_n\} \\
\text{else if } (C_5 \text{ OR } C_6 \text{ OR } C_7) \ \{\omega = 20\%, v = v_n\} \\
\text{else if } (C_8) \ \{\omega = 0\%, v = 0\} \\
\text{else } (C_9) \ \{\omega = 0\%, v = v_n\}
\end{aligned} \tag{49}$$

where conditions $C_1$–$C_9$ are defined by Equations (50)–(58)

$$\text{if } (ll = 0 \text{ AND } fl > 0 \text{ AND } fr > 0 \text{ AND } lr = 0)$$
$$\{C_1 = true\} \tag{50}$$

$$\text{if } (ll = 0 \text{ AND } fl > 0 \text{ AND } fr = 0 \text{ AND } lr = 0)$$
$$\{C_2 = true\} \tag{51}$$

$$\text{if } (ll > 0 \text{ AND } fl = 0 \text{ AND } fr = 0 \text{ AND } lr = 0)$$
$$\{C_3 = true\} \tag{52}$$

$$\text{if } (ll > 0 \text{ AND } fl > 0 \text{ AND } fr = 0 \text{ AND } lr = 0)$$
$$\{C_4 = true\} \tag{53}$$

$$\text{if } (ll = 0 \text{ AND } fl = 0 \text{ AND } fr > 0 \text{ AND } lr = 0)$$
$$\{C_5 = true\} \tag{54}$$

$$\text{if } (ll = 0 \text{ AND } fl = 0 \text{ AND } fr = 0 \text{ AND } lr > 0)$$
$$\{C_6 = true\} \tag{55}$$

$$\text{if } (ll = 0 \text{ AND } fl = 0 \text{ AND } fr > 0 \text{ AND } lr > 0)$$
$$\{C_7 = true\} \tag{56}$$

$$\text{if } (ll > 0 \text{ AND } fl > 0 \text{ AND } fr > 0 \text{ AND } lr > 0)$$
$$\{C_8 = true\} \tag{57}$$

$$\text{if } (ll = 0 \text{ AND } fl = 0 \text{ AND } fr = 0 \text{ AND } lr = 0)$$
$$\{C_9 = true\} \tag{58}$$

In Figure 11, the orange arrows represent the discrete-time solution and the green arrows the proposed event-based control algorithm. This algorithm is parametrized by the linear speed $v$ and the angular speed $\omega$. Theses magnitudes are transformed into wheel speeds $sl$ and $sr$ by

$$sl = v - \omega \frac{L}{2} \omega_{max} \tag{59}$$

$$sr = v + \omega \frac{L}{2} \omega_{max} \tag{60}$$

where $\omega$ is expressed as a percentage of the angular speed and $\omega_{max}$ is the maximum angular speed of the robot. Notice that, in this control algorithm, condition $C_9$ is only executed in the discrete-time implementation.

As presented in Figure 11, the algorithm receives from the robot the sensor information $(fl, fr, ll, lr)$. In the discrete-time implementation, this information is received every period of time $T = 1/f_s$ and in the event-based solution when an event is generated in the robot. Depending on the values of the sensors (conditions $C_1$ to $C_9$), different control actions are taken. In this implementation, $\omega > 0$ represents a turn of the robot to the left, on the contrary $\omega < 0$ does the robot turns to the right. For example, if the right sensors detect an obstacle (the conditions $C_5$ OR $C_6$ OR $C_7$ is fulfilled) the robot must turn to the left. In this case, the control law is defined as $\omega = 20\%$ (turn to left) and $v = v_n$ (maintain constant linear velocity).

6.1.2. Wall Following Algorithm

Two variants of the WF algorithm are usually implemented: clockwise (CW) or counter-clockwise (CCW) [45]. In this work, the second option has been selected. The event condition in the algorithm is defined by

$$\text{if } ((fl > 0) \text{ OR } (abs(ll - w) > \bar{e}_{WF})$$
$$\{event = true\} \quad \quad (61)$$
$$\text{else } \{event = false\}$$

where $\bar{e}_{WF}$ represents the event-threshold and the parameter $w$ represents the target distance between the robot and the wall. As presented in Equation (61), if there is an obstacle in front of the robot ($fl > 0$) or the distance from the robot to the wall exceeds the event threshold ($abs(ll - w) > \bar{e}_{WF}$), an event is generated in the robot.

The controller algorithm, which is the same in both architectures (discrete-time and event-based), is defined by

$$\text{if } ((fl > 0)$$
$$\{sl = v_n, sr = 0\} \quad \quad (62)$$
$$\text{else } \{sl = (ll - w)v_w + v_n, sr = v_n\}$$

where $v_w$ represents the approach speed to the wall. In this case, the control law is the same for both implementations (discrete-time and event-based). In this algorithm, when there is an obstacle in front of the robot ($fl >= 0$), the left wheel rotates with a constant speed ($sl = v_n$) and the right wheel stops ($sr = 0$). In any other situation, the speed of the right wheel remains constant ($sr = v_n$) and the right wheel is modulated according to $v_w$ to maintain a constant distance $w$ to the wall.

*6.2. System Activity*

In this work, two experiments have been set up, the OA algorithm has been checked in experiment 1 and the FW in experiment 2. In both experiments, the discrete-time architecture was implemented in robot number 1 ($rn = 1$) and the event-based one in robot number 2 ($rn = 2$).

The discrete-time system works with a sampling frequency of 10 Hz, and the event-based system uses the send-on-delta sampling method. The obstacles sensors have a maximum change rate ($mcr$) of 3 cm/s. To compare the efficiency of both systems (discrete-time and event-based), they must have the same accuracy. In other words, the precision of the discrete-time system $\bar{e}_P$ has to be the same as $\bar{e}_S$ (see Section 3.3). In this case, $\bar{e}_P = mcr/f_s$ and $\bar{e}_S$ is defined as $\bar{e}_{OA}$ for the OA algorithm and as $\bar{e}_{WF}$ for the WF algorithm. Taking into account the previous assumptions, the event thresholds were set up to $\bar{e}_{OA} = \bar{e}_{WF} = \bar{e}_P = mcr/f_s = 0.3$ cm.

The parameters of the algorithms are presented in Tables 1 and 2.

**Table 1.** Experiment 1: obstacles avoidance algorithm.

| Architecture | $rn$ | $v_n$ (cm/s) | $\bar{e}_{OA}$ (cm) |
|---|---|---|---|
| Discrete-time | 1 | 12 | - |
| Event-based | 2 | 12 | 0.3 |

**Table 2.** Experiment 2: wall following algorithm.

| Architecture | $rn$ | $v_n$(cm/s) | $w$ (cm) | $v_w$ (1/s) | $\bar{e}_{WF}$ (cm) |
|---|---|---|---|---|---|
| Discrete-time | 1 | 12 | 1.5 | - | - |
| Event-based | 2 | 12 | 1.5 | 6 | 0.3 |

In the experiments, the responses of the systems were analysed during fifteen minutes ($\Delta T = 15$ min). To analyse the activity of the control scheme, the number of samples $N_{per}$ in the discrete-time system and the number of events $N_{eve}$ in the proposed architecture have been measured.

In Figures 12 and 13, some snapshots of the experiment are shown. In both experiments, the robots show a stable behaviour. The control scheme solve the navigation problem in the discrete-time solution $rn = 1$ as in the event-based implementation $rn = 2$.

In Figure 14, the activity of the systems is presented.

In both experiments, the number of samples $N_{per}$ in the discrete-time architecture is the same (Figure 14a,b). On the other hand, the number of events $N_{eve}$ is always smaller than the number of samples $N_{per}$ in Equation (17) (Figure 14c,d), as demonstrated in Section 3.3. In this case, the event efficiency $\eta_N$ for the 15 min experiment is 63% for the OA and 17% for WF algorithm, as presented in Figure 14e,f and in Table 3.

**Table 3.** Algorithm activity by architecture, number of samples, number of events, ratio of events and event efficiency.

| Algorithm | $N_{per}$ | $N_{eve}$ | $N_N$ | $\eta_N$ |
|---|---|---|---|---|
| Obstacles Avoidance | 9544 | 3548 | 0.37 | 63% |
| Wall Following | 9544 | 7889 | 0.83 | 17% |

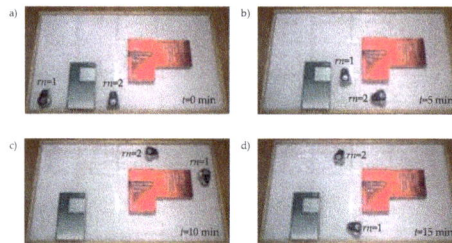

**Figure 12.** Snapshots of experiment 1. Positions of the robots at: (**a**) $t = 0$ min; (**b**) $t = 5$ min; (**c**) $t = 10$ min; and (**d**) $t = 15$ min.

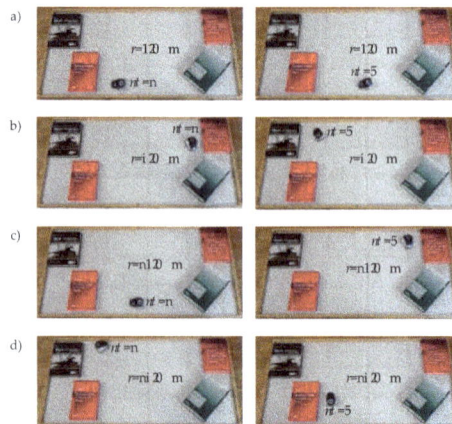

**Figure 13.** Snapshots of experiment 2. Positions of the robots at: (**a**) $t = 0$ min; (**b**) $t = 5$ min; (**c**) $t = 10$ min; and (**d**) $t = 15$ min.

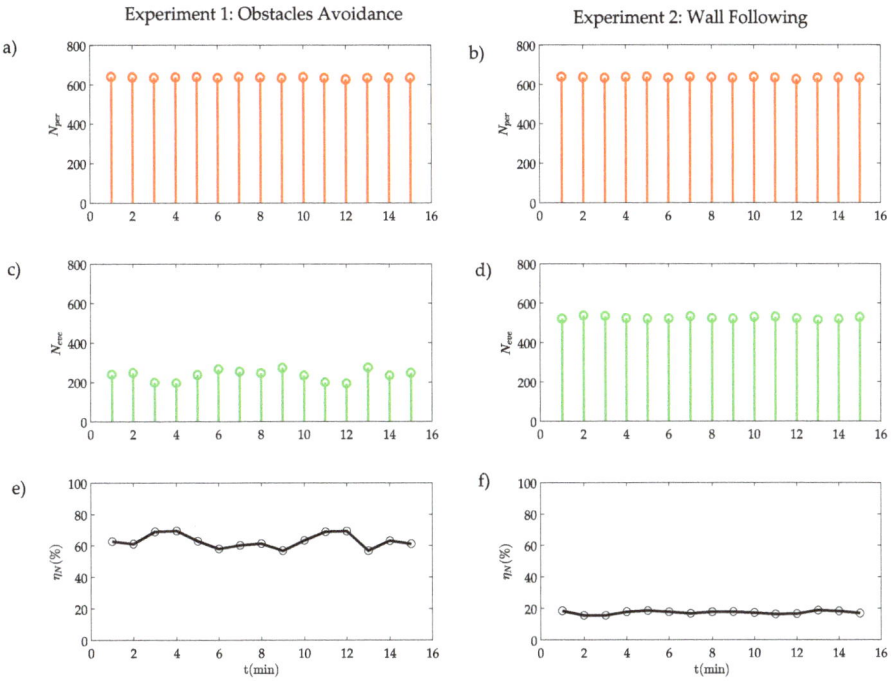

**Figure 14.** Activity of the systems: number of samples $N_{per}$ (**a**,**b**); number of events $N_{eve}$ (**c**,**d**); and event efficiency $\eta_N$ (**e**,**f**).

## 6.3. Resource Efficiency

In this section, the RF bandwidth, the computational load and the energy consumption in both architectures have been analysed. At the same time, the experimental results are discussed.

### 6.3.1. RF bandwidth

In each sensor and in each actuator, the transmitted information is a 8 bit code. Taking into account this assumption, the RF uplink (from controller to robot) uses 16 bits and the RF downlink (from robot to controller) needs 48 bits to send a packet of information (see Figure 10). Some additional bits also have to be included to manage the radio interface. In this case, the downlink is the critical link because it needs most of the bandwidth.

The downlink (DL) bandwidth for the two experiments is depicted in Figure 15a,b.

As demonstrated in Section 5.1, the efficiency in the bandwidth usage $\eta_{BW}$ (Figure 15e,f) is the same as the event efficiency $\eta_N$ (32) (Figure 15c,d). The average bandwidths for each architecture $((\overline{BW}_{DL})_{dis}, (\overline{BW}_{DL})_{eve})$, the bandwidth usage ratio $N_{BW}$ and the bandwidth usage efficiency $\eta_{BW}$ are presented in Table 4.

**Table 4.** Bandwidth efficiency.

| Algorithm | $(\overline{BW}_{DL})_{dis}$ | $(\overline{BW}_{DL})_{eve}$ | $N_{BW}$ | $\eta_{BW}$ |
|---|---|---|---|---|
| Obstacles Avoidance | 509 bps | 189 bps | 0.37 | 63% |
| Wall Following | 509 bps | 421 bps | 0.83 | 17% |

Taking into account these results, the bandwidth efficiency is 63% for the OA algorithm and 17% for the WF one.

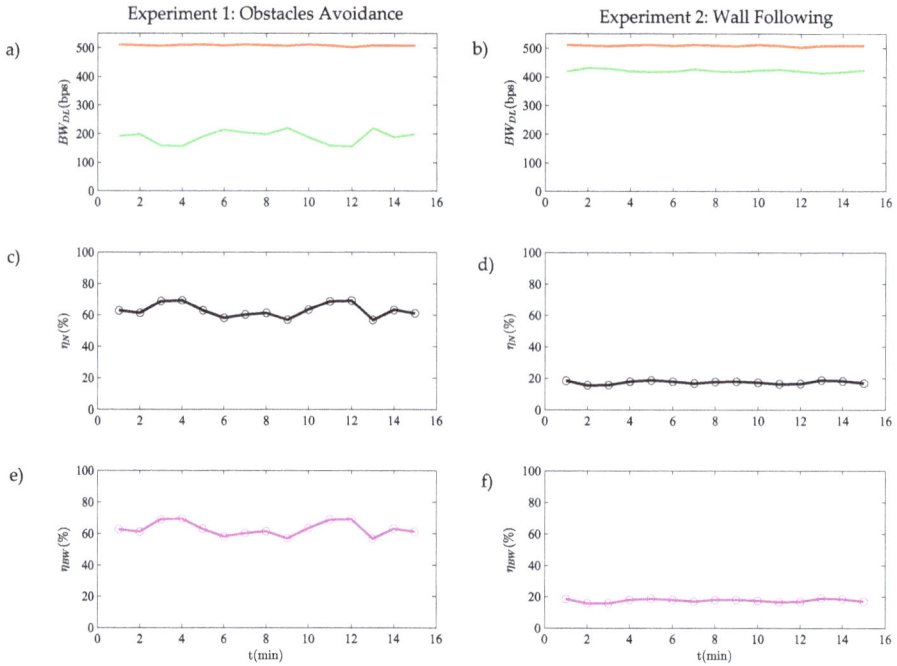

**Figure 15.** Bandwidth efficiency: (**a,b**) download bandwidth, where orange line represents the discrete-time architecture and the green line the event-based architecture; (**c,d**) event efficiency $\eta_N$; and (**e,f**) download bandwidth efficiency $\eta_{BW}$.

### 6.3.2. Computational Load

To estimate the computational load in both experiments, a high level analysis of the algorithms has been performed. In Table 5, the number of operations for each algorithm is presented.

**Table 5.** Algorithm operations.

| a) OA Algorithm Architecture | $no_C$(ops) | $no_R$(ops) | $no_{Pd}$(ops) | $no_{Pe}$(ops) |
|---|---|---|---|---|
| Discrete-time | 11 | 4 | 5 | - |
| Event-based | 11 | 10 | - | 3 |
| b) WF Algorithm Architecture | $no_C$(ops) | $no_R$(ops) | $no_{Pd}$(ops) | $no_{Pe}$(ops) |
| Discrete-time | 5 | 4 | 5 | - |
| Event-based | 5 | 10 | - | 3 |

The parameters $no_C$, $no_R$, $no_{Pd}$ and $no_{Pe}$ have been calculated taking into account the high level representation of the algorithms of the robots and the controllers (see Figure 5 and Appendix A). Considering Table 5 and the activity of the system ($N_{per}$ and $N_{eve}$), the computational load for each experiment can be calculated by Equations (33)–(35).

The results for experiments 1 and 2 are depicted in Figure 16.

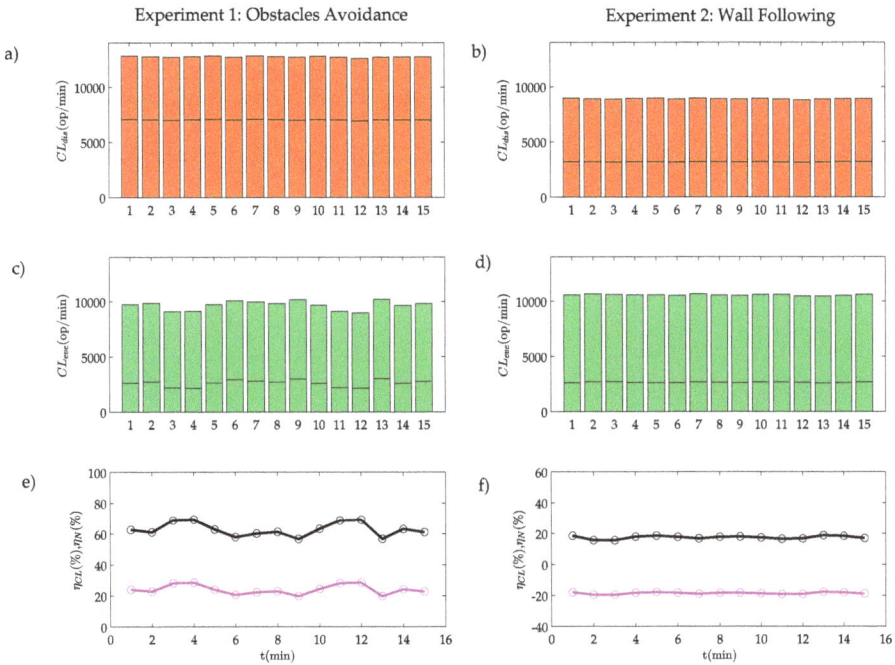

**Figure 16.** Computational load: (**a,b**) Computational load in the discrete-time implementation, where the bottom bars represent the controller and the top bars the robot; (**c,d**) computational load in the event-based implementation, where the bottom bars represent the controller and the top bars the robot; and (**e,f**) event efficiency $\eta_N$ (black line) and computational load efficiency $\eta_{CL}$ (pink line).

In experiment 1, the computational load is lower in the event-based implementation than in the discrete-time one (Figure 16a,c). In this case, the computational load efficiency $\eta_{CL}$ reaches 24% (Figure 16e and Table 6). On the other hand, in experiment 2, the computational load in the event-based solution is higher than the discrete-time one (Figure 16b,d). In this case, the computational load efficiency is −19% (Figure 16f and Table 6).

**Table 6.** Computational load efficiency.

| Algorithm | $\overline{(CL)}_{dis}$ | $\overline{(CL)}_{eve}$ | $N_{CL}$ | $\eta_{CL}$ |
|---|---|---|---|---|
| Obstacles Avoidance | 12,725 ops/min | 9674 ops/min | 0.76 | 24% |
| Wall Following | 8907 ops/min | 10,570 ops/min | 1.19 | −19% |

As discussed in Section 5.2, the computational load has a high dependence on the ratio of events $N_N$ and the polling algorithms. In these experiments, the polling algorithms have a small computational load (see Table 5), but the ratio of events is very large especially in the WF algorithm (see Table 3). This is the main reason why the efficiency in WF algorithm is negative. Therefore, the proposed system does not present good results for computational load in the WF algorithm. In this case, the solution could be to modify the event threshold $\bar{e}_{WF}$ to improve the efficiency with the inconvenience of reducing the accuracy of the system.

### 6.3.3. Energy Consumption

In these experiments, the controller has been implemented in a laptop with Windows OS. In this system, it is very complicated to measure the consumed energy by the controller and therefore this consumption has not been considered. On the other hand, in the mOway robot measuring this energy is easy by using the battery level sensor *bl*.

The results of the experiments are depicted in Figure 17.

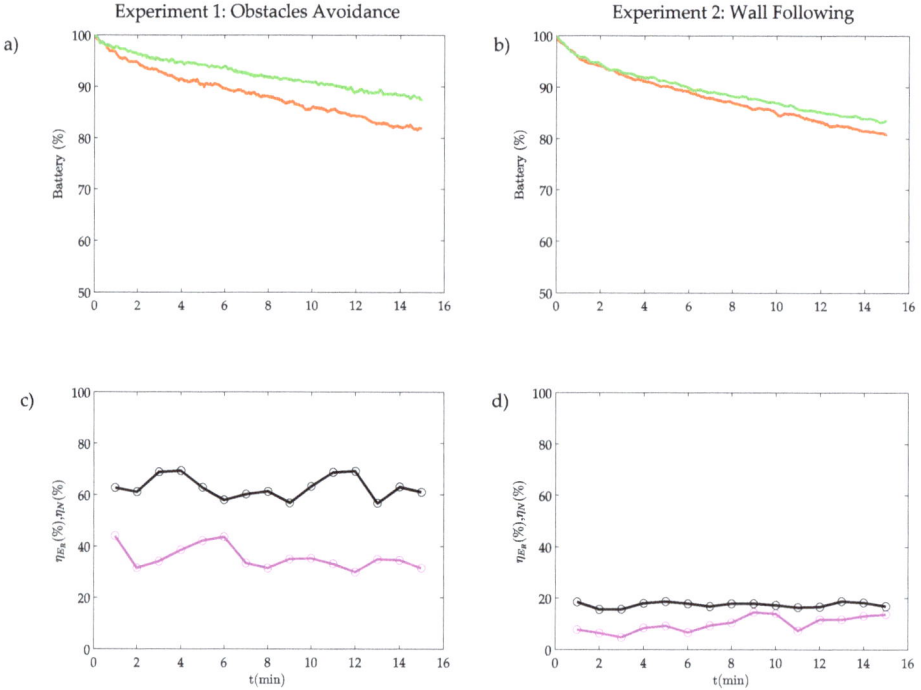

**Figure 17.** Energy consumption: (**a**,**b**) battery level, where the orange line represents the battery level in the discrete-time robot, and the green line in the event-based robot; and (**c**,**d**) event efficiency $\eta_N$ (black line) and robot energy efficiency $\eta_{CL}$ (pink line).

As shown in Figure 17a,b, when both experiments end, the battery level in the proposed event-based architecture is higher than in the discrete-time implementation, which means that the proposed system uses less energy than the classical discrete-time solution. The energy used in the robot can be estimated in the discrete-time solution as

$$(ER)_{dis} = 100\% - (Battery(\%))_{dis} \tag{63}$$

and in the event-based architecture by

$$(ER)_{eve} = 100\% - (Battery(\%))_{eve} \tag{64}$$

where $(Battery(\%))_{dis}$ denotes the level of battery measured in sensor *bl* for the discrete-time solution and $(Battery(\%))_{eve}$ for the event-based one.

In these experiments, the energy ratio in the robot can be obtained directly by $N_{ER} = \frac{(ER)_{eve}}{(ER)_{dis}}$ and energy efficiency by $\eta_{ER} = (1 - N_{ER})100$. The results are presented in Figure 17c,d and Table 7.

**Table 7.** Energy efficiency.

| Algorithm | $(ER)_{dis}$ | $(ER)_{eve}$ | $N_{ER}$ | $\eta_{ER}$ |
|---|---|---|---|---|
| Obstacles Avoidance | 18.09% | 12.48% | 0.69 | 31% |
| Wall Following | 19.18% | 16.56% | 0.86 | 14% |

In this case, the energy efficiency is 31% for experiment 1 and 14% for experiment 2. The energy consumed by the robot could be directly obtained using the measures of the sensor *bl* and this has allowed to calculate the energy efficiency directly. On the other hand, using the model developed in Section 5.3, the power ratio $\gamma$ can be obtained. In these experiments $\gamma < 1\%$, this means that the modem of the robot consumes less than 1% of the robots energy. This value is extremely small because the modem has very little power ($< 1$ mW) and consequently a short range (less than 20 m).

As analysed previously, if the power of the modem increases, the efficiency also increases. In these examples, if the power of the modem is increased (e.g., 100 mW which implies a range of 1 km, $\gamma = 63\%$) the efficiency in the OA algorithm increases from 31% to 62% and in the WF algorithm from 14% to 17%.

## 7. Conclusions and Future Work

The event-based control architectures presented in this work can be an alternative to the classical discrete-time control systems. The features of these new control schemes help to manage the resources of the system under optimal conditions because they present high efficiency in the resource usage. In a general way, by using a sampling criterion such as the send-on-delta or the integral criterion, the activity of the presented solution can be reduced. In this paper, some new methods have been proposed to analyse the resources usage and the criteria to minimize their consumption. Thus, it can be concluded that these new schemes use fewer resources such as bandwidth, computational load and energy than the classical ones. The ideas presented in this work have been applied to an NCS for mobile robots as a practical approach. The navigation problem was solved using this new control paradigm and it has been compared with a classical discrete-time solution. Finally, the experimental results have demonstrated the stability and the efficiency in the resources usage of the proposed event-based control architecture if it is compared with classical control schemes.

In this work, the effects of the delays, the packet dropouts and the packet disorders in the network have not been analysed. To avoid these undesirable effects that the network produces in the proposed control system a free RF channel is selected to communicate the robots and the controller. This reduces the interference that other agents using the same frequency can produce. To improve the proposed system, a logic-like trigger [49] or other similar ideas could be applied in a simple way. To apply these methods, two main aspects must be previously developed. First, it is necessary to find a stochastic model of the mobile robots. Then, an exhaustive analysis of the network with different levels of congestion must be performed. With this model, an estimation of the delay between the packets and the volume of the disordered packets can be obtained. Finally, once these models have been developed a network-based H∞ filtering using a logic jumping-like trigger could be applied.

As a future work, the proposed strategy and the ideas presented in this paper will be analysed in systems such as Unmanned Aerial Vehicles (UAVs), Autonomous Underwater Vehicles (AUVs) or Legged Mobile Robots (LMRs). Furthermore, new strategies to ensure the stability and the efficiency of these systems will be investigated.

**Acknowledgments:** This work was supported in part by the UNED project GID2016-6-1, the Spanish Ministry of Economy and Competitiveness under Projects DPI2014-55932-C2-2-R and ENE2015-64914-C3-2-R and FEDER funds.

**Author Contributions:** Rafael Socas designed the system, performed the experiments, analysed the results and prepared the first draft of the manuscript. Raquel Dormido performed the theoretical conception and planning of the system. The manuscript was revised and modified by Sebastián Dormido.

**Conflicts of Interest:** The authors declare no conflict of interest.

## Appendix A. Pseudocodes of Control Algorithms

In this appendix, the pseudocodes of the following algorithms are presented:

- Algorithm 1: Code for the discrete-time robot, which is used for Obstacle Avoidance and Wall Following navigation algorithms.
- Algorithm 2: Code for the event-based robot, which is used for Obstacle Avoidance and Wall Following navigation algorithms.
- Algorithm 3: Code for the Obstacle Avoidance navigation algorithm, which is used in both architectures (discrete-time and event-based).
- Algorithm 4: Code for the Wall Following navigation algorithm, which is used in both architectures (discrete-time and event-based).

---

**Algorithm 1** Pseudocode in the discrete-time robot

---

**assign** $sl \leftarrow v_n$
**assign** $sr \leftarrow v_n$
**while** (true) **do**
{
1: **assign** *left wheel speed* $\leftarrow sl$
2: **assign** *right wheel speed* $\leftarrow sr$
3: **measure** *obstacle front left sensor* $\rightarrow fl$
4: **measure** *obstacle front right sensor* $\rightarrow fr$
5: **measure** *obstacle lateral left sensor* $\rightarrow ll$
6: **measure** *obstacle lateral right sensor* $\rightarrow lr$
7: **measure** *battery level* $\rightarrow bl$
8: **transmit** *sensor information* $(rn, fl, fr, ll, lr, bl)$
9: **receive** *control information* $(sl, sr)$
}

---

**Algorithm 2** Pseudocode in the event-based robot

---

**assign** $sl \leftarrow v_n$
**assign** $sr \leftarrow v_n$
**while** (true) **do**
{
1: **assign** *left wheel speed* $\leftarrow sl$
2: **assign** *right wheel speed* $\leftarrow sr$
3: **measure** *obstacle front left sensor* $\rightarrow fl$
4: **measure** *obstacle front right sensor* $\rightarrow fr$
5: **measure** *obstacle lateral left sensor* $\rightarrow ll$
6: **measure** *obstacle lateral right sensor* $\rightarrow lr$
7: **measure** *battery level* $\rightarrow bl$
8:  **if** (event condition==true)
   {
   event_code()
   }
   else
   {
   9: **assign** $sl \leftarrow v_n$
   10: **assign** $sr \leftarrow v_n$
   }
}

event_code()
{
1: **transmit** *sensor information* $(rn, fl, fr, ll, lr, bl)$
2: **receive** *control information* $(sl, sr)$
3: **return**()
}

---

---

**Algorithm 3** Pseudocode of the OA algorithm in the controller

---

1: **receive** *obstacle sensor information* $(fl, fr, ll, lr)$
2: **check** *condition* $C_1$
3: **check** *conditions* $C_2, C_3, C4$
4: **check** *conditions* $C_5, C_6, C7$
5: **check** *condition* $C_8$
6: **check** *condition* $C_9$ // only in the discrete-time scheme
7: **calculate** $\omega$
8: **calculate** $v$
9: **transform** $(\omega, v)$ *to sl*
10: **transform** $(\omega, v)$ *to sr*
11: **transmit control information** $(sl, sr)$

---

---

**Algorithm 4** Pseudocode of the FW algorithm in the controller

---

1: **receive** *obstacle sensor information* $(fl, fr, ll, lr)$
2: **check** *condition* $(fl > 0)$
3: **calculate** *sl*
4: **calculate** *sr*
5: **transmit** *control information* $(sl, sr)$

---

## References

1. Guinaldo, M.; Sánchez, J.; Dormido, S. Control en red basado en eventos: De lo centralizado a lo distribuido. *Revista Iberoamericana de Automática e Informática Industrial RIAI* **2017**, *14*, 16–30.
2. Mansano, R.K.; Godoy, E.P.; Porto, A.J. The Benefits of Soft Sensor and Multi-Rate Control for the Implementation of Wireless Networked Control Systems. *Sensors* **2014**, *14*, 24441–24461.
3. Roohi, M.H.; Ghaisari, J.; Izadi, I.; Saidi, H. Discrete-time event-triggered control for wireless networks: Design and network calculus analysis. In Proceedings of the 2015 International Conference on Event-Based Control, Communication, and Signal Processing (EBCCSP), Krakow, Poland, 17–19 June 2015; pp. 1–8.
4. Cloosterman, M.B.; Hetel, L.; Van de Wouw, N.; Heemels, W.; Daafouz, J.; Nijmeijer, H. Controller synthesis for networked control systems. *Automatica* **2010**, *46*, 1584–1594.
5. Montestruque, L.A.; Antsaklis, P.J. State and output feedback control in model-based networked control systems. In Proceedings of the 41st IEEE Conference on Decision and Control, Las Vegas, NV, USA, 10–13 December 2002; Volume 2, pp. 1620–1625.
6. Montestruque, L.A.; Antsaklis, P.J. On the model-based control of networked systems. *Automatica* **2003**, *39*, 1837–1843.
7. Sun, Y.; El-Farra, N.H. Quasi-decentralized model-based networked control of process systems. *Comput. Chem. Eng.* **2008**, *32*, 2016–2029.
8. Tabuada, P. Event-triggered real-time scheduling of stabilizing control tasks. *IEEE Trans. Autom. Control* **2007**, *52*, 1680–1685.
9. Wang, X.; Lemmon, M.D. Event-triggering in distributed networked control systems. *IEEE Trans. Autom. Control* **2011**, *56*, 586–601.
10. Garcia, E.; Antsaklis, P.J. Decentralized model-based event-triggered control of networked systems. In Proceedings of the 2012 American Control Conference (ACC), Montreal, QC, Canada, 27–29 June 2012; pp. 6485–6490.
11. Guinaldo, M.; Dimarogonas, D.V.; Johansson, K.H.; Sánchez, J.; Dormido, S. Distributed event-based control strategies for interconnected linear systems. *IET Control Theory Appl.* **2013**, *7*, 877–886.
12. Romero, J.A.; Pascual, N.J.; Peñarrocha, I.; Sanchis, R. Event-Based PI controller with adaptive thresholds. In Proceedings of the 2012 4th International Congress on Ultra Modern Telecommunications and Control Systems and Workshops (ICUMT), St. Petersburg, Russia, 3–5 October 2012; pp. 219–226.

13. Molin, A.; Hirche, S. Adaptive event-triggered control over a shared network. In Proceedings of the 2012 IEEE 51st IEEE Conference on Decision and Control (CDC), Maui, HI, USA, 10–13 December 2012; pp. 6591–6596.

14. Socas, R.; Dormido, S.; Dormido, R. Event-based controller for noisy environments. In Proceedings of the 2014 Second World Conference on Complex Systems (WCCS), Agadir, Morocco, 10–12 November 2014; pp. 280–285.

15. Anta, A.; Tabuada, P. To sample or not to sample: Self-triggered control for nonlinear systems. *IEEE Trans. Autom. Control* **2010**, *55*, 2030–2042.

16. Wang, X.; Lemmon, M.D. Event-triggering in distributed networked systems with data dropouts and delays. In *International Workshop on Hybrid Systems: Computation and Control*; Springer: Berlin, Germany, 2009; pp. 366–380.

17. Mazo, M.; Tabuada, P. Decentralized event-triggered control over wireless sensor/actuator networks. *IEEE Trans. Autom. Control* **2011**, *56*, 2456–2461.

18. Dimarogonas, D.V.; Johansson, K.H. Event-triggered control for multi-agent systems. In Proceedings of the 48th IEEE Conference on Decision and Control, 2009 Held Jointly with the 2009 28th Chinese Control Conference, Shanghai, China, 15–18 December 2009; pp. 7131–7136.

19. Liu, J.; Yue, D. Event-triggering in networked systems with probabilistic sensor and actuator faults. *Inf. Sci.* **2013**, *240*, 145–160.

20. Zhang, X.M.; Han, Q.L.; Zhang, B.L. An overview and deep investigation on sampled-data-based event-triggered control and filtering for networked systems. *IEEE Trans. Ind. Inform.* **2017**, *13*, 4–16.

21. Dormido, S.; Sánchez, J.; Kofman, E. Muestreo, control y comunicación basados en eventos. *Revista Iberoamericana de Automática e Informática Industrial RIAI* **2008**, *5*, 5–26.

22. Socas, R.; Dormido, S.; Dormido, R. Event-based control strategy for the guidance of the Aerosonde UAV. In Proceedings of the 2015 European Conference on Mobile Robots (ECMR), Lincoln, UK, 2–4 September 2015; pp. 1–6.

23. Nguyen, V.H.; Suh, Y.S. Networked estimation with an area-triggered transmission method. *Sensors* **2008**, *8*, 897–909.

24. Pawlowski, A.; Guzman, J.L.; Rodríguez, F.; Berenguel, M.; Sánchez, J.; Dormido, S. Simulation of greenhouse climate monitoring and control with wireless sensor network and event-based control. *Sensors* **2009**, *9*, 232–252.

25. Miskowicz, M. Efficiency of event-based sampling according to error energy criterion. *Sensors* **2010**, *10*, 2242–2261.

26. Li, S.; Sauter, D.; Xu, B. Fault isolation filter for networked control system with event-triggered sampling scheme. *Sensors* **2011**, *11*, 557–572.

27. Sánchez, J.; Visioli, A.; Dormido, S. Event-based PID control. In *PID Control in the Third Millennium*; Springer: Berlin, Germany, 2012; pp. 495–526.

28. Åström, K.J. Event based control. *Anal. Des. Nonlinear Control Syst.* **2008**, *3*, 127–147.

29. Lunze, J.; Lehmann, D. A state-feedback approach to event-based control. *Automatica* **2010**, *46*, 211–215.

30. Miskowicz, M. Send-on-delta concept: An event-based data reporting strategy. *Sensors* **2006**, *6*, 49–63.

31. Miskowicz, M. Analytical approximation of the uniform magnitude-driven sampling effectiveness. In Proceedings of the 2004 IEEE International Symposium on Industrial Electronics, Ajaccio, France, 4–7 May 2004; Volume 1, pp. 407–410.

32. Miskowicz, M. Asymptotic effectiveness of the event-based sampling according to the integral criterion. *Sensors* **2007**, *7*, 16–37.

33. Zhu, W.; Jiang, Z.P. Event-based leader-following consensus of multi-agent systems with input time delay. *IEEE Trans. Autom. Control* **2015**, *60*, 1362–1367.

34. Vasyutynskyy, V.; Kabitzsch, K. Simple PID control algorithm adapted to deadband sampling. In Proceedings of the 2007 ETFA. IEEE Conference on Emerging Technologies and Factory Automation, Patras, Greece, 25–28 September 2007; pp. 932–940.

35. Otanez, P.G.; Moyne, J.R.; Tilbury, D.M. Using deadbands to reduce communication in networked control systems. In Proceedings of the 2002 American Control Conference, Anchorage, AK, USA, 8–10 May 2002; Volume 4, pp. 3015–3020.

36. Miskowicz, M. Improving the performance of the networked control system using event-triggered observations. *PDS'2004* **2004**, *37*, 53–58.

37. Pawlowski, A.; Guzmán, J.L.; Rodríguez, F.; Berenguel, M.; Sánchez, J.; Dormido, S. The influence of event-based sampling techniques on data transmission and control performance. In Proceedings of the 2009 ETFA 2009. IEEE Conference on Emerging Technologies & Factory Automation, Mallorca, Spain, 22–25 September 2009; pp. 1–8.

38. Lian, F.L.; Yook, J.K.; Tilbury, D.M.; Moyne, J. Network architecture and communication modules for guaranteeing acceptable control and communication performance for networked multi-agent systems. *IEEE Trans. Ind. Inform.* **2006**, *2*, 12–24.

39. Nguyen, V.H.; Suh, Y.S. A modified multirate controller for networked control systems with a send-on-delta transmission method. In *Advanced Intelligent Computing Theories and Applications. With Aspects of Theoretical and Methodological Issues*; Springer: Berlin, Germany, 2007; pp. 304–315.

40. Yook, J.K.; Tilbury, D.M.; Soparkar, N.R. Trading computation for bandwidth: Reducing communication in distributed control systems using state estimators. *IEEE Trans. Control Syst. Technol.* **2002**, *10*, 503–518.

41. Valera, A.; Soriano, A.; Vallés, M. Plataformas de Bajo Coste para la Realización de Trabajos Prácticos de Mecatrónica y Robótica. *Revista Iberoamericana de Automática e Informática Industrial RIAI* **2014**, *11*, 363–376.

42. Siegwart, R.; Nourbakhsh, I.R.; Scaramuzza, D. *Introduction to Autonomous Mobile Robots*; MIT Press: Cambridge, MA, USA, 2011.

43. Nehmzow, U. *Robot Behaviour: Design, Description, Analysis and Modelling*; Springer Science & Business Media: Berlin, Germany, 2008.

44. Nehmzow, U. *Mobile Robotics: A Practical Introduction*; Springer Science & Business Media: Berlin, Germany, 2012.

45. Socas, R.; Dormido, S.; Dormido, R.; Fabregas, E. Event-based control strategy for mobile robots in wireless environments. *Sensors* **2015**, *15*, 30076–30092.

46. Socas, R.; Dormido, S.; Dormido, R. Optimal Threshold Setting for Event-Based Control Strategies. *IEEE Access* **2017**, *5*, 2880–2893.

47. Socas, R.; Dormido, S.; Dormido, R.; Fábregas, E. 3D positioning algorithm for low cost mobile robots. In Proceedings of the 2015 12th International Conference on Informatics in Control, Automation and Robotics (ICINCO), Colmar, France, 21–23 July 2015; Volume 2, pp. 5–14.

48. Socas, R.; Dormido, S.; Dormido, R.; Fabregas, E. Improving the 3D positioning for low cost mobile robots. In Proceedings of the Informatics in Control, Automation and Robotics 12th International Conference, Colmar, France, 21–23 July 2015; *Revised Selected Papers*; Springer: Berlin, Germany, 2016; pp. 97–114.

49. Zhang, X.M.; Han, Q.L. Network-based H∞ filtering using a logic jumping-like trigger. *Automatica* **2013**, *49*, 1428–1435.

*Article*

# Textile Concentric Ring Electrodes for ECG Recording Based on Screen-Printing Technology

José Vicente Lidón-Roger [1], Gema Prats-Boluda [2], Yiyao Ye-Lin [2], Javier Garcia-Casado [2] and Eduardo Garcia-Breijo [1,*]

[1]   Instituto Interuniversitario de Investigación de Reconocimiento Molecular y Desarrollo Tecnológico (IDM), Universitat Politècnica de València, Universitat de València, Valencia 46022, Spain; jvlidon@eln.upv.es
[2]   Centro de Investigación e Innovación en Bioingeniería, Universitat Politècnica de València, Valencia 46022, Spain; gprats@ci2b.upv.es (G.P.B.); yiye@eln.upv.es (Y.Y.L.); jgarciac@ci2b.upv.es (J.G.C.)
*   Correspondence: egarciab@eln.upv.es; Tel.: +34-96-387-7608

Received: 29 December 2017; Accepted: 17 January 2018; Published: 21 January 2018

**Abstract:** Among many of the electrode designs used in electrocardiography (ECG), concentric ring electrodes (CREs) are one of the most promising due to their enhanced spatial resolution. Their development has undergone a great push due to their use in recent years; however, they are not yet widely used in clinical practice. CRE implementation in textiles will lead to a low cost, flexible, comfortable, and robust electrode capable of detecting high spatial resolution ECG signals. A textile CRE set has been designed and developed using screen-printing technology. This is a mature technology in the textile industry and, therefore, does not require heavy investments. Inks employed as conductive elements have been silver and a conducting polymer (poly (3,4-ethylenedioxythiophene) polystyrene sulfonate; PEDOT:PSS). Conducting polymers have biocompatibility advantages, they can be used with flexible substrates, and they are available for several printing technologies. CREs implemented with both inks have been compared by analyzing their electric features and their performance in detecting ECG signals. The results reveal that silver CREs present a higher average thickness and slightly lower skin-electrode impedance than PEDOT:PSS CREs. As for ECG recordings with subjects at rest, both CREs allowed the uptake of bipolar concentric ECG signals (BC-ECG) with signal-to-noise ratios similar to that of conventional ECG recordings. Regarding the saturation and alterations of ECGs captured with textile CREs caused by intentional subject movements, silver CREs presented a more stable response (fewer saturations and alterations) than those of PEDOT:PSS. Moreover, BC-ECG signals provided higher spatial resolution compared to conventional ECG. This improved spatial resolution was manifested in the identification of P1 and P2 waves of atrial activity in most of the BC-ECG signals. It can be concluded that textile silver CREs are more suitable than those of PEDOT:PSS for obtaining BC-ECG records. These developed textile electrodes bring the use of CREs closer to the clinical environment.

**Keywords:** textile electrode; concentric ring electrode (CRE); Laplacian electrocardiogram; PEDOT:PSS

---

## 1. Introduction

The recording of electrophysiological signals in its simplest form—that is, through contact electrodes attached to the skin—is subject to continuous studies both to optimize these records and in the search for new technologies that improve the measurement process. Today the diagnosis, therapy, and monitoring of health are based to a large extent on the measurement of signals from the brain, heart, and muscles. Even so, most of the recording systems of these signals continue to have a traditional approach, using monopolar disk electrodes (mainly Ag or AgCl). In recent years, an effort has been made to look for alternative geometries and new technologies for the manufacture of contact electrodes that allow signals to be obtained that are of better quality and/or have more precise information.

*Sensors* **2018**, *18*, 300

In addition, systems with multi-electrodes that allow the recording of several signals simultaneously are being imposed. Finally, the integration of electrodes into clothes is being sought, which would lead to medical control beyond the clinical work environment. The use of textile-based electrodes entails a series of special characteristics such as ultra-thinness, light-weighted, high flexibility, stretchability, and conformity [1]. There are two tendencies for the realization of generic textile-based electrodes: printing the electrodes on the textile using different types of inks and printing techniques [2–6] or using fibers and weaving or sewing the electrodes [7–10]. In particular, on the use of textile-based electrodes for the control of health, there are several very interesting reviews in the literature on this subject [1,11–13] that confirm the trends mentioned above.

Two important aspects to consider when designing and using electrodes for measuring bioelectric signals are the materials to be used and where to place them. As for the material, you can find electrodes that are of metal inks, conductive polymer inks or are directly of conductive textile. Each of these materials must provide flexibility to improve contact during movement of the individual. Based on this flexibility, designs can be found with conductive foam [14], conducting polymers (poly (3,4-ethylenedioxythiophene) polystyrene sulfonate (PEDOT:PSS)) and polymers with conductive particles (silver) [15], nanoparticles [16], or carbon nanotubes [17]. As for the location of the electrodes, there are works published in the literature with electrodes arranged in different positions on T-shirts, vests, girdles, and swimsuits [18–20].

As mentioned, one of the techniques for manufacturing electrodes for capturing bioelectric signals is direct printing on a substrate. In recent years, works have been developing printing techniques based on graphic arts such as screen printing, gravure, or inkjet for the manufacture of these electrodes on flexible substrates and more specifically on textiles [21,22]. The screen-printing technology is the most used and mature printing technology and has been used for decades in the manufacture of electronic systems. The revolution in the use of screen-printing techniques on flexible substrates occurred with the development of polymer-based inks, which allow low curing temperatures compatible with textile substrates [2].

On the other hand, one of the main limitations of surface bioelectric recording by means of conventional disc electrodes is the poor spatial resolution, mainly originating from the blurring effect due to different conductivities of the body volume conductor [23,24]. To overcome this limitation, surface Laplacian potential records have been proposed [25]. Literature has confirmed that Laplacian records are able to mitigate this effect and provide enhanced spatial resolution surface potential recordings—i.e., they are able to improve the detection of the bioelectric dipole sources closest to the recording electrodes, rejecting the contribution of distant bioelectric dipole sources—when compared to bipolar records made with disk electrodes [24]. First, surface Laplacian potentials were estimated using monopolar disc electrodes and applying discretization techniques [26–28]. Subsequently, body surface Laplacian potentials, such as Laplacian electrocardiography (LECG), were obtained by designing and implementing concentric ring electrodes in several configurations (bipolar, quasi-bipolar, and tripolar). Concentric ring electrodes were initially implemented in rigid substrates, mainly printed circuit boards (PCBs) [29]. In that context, Besio et al. [30] developed concentric ring electrodes on PCBs to compare the uptake capacity and spatial sensitivity of different electrode configurations—bipolar conventional (disc electrodes), bipolar concentric, and tripolar—when recording surface electrocardiographic signals. Once the capacity of the ring electrodes to detect the electrocardiographic signal and its higher spatial resolution with respect to the bipolar registers with conventional monopolar electrodes had been demonstrated, Garcia-Breijo et al. [2] compared different printing technologies to make the concentric ring electrodes on flexible plastic substrates (serigraphy, inject-printing, gravure). They concluded that the electrodes with higher reproducibility and better properties for surface bioelectric recordings were those implemented by serigraphy. Then, ring electrodes were developed on flexible plastic substrates with the aim of determining the best dimensions and CRE location to pick up electrocardiographic activity [31–33]. Flexible CREs were also developed on plastic substrates to detect uterine electrical activity [34]. Other research groups have also developed flexible electrodes on plastic substrates for capturing different bioelectrical records such as those in electrocardiography (ECG), electroencephalography (EEG) and, to a minor extent, electromyography (EMG) [35–37].

Despite the improvements introduced by the implementation of the CRE on flexible substrates, CRE use has not been transferred to the clinical environment. In order to facilitate this, we have worked on the design and validation of a set of two CREs developed on a textile substrate that will improve patient comfort during recording, especially for long-term recordings, and that will enable the detection of bioelectric signals with a similar quality to that of conventional bipolar recordings and with enhanced spatial resolution. In the present work, a set of two concentric ring electrodes screen printed onto a textile substrate has been designed and their features have been compared (electrical characteristics and bioelectric signals quality). Two different types of inks were used: one based on silver and another on a conducting polymer (poly (3,4-ethylenedioxythiophene) polystyrene sulfonate, PEDOT:PSS).

This study is structured in the following way: Section 2 includes the material and methods, describing the design of the CREs, the manufacturing processes, their characterization, the ECG recording protocol and parameters to assess signals quality; Section 3 presents the results corresponding to the CREs' characterization, ECG signals recorded, and CRE performance; in Section 4 the results are discussed and finally a conclusion is presented in Section 5.

## 2. Materials and Methods

### 2.1. Textile Concentric Ring Electrodes (CRE): Design and Development

The sensing part consists of a set of two concentric ring electrodes, each one made up of an inner disc electrode (Figure 1a) and an outer ring. Although the recording areas of the central disc and the outer ring are not equal, CREs will be connected to commercial bioamplifiers (P511, Grass Technologies, Warwick, RI, USA) with input impedances high enough to disregard the imbalance between the impedances of both poles of the CREs. Taking into account that the CRE's external diameter should be approximately at the distance between the body surface and the bioelectric sources to be recorded [29,38], the ring's external diameter was set to 5 cm since the distance between the torso surface and the heart is between 3.5 and 5.0 cm [30]. In addition, textile electrodes could provide a higher impedance electrode with respect to the CRE implemented on plastic substrates [32,33], together with the signal recording being under dry conditions (without electrolytic gel), which could hinder their capability of detecting the ECG signal, we decided to increase the recording area of these textile CREs.

The CRE dimensions are shown in Table 1. Furthermore, the distance between the CREs has been set taking into consideration that it is desired to record ECG signals in positions that are as close as possible to the standard recording positions CMV1 (position comparable to precordial V1 near the right atrium) and CMV2 (comparable to precordial V2 near the left atrium); see Figure 1b.

**Table 1.** Concentric ring electrode (CRE) dimensions and distance.

| Parameter | Units (mm) |
|---|---|
| Inner disc diameter | 16 |
| Ring internal diameter | 36 |
| Ring external diameter | 50 |
| Distance (between the discs' centers) | 120 |

Manufacturing technology used to implement this type of sensor was based on serigraphic technology of thick film. The screen-printing process consists of forcing inks of different characteristics over a substrate through some screens using squeegees. Openings in the screen define the pattern that will be printed on the substrate by serigraphy. The final thickness of the inks can be adjusted by varying the thickness of the screens. Specifically, textile CREs were produced by screen-printing technology using a four-layer design as shown in Figure 2. The first layer corresponds to the disc electrode (conductor layer). The second layer insulates the connection line that joins the inner disc to the connector, preventing a short circuit with the ring electrode and preserving the shape of the disc electrode. The concentric ring electrode is implemented in the third layer (conductor layer). The fourth layer, similar to the second layer, insulates the connection line that joins the concentric ring to the connector and the skin.

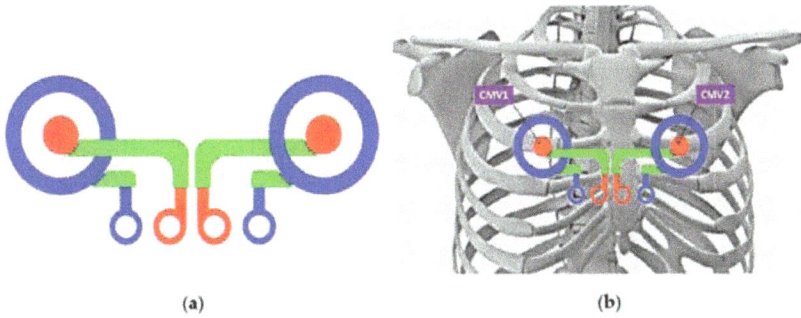

(a)                                             (b)

**Figure 1.** (a) Graphic representation of a concentric ring electrode (CRE); (b) CRE locations coincide as far as possible with the precordial registration positions CMV1 and CMV2.

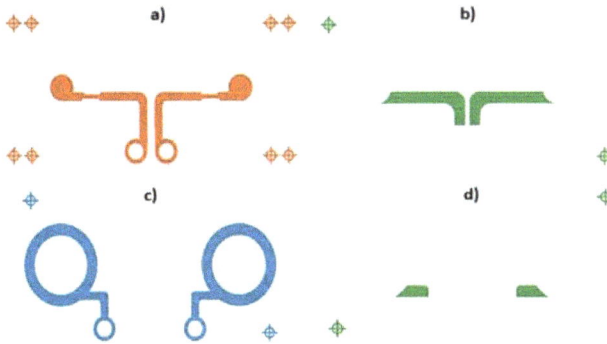

**Figure 2.** Screen patterns used. (a) The first layer corresponds to the disc electrode (conductor layer); (b) dielectric that insulates the connection line that joins the inner disc to the connector; (c) concentric ring electrode is implemented in the third layer (conductor layer); (d) fourth layer, dielectric, insulates the connection line that joins the concentric ring electrode to the connector and the skin.

The screen for the conductors was a 230 mesh of polyester material (PET 1500 90/230-48, Sefar, Thal, Switzerland) and the screen for the dielectric layer was a 175 mesh polyester material (PET 1500 68/175-64 PW, Sefar). In order to transfer the stencil to the screen mesh, a UV film Dirasol 132 (Fujifilm, Tokyo, Japan) was used. The final screen thickness was 10 μm for the screen for conductors and 15 μm for the screen for the dielectric. The patterns were transferred to the screen by using a UV light source unit. The materials used were the textile Mediatex TT ACQ 120 μm (Junkers&Muellers gmbh, Mönchengladbach, Germany) for the substrate, C2131014D3 Silver ink 59.75% (Gwent Group, Pontypool, UK) and C2100629D1 PEDOT:PSS (Gwent Group, Pontypool, UK) as the conductive inks and D2081009D6 polymer dielectric (Gwent Group, Pontypool, UK) as the dielectric ink. Flexibility is one of the most important characteristics of these inks in order to use them with textiles. Their main characteristics are shown in Table 2. Sheet resistivity (Ω/sq) measured was 76 mΩ/sq for Ag and 268 Ω/sq in the case of PEDOT:PSS for final thickness obtained. Printing was carried out using an Ekra E2 XL screen-printer (ASYS Group GmbH, Dornstadt, Germany) with a 750 shore squeegee hardness,

3.5 bar force, and 8 mm/s. After depositing the inks, they were cured in an air oven (UNB-100 Memmert GmbH+Co.KG, Schwabach, Germany) at 130 °C for 3 min (Ag ink) and 15 min (PEDOT:PSS ink).

**Table 2.** Inks' parameters.

| Property | Ag C2131014D3 | PEDOT:PSS C2100629D1 |
|---|---|---|
| Solids Content (%) | 57.00–59.75 | - |
| Vicosity (Pa.s) | 6.5–13.5 | 0.5–2.0 |
| Curing condition (°C) | 130°/3 min | 130°/15 min |
| Sheet resistivity (25 µm) | 100 mΩ/sq | 500–700 Ω/sq |

Figure 3 is a photograph of the CREs implemented with silver (Figure 3a) and with PEDOT:PSS (Figure 3b) inks. To facilitate the electrical connection with the measuring system, a snap (Sparkfun) was incorporated at each terminal of the electrodes as shown. Due to the relationship between the electrode-skin contact and signal quality, in this first prototype, the textile CREs were integrated into an adjustable belt (see Figure 3c) that exerted a certain pressure on the chest contour so as to guarantee the electrode-skin contact. In this sense, it is well known that as pressure on the chest contour increases, the skin-electrode impedance is reduced and, therefore, a better signal quality is obtained [39,40]. This belt is to be placed in a supramamarian position, making the electrodes coincide as far as possible with the registration positions CMV1 and CMV2.

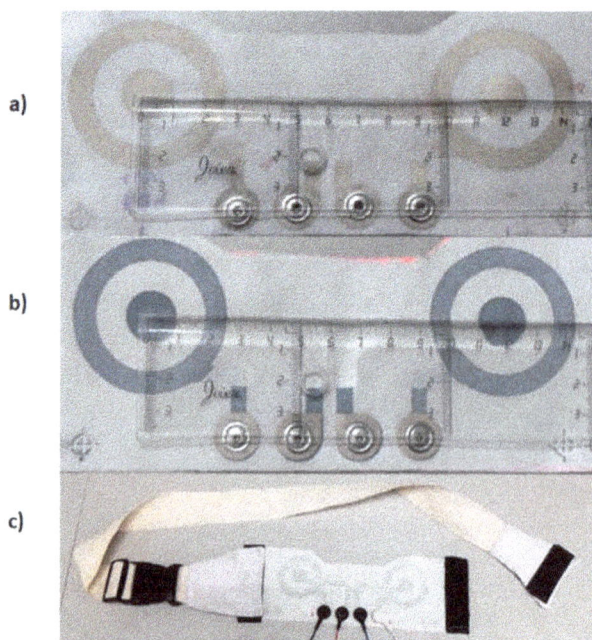

**Figure 3.** Photograph of the CRE set implemented: (**a**) corresponds to the silver electrode, (**b**) corresponds to the PEDOT:PSS (poly (3,4-ethylenedioxythiophene) polystyrene sulfonate) electrode and (**c**) CRE integrated with an adjustable belt.

## 2.2. Physical and Electrical Electrode Characterization

A physical characterization was made by measuring the final layer thickness with Profilm3D (Filmetrics) with a 20× Mirau objective.

An electrical characterization was carried out by measuring the magnitude of the impedance and the angle of the phase by using electrochemical impedance spectroscopy, which was taken with a potentiostat (Bio-Logic SP-300, Bio-Logic Science Instruments, Seyssinet-Pariset, France) in a two-electrode configuration supplying a sinusoidal signal of 1 V without any dc bias. The measurement was made between the two terminals of one of the electrodes (pole-to-pole impedance), both with skin contact.

The skin-electrode impedance of each CRE pole was carried out using the EIM-105 Prep-Check (General Devices Co Inc., Indianapolis, IN, USA) in a three-electrode configuration at 10 Hz.

*2.3. Recording Protocol*

In total, ten recording sessions were performed with 7 male and 3 female healthy subjects aged between 20 and 41 years and with body mass indices between 19.6 and 27.4 kg/m$^2$. The recordings were carried out with the subjects lying on a stretcher. This study was approved by the Polytechnic University of Valencia Ethics Committee and adhered to the Declaration of Helsinki. The volunteers were informed about the nature of the study and briefed on the recording protocol before they signed a consent form.

To reduce contact impedance, the skin area on which the electrodes (conventional and concentric) were placed was previously gently exfoliated (Nuprep, Weaver and Company, Aurora, IL, USA) and was also shaved in the case of subjects with excess hair. Two disposable Ag/AgCl electrodes (Kendall 100 series Foam electrodes, Medtronic, Minneapolis, MN, USA) were positioned on the left leg and right arm for obtaining standard lead-II ECG signals with the ground electrode being on the right leg. First, the silver CRE, previously cleaned with alcohol, was attached to the chest with an adjustable belt that exerted a certain pressure on the skin. The right electrode was positioned as close as possible to CMV1 (position comparable to V1 near to the right atrium) (see Figure 4). Immediately after placing the electrodes, the skin-electrode impedance of each CRE pole was measured as indicated in the previous section. Subsequently, two bipolar concentric ECG (BC-ECG) signals using silver CRE and standard lead-II ECG signals were recorded for 1 minute with the subject at rest. For obtaining and conditioning the two BC-ECG and lead II ECG, commercial instrumentation amplifiers (Grass Technologies P511, AstroNova, Inc., West Warwick, RI, USA) were used. Since the main information of ECG signals is distributed in the bandwidth 0.1–100 Hz [41], signals were band-pass filtered in this bandwidth and then acquired with a sampling rate of 1000 Hz. Subsequently, the electrodes' sensitivity to possible movements was analyzed. For this purpose, 1-minute of all BC-ECGs and lead II signals were recorded with the subject performing the movements listed for 10 s: lateral head movement, vertical arm movement, vertical leg movement, laughing, and deep breathing. Subsequently, the silver CRE was replaced with the PEDOT electrode and the skin-electrode impedance measurement and ECG signal acquisition during rest and during motion were repeated.

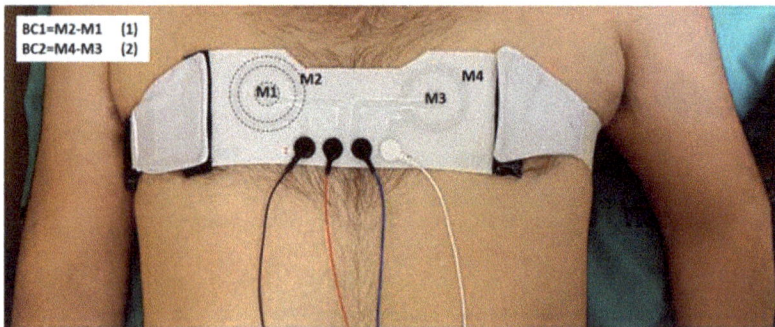

**Figure 4.** Attachment of the concentric ring electrode to the chest for obtaining two BC-ECG recordings simultaneously. M1: inner disc of the patient's right electrode, M2: outer ring of the patient's right electrode, M3: inner disk of the patient's left electrode and M4: outer ring of the patient's left electrode.

### 2.4. ECG Analysis

ECG can be corrupted by background noise and different types of interferences, such as baseline drift and power line and abdominal muscle interference. First, the ECG signals were digitally filtered with a fifth-order Butterworth high pass filter with the cutoff frequency being at 0.3 Hz. ECG fiducial points were then obtained by detecting the R wave of the ECG signal with the algorithm proposed by Pan and Tompkins [42] and slightly modified by Hamilton and Tompkins [43]. Then, the averaged beat ($E\tilde{C}G$) extending from 275 ms prior to the R wave to 425 ms after it was computed.

To compare the ECG signals recorded by the silver and PEDOT:PSS CREs during rest, the peak-peak amplitude of the averaged beat ($E\tilde{C}G$) and the signal-to-noise ratio (SNR) were worked out. The latter is defined as the ratio of the root mean square (rms) value of the average beat ($E\tilde{C}G$) and that of the noise during the isoelectric period between beats.

The ECG signals sensed during intentional movement was analyzed in order to quantify the motion artifact sensitivity of these electrodes. For this purpose, the time percentage in which the ECG signal presented alterations with respect to the sensed signal during rest and/or signal saturation due to each movement was annotated for the signals sensed in each position (left and right). In the present study, signal alteration was considered as any visually appreciable variation of the ECG signal recorded during movement with respect to that obtained at rest. Alterations consisted mainly of baseline changes. By contrast, signal saturation was referred to as amplified signals that reached the maximum or minimum output voltage allowed for conditioning the system, made up by the commercial P511 bioamplifiers connected to the DAQ NI USB 6229 (National Instruments, Austin, TX, USA) with the saturation voltage being ±5 V. Then, the mean time percentage of all the involved subjects was computed for each movement and each position.

## 3. Results

### 3.1. CREs Physical and Electrical Characteristics

A magnified view of the two designs is shown in Figure 5 (Figure 5a for PEDOT:PSS and Figure 5b for Ag). As the PEDOT:PSS was embedded in the fabric while the silver remained on the fabric, the different effective thicknesses were obtained in both cases. In the case of PEDOT:PSS, the average thickness obtained was 15 μm (Figure 6a); in the case of the silver, the average thickness obtained was 40 μm (Figure 6b).

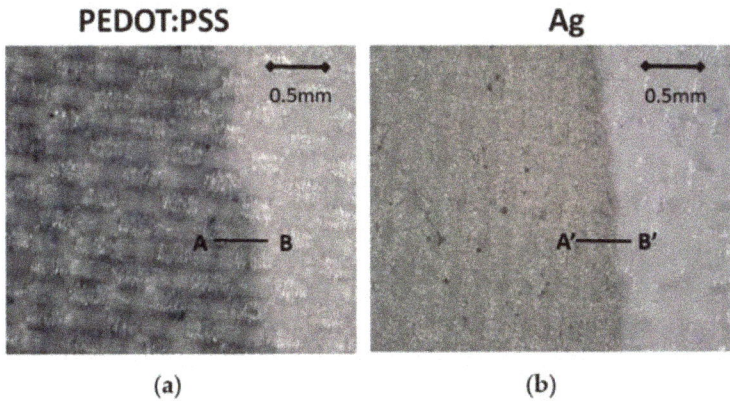

**Figure 5.** (**a**) Detail of the PEDOT:PSS on the substrate; the PEDOT:PSS is embedded in the fabric pattern; (**b**) Detail of the Ag on the substrate.

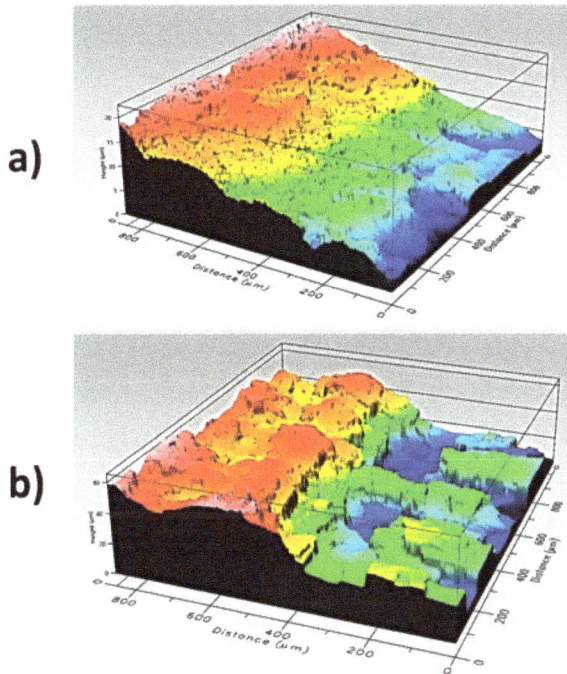

**Figure 6.** (**a**) Thickness of the PEDOT:PSS on the substrate (view A–B from Figure 5a); (**b**) thickness of the Ag on the substrate (view A'–B' from Figure 5b).

Regarding the electrical characterization, Figure 7 shows the impedance and phase of the electrodes between 0.1 and 200 Hz—frequency bandwidth of ECG—both in the case of electrode-skin contact.

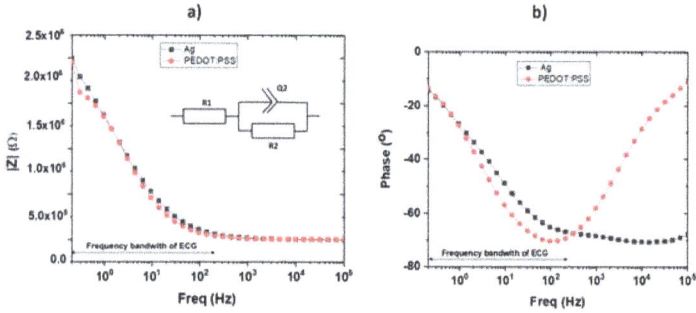

**Figure 7.** External ring: (**a**) electrode–skin impedance (pole-to-pole) magnitude and (**b**) phase angle.

The electrode impedance characteristics were analyzed using a modified Ershler-Randles equivalent circuit called ZARC—shown in the inset of Figure 7—the equation for which is shown in (3) where $R_1$ is the series resistance, $R_2$ is the charge-transfer resistance, $Q_2$ is the constant phase element, and the exponent $\alpha$ determines the character of frequency dependence. $R_1$ represents the resistance between electrodes through the skin, $R_2$ represents the transfer of electrons by redox between the electrodes and the skin and $Q_2$ can vary by the roughness, thickness, and composition of the material. Table 3 summarizes the data of these variables for the case of using the silver or the PEDOT:PSS.

$$Z(f) = R_1 + \frac{R_1}{R_2 \cdot Q_2 (i \cdot \pi \cdot f)^{\alpha} + 1}$$

**Table 3.** Values of equivalent circuit model.

| Property | Ag C2131014D3 | PEDOT:PSS C2100629D1 |
|---|---|---|
| $R_1$ ($\Omega$) | $4.80 \times 10^2$ | $2.73 \times 10^3$ |
| $R_2$ ($\Omega$) | $1.81 \times 10^6$ | $1.85 \times 10^6$ |
| $Q_2$ (F·s$^{(\alpha-1)}$) | $69.79 \times 10^{-9}$ | $79.96 \times 10^{-9}$ |
| $\alpha$ | 0.76 | 0.79 |

The behavior of the two electrodes in the working frequency range is very similar. The difference in the value of Q2 makes sense since they are two different materials with different thickness and roughness. It must be taken into account that PEDOT:PSS is permeable to cations and a redox process occurs in its presence.

The skin-electrode impedance measurements during the recording sessions are shown in Table 4. The skin-electrode impedance values presented a high variability among the subjects and were relatively higher than the conventional pre-gelled Ag/AgCl electrodes, which usually provide a skin-electrode impedance lower than 10 kΩ; however, they were still within the admissible limit for bioelectrical signal acquisition. They were also higher than that obtained by the silver CRE implemented on other flexible substrates: Valox, Melinex, Ultem [31]. In comparison to the PEDOT:PSS electrode, the silver electrode impedance was generally lower.

**Table 4.** Skin-electrode impedance measurements for both external ring and inner disc of the CREs.

| | Left Ag | PEDOT:PSS | Right Ag | PEDOT:PSS |
|---|---|---|---|---|
| External ring impedance (kΩ) | $18.3 \pm 20.5$ | $27.3 \pm 22.3$ | $21.3 \pm 22.3$ | $32.0 \pm 21.0$ |
| Inner disc impedance (kΩ) | $25.0 \pm 20.1$ | $25.3 \pm 24.0$ | $24.0 \pm 19.3$ | $32.0 \pm 20.1$ |

## 3.2. ECG Analysis

Figure 8 shows 5 seconds of the simultaneous recordings of standard lead II (trace c.1 and f.1) and two BC-ECG recordings acquired using the silver electrode (trace a.1 and b.1) and the PEDOT:PSS electrode (trace d.1 and e.1) with the subject at rest. Their corresponding averaged beats are shown on the right side. Initially, fiducial points of ECG signals can be clearly identified in all BC-ECG recordings using both silver and PEDOT:PSS electrodes, being of the BC-ECG signal quality. In addition, the P1 and P2 waves corresponding to the depolarization of the right and left atria can be clearly identified in the BC-ECG averaged beat at the right position (CMV1, trace a.2 and d.2), regardless of the CRE conductor material (silver or PEDOT:PSS). The relative amplitude of the P wave with respect to the QRS complex of BC-ECG acquired at the right position (CMV1) was much higher than that of the standard lead-II ECG signal. By contrast, the signal amplitudes recorded at the left position was higher than that sensed at the right position, although the P wave associated with atrial activity was not appreciated (see traces b and e) since the CRE was positioned away from the atrium. When comparing the BC-ECG signals acquired with the silver and PEDOT:PSS electrodes, no significant morphology change was observed except for the signal amplitude, which may be due to the fact that the silver and PEDOT:PSS electrodes were not positioned exactly in the same position.

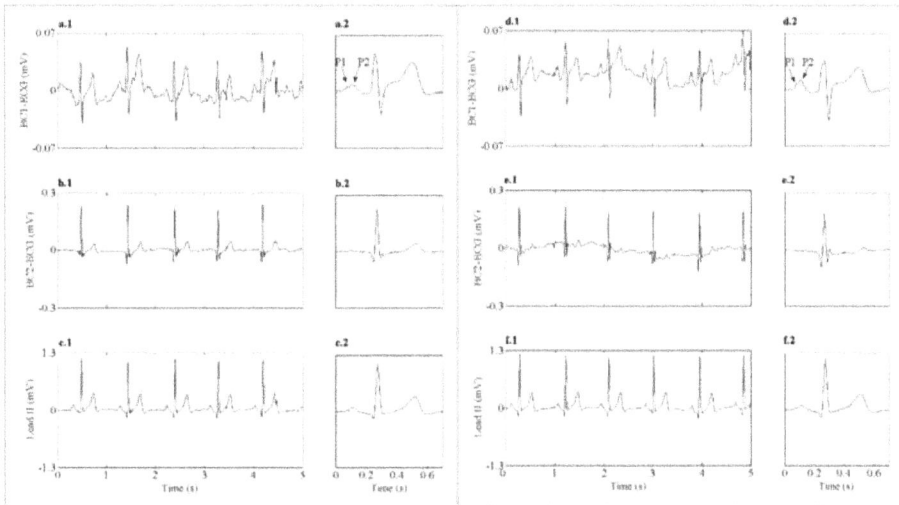

**Figure 8.** Five seconds of raw ECG signals and its corresponding averaged beat at its right. (**a.1**) BC1-ECG acquired with the silver CRE at the right position (CMV1). (**b.1**) BC2-ECG acquired with the silver CRE at the left position. (**d.1**) BC1-ECG acquired with the PEDOT:PSS CRE in the right position (CMV1). (**e.1**) BC2-ECG acquired with the PEDOT:PSS CRE in the left position. (**c.1,f.1**) Standard lead II simultaneously recorded with the two BC-ECG signals sensed by the silver and PEDOT:PSS CREs, respectively. (**a.2–f.2**) averaged beats of BC-ECG signals shown in traces (**a.1–f.1**) respectively.

Regarding the ECG signals detected during rest, it can be observed in Table 5 that signal amplitude presented a high inter-subject variability with higher ECG signals being detected in the left position. In addition, the amplitude of the signals sensed by the PEDOT:PSS electrode was slightly higher than that obtained by the silver electrode. In contrast, similar signal to noise ratio (SNR) values (around of 21 dB) were obtained for both electrodes regardless of their position.

**Table 5.** Main characteristics of the sensed BC-ECG signal at rest.

| | Left | | Right | |
|---|---|---|---|---|
| | Ag | PEDOT:PSS | Ag | PEDOT:PSS |
| Peak to peak amplitude (µV) | 330.2 ± 126.2 | 363.9 ± 148.0 | 124.8 ± 133.6 | 143.8 ± 128.3 |
| SNR (dB) | 22.4 ± 6.3 | 20.1 ± 6.8 | 20.8 ± 6.6 | 20.2 ± 3.9 |

To quantify motion artifact sensitivity of both CREs, ECG signals were acquired during intentional movements and analyzed. Table 6 shows the mean time percentage in which ECG signals were altered and/or saturated for the total of patients. The last row shows the mean and deviation of the mean time percentage for all of the intentional movements. The mean time percentage of the saturated signal for the silver electrode was lower than 2% regardless of its position. In addition, with respect to the signal sensed during rest, the ECG signal presented alterations almost all of the time in which the intentional movements were generated (~16%). In contrast, using the PEDOT:PSS electrode, the signals remained altered for a longer period (25–30%), even after the movement had finished. These electrodes were in general more sensitive to motion artifact with the mean time percentage of the altered signal and saturated signal being higher than that of the silver CRE. When comparing the different types of movements, both dry electrodes seem to be less sensitive to horizontal head movement. By contrast, the silver CRE seems to be more sensitive to laughing and deep breathing, which entail rib cage movement.

**Table 6.** Mean time percentage in which ECG signal was altered and/or saturated due to intentional movements of the total patients for both electrodes (Ag and PEDOT:PSS) in different positions.

| | Left | | | | Right | | | |
|---|---|---|---|---|---|---|---|---|
| | Altered (%) | | Saturated (%) | | Altered (%) | | Saturated (%) | |
| | Ag | PEDOT:PSS | Ag | PEDOT:PSS | Ag | PEDOT:PSS | Ag | PEDOT:PSS |
| Head | 3.2 | 10.1 | 0.0 | 3.1 | 5.9 | 9.6 | 0.0 | 1.1 |
| Arm | 13.9 | 28.8 | 0.9 | 17.7 | 13.6 | 31.6 | 0.9 | 12.2 |
| Leg | 18.9 | 36.9 | 0.1 | 13.1 | 24.2 | 28.1 | 0.0 | 10.7 |
| Laughing | 18.1 | 41.4 | 2.1 | 26.9 | 22.9 | 24.9 | 4.2 | 7.1 |
| Breathing | 13.9 | 44.5 | 6.1 | 36.1 | 25.5 | 29.9 | 3.5 | 13.3 |
| $\mu \pm \sigma$ | 13.6 ± 6.2 | 32.3 ± 13.8 | 1.8 ± 2.6 | 19.4 ± 12.7 | 18.4 ± 8.4 | 24.9 ± 8.9 | 1.7 ± 2.0 | 8.9 ± 5.0 |

## 4. Discussion

An electrode for comfortable high spatial resolution surface ECG recording has been manufactured using a textile substrate and with two types of conductive materials, namely silver and PEDOT:PSS. So far, metals such as Ag or AgCl have been used for the construction of these electrodes, but these materials can cause allergies in some individuals. Therefore, the use of totally biocompatible materials, such as some conducting polymers, is a desired alternative. PEDOT:PSS used in this work is a commercial product to which no compound has been added to improve its conductivity or to avoid further problems. However, in view of the results, it would be interesting to add an organic compound such as dimethyl sulfoxide (DMSO) or ethylene glycol to enhance the conductivity one to three orders of magnitude. It should also be noted that some of the manufactured electrodes using PEDOT:PSS had to be discarded since the material had not been deposited uniformly enough to ensure good conductivity and contact area. In this context, some changes to the manufacturing process are required to enhance its reproducibility in terms of correct and uniform deposition of the material. A possible way to improve the response is to first deposit a layer of silver and then on top of this to deposit a layer of PEDOT:PSS [34].

The developed electrodes have been tested in healthy volunteers in order to assess its capability of detecting cardiac activity in patients at rest as the first step toward transferring CREs to clinical applications. It should be noted that in the present work, compared to previous publications [33,34,43], larger CREs were designed. It was taken into account that CREs would be implemented on textile substrates, which could compromise the ability to capture cardiac signals compared to smaller CREs developed on flexible plastic substrates [32,33,39]. Nonetheless, in this regard, it should be noted that the amplitudes of the BC-ECG signals captured with the textile CRE have been higher (more than two times) than those of smaller flexible CREs implemented on plastic substrates [33,35,39]. It has also been verified—as in previous studies—that the amplitudes of the ECG signals captured in position CMV1 are considerably lower than those associated with position CMV2, located on the left side of the patient and closer to the central part of the patient's heart [33]. Regarding the CREs' location on the chest for obtaining BC-ECG signals, previous studies carried out by the present research group revealed the superiority of CREs located in chest positions comparable to precordial V1 (CMV1) and precordial V2 (CMV2) for picking up cardiac activity compared to positions comparable to precordial V4R (CMV4R) or comparable to precordial V5 (CMV5) [33]. To simultaneously record cardiac signals in CMV1 and CMV2, an inter-CRE distance of 120 mm was chosen for the CREs' set design. This distance permits the correct placing of the two CREs in the desired positions on a torso of a medium-build subject. In the case of subjects with a very thin or wide frame, placement of the electrode on the CMV1 position is prioritized. Once the ability of textile CREs to capture cardiac activity has been proven, future work aims to improve the ability of the system to adapt to the patient's anatomy.

Regarding spatial resolution, previous work that used smaller CREs implemented on flexible plastic substrates [33] reported the identification of more 'local cardiac activity' as is the case of P1 and P2 atrial waves. Despite the spatial resolution decreasing with increasing size of the CRE [44], the P1 and P2 atrial waves could also be identified in the BC-ECG records performed in the present work with the larger electrodes in this work. These waves are not identifiable in conventional recordings with disc electrodes. At present, there has not yet been a translation of surface high spatial resolution ECG techniques, either with disk electrodes and the application of interpolation techniques or through the use of CRE to clinical practice. In the first case, the use of a large number of monopolar electrodes to carry out the ECG mapping entails the use of complicated recording systems that are not viable in clinical use. The use of CREs could reduce the number of required electrodes. Nonetheless, the best locations to obtain useful diagnostic information should still be studied. Furthermore, the information provided by surface bioelectric signals from CREs has not been analyzed and compared with that of conventional records from standard derivations.

With this work, we aimed to make a prototype that brings the use of CREs closer to clinical use. To do this, systems must still be developed that are comfortable for the patient, are easy to use, and that provide information that is easy for the physician to interpret. In future work, we propose to analyze the behavior of this type of electrode in stress tests and ambulatory recording systems.

## 5. Conclusions

- The use of silver on textiles presents better characteristics than PEDOT:PSS. Even so, techniques that improve the transfer of PEDOT:PSS to textiles could be used.
- Both textile silver and PEDOT:PSS CREs implemented on textile substrates are able to detect surface electrocardiographic activity in standard precordial recording positions, similar to V1 and V2 (CMV1 and CMV2).
- The amplitudes the BC-ECG signals detected with the textile CREs developed are of hundreds of microvolts, which is slightly lower than those of conventional bipolar ECG signals in precordial positions.
- BC-ECGs recorded with silver and PEDOT:PSS textile CREs presented similar signal to noise ratios with these values being similar to those of BC-ECGs from CREs implemented on plastic substrates published in the literature.
- Regarding the saturation and alterations of the BC-ECGs associated with movement of the subject, textile silver CRE showed a more stable response (fewer saturations and alterations) than PEDOT:PSS.

- Regardless the use of silver or PEDOT:PSS, BC-ECG signals captured with the developed textile CREs have a better spatial resolution than that of conventional recordings (lead II). Specifically, BC-ECG signals have an improved capability of recording atrial activity on the surface, with P1 and P2 waves being associated with the activity of the left and right atria identified in most BC-ECG signals.

To sum up, surface ECG records of high spatial resolution could be obtained with a comfortable and simple system using CREs of the proposed dimensions, implemented using screen printing techniques on textile substrates and with Ag ink as the conductive material.

**Acknowledgments:** Grant from the Ministerio de Economía y Competitividad y del Fondo Europeo de Desarrollo Regional. DPI2015-68397-R (MINECO/FEDER). This work was also supported by the Spanish Government/FEDER funds (grant number MAT2015-64139-C4-3-R (MINECO/FEDER)).

**Author Contributions:** Gema Prats-Boluda, Yiyao Ye-Lin, and Javier Garcia-Casado designed the electrode; José Vicente Lidón-Roger and Eduardo García-Breijo manufactured the electrodes and designed the experiments for the physical and electrical characterization; Gema Prats-Boluda, Yiyao Ye-Lin, and Javier Garcia-Casado designed the experiments for recording the electrophysiological signals and analyzed the signals; Gema Prats-Boluda, Yiyao Ye-Lin, and Eduardo García-Breijo wrote the paper, which was enriched by the rest of the authors.

**Conflicts of Interest:** The authors declare no conflict of interest. The founding sponsors had no role in the design of the study; in the collection, analyses, or interpretation of data; in the writing of the manuscript, or in the decision to publish the results.

## References

1. Trung, T.Q.; Lee, N.E. Flexible and stretchable physical sensor integrated platforms for wearable human-activity monitoring and personal healthcare. *Adv. Mater.* **2016**, *28*, 4338–4372. [CrossRef] [PubMed]
2. Garcia-Breijo, E.; Prats-Boluda, G.; Lidon-Roger, J.V.; Ye-Lin, Y.; Garcia-Casado, J. A comparative analysis of printing techniques by using an active concentric ring electrode for bioelectrical recording. *Microelectron. Int.* **2015**, *32*, 103–107. [CrossRef]
3. Linti, C.; Horter, H.; Österreicher, P.; Planck, H. Sensory baby vest for the monitoring of infants. In Proceedings of the International Workshop on Wearable Implantable Body Sensor Networks BSN 2006, Cambridge, MA, USA, 3–5 April 2006; pp. 135–137. [CrossRef]
4. Wei, Y.; Torah, R.; Li, Y.; Tudor, J. Dispenser printed capacitive proximity sensor on fabric for applications in the creative industries. *Sens. Actuators A Phys.* **2016**, *247*, 239–246. [CrossRef]
5. Takamatsu, S.; Lonjaret, T.; Ismailova, E.; Masuda, A.; Itoh, T.; Malliaras, G.G. Wearable keyboard using conducting polymer electrodes on textiles. *Adv. Mater.* **2016**, *28*, 4485–4488. [CrossRef] [PubMed]
6. Kim, D.K.; Kim, J.H.; Kwon, H.J.; Kwon, Y.H. A touchpad for force and location sensing. *ETRI J.* **2010**, *32*, 722–728. [CrossRef]
7. Catrysse, M.; Puers, R.; Hertleer, C.; Van Langenhove, L.; Van Egmond, H.; Matthys, D. Fabric sensors for the measurement of physiological parameters. In Proceedings of the 12th International Conference on Transducers Solid-State Sensors, Actuators and Microsystems, Boston, MA, USA, 8–12 June 2003; pp. 1758–1761. [CrossRef]
8. Catrysse, M.; Puers, R.; Hertleer, C.; Van Langenhove, L.; Van Egmond, H.; Matthys, D. Towards the integration of textile sensors in a wireless monitoring suit. *Sens. Actuators A Phys.* **2004**, *114*, 302–311. [CrossRef]
9. Chi, Y.M.; Deiss, S.R.; Cauwenberghs, G. Non-contact low power EEG/ECG electrode for high density wearable biopotential sensor networks. In Proceedings of the 6th International Workshop on Wearable Implantable Body Sensor Networks BSN 2009, Berkeley, CA, USA, 3–5 June 2009. [CrossRef]
10. Al-huda Hamdan, N.; Heller, F.; Wacharamanotham, C.; Thar, J.; Borchers, J. Grabrics: A foldable two-dimensional textile input controller. *CHI Ext. Abstr. Hum. Factors Comput. Syst.* 2016. [CrossRef]
11. Baig, M.M.; Gholamhosseini, H.; Connolly, M.J. A comprehensive survey of wearable and wireless ECG monitoring systems for older adults. *Med. Biol. Eng. Comput.* **2013**, *51*, 485–495. [CrossRef] [PubMed]

12. Zheng, Y.L.; Ding, X.R.; Poon, C.C.Y.; Lo, B.P.L.; Zhang, H.; Zhou, X.L.; Yang, G.Z.; Zhao, N.; Zhang, Y.T. Unobtrusive sensing and wearable devices for health informatics. *IEEE Trans. Biomed. Eng.* **2014**, *61*, 1538–1554. [CrossRef] [PubMed]

13. Sun, Y.; Yu, X.B. Capacitive biopotential measurement for electrophysiological signal acquisition: A review. *IEEE Sens. J.* **2016**, *16*, 2832–2853. [CrossRef]

14. Lin, C.T.; Liao, L.D.; Liu, Y.H.; Wang, I.J.; Lin, B.S.; Chang, J.Y. Novel dry polymer foam electrodes for long-term EEG measurement. *IEEE Trans. Biomed. Eng.* **2011**, *58*, 1200–1207. [CrossRef] [PubMed]

15. Hoffmann, K.P.; Ruff, R.; Poppendieck, W. Long-term characterization of electrode materials for surface electrodes in biopotential recording. *Annu. Int. Conf. IEEE Eng. Med. Biol.-Proc.* **2006**. [CrossRef]

16. Pylatiuk, C.; Müller-Riederer, M.; Kargov, A.; Schulz, S.; Schill, O.; Reischl, M.; Bretthauer, G. Comparison of surface EMG monitoring electrodes for long term use in rehabilitation device control. In Proceedings of the 2009 IEEE 11th International Conference on Rehabilitation Robotics, Kyoto, Japan, 23–26 June 2009; pp. 300–304.

17. Jung, H.; Moon, J.; Baek, D.; Lee, J.; Choi, Y.; Hong, J. CNT/PDMS composite flexible dry electrodes for long-term ECG monitoring. *IEEE Trans. Biomed. Eng.* **2012**, *59*, 1472–1479. [CrossRef] [PubMed]

18. Guo, Y.; Otley, M.T.; Li, M.; Zhang, X.; Sinha, S.K.; Treich, G.M.; Sotzing, G.A. PEDOT:PSS "wires" printed on textile for wearable electronics. *ACS Appl. Mater. Interfaces.* **2016**, *8*, 26998–27005. [CrossRef] [PubMed]

19. Papaiordanidou, M.; Takamatsu, S.; Rezaei-Mazinani, S.; Lonjaret, T.; Martin, A.; Ismailova, E. Cutaneous recording and stimulation of muscles using organic electronic textiles. *Adv. Healthc. Mater.* **2016**, *5*, 2001–2006. [CrossRef] [PubMed]

20. Rivnay, J.; Leleux, P.; Ferro, M.; Sessolo, M.; Williamson, A.; Koutsouras, D.A.; Khodagholy, D.; Ramuz, M.; Strakosas, X.; Owens, R.M.; et al. High-performance transistors for bioelectronics through tuning of channel thickness. *Sci. Adv.* **2015**, *1*, e1400251. [CrossRef] [PubMed]

21. Takamatsu, S.; Lonjaret, T.; Crisp, D.; Badier, J.-M.; Malliaras, G.G.; Ismailova, E. Direct patterning of organic conductors on knitted textiles for long-term electrocardiography. *Sci. Rep.* **2015**, *5*, 15003. [CrossRef] [PubMed]

22. Pandian, P.S.; Mohanavelu, K.; Safeer, K.P.; Kotresh, T.M.; Shakunthala, D.T.; Gopal, P.; Padaki, V.C. Smart vest: wearable multi-parameter remote physiological monitoring system. *Med. Eng. Phys.* **2008**, *30*, 466–477. [CrossRef] [PubMed]

23. Bradshaw, L.A.; Richards, W.O.; Wikswo, J.P. Volume conductor effects on the spatial resolution of magnetic fields and electric potentials from gastrointestinal electrical activity. *Med. Biol. Eng. Comput.* **2001**, *39*, 35–43. [CrossRef] [PubMed]

24. Besio, W.; Aakula, R.; Dai, W. Comparison of bipolar vs. tripolar concentric ring electrode Laplacian estimates. *Annu. Int. Conf. IEEE Eng. Med. Biol. Soc.* **2004**, *3*, 2255–2258. [CrossRef]

25. Makeyev, O.; Ding, Q.; Besio, W.G. Improving the accuracy of Laplacian estimation with novel multipolar concentric ring electrodes. *Measurement* **2016**, *80*, 44–52. [CrossRef] [PubMed]

26. Hjorth, B. An online transformation of EEG scalp potentials into orthogonal source derivations. *Electroencephalogr. Clin. Neurophysiol.* **1975**, *39*, 526–530. [CrossRef]

27. Wu, D.; Tsai, H.C.; He, B. On the estimation of the Laplacian electrocardiogram during ventricular activation. *Ann. Biomed. Eng.* **1999**, *27*, 731–745. [CrossRef] [PubMed]

28. Tandonnet, C.; Burle, B.; Hasbroucq, T.; Vidal, F. Spatial enhancement of EEG traces by surface Laplacian estimation: comparison between local and global methods. *Clin. Neurophysiol.* **2005**, *116*, 18–24. [CrossRef] [PubMed]

29. Lu, C.C.; Tarjan, P.P. An ultra high common mode rejection ratio (CMRR) AC instrumentation amplifier for Laplacian electrocardiographic measurement. *Biomed. Instrum. Technol.* **1999**, *33*, 76–83. [PubMed]

30. Besio, W.; Chen, T. Tripolar Laplacian electrocardiogram and moment of activation isochronal mapping. *Physiol. Measurement* **2007**, *28*, 515–529. [CrossRef] [PubMed]

31. Prats-Boluda, G.; Ye-Lin, Y.; Garcia-Breijo, E.; Ibañez, J.; Garcia-Casado, J. Active flexible concentric ring electrode for non-invasive surface bioelectrical recordings. *Meas. Sci. Technol.* **2012**, *23*, 125703. [CrossRef]

32. Prats-Boluda, G.; Ye-Lin, Y.; Bueno Barrachina, J.M.; Senent, E.; Rodriguez de Sanabria, R.; Garcia-Casado, J. Development of a portable wireless system for bipolar concentric ECG recording. *Meas. Sci. Technol.* **2015**, *26*, 75102. [CrossRef]

33. Prats-Boluda, G.; Ye-Lin, Y.; Bueno-Barrachina, J.; Rodriguez de Sanabria, R.; Garcia-Casado, J. Towards the clinical use of concentric electrodes in ECG recordings: influence of ring dimensions and electrode position. *Meas. Sci. Technol.* **2016**, *27*, 25705. [CrossRef]

34. Ye-Lin, Y.; Alberola-Rubio, J.; Prats-Boluda, G.; Perales, A.; Desantes, D.; Garcia-Casado, J. Feasibility and analysis of bipolar concentric recording of electrohysterogram with flexible active electrode. *Ann. Biomed. Eng.* **2014**, *4*, 968–976. [CrossRef] [PubMed]

35. Wang, K.; Parekh, U.; Pailla, T. Stretchable dry electrodes with concentric ring geometry for enhancing spatial resolution in electrophysiology. *Adv. Healthc. Mater.* **2017**, *6*, 1700552. [CrossRef] [PubMed]

36. Junwei, M.; Han, Y.; Sunderam, S.; Besio, W.; Lei, D. Computation of surface Laplacian for tri-polar ring electrodes on high-density realistic geometry head model. In Proceedings of the 39th Annual International Conference of the IEEE Engineering in Medicine and Biology Society (EMBC), Seogwipo, South Korea, 13–15 July 2017. [CrossRef]

37. Besio, W.G.; Martínez-Juárez, I.E.; Makeyev, O.; Gaitanis, J.N.; Blum, A.S.; Fisher, R.S.; Medvedev, A.V. High-frequency oscillations recorded on the scalp of patients with epilepsy using tripolar concentric ring electrodes. *IEEE J. Transl. Eng. Health Med.* **2014**, *2*, 2000111. [CrossRef] [PubMed]

38. Kaufer, M.; Rasquinha, L.; Tarjan, P. Optimization of multi-ring sensing electrode set. In Proceedings of the Annual Conference on Engineering in Medicine and Biology, Philadelphia, PA, USA, 1–4 November 1990; pp. 612–613.

39. Besio, W.; Prasad, A. Analysis of skin-electrode impedance using concentric ring electrode. In Proceedings 28th Annual International Conference of the IEEE Engineering in Medicine and Biology Society, New York, NY, USA , 30 August 2006. [CrossRef]

40. O'Mahony, C.; Grygoryev, K.; Ciarlone, A.; Giannoni, G.; Kenthao, A.; Galvin, P. Design, fabrication and skin-electrode contact analysis of polymer microneedle-based ECG electrodes. *J. Micromech. Microeng.* **2016**, *26*, 1–11. [CrossRef]

41. Carr, J.J.; Brown, J.M. *Introduction to Biomedical Equipment Technology*, 4th ed.; Prentice Hall: Upper Saddle River, NJ, USA, 2001; ISBN 0-13-010492-2.

42. Pan, J.; Tompkins, W.J. A real-time qrs detection algorithm. *IEEE Trans. Biomed. Eng.* **1985**, *32*, 230–236. [CrossRef] [PubMed]

43. Hamilton, P.S.; Tompkins, W.J. Quantitative investigation of QRS detection rules using the MIT/BIH arrhythmia database. *IEEE Trans. Biomed. Eng.* **1986**, *33*, 1157–1165. [CrossRef] [PubMed]

44. Ye-Lin, Y.; Bueno-Barrachina, J.M.; Prats-boluda, G.; de Sanabria, R.R.; Garcia-Casado, J. Wireless sensor node for non-invasive high precision electrocardiographic signal acquisition based on a multi-ring electrode. *Measurement* **2017**, *97*, 195–202. [CrossRef]

**sensors**

MDPI

*Article*

# A Flexible Multiring Concentric Electrode for Non-Invasive Identification of Intestinal Slow Waves

Victor Zena-Giménez [1], Javier Garcia-Casado [1,*], Yiyao Ye-Lin [1], Eduardo Garcia-Breijo [2] and Gema Prats-Boluda [1]

[1]   Centro de Investigación e Innovación en Bioingeniería, Universitat Politècnica de València, Valencia 46022, Spain; vfzena@ci2b.upv.es (V.Z.-G.); yiye@ci2b.upv.es (Y.Y.-L.); gprats@ci2b.upv.es (G.P.-B.)
[2]   Instituto Interuniversitario de Investigación de Reconocimiento Molecular y Desarrollo Tecnológico, Universitat Politècnica de València, Valencia 46022, Spain; egarciab@eln.upv.es
*   Correspondence: jgarciac@ci2b.upv.es; Tel.: +34-963-877-007 (ext. 76027 & 67041)

Received: 28 December 2017; Accepted: 26 January 2018; Published: 30 January 2018

**Abstract:** Developing new types of optimized electrodes for specific biomedical applications can substantially improve the quality of the sensed signals. Concentric ring electrodes have been shown to provide enhanced spatial resolution to that of conventional disc electrodes. A sensor with different electrode sizes and configurations (monopolar, bipolar, etc.) that provides simultaneous records would be very helpful for studying the best signal-sensing arrangement. A 5-pole electrode with an inner disc and four concentric rings of different sizes was developed and tested on surface intestinal myoelectrical recordings from healthy humans. For good adaptation to a curved body surface, the electrode was screen-printed onto a flexible polyester substrate. To facilitate clinical use, it is self-adhesive, incorporates a single connector and can perform dry or wet (with gel) recordings. The results show it to be a versatile electrode that can evaluate the optimal configuration for the identification of the intestinal slow wave and reject undesired interference. A bipolar concentric record with an outer ring diameter of 30 mm, a foam-free adhesive material, and electrolytic gel gave the best results.

**Keywords:** ring electrodes; surface recording; Laplacian recordings; intestinal slow wave

## 1. Introduction

### 1.1. Biosignal Recordings on the Body Surface

The non-invasive recording of electrophysiological signals is commonly used in several fields of applied medicine, such as electrocardiography, electroencephalography and electromyography of many other organs and muscles. Although these recordings are usually made with conventional disc electrodes, these yield poor spatial resolution, caused in part by the blurring effect of the different conductivities of the volume conductor [1,2].

The Laplacian potential has been demonstrated to reduce this effect and provide better spatial resolution. Surface Laplacian potentials can be estimated by applying discretization techniques [3,4] and combining the surface potentials picked up by a set of monopolar electrodes, usually with a central electrode and four additional electrodes in the form of a square [3], or alternatively three electrodes in a triangle [4]. On the other hand, concentric ring electrodes of different sizes and configurations (bipolar, tripolar, tripolar connected in bipolar configuration) have been designed and tested to obtain a direct estimation of the Laplacian potential [5–7]. However, as these electrodes are mostly developed on rigid substrates, they cannot properly adapt to the body curvature and so can cause discomfort to the patient and also require to be fixed in place by external adhesive. The tripolar recording configuration is more spatially sensitive with a smaller amplitude than the bipolar. When signal amplitude is more important than spatial specificity, then the bipolar configuration appears superior [2,8]. Regarding electrode dimensions, it is known that the larger the electrode the wider the recording area, the deeper the sources that can be sensed and the higher the signal amplitude, but also the poorer the spatial selectivity [8]. The electrode dimensions should therefore be optimized for specific applications. In this regard, one of the objectives of this work was to develop a versatile electrode that admits multiple configurations with different ring sizes to study the best sensing setup for different bioelectrical recording applications. To facilitate its clinical use, the developed electrode is implemented onto a flexible substrate, is self-adhesive (two adhesive materials were tested with and without foam), incorporates a single connector and can perform dry or wet (with gel) recordings.

### 1.2. Intestinal Myoelectrical Activity

The electroenterogram (EEnG) is a record of the myoelectric activity of the small intestine. It has two components: the slow wave (SW) and the spike bursts (SB). The SW is associated with the slow and periodic oscillations of the basal signal. It is the intestinal pacemaker and determines the maximum frequency of intestinal contractions associated with the SB. The frequency of the SW varies along the small intestine, from around 12 cycles per minute (cpm) in the duodenum to 9 and 11 cpm in the jejunum and about 8 cpm in the ileum [9–11]. Although abnormal SW patterns are related with diabetes [12] and several intestinal pathologies, such as mechanical intestinal obstruction, irritable bowel syndrome, paralytic ileus, and intestinal ischemia [13,14], surface EEnG recording is not yet used in clinical practice. One of the main reasons for this is the poor quality of the surface signals. Surface EEnG studies [3,15–19], reported low amplitude of the intestinal signal picked up on the abdominal surface (above and below navel) and the presence of different types of interference such as low frequency drifts, respiratory interference (whose spectrum can sometimes overlap with that of the SW [16,18]) and cardiac interference (whose spectrum can overlap with that of the SB). In addition, all these studies focus on the SW, since the energy of the SB is more attenuated due to the effect of low-pass filtering associated with the abdominal layers and the distance between the signal source and the pickup electrodes.

It should be highlighted that, as in other bioelectrical recording applications, the signals are sensed by conventional disc electrodes in most cases. As mentioned in the previous section, the use of concentric ring electrodes can enhance the spatial resolution and signal quality. In fact, in surface EEnG studies these electrodes have proven to be more immune to cardiac and respiratory interference [3,18,19]. However, there is no agreed criteria on the size of the electrodes: ring electrodes have been tested with an external diameter of 24 mm and 37 mm [18,19]. Both these studies took dry recordings and reported high contact impedance and the influence of signal drifts, indicating the need for electrolytic gel in future studies. The electrodes should also be self-adhesive to facilitate their use and to provide a 'track' for gel deposition. Regarding the configuration of the concentric ring electrode, given the low amplitude of intestinal signals on the abdominal surface, bipolar or even monopolar records may be recommended (the latter not being a direct estimation of the Laplacian potential). The second goal of this work was to determine the best recording setup for intestinal SW identification in terms of: configuration (monopolar/bipolar records), ring electrode size, adhesive material and electrode location.

This paper is organized as follows: Section 2 describes the materials and methods, the concentric electrode developed, the signal recording protocol, the parameters calculated and the proposed methodology for comparing and selecting the best combination of recording factors. Section 3 gives the results, recorded signals and characteristic parameters. In Section 4 the results are discussed and the conclusions are given in Section 5.

## 2. Materials and Methods

### 2.1. Sensing Part

A multi-ring concentric ring electrode (multi-CRE) was designed as shown in Figure 1. The sensing electrode consists of a set of four hook-shaped electrodes and an inner circular electrode. This design allows direct connection of the inner disk and outer rings to the connector to the signal conditioning unit in a single conductive layer, and presents little deviance in spatial sensitivity to that of closed rings [20]. In this work, the diameter of the inner disc was set to 10 mm. The external ring diameter ranged from 20 mm to 50 mm, as shown in Figure 1b. The width of the rings was set to 2 mm. This provides sufficient contact area between the electrode and the skin and means the inter-ring distance can be set to 3 mm, which reduces the risk of undesired shortcuts between poles due to possible scattering of electrolytic gel in wet recordings.

In order to facilitate the adaptation of the electrode to a curved body surface and to make it more robust to interference and movement artifacts [21], the multi-CRE was implemented onto a flexible polyester substrate (Melinex ST506, 175 µm, DuPont, Chester, VA, USA).

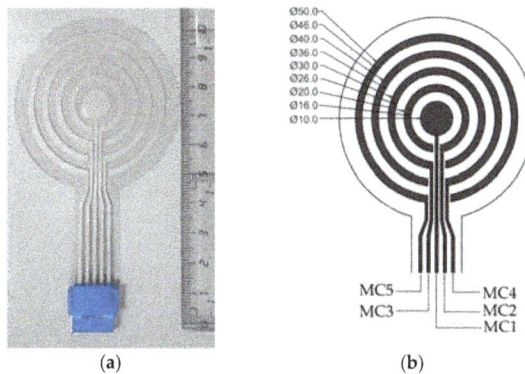

| (a) | (b) |

**Figure 1.** (a) Multi-ring concentric electrode; (b) Dimensions (in mm) of poles and annotation of signals sensed by each pole.

Thick film serigraphic technology was used to produce the sensor. The screen printing process consists of forcing pastes of different characteristics through squeegees onto a substrate through screens. The gaps in the screen define the pattern that will be printed on the substrate by serigraphy. The final thickness of the pastes can be adjusted by varying the thickness of the screens. It is necessary to manufacture frames with screen mesh for each layer of the design. To build the electrode, one screen was made with 230 mesh polyester material (PET 1500 90/230-48 from Sefar, Heiden, Switzerland). A UV film Dirasol 132 (Fujifilm, Kansas, KS, USA) was used to transfer the stencil onto the screen mesh. Final screen thickness was 10 µm. The pattern was transferred onto the screen by a UV light source. The flexible electrodes were screen-printed with a biocompatible 80/20 Silver/Silver Chloride mix of Ag/Ag-Cl paste (C2130429D3, Gwent, Pontypool, UK). The main characteristics of this paste are shown in Table 1.

**Table 1.** C2130429D3 ink's parameters.

| Property | Value |
|---|---|
| Solids Contents (%) | 81.14–83.14 |
| Vicosity (Pa.s) | 2.0–5.5 |
| Curing condition (°C) | 80 °C/10 min |
| Sheet resistivity (25 µm) | < 30 mΩ/sq |

Printing was carried out by a high precision screen stencil printer (AUREL 900, Modigliana FC, Italy) with a 75° shore squeegee hardness, 3.5 bar force, and 8 mm/s. After paste depositing, this was cured in an air oven (UNB-100, Memmert GmbH+Co. KG, Schwabach, Germany) at 80 °C for 10 min. The ink curing period was 130 °C for 10 min. Figure 2 shows details of the results of the printing process.

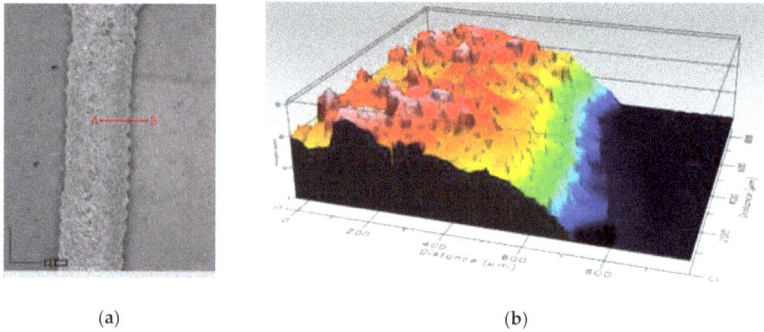

(a)                                                         (b)

**Figure 2.** (a) Magnification view of a track of electrode; (b) Right: thickness of view A–B; average thickness is 20 μm measured by Profilm 3D (Filmetrics, San Diego, CA, USA) with 20× lens.

This electrode design can yield signals with different configurations such as: monopolar concentric (MC), bipolar concentric (BC), tested in this work, and some others such as tripolar or quasi-bipolar configurations [2]. MC signals are the monopolar records obtained with the 5 poles of the electrode (see Figure 1b), with respect to a distant reference electrode. BC signals derived from multi-CRE are given by:

$$BC_n = MC_{n+1} - MC_1 \tag{1}$$

where, $MC_{n+1}$, $n = 1 \ldots 4$ are the biopotentials picked up by the four rings from the inside out; $MC_1$ is the biopotential picked up by the inner disc.

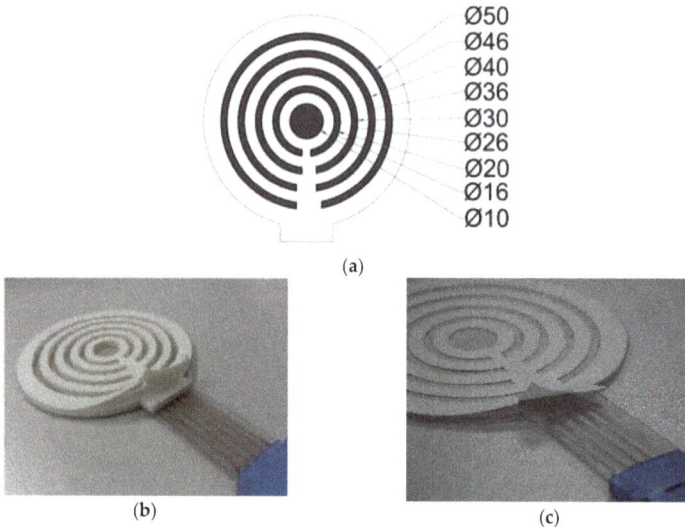

(a)

(b)                                                         (c)

**Figure 3.** (a) Template design, size in mm; (b) Multi-CRE with foam; (c) Multi-CRE with adhesive.

In order to help the multi-CRE to stick to the skin and to create a gap for the deposition of the conductive gel between the electrode and skin, two types of solutions were used: adhesive only and adhesive+foam (hereinafter 'adhesive' and 'foam'). The foam (ethylvinyl acetate) is a non-toxic polymer widely used in electrodes for the uptake of biosignals [22,23]. It can provide the appropriate stiffness while adapting to the curvature of the body and provides electrode stability to sudden movements (coughing, sneezing, involuntary movements of the subject) [23–25]. Its 1 mm thickness can deposit a thicker gel layer. The foam material does not have adherent properties; this is given by a second adhesive material, a double-sided adhesive (TM8710, MacTac, Soignies, Belgium) made of polyethylene terephthalate (PET) designed to adhere medical products to human skin. The template designed for the die-cutting of foam and adhesive materials and the implemented electrode with both adhesive materials are shown in Figure 3.

### 2.2. Recording Sessions

This study was approved by the Polytechnic University Ethics Committee and adhered to the Declaration of Helsinki. The subjects were informed of the nature of the study, briefed on the recording protocol and signed a consent form. Twenty recording sessions were performed on twenty healthy volunteers, ten for each type of foam and adhesive material. To study the effect of the recording position, EEnG records were made at both the above navel (AN) and below navel (BN) sites, with each volunteer in a supine position (30° from the horizontal). Firstly, basic biostatistical parameters such as weight, height, sex, surgical interventions relevant to the study and any history of gastrointestinal, cardiovascular or infectious disorders was noted. The skin was then prepared (exfoliation using abrasive pads and cleaning with 96% alcohol) in the areas under the electrodes at the level of the right and left middle clavicle (ECG recordings), AN and BN abdominal region (EEnG recordings) and the right ankle and left hip (reference and ground electrodes). The subjects were also shaved if hairs were present in these areas.

The conductive gel was spread on the multi-CRE without removing the adhesive backing of either material, spreading the gel carefully over all the rings. The adhesive backing was then removed and the electrode placed 2.5 cm above or below the navel (AN or BN positions, respectively), as shown in Figure 4a,b. Two monopolar Ag/Ag-Cl 8 mm recording diameter disk electrodes with 2.5 cm separations were placed 2.5 cm above or below the umbilicus to obtain one bipolar conventional EEnG recording. The duration of the recording sessions was 1 h with the multi-CRE in the BN (best expected location) and 10 min in the AN position.

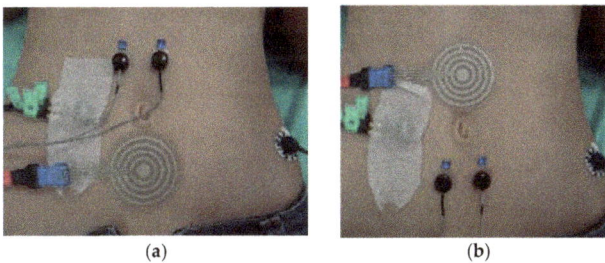

(a)                    (b)

**Figure 4.** Location of electrodes and accelerometer (covered by adhesive tape). (**a**) Multi-CRE in BN position and conventional bipolar recording in AN position, reference electrode for monopolar measurements on left hip; (**b**) multi-CRE in AN position and conventional bipolar recording in BN position.

The main sources of common types of physiological interference in surface EEnG recordings were also noted: ECG was recorded in the ML-Lead-I using disposable electrodes; respiration was measured with an airflow transducer (1401G Grass Technologies, Warwick, RI, USA) and movements were sensed by a triaxial accelerometer (ADXL 335, Analog Devices, Norwood, MA, USA). In order to help the multi-CRE to stick to the skin and to create a gap for the deposition of the conductive gel between the electrode and skin, two types of solutions were used: adhesive only and adhesive+foam (hereinafter 'adhesive' and 'foam'). The foam (ethylvinyl acetate) is a non-toxic polymer widely used in electrodes for the uptake of biosignals [22,23]. It can provide the appropriate stiffness while adapting to the curvature of the body and provides electrode stability to sudden movements (coughing, sneezing, involuntary movements of the subject) [23–25]. Its 1 mm thickness can deposit a thicker gel layer. The foam material does not have adherent properties; this is given by a second adhesive material, a double-sided adhesive (TM8710, MacTac, Soignies, Belgium) made of polyethylene terephthalate (PET) designed to adhere medical products to human skin.

The template designed for the die-cutting of foam and adhesive materials and the implemented electrode with both adhesive materials are shown in Figure 3. A disposable electrode was placed on the right ankle and used as the bioelectric ground. Another electrode was placed on the left hip and used as reference potential for EEnG monopolar records.

All signals except those from the accelerometer were amplified and band-pass filtered (0.1–100 Hz) by means of conventional bioamplifiers (P511, Grass Technologies). Signals were simultaneously recorded at a sampling rate of 1 kHz.

### 2.3. Signal Analysis

In order to study the effect of the material, position, configuration and dimensions of the electrode rings on the detection of SW, ten EEnG signals were analyzed in each session: five monopolar concentric (MC-EEnG), four bipolar concentric (BC-EEnG) and one conventional bipolar (BIP). Artifacted signal segments were identified by accelerometer signals and visual inspection and were removed from the analysis (<3% of total windows). The EEnG signals and respiration signal were digitally low-pass filtered (fc = 0.5 Hz) and resampled at 4 Hz. The power spectral density (PSD) of these signals was estimated by means of autoregressive parametric techniques (AR, order 120) calculated in moving 120 s windows every 15 s. The dominant frequency (DF) was calculated in each moving window, being defined as the frequency of the maximum energy peak above 6 cpm. The signal quality in terms of respiration interference and low frequency components was also evaluated, calculating the Welch Periodogram for each moving window so as to compute subband energies. The following parameters were calculated from the obtained PSDs [19]:

- %DF$_{TFSW}$: defined as the ratio of analyzed windows whose DF is inside the typical frequency range of intestinal SW (8–12 cpm).
- %DF$_{RESP}$: defined as the ratio between the number of windows in which the DF of the surface signal is within the DF of respiration (DF$_{RESP}$) ±1 cpm and the total number of windows
- PR$_{SW/RESP}$: defined as the ratio between the power within the SW frequency range and the power in the respiratory bandwidth, calculated as follows:

$$\text{PR}_{SW/RESP}(\text{dB}) = 10 \cdot \log \left( \frac{\text{Power(EEnG)}|_{8\ \text{cpm}}^{12\ \text{cpm}}}{\text{Power(EEnG)}|_{DF_{RESP}-1\ \text{cpm}}^{DF_{RESP}+1\ \text{cpm}}} \right) \tag{2}$$

- %DF$_{LF}$: defined as the ratio between the number of windows in which the DF is within 6–8 cpm and the total number of windows.
- PR$_{SW/LF}$: defined as the ratio between the power within the SW frequency range and the power in the low frequency bandwidth, calculated as follows:

$$\text{PR}_{SW/LF}(\text{dB}) = 10 \cdot \log \left( \frac{\text{Power(EEnG)}|_{8\ \text{cpm}}^{12\ \text{cpm}}}{\text{Power(EEnG)}|_{6\ \text{cpm}}^{8\ \text{cpm}}} \right) \tag{3}$$

- %DF$_{OTHERS}$: defined as the ratio between the number of other cases and the total number of windows. Ideally, this parameter should be 0%.
- %DF$_{SW}$: defined as the ratio of analyzed windows whose DF after discarding peaks on the low frequency and respiration bandwidth is in the SW range.
- PR$_{ECG}$: defined as the ratio between the difference between the power of the raw signal (Power$_{EEnG}$) and the power of the ECG interference estimation (Power$_{I_{ECG}}$). The power of the ECG interference estimation was worked out as described in [26,27]:

$$PR_{ECG}(dB) = 10 \cdot \log \left( \frac{Power_{EEnG} - Power_{I_{ECG}}}{Power_{I_{ECG}}} \right) \tag{4}$$

- %RS (rhythm stability): which is the percentage of the recording session windows in which DF are within the range of $mean(DF) \pm 2cpm$. This interval is considered an acceptable frequency change for the SW component of the EEnG in humans [3,28].
- MV (mean variability): defined as the average of the dominant frequency difference between consecutive windows in the range of 8 to 12 cpm, where '$R_i$' is the dominant frequency of the EEnG signal in the window '$i$' and '$N$' is the number of analyzed windows in the session:

$$MV = \frac{1}{N} \sum_{i=1}^{N} |R_{i+1} - R_i| \tag{5}$$

*2.4. Selecting the Best Combination of Recording Factors of CRE*

Eight combinations of three signal recording factors were analyzed: configuration (MC, BC), material (adhesive, foam) and position (AN, BN). It should be noted that for the MC and BC configurations, the parameters obtained for the five MC and four BC channels were averaged, respectively. The influence of the electrode size was studied in a subsequent approach.

In order to determine the best combination of factors for recording the EEnG signal on the body surface, an Improvement Ratio of a given parameter 'Y' (IR$_Y$) was proposed and calculated, defined in such a way that its value varies between '1' for the combination of factors that yields the 'best' average (N = 10 subjects) value of 'Y' parameter, and '0' for the worst of the 8 factor combinations. The 'best' result for a parameter is its maximum value if it is a beneficial parameter i.e., a higher signal/interference ratio, and is its minimum value if it is a detrimental to signal quality, i.e., a higher percentage of dominant frequencies associated with an interference. According to this, the improvement ratio of a given parameter is calculated as follows:

$$IR_{Yi} = \frac{M_{Yi} - \min\{M_{Yi}\}}{\max\{M_{Yi}\} - \min\{M_{Yi}\}} \tag{6}$$

where Y = 1...6 are beneficial parameters (see Table 2); i = 1...8, 'i' being each combination of factors; $M_{Yi}$ is the mean of parameter 'Y' for combination 'i';

$$IR_{Yi} = \frac{\max\{M_{Yi}\} - M_{Yi}}{\max\{M_{Yi}\} - \min\{M_{Yi}\}} \tag{7}$$

where Y = 7...10 are detrimental parameters (see Table 2); i = 1...8, 'i' being each combination of factors; $M_{Yi}$ is the mean of parameter 'Y' for combination 'i'.

In order to determine whether a combination of factors globally improves or worsens EEnG signal quality and the capacity to detect intestinal SW, a weight (W) was assigned to each improvement ratio parameter (IR$_{Yi}$). The value of the weight was assigned according to the research group's experience of the greater or lesser relevance of the parameter on EEnG recordings. The weights associated with each parameter are shown in Table 2. The highest weight (0.25) is assigned to %DF$_{TFSW}$, since it is the most frequently used parameter in the literature for the identification of intestinal SW activity [16]. The main types of interference, such as respiration and low frequency, were equally weighted (0.125), 0.0625 then being the weight of each of the two parameters that value each type of interference (%DF$_{RESP}$, PR$_{SW/RESP}$, %DF$_{LF}$, PR$_{SW/LF}$). The same weight (0.125) was assigned to parameters that assess the rhythmicity of the dominant frequency, 0.0625 for %RS and for MV. The parameter %DF$_{SW}$ was assigned with half of the %DF$_{TFSW}$ weight (0.125) because its value is conditioned by discarding interferences; %DF$_{OTHERS}$ is also weighted by the same value (0.125) since the combinations with the smallest values of this parameter are of interest. Finally, PR$_{ECG}$ was weighted by 0.125. Although the ECG is outside the range of the SW and the weights are focused on SW detection, using the electrode to detect intestinal bursts is not ruled out. In this case it may also be important that the electrode attenuate cardiac interference.

Finally, for each combination of factors 'i', a global improvement index IRglobal$_i$ is computed as a weighted sum of the IR of the 10 parameters studied:

$$IRglobal_i = \sum_{Y=1}^{10} (W_Y * IR_{Yi}) \qquad (8)$$

IR$_{global}$ will assess how close combination 'i' would be to an ideal case using the optimal combination of each of the parameters tested. '1' would be optimal and '0' the worst, i.e., always use the worst possible combination for each parameter.

**Table 2.** Weights for the improvement ratios of parameters involved in the selection of the most appropriate combination of factors.

| Y | Improvement Ratio (IR$_Y$) | Weight (W$_Y$) |
|---|---|---|
| 1 | %DF$_{TFSW}$ | 0.25 |
| 2 | %DF$_{SW}$ | 0.125 |
| 3 | PR$_{ECG}$ | 0.125 |
| 4 | PR$_{SW/RESP}$ | 0.0625 |
| 5 | PR$_{SW/LF}$ | 0.0625 |
| 6 | %RS | 0.0625 |
| 7 | %DF$_{RESP}$ | 0.0625 |
| 8 | %DF$_{LF}$ | 0.0625 |
| 9 | %DF$_{OTHERS}$ | 0.125 |
| 10 | MV | 0.0625 |

*2.5. Selecting the Best CRE Ring Size and Comparison with Conventional Bipolar Recording*

After determining the most suitable combination of factors for picking up the surface EEnG signal by CRE had been determined, the effect of ring size was studied. The parameters associated with the bipolar recording obtained with conventional disc electrodes (BIP) were also part of the group studied to look for the most appropriate way to record EEnG signals. Signal quality and capacity to identify SW frequency was evaluated and compared by means of improvement ratios, as described above.

## 3. Results

*3.1. Signal Acquisition and Parameters*

Figure 5 shows an example of the signals recorded and their PSDs. It can be seen that the amplitude of the MC-EEnG signals does not change significantly for different ring sizes. However, the larger the size of the ring, the larger the amplitude of the BC-EEnG signal. It can also be seen that ECG interference was higher in MC-EEnG and BIP than BC-EEnG signals, as corroborated by the results of the PR$_{ECG}$ parameter shown in Tables 3 and 4.

Visual identification of the slow wave in the time domain is not clear, mainly due to cardiac and respiratory interference in MC-EEnG and BIP records, and to the low amplitude in BC-EEnG records. Figure 5b shows the PSD of the corresponding temporal signals in analysis windows of 120 s after low-pass filtering and resampling. It can be seen that in BC-EEnG records the highest energy peaks and the associated DFs are clearly in the SW frequency range (8–12 cpm), while the respiratory interference is more manifest in MC-EEnG and BIP and maximum energy peaks could be obtained in the respiratory frequency. The better immunity of bipolar concentric configuration to respiratory interference is confirmed by the results from parameters $\%DF_{RESP}$ and $PR_{SW/RESP}$, as shown in Tables 3 and 4. Higher signal-to-respiratory interference ratios were obtained for BC-EEnG than MC-EEnG and BIP; the DF frequencies were also associated with respiratory interference in a smaller number of cases. However, the opposite behavior is seen in the influence of low frequency interference. The signal-to-low frequency interference ratio was lower and $\%DF_{LF}$ higher, for BC-EEnG than for MC-EEnG or BIP, as shown in Tables 3 and 4. Overall, the best $\%DF_{SW}$ results were obtained with the BC configuration, at mean values of 93% in the BN position, which drops to 85–87% in the AN position. In the RS and MV rhythmicity parameters, no clear superiority of any configuration or material was observed, with the exception of MC in BN and adhesive material, which proved to be the best combination in this context.

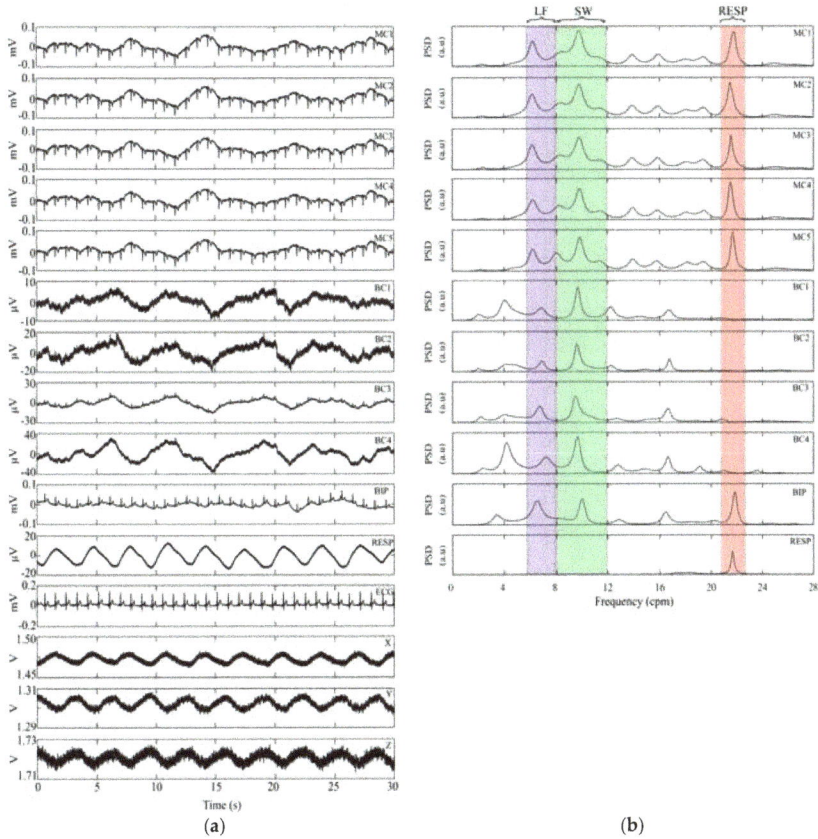

**Figure 5.** (**a**) Thirty seconds of signals: monopolar concentric (MC1-5), bipolar concentric (BC1-4) with adhesive material below navel position; bipolar (BIP), respiration (RESP), ECG ML-Lead I and accelerometer signals (X, Y, Z); (**b**) Power spectral density of signals shown on the left.

**Table 3.** Mean and standard deviation of the parameters defined for SW identification and interference quantification of surface abdominal signals picked up with different recording factors. Signals recorded in BN position in 10 volunteers.

| | MC-EEnG (*n* = 50) | | BC-EEnG (*n* = 40) | | BIP (*n* = 10) | |
|---|---|---|---|---|---|---|
| | Foam | Adh | Foam | Adh | Foam | Adh |
| %DF$_{TFSW}$ (%) | 56.0 ± 12.6 | 56.1 ± 9.5 | 54.1 ± 8.3 | 57.3 ± 12.7 | 53.4 ± 17.2 | 58.7 ± 11.5 |
| %DF$_{RESP}$ (%) | 19.7 ± 14.1 | 16.1 ± 9.2 | 14.6 ± 9.8 | 14.4 ± 11.2 | 23.9 ± 22.1 | 21.7 ± 9.5 |
| PR$_{LF}$ (dB) | 5.01 ± 1.82 | 4.74 ± 1.74 | 5.59 ± 1.76 | 5.19 ± 2.20 | 3.81 ± 2.03 | 4.07 ± 2.21 |
| %DF$_{LF}$ (%) | 16.6 ± 7.1 | 16.7 ± 7.6 | 24.6 ± 10.0 | 21.6 ± 9.3 | 12.5 ± 11.3 | 9.6 ± 3.8 |
| PR$_{SW/LF}$ (dB) | 3.78 ± 1.44 | 3.86 ± 0.92 | 2.86 ± 1.45 | 3.55 ± 1.25 | 4.25 ± 1.58 | 4.97 ± 1.47 |
| %DF$_{OTHERS}$ (%) | 7.9 ± 4.3 | 11.1 ± 5.5 | 6.7 ± 5.3 | 6.7 ± 3.9 | 10.2 ± 12.9 | 9.9 ± 8.2 |
| PR$_{ECG}$ (dB) | 6.39 ± 3.89 | 6.13 ± 1.82 | 11.17 ± 4.32 | 19.77 ± 4.90 | 7.93 ± 4.91 | 6.57 ± 3.55 |
| %DF$_{SW}$ (%) | 92.1 ± 4.2 | 88.8 ± 5.5 | 93.2 ± 5.3 | 93.2 ± 3.9 | 87.0 ± 12.1 | 89.6 ± 8.5 |
| DF$_{SW}$ (cpm) | 9.97 ± 0.26 | 9.84 ± 0.35 | 9.99 ± 0.21 | 9.72 ± 0.49 | 9.49 ± 0.48 | 9.64 ± 0.20 |
| RS (%) | 58.9 ± 15.5 | 54.5 ± 21.7 | 58.8 ± 14.3 | 56.4 ± 22.8 | 60.9 ± 26.3 | 72.6 ± 28.8 |
| MV (cpm) | 0.38 ± 0.07 | 0.39 ± 0.10 | 0.41 ± 0.09 | 0.40 ± 0.11 | 0.38 ± 0.23 | 0.30 ± 0.26 |

**Table 4.** Mean and standard deviation of the parameters defined for SW identification and interference quantification of surface abdominal signals picked up with different recording factors. Signals recorded in AN position in 10 volunteers.

| | MC-EEnG (*n* = 50) | | BC-EEnG (*n* = 40) | | BIP (*n* = 10) | |
|---|---|---|---|---|---|---|
| | Foam | Adh | Foam | Adh | Foam | Adh |
| %DF$_{TFSW}$ (%) | 45.9 ± 21.0 | 57.8 ± 14.9 | 32.9 ± 11.4 | 57.0 ± 12.5 | 49.3 ± 10.7 | 50.1 ± 10.2 |
| %DF$_{RESP}$ (%) | 25.9 ± 23.0 | 19.6 ± 18.1 | 24.8 ± 18.4 | 12.3 ± 11.1 | 29.8 ± 14.3 | 25.6 ± 14.0 |
| PR$_{SW/RESP}$ (dB) | 4.23 ± 2.92 | 4.14 ± 2.13 | 2.58 ± 1.77 | 4.98 ± 2.18 | 3.83 ± 1.83 | 3.51 ± 1.94 |
| %DF$_{LF}$ (%) | 18.8 ± 12.4 | 8.8 ± 5.9 | 29.8 ± 16.3 | 15.9 ± 10.1 | 13.5 ± 8.1 | 14.1 ± 8.6 |
| PR$_{SW/LF}$ (dB) | 3.69 ± 1.90 | 5.15 ± 0.81 | 2.11 ± 1.55 | 4.16 ± 1.84 | 5.16 ± 1.26 | 4.63 ± 1.37 |
| %DF$_{OTHERS}$ (%) | 9.40 ± 8.6 | 13.8 ± 7.8 | 12.4 ± 8.4 | 14.8 ± 10.2 | 7.4 ± 4.4 | 10.2 ± 8.8 |
| PR$_{ECG}$ (dB) | 7.82 ± 3.96 | 4.44 ± 2.94 | 11.83 ± 6.65 | 14.43 ± 6.17 | 6.13 ± 4.49 | 4.52 ± 2.03 |
| %DF$_{SW}$ (%) | 90.5 ± 8.6 | 86.1 ± 7.8 | 87.5 ± 8.4 | 85.1 ± 10.2 | 89.4 ± 4.3 | 88.5 ± 8.8 |
| DF$_{SW}$ (cpm) | 9.98 ± 0.32 | 10.29 ± 0.70 | 9.69 ± 0.51 | 9.69 ± 0.55 | 10.04 ± 0.26 | 9.59 ± 0.28 |
| RS (%) | 52.8 ± 29.2 | 80.1 ± 20.1 | 40.3 ± 30.0 | 54.5 ± 30.4 | 56.1 ± 10.7 | 47.5 ± 23.6 |
| MV (cpm) | 0.35 ± 0.16 | 0.34 ± 0.21 | 0.46 ± 0.11 | 0.44 ± 0.15 | 0.49 ± 0.11 | 0.37 ± 0.08 |

*3.2. Selection of the Best Combination of CRE Recording Factors*

Table 5 shows the improvement ratio of each combination of factors for the different parameters. The BN provides clearly better results than AN for the ratio of cases in which 'raw' signal DF could be attributed to the SW frequency (%DF$_{TFSW}$) when foam was used and very similar ones for adhesive. BN is also clearly better after discarding the energy peaks associated with possible interference (%DF$_{TFSW}$). The highest immunity to respiratory interference is obtained by BN plus adhesive. The best behavior in terms of robustness to low frequency and cardiac interference is obtained by the MC and BC configurations, respectively.

The color scheme in Table 5 shows the best and worst combinations when considering all parameters in a global approach. This is clearest in the IR$_{Global}$ values given in Figure 6. It can be seen that in general, the lowest IRs and IR$_{Global}$ are from the records from AN with foam for both MC and BC (BCFoAN and MCFoAN). The best IRs seem to concentrate in the combination of BC with adhesive and BN ition(BCAdBN), which obtained the highest IR$_{Global}$ values (0.7), followed by MCAdBN and BCFoBN, with values around 0.6.

**Table 5.** IR values for the eight combinations of recordings factors and for the 10 parameters defined.

| | Below Navel | | | | Above Navel | | | |
|---|---|---|---|---|---|---|---|---|
| | Foam | | Adhesive | | Foam | | Adhesive | |
| | BC | MC | BC | MC | BC | MC | BC | MC |
| | BCFoBN | MCFoBN | BCAdBN | MCAdBN | BCFoAN | MCFoAN | BCAdAN | MCAdAN |
| $\%DF_{TFSW}$ | 0.84 | 0.92 | 0.97 | 0.93 | 0.00 | 0.51 | 0.96 | 1.00 |
| $\%DF_{RESP}$ | 0.83 | 0.46 | 0.84 | 0.72 | 0.08 | 0.00 | 1.00 | 0.86 |
| $PR_{SW/RESP}$ | 1.00 | 0.81 | 0.86 | 0.72 | 0.00 | 0.54 | 0.79 | 0.52 |
| $\%DF_{LF}$ | 0.24 | 0.62 | 0.39 | 0.62 | 0.00 | 0.52 | 0.66 | 1.00 |
| $PR_{SW/LF}$ | 0.24 | 0.54 | 0.47 | 0.57 | 0.00 | 0.52 | 0.67 | 1.00 |
| $\%DF_{OTHERS}$ | 1.00 | 0.85 | 0.99 | 0.45 | 0.29 | 0.66 | 0.00 | 0.11 |
| $PR_{ECG}$ | 0.44 | 0.12 | 1.00 | 0.11 | 0.48 | 0.22 | 0.65 | 0.00 |
| $\%DF_{SW}$ | 1.00 | 0.87 | 0.99 | 0.45 | 0.29 | 0.66 | 0.00 | 0.11 |
| RS | 0.55 | 0.59 | 0.41 | 0.36 | 0.00 | 0.31 | 0.35 | 1.00 |
| MV (cpm) | 0.24 | 0.36 | 0.27 | 0.34 | 0.00 | 0.51 | 0.11 | 1.00 |
| $IR_{Global}$ | 0.58 | 0.56 | 0.69 | 0.51 | 0.10 | 0.39 | 0.54 | 0.60 |

Best (1) Poor (0)

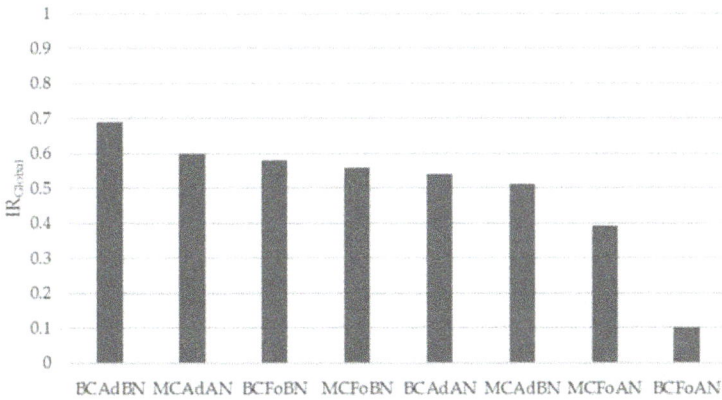

**Figure 6.** $IR_{Global}$ for each combination of factors: configuration (MC: monopolar concentric; BC: bipolar concentric), material (Ad: adhesive; Fo: foam) and position (AN: above navel; BN: below navel).

### 3.3. Selection of the Best Ring Size of CRE and Comparison with Conventional Bipolar Recording

After identifying the best combination of CRE recording factors to capture the SW, the next step was to study the influence of ring dimensions on the values of the characteristic EEnG parameters. The influence of the electrode dimensions was analyzed for the BC-EEnG configuration in BN position plus adhesive. The characteristic parameters of the four bipolar concentric signals (BCn-EEnG, $n = 1.4$) of $\varnothing20$, $\varnothing30$, $\varnothing40$, $\varnothing50$ mm outer ring diameters, respectively, were studied and the best choice was determined in the same way as in the previous step. The results obtained were compared with those of conventional bipolar registers by means of cup disc electrodes at the same BN position.

Table 6 shows the IR values of each BC-EEnG signal and BIP. It can be seen that the IRs of the BIP were lowest for most parameters, except for those associated with immunity to low-frequency interference. It is also noticeable that for BC-EEnG signals, the larger the electrode the worse the immunity to respiratory interference, and the better to low frequency interference. It seems that the best compromise is found for BC2-EEnG ($\varnothing30$ mm), whose IR values tend to be highest. These results agree with those given in Figure 7, which represents the $IR_{Global}$ values. According to the selection criteria, the optimal diameter of the concentric electrode ring for SW identification in surface EEnG signals is 30 mm; which belonged to BC2-EEnG with an $IR_{Global}$ of 0.80.

**Table 6.** IR values for the concentric bipolar records (BC1-EEnG, BC2-EEnG, BC3-EEnG, BC4-EEnG) in BN position and using adhesive material and conventional bipolar (BIP) in BN position, for the 10 characteristic signal parameters.

| | BC1-EEnG | BC2-EEnG | BC3-EEnG | BC4-EEnG | BIP |
|---|---|---|---|---|---|
| %$DF_{TFSW}$ | 0.84 | 1.00 | 0.00 | 0.21 | 0.68 |
| %$DF_{RESP}$ | 1.00 | 0.83 | 0.35 | 0.00 | 0.05 |
| $PR_{RESP}$ | 1.00 | 0.91 | 0.19 | 0.05 | 0.00 |
| %$DF_{LF}$ | 0.00 | 0.09 | 0.19 | 0.59 | 1.00 |
| $PR_{LF}$ | 0.00 | 0.31 | 0.42 | 0.53 | 1.00 |
| %$DF_{OTHERS}$ | 0.51 | 1.00 | 0.45 | 0.59 | 0.00 |
| $PR_{ECG}$ | 0.67 | 0.91 | 0.47 | 1.00 | 0.00 |
| %$DF_{SW}$ | 0.53 | 1.00 | 0.49 | 0.62 | 0.00 |
| %RS | 0.14 | 0.38 | 1.00 | 0.92 | 0.00 |
| MV | 0.40 | 0.52 | 0.51 | 0.00 | 1.00 |
| $IR_{Global}$ | 0.58 | 0.80 | 0.34 | 0.46 | 0.36 |

Best (1) Poor (0)

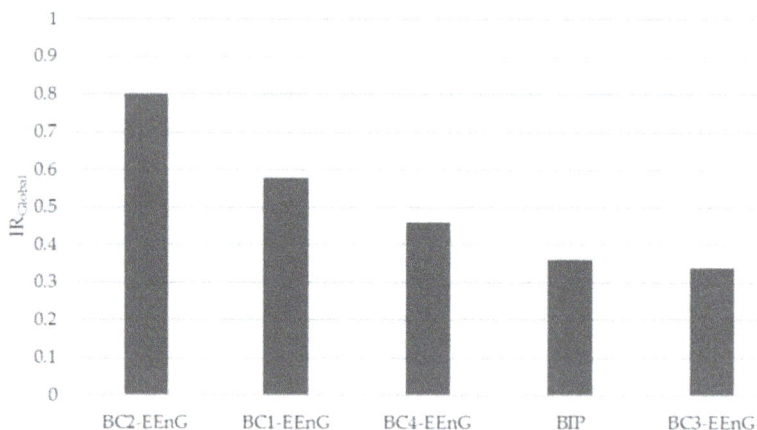

**Figure 7.** $IR_{Global}$ for each concentric bipolar configuration (BC1-EEnG, BC2-EEnG, BC3-EEnG and BC4-EEnG) and conventional bipolar (BIP).

## 4. Discussion

The multi-CRE developed in this work was implemented onto a flexible substrate and had an adhesive layer to fix it to the abdomen. It was comfortable for the patient, adapted to fit the body surface and could be easily handled by the technician. The fact that it had only a single connector reduced the possibility of incorrect connections and simplified the protocol.

The electrode was produced by screen printing technology. At present, several different graphic art printing techniques are available to make electronic devices, sensors or electrodes. The most widely used methods are screen printing, inkjet printing, stamping/nanoimprinting, and gravure printing. Previous studies compared some of these techniques for the implementation of flexible CRE [29]. The results obtained for ink thickness, resistivity, electrode resistance, and other performance parameters derived from electrocardiographic signal recording tests suggest that screen-printing and inkjets are suitable for non-invasive bioelectric signal acquisition. In the present work, the inkjet printing option was discarded since it is less robust in terms of conductive ink fixation, i.e., the ink tends to peel off when removing the adhesive.

The use of the developed multi-CRE for non-invasive EEnG recording permitted the study of the influence of the factors affecting recording conditions on different signal quality parameters of this specific application. In a preliminary work, we computed a reduced set of parameters (%$DF_{TFSW}$, $PR_{RESP}$, $PR_{LF}$ and %$DF_{SW}$) for different ring sizes and configurations with multi-CRE in below navel position and adhesive material [30]. In the present work, factors such as the recording position and the type of adhesive material have also been considered, and its effects has been studied for a more comprehensive set of 10 parameters. It should be noted that we did not specifically study the influence of recording factors on the signal amplitude, since it is strongly influenced by interference such as respiration and the ECG. As these types of interference can be greater than the intestinal signal itself [16,18,31], the amplitudes captured on the surface cannot be attributed solely to intestinal myoelectric activity, which would need to be eliminated in order to determine the EEnG amplitude. It was observed that amplitude usually increases with ring size for monopolar and bipolar concentric records, especially with adhesive and regardless of the position. This agrees with previous studies on other bioelectric signals that report larger amplitudes for larger rings, although without significant differences [32–35] and in finite element models [8].

The best results were obtained from the BN position as AN was more affected by respiratory and cardiac interference. In this position the electrode is closer to the thoracic cage and the diaphragm, and thus more liable to respiratory movement. It is also closer to the heart, and therefore cardiac interference is also stronger. In the BN there were fewer cases where the DF did not correspond to the range defined for SW or its main types of interference. All this leads to better SW identification in the BN, especially after discarding undesired energy peaks, in agreement with previous studies [18], given that there is a greater concentration of intestinal loops in this area and it is further away from the stomach and diaphragm, which can generate high gastric and respiratory interference.

Of the materials used to fix the electrode in place, i.e., adhesive only or adhesive with foam, the former showed better behavior in terms of robustness to respiratory and low frequency interference. As seen in other studies, respiratory interference in surface records is due to a great extent to the potential difference between the electrodes due to physical movements [16,18,31], so that the greater rigidity introduced by the foam could be responsible for this higher respiratory interference. It could also lead to poor skin contact, increasing electrode-skin impedance [36,37] and in turn increasing low frequency interference. Also, foam leaves a larger volume of gel on the multi-CRE, which may not be evenly distributed and thus could lead to poorer results. However, this could be solved by using solid electrolyte gel or hydrogel, which would achieve two important improvements: better skin contact and the fixation would be self-adhesive, since many solid conductor gels are also adhesives, although it would also increase costs.

With respect to the configuration of the signal recording, BC proved to be better than MC in terms of attenuation of respiratory and cardiac interference. Both types of interference are coupled to the signal as common mode interference, i.e., with similar morphology and amplitude in the different poles of the multi-CRE. Therefore, they are significantly reduced in BCs, which are obtained by subtracting the potential sensed by the central disk with that of the outer rings. In fact, most studies with concentric ring electrodes perform bipolar records [18,19,38], while it has been found that low frequency interference affects the EEnG signals in the BC configuration more than MC regardless of the electrode position and material. This seems to indicate that the low frequency interference, like fluctuations of the contact potential or baseline are not similar (common mode component) between both poles of the record or do not cancel, in fact, they increase slightly, when computing the potential difference between both. The different behavior of the BC signals could also be due to the fact that the amplitude of the signals are smaller, increasing the effect of this interference.

It should be mentioned that some authors have also proposed other recording configurations with multipole concentric ring electrodes for a more accurate estimation of the Laplacian potential, such as a tripolar configuration (experimentally tested) [39] and even a generalized approach for n rings (at a theoretical level) [40]. These other configurations lie beyond the scope of the present work, but they could be worked out with the electrode developed in the present study.

All in all, the proposed methodology based on the definition of improvement ratios found that when considering all parameters in a global approach the best combination of factors was a bipolar concentric configuration with adhesive material in the BN position. With this combination of factors, the influence of the ring size was studied and the results were compared with conventional bipolar recordings with disk electrodes. Theoretically, smaller rings capture dipole sources closer to the body surface and the larger the ring the deeper the activity that can be sensed. However, the larger the diameter, the greater the averaging effect of different dipole sources and the lower the spatial resolution.

It was observed that larger ring sizes have less low frequency interference. Since all the rings were 2 mm wide, this behavior could be due to the larger skin-to-electrode contact area and the associated reduced contact impedance. On the other hand, the influence of respiratory interference on the EEnG record is greater with increasing ring size, probably because the possible movements of the central disk and the outer ring caused by respiration are less temporarily synchronized by the larger outer ring [30]. The best results in terms of intestinal SW identification and global signal quality were obtained with the 30 mm outer diameter ring, which manages to balance the previously indicated factors. Some authors have proposed the used of ring electrodes to pick up other bioelectric surface signals with similar dimensions to those analyzed in the present work. For instance Ye-Lin et al. pointed out the feasibility of capturing uterine myoelectric activity in at-term pregnant women by 24 mm diameter concentric bipolar electrodes [32]. Also in the field of electrocardiography, Prats et al. and Lu et al. observed that with concentric rings of 33.5 mm and 36 mm a compromise was reached between the better usability and spatial resolution of small rings and the larger signal amplitudes achieved by larger electrodes [5,41].

Lastly, comparing the signals picked up by the multi-CRE with those from conventional bipolar recordings by disc electrodes, bipolar concentric records presented greater attenuation of cardiac interference. These results agree with other studies in which bipolar configuration concentric ring electrodes were used to sense surface myoelectric activity from different organs and muscles, such as the small bowel (in quasi-bipolar configuration) [18,19], the uterus [38], the diaphragm [42] and the sternocleidomastoid [43]. This behavior could be mainly attributed to the enhanced spatial resolution of bipolar concentric electrodes, which are more sensitive to the activity of dipole sources close to the electrode and less sensitive to distant dipoles such as those of cardiac origin. On the other hand, unlike conventional bipolar records, BC electrodes attenuate respiratory interference. This is probably due to the more homogeneous movement of the central disc and the outer ring combined on a single electrode, than that of the two independent disc electrodes used for conventional bipolar records. However, the latter outperform the BC in terms of the influence of low frequency interference. All in all, in identifying the SW frequency and obtaining surface EEnG signals, the multi-CRE developed for this study in general achieved better results.

## 5. Conclusions

In this work a multipole concentric ring electrode was developed to perform simultaneous recordings with different ring sizes and configurations (monopolar, bipolar, tripolar, etc.). To facilitate its clinical use and improve its body contact and comfort, it was implemented onto a flexible substrate and incorporated an adhesive layer and a single output connector.

This electrode can provide the best signal-sensing arrangement for each surface bioelectrical recording application. The factors evaluated were: the effect of the electrode position, adhesive material, configuration and ring size on signal quality and intestinal SW detection capacity parameters derived from surface EEnG recordings.

The proposed methodology for the evaluation of the different signal sensing arrangements based on the definition of improvement ratios found that for this specific application the best combination of factors was: bipolar concentric recordings below the navel with an adhesive material without foam,

*Sensors* **2018**, *18*, 396

and a ring size of 30mm outer diameter. The signals derived from the multi-CRE yielded signals of better quality and were better able to identify intestinal SW than conventional bipolar records with disc electrodes.

The developed multi-CRE and proposed methodology for the comparison of alternatives could thus be used to find the most appropriate signal sensing arrangement for other biomedical applications providing non-invasive bioelectrical recordings from other organs and muscles, such as the heart, brain, uterus, respiratory muscles, etc.

**Acknowledgments:** Grant from the Ministerio de Economía y Competitividad y del Fondo Europeo de Desarrollo Regional. DPI2015-68397-R (MINECO/FEDER).

**Author Contributions:** J.G.-C., Y.Y.-L. and G.P.-B. conceived and designed the multi-CRE, the experiments and data analysis, E.G.-B. built the multi-CRE and performed preliminary tests, V.Z.-G. performed the experiments, analyzed the data and wrote the manuscript, which was enriched by all the rest of the authors.

**Conflicts of Interest:** The authors declare no conflict of interest.

## References

1. Bradshaw, L.A.; Richards, W.O.; Wikswo, J.P. Volume conductor effects on the spatial resolution of magnetic fields and electric potentials from gastrointestinal electrical activity. *Med. Biol. Eng. Comput.* **2001**, *39*, 35–43. [CrossRef] [PubMed]

2. Besio, W.; Aakula, R.; Dai, W. Comparison of bipolar vs. tripolar concentric ring electrode Laplacian estimates. *Annu. Int. Conf. IEEE Eng. Med. Biol. Soc.* **2004**, *3*, 2255–2258.

3. Prats-Boluda, G.; Garcia-Casado, J.; Martinez-de-Juan, J.L.; Ponce, J.L. Identification of the slow wave component of the electroenterogram from Laplacian abdominal surface recordings in humans. *Physiol. Meas.* **2007**, *28*, 1115–1133. [CrossRef] [PubMed]

4. Fukuoka, Y.; Miyazawa, K.; Mori, H.; Miyagi, M.; Nishida, M.; Horiuchi, Y.; Ichikawa, A.; Hoshino, H.; Noshiro, M.; Ueno, A. Development of a compact wireless laplacian electrode module for electromyograms and its human interface applications. *Sensors* **2013**, *13*, 2368–2383. [CrossRef] [PubMed]

5. Lu, C.C.; Feng, W.Y.; Tarjan, P.P. Laplacian electrocardiograms with active electrodes for arrhythmia detection. In Proceedings of the 19th Annual International Conference of the IEEE Engineering in Medicine and Biology Society, Chicago, IL, USA, 30 October–2 November 1997; Volume 1.

6. Besio, W.; Prasad, A. Analysis of skin-electrode impedance using concentric ring electrode. *Conf. Proc. IEEE Eng. Med. Biol. Soc.* **2006**, *1*, 6414–6417. [PubMed]

7. Boudria, Y.; Feltane, A.; Besio, W. Significant improvement in one-dimensional cursor control using Laplacian electroencephalography over electroencephalography. *J. Neural Eng.* **2014**, *11*, 35014. [CrossRef] [PubMed]

8. Kaufer, M.; Rasquinha, L.; Tarjan, P. Optimization of Multi-ring Sensing Electrode Set. In Proceedings of the Twelfth Annual International Conference of the IEEE Engineering in Medicine and Biology Society, Philadelphia, PA, USA, 1–4 November 1990; pp. 612–613.

9. Fleckenstein, P.; Oigaard, A. Electrical spike activity in the human small intestine. A multiple electrode study of fasting diurnal variations. *Am. J. Dig. Dis.* **1978**, *23*, 776–780. [CrossRef] [PubMed]

10. Quigley, E.M. Disturbances in small bowel motility. *Baillieres. Best Pract. Res. Clin. Gastroenterol.* **1999**, *13*, 385–395. [CrossRef] [PubMed]

11. Vantrappen, G. Small intestinal motility and bacteria. In *Gastrointestinal Motility*; Heidt, P.J., Rusch, V., Eds.; Old Herborn University: Herborn, Germany, 1997; Volume 67, pp. 53–67.

12. Ouyang, X.; Li, S.; Foreman, R.; Farber, J.; Lin, L.; Yin, J.; Chen, J.D.Z. Hyperglycemia-induced small intestinal dysrhythmias attributed to sympathovagal imbalance in normal and diabetic rats. *Neurogastroenterol. Motil.* **2015**, *27*, 406–415. [CrossRef] [PubMed]

13. Quigley, E.M. Gastric and small intestinal motility in health and disease. *Gastroenterol. Clin. N. Am.* **1996**, *25*, 113–145. [CrossRef]

14. Somarajan, S.; Muszynski, N.D.; Cheng, L.K.; Bradshaw, L.A.; Naslund, T.C.; Richards, W.O. Noninvasive biomagnetic detection of intestinal slow wave dysrhythmias in chronic mesenteric ischemia. *Am. J. Physiol. Gastrointest. Liver Physiol.* **2015**, *309*, G52–G58. [CrossRef] [PubMed]

15. Bradshaw, L.; Allos, S.; Wikswo, J.P., Jr.; Richards, W. Correlation and comparison of magnetic and electric detection of small intestinal electrical activity. *Am. J.* **1997**, *272*, G1159–G1167. [CrossRef] [PubMed]

16. Chen, J.D.; Schirmer, B.D.; McCallum, R.W. Measurement of electrical activity of the human small intestine using surface electrodes. *IEEE Trans. Biomed. Eng.* **1993**, *40*, 598–602. [CrossRef] [PubMed]

17. Chen, J.D.; Lin, Z. Adaptive cancellation of the respiratory artifact in surface recording of small intestinal electrical activity. *Comput. Biol. Med.* **1993**, *23*, 497–509. [CrossRef]

18. Prats-Boluda, G.; Garcia-Casado, J.; Martinez-de-Juan, J.L.; Ye-Lin, Y. Active concentric ring electrode for non-invasive detection of intestinal myoelectric signals. *Med. Eng. Phys.* **2011**, *33*, 446–455. [CrossRef] [PubMed]

19. Garcia-Casado, J.; Zena-Gimenez, V.; Prats-Boluda, G.; Ye-Lin, Y. Enhancement of non-invasive recording of electroenterogram by means of a flexible array of concentric ring electrodes. *Ann. Biomed. Eng.* **2014**, *42*, 651–660. [CrossRef] [PubMed]

20. Garcia-Casado, J.; Zena, V.; Perez, J.J.; Prats-Boluda, G.; Ye-Lin, Y.; Garcia-Breijo, E. Opened-Ring Electrode Array for Enhanced Non-invasive Monitoring of Bioelectrical Signals: Application to Surface EEnG Recording. In *Biomedical Engineering Systems and Technologies*; Fernández-Chimeno, M., Fernandes, P.L., Alvarez, S., Stacey, D., Solé-Casals, J., Fred, A., Gamboa, H., Eds.; Springer: Berlin/Heidelberg, Germany, 2014; Volume 452, pp. 1–15.

21. Prats-Boluda, G.; Ye-Lin, Y.; Garcia-Breijo, E.; Ibañez, J.; Garcia-Casado, J. Active flexible concentric ring electrode for non-invasive surface bioelectrical recordings. *Meas. Sci. Technol.* **2012**, *23*, 125703. [CrossRef]

22. Webster, J.G. *Medical Instrumentation: Application and Design*, 4th ed.; John Wiley & Sons, Inc.: New York, NY, USA, 2010; pp. 189–215.

23. Lin, C.-T.; Liao, L.-D.; Liu, Y.-H.; Wang, I.-J.; Lin, B.-S.; Chang, J.-Y. Novel dry polymer foam electrodes for long-term EEG measurement. *IEEE Trans. Biomed. Eng.* **2011**, *58*, 1200–1207. [PubMed]

24. Gruetzmann, A.; Hansen, S.; Müller, J. Novel dry electrodes for ECG monitoring. *Physiol. Meas.* **2007**, *28*, 1375. [CrossRef] [PubMed]

25. Meziane, N.; Webster, J.G.; Attari, M.; Nimunkar, A.J. Dry electrodes for electrocardiography. *Physiol. Meas.* **2013**, *34*, R47–R69. [CrossRef] [PubMed]

26. Garcia-Casado, J.; Martinez-de-Juan, J.L.; Ponce, J.L. Adaptive filtering of ECG interference on surface EEnGs based on signal averaging. *Physiol. Meas.* **2006**, *27*, 509–527. [CrossRef] [PubMed]

27. Ye-Lin, Y.; Garcia-Casado, J.; Martinez-de-Juan, J.L.; Prats-Boluda, G.; Ponce, J.L. The detection of intestinal spike activity on surface electroenterograms. *Phys. Med. Biol.* **2010**, *55*, 663–680. [CrossRef] [PubMed]

28. Hou, X.; Yin, J.; Liu, J.; Pasricha, P.J.; Chen, J.D.Z. In vivo gastric and intestinal slow waves in W/Wv mice. *Dig. Dis. Sci.* **2005**, *50*, 1335–1341. [CrossRef] [PubMed]

29. Garcia-Breijo, E.; Prats-Boluda, G.; Lidon-Roger, J.V.; Ye-Lin, Y.; Garcia-Casado, J. A comparative analysis of printing techniques by using an active concentric ring electrode for bioelectrical recording. *Microelectron. Int.* **2015**, *32*, 103–107. [CrossRef]

30. Zena-Gimenez, V.; Garcia-Casado, J.; Prats-Boluda, G.; Ye-Lin, Y. Effects of Configuration and Dimension of Conecnetric Ring Electrodes in EEnG Recordin Applications. In Proceedings of the 10th International join Conference on Biomedical Engineering Systems and Technologies (BIOSTEC 2017), Porto, Portugal, 21–23 February 2017; Volume 1, pp. 32–37.

31. Chang, F.Y.F. Electrogastrography: Basic knowledge, recording, processing and its clinical applications. *J. Gastroenterol. Hepatol.* **2005**, *20*, 502–516. [CrossRef] [PubMed]

32. Ye-Lin, Y.; Alberola-Rubio, J.; Prats-Boluda, G.; Barrachina, J.M.B.; Perales, A.; Valero, J.; Desantes, D.; Garcia-Casado, J. Non-invasive electrohysterogram recording using flexible concentric ring electrode. In Proceedings of the 2014 36th Annual International Conference of the IEEE Engineering in Medicine and Biology Society, Chicago, IL, USA, 26–30 August 2014; pp. 4050–4053.

33. Ye-Lin, Y.; Alberola-Rubio, J.; Prats-Boluda, G.; Perales, A.; Desantes, D.; Garcia-Casado, J. Feasibility and Analysis of Bipolar Concentric Recording of Electrohysterogram with Flexible Active Electrode. *Ann. Biomed. Eng.* **2014**, *43*, 968–976. [CrossRef] [PubMed]

34. Ye-Lin, Y.; Bueno-Barrachina, J.M.; Prats-boluda, G.; Rodriguez de Sanabria, R.; Garcia-Casado, J. Wireless sensor node for non-invasive high precision electrocardiographic signal acquisition based on a multi-ring electrode. *Meas. J. Int. Meas. Confed.* **2017**, *97*, 195–202. [CrossRef]

35. Prats-Boluda, G.; Ye-Lin, Y.; Bueno Barrachina, J.M.; Senent, E.; Rodriguez de Sanabria, R.; Garcia-Casado, J. Development of a portable wireless system for bipolar concentric ECG recording. *Meas. Sci. Technol.* **2015**, *26*, 75102. [CrossRef]

36. Huigen, E.; Peper, A.; Grimbergen, C.A. Investigation into the origin of the noise of surface electrodes. *Med. Biol. Eng. Comput.* **2002**, *40*, 332–338. [CrossRef] [PubMed]

37. Verhagen, M.A.M.T.; Van Schelven, L.J.; Samsom, M.; Smout, A.J.P.M. Pitfalls in the analysis of electrogastrographic recordings. *Gastroenterology* **1999**, *117*, 453–460. [CrossRef] [PubMed]

38. Alberola-Rubio, J.; Prats-Boluda, G.; Ye-Lin, Y.; Valero, J.; Perales, A.; Garcia-Casado, J. Comparison of non-invasive electrohysterographic recording techniques for monitoring uterine dynamics. *Med. Eng. Phys.* **2013**, *35*, 1736–1743. [CrossRef] [PubMed]

39. Besio, W.; Aakula, R.; Koka, K.; Dai, W. Development of a tri-polar concentric ring electrode for acquiring accurate Laplacian body surface potentials. *Ann. Biomed. Eng.* **2006**, *34*, 426–435. [CrossRef] [PubMed]

40. Makeyev, O.; Ding, Q.; Besio, W.G. Improving the accuracy of Laplacian estimation with novel multipolar concentric ring electrodes. *Meas. J. Int. Meas. Confed.* **2016**, *80*, 44–52. [CrossRef] [PubMed]

41. Prats-Boluda, G.; Ye-Lin, Y.; Bueno-Barrachina, J.; Rodriguez de Sanabria, R.; Garcia-Casado, J. Towards the clinical use of concentric electrodes in ECG recordings: Influence of ring dimensions and electrode position. *Meas. Sci. Technol.* **2016**, *27*, 25705. [CrossRef]

42. Estrada, L.; Torres, A.; Garcia-Casado, J.; Prats-Boluda, G.; Ye-Lin, Y.; Jane, R. Evaluation of Laplacian diaphragm electromyographic recording in a dynamic inspiratory maneuver. *Conf. Proc. IEEE Eng. Med. Biol. Soc.* **2014**, *2014*, 2201–2204. [PubMed]

43. Estrada, L.; Torres, A.; Sarlabous, L.; Jané, R. Evaluation of sternocleidomastoid muscle activity by electromyography recorded with concentric ring electrodes. In Proceedings of the XXXIII Congreso Anual la Socieded Española Ingeniería Biomédica (CASEIB 2015), Madrid, Spain, 4–6 November 2015; pp. 183–186.

*sensors*

MDPI

*Article*

# VLSI Design of Trusted Virtual Sensors

**Macarena C. Martínez-Rodríguez \*, Miguel A. Prada-Delgado, Piedad Broxand Iluminada Baturone**

Instituto de Microelectrónica de Sevilla IMSE-CNM, CSIC, Universidad de Sevilla, Américo Vespucio, 41092 Sevilla, Spain; prada@imse-cnm.csic.es (M.A.P.-D.); brox@imse-cnm.csic.es (P.B.); lumi@imse-cnm.csic.es (I.B.)
* Correspondence: macarena@imse-cnm.csic.es; Tel.: +34-954-466-666

Received: 30 December 2017; Accepted: 22 January 2018; Published: 25 January 2018

**Abstract:** This work presents a Very Large Scale Integration (VLSI) design of trusted virtual sensors providing a minimum unitary cost and very good figures of size, speed and power consumption. The sensed variable is estimated by a virtual sensor based on a configurable and programmable PieceWise-Affine hyper-Rectangular (PWAR) model. An algorithm is presented to find the best values of the programmable parameters given a set of (empirical or simulated) input-output data. The VLSI design of the trusted virtual sensor uses the fast authenticated encryption algorithm, AEGIS, to ensure the integrity of the provided virtual measurement and to encrypt it, and a Physical Unclonable Function (PUF) based on a Static Random Access Memory (SRAM) to ensure the integrity of the sensor itself. Implementation results of a prototype designed in a 90-nm Complementary Metal Oxide Semiconductor (CMOS) technology show that the active silicon area of the trusted virtual sensor is 0.86 mm$^2$ and its power consumption when trusted sensing at 50 MHz is 7.12 mW. The maximum operation frequency is 85 MHz, which allows response times lower than 0.25 μs. As application example, the designed prototype was programmed to estimate the yaw rate in a vehicle, obtaining root mean square errors lower than 1.1%. Experimental results of the employed PUF show the robustness of the trusted sensing against aging and variations of the operation conditions, namely, temperature and power supply voltage (final value as well as ramp-up time).

**Keywords:** virtual sensors, CMOS integrated circuits; data security; hardware security; Physical Unclonable Function (PUF); piecewise linear approximation

## 1. State of the Art

A virtual sensor estimates the value of a variable that is very difficult or costly to measure physically by modelling the relation between that variable and others that can be measured easily with low-cost commercial sensors. The use of virtual sensors has increased continuously since the early 1980s in a wide number of industrial applications, such as building monitoring [1], robotics [2], process control [3,4], or automotive engineering [5]. In the latter case, for example, virtual sensors are employed to monitor vehicle and driving status as well as road conditions and even communication between vehicles [5–7]. The model that relates input and output variables can be derived from fundamental physical laws using adjustable parameters, from empirical data of the input and output variables without any knowledge of the physical process (usually referred to as black-box models), or a combination of physical and empirical knowledge (a gray-box model). Neural networks and fuzzy logic techniques are used widely to obtain black- and gray-box models [6,7]. PieceWise-Affine (PWA) virtual sensors are also employed to provide black-box models [8–10].

Virtual sensors are usually implemented as software installed in the electronic control units [6–8]. However, faster responses, smaller sizes and lower power consumption are achieved if virtual sensors are implemented in hardware. Solutions that employ PWA Simplicial (PWAS) models implemented in

Field Programmable Gate Arrays (FPGAs) were proposed in [9,10]. PWAS-based virtual sensors are simpler to implement in FPGAs than neural-network-based sensors and provide higher computation speed, as shown in [10]. FPGA implementations of PWA functions based on hyper-rectangular partitions (PWAR-based models) are further simpler than PWAS-based models, as shown in [11]. Taking into account that the sensor market size is growing (in the automotive industry, for example, approximately 22 billion sensors are estimated to be used per year by 2020 [12]), a CMOS Integrated Circuit (IC) solution for virtual sensors is very interesting since it can provide a minimum unitary cost with high performance in terms of area, power and speed. This is why this paper focuses on virtual sensor design into ICs. PWAR-based models are selected since they are very suitable to be realized by programmable ICs that can be adjusted to different applications.

Considering that virtual sensors are usually part of ubiquitous and distributed networks, security is becoming increasingly important [13,14]. The sensing data must be trusted by the receiver, hence the integrity of the output data must be ensured. However, the data can be authenticated by an impostor sensor. The counterfeit problem is of such magnitude that the Semiconductor Industry Association (SIA) keeps an anti-counterfeiting task force [15]. The trusted sensing significance has increased up to the point that it is considered not only by the sensors [14,16,17] but also by well-established protocols that interconnect them [18]. To solve the problem of the IC integrity, the key is not stored in the IC. It is recovered by using a Physical Unclonable Function (PUF) inside the sensor hardware [19]. Consequently, the trusted sensor can recover the cryptographic key, while any impostor sensor is unable to do that. To solve the problem of data integrity, Message Authentication Codes (MACs) are usually employed. MACs can be obtained from block ciphers (such as AES-CMAC) or from hash functions (such as HMAC). In the case of AES-CMAC, the block cipher AES encrypts the data and then AES-CMAC is applied to the resulting ciphertext to generate the authentication tag. Due to the significance of the standardization in any industrial application the standard Keyed-Hash Message Authentication Code (HMAC) was selected in a previous work [20]. The HMAC standard is essentially a two-pass hash-based MAC [21]. In [20], a PHOTON-based HMAC was implemented [22]. In this paper a new solution based on the authenticated encryption algorithm called AEGIS [23] is proposed. AEGIS algorithm is one of the third-round candidates of the Competition for Authenticated Encryption: Security, Applicability, and Robustness (CAESAR) [24]. It is recommended for lightweight solutions, providing small response times because symmetric key encryption and MAC are combined efficiently to share part of the computation. AEGIS not only authenticates the virtual sensor data but also encrypts it, ensuring confidentiality.

This paper describes the VLSI design of trusted virtual sensors. The paper is organized as follows. Firstly, the main features of the proposed trusted virtual sensor are described in Section 2. Then, an architectural description of the sensor is provided in Section 3. In Section 4, implementation results of the proposal in a 90-nm CMOS technology are shown and an application example in the automotive domain is described. Finally, conclusions are given in Section 5.

## 2. Features of the Proposed CMOS Sensor

The proposed sensor provides the encrypted value of a variable that is virtually measured. In addition, it provides an authentication code that ensures the integrity, confidentiality, and authenticity of the sensor data.

The virtual measurement is estimated from a PWA-based model. Both PWAR and PWAS forms are able to approximate any function and extract any black-box model. The PWAS form has been widely explored for virtual sensors [9,10]. However, the PWAR form is selected herein since its implementation is simpler than PWAS implementation.

The algorithm AEGIS is selected to encrypt and authenticate the virtual measurement providing confidentiality and authenticity to the virtual sensor. AEGIS offers a high security since it is not possible to recover the state and the key faster than exhaustive key search provided that a non-reused nonce is used and assuming that forgery attacks are not successful. The output provided by the proposed trusted virtual sensor are the used nonce, the resulting ciphertext, and the authentication tag.

The integrity of the virtual sensor is ensured if the key employed by AEGIS is not stored but recovered whenever needed by using PUFs. The trusted sensor is able to recover the cryptographic key shared with the receiver of the sensing data, while any impostor is unable due to the uniqueness provided by the start-up values of the SRAM in the sensor, which is exploited as a PUF. Non-sensitive Helper Data, $H$, are stored to recover the key with a Helper Data Algorithm (HDA) based on an Error Correcting Code (ECC) [25]. Helper Data do not reveal anything about the cryptographic key because the start-up values of SRAM cells obfuscate it. Similarly, Helper Data do not reveal anything about the intrinsic nature of the sensor because the cryptographic key obfuscates it.

Hence, two main components are differentiated in the proposed trusted virtual sensor. One part is associated with the generation of the virtual measurement and the other one provides the security of the measurement. They are both detailed in the following.

## 2.1. Virtual Sensing Based on PWAR approach

A black-box model establishes the relation between input variables and the empirical or simulated output. The virtual sensor is obtained using a black-box identification algorithm by assuming that the virtually measured output, $y$, is set as a PWAR function of the input variables, $x = \{x_1, \ldots, x_n\}$.

A generic PWA function with multiple inputs and one output $y(x) : D \subset \mathbb{R}^n \to \mathbb{R}$ is represented as

$$y(x) = \sum_{j=1}^{n} f_{ij} \cdot x_j + f_{i0} \ \forall \ x \in P_i \ (i = 1, \ldots, P) \tag{1}$$

where $f_i = [f_{i0} \cdots f_{in}] \in \mathbb{R}^{n+1}$, and $P_i \subset D$ are $P$ non-overlapping regions ( $P_i \cap P_j = \emptyset \ \forall i \neq j$ ), called polytopes, which form a polyhedral partition of the domain, $D$, so that $\bigcup_{i=1}^{p} P_i = D$.

In the case of regular PWAR functions, the domain is partitioned into hyper-rectangular polytopes by dividing each $k$ dimension of the domain into $L_k = 2^{p_k}$ intervals with the same amplitude, thus resulting $P = \prod_{k=1}^{n} L_k = \prod_{k=1}^{n} 2^{p_k} = 2^{\sum_{k=1}^{n} p_k}$ polytopes. Hence, a PWAR function is defined by its partition, $L = \{p_1, \ldots, p_n\}$, and the coefficients and the offset of each affine function, $F = \{f_1, \ldots, f_P\}$ with $P = 2^{\sum_{k=1}^{n} p_k}$. Figure 1 shows a bi-dimensional example of a regular PWAR function with its domain partitioned into $32 = 16 \times 2 = 2^4 \times 2^1$ rectangles, that is $L = \{4, 1\}$.

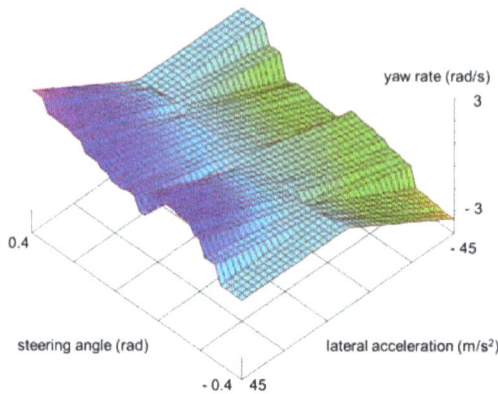

**Figure 1.** Vehicle yaw rate estimation by a PWAR-based virtual sensor.

The partition and affine functions of a PWAR function can be adjusted conveniently to approximate the relation between input variables and sensing variable to estimate. An algorithm has been developed to find the parameters $\{L, F\}$ that minimize the mean square error (MSE) between the PWAR function output (variable to sense, $y$) and the empirical or simulated output data, $y_\mu$. The algorithm takes into account the constraints imposed by digital implementations, which are the following.

- Maximum number of input variables: $n$.
- Maximum number of hyper-rectangles: $P = 2^p$.
- Maximum number of intervals per input: $Q = 2^q$, being $L_k = 2^{p_k}$ the number of intervals per input, so that, $q \geq p_k \ \forall k$.

The algorithm determines the hyper-rectangular partition of the input domain , $L = \{p_1, \ldots, p_n\}$, and the value of the affine functions, $F = \{f_1, \ldots, f_P\}$.

First, the partition is established, and then an off-line optimization algorithm extracts the $F$ parameters using a set of simulated or empirical values of the output and its corresponding input data. All input data should belong to the domain $D$. At least, a given percentage, $C$, of the hyper-rectangles in a partition should be covered by the data in order to proceed with optimization. Otherwise, such high resolution is not required and the partition is not considered. The result of the approximation algorithm for a given partition, $L$, is not only the value of the $F$ parameters but also the MSE between the simulated or empirical output data and the output of the PWAR function (the MSE taking into account training and testing data). Then, the MSE of all possible partitions are compared, and the partition with the lowest MSE (with the MSE that allows the best trade-off taking into account training and testing data) is selected.

The algorithm steps are detailed in the pseudo-code of Algorithm 1. Its inputs are a set, $B$, of output data, $y_\mu$, with $\mu = 1, \ldots, M$, corresponding to the input data $\{x_{1\mu}, \ldots, x_{n\mu}\}$; the value of $p$ that fixes the maximum number of hyper-rectangles; the value $q$ that fixes the maximum number of intervals per input; and the minimum percentage, $C$, of hyper-rectangles that should be covered by input data. Data set B is divided into two subsets, one used as training file and the other as test (validation) file.

The partitions with high resolution, that is, which verify that $p = \sum_{k=1}^{m} p_k$, are firstly explored with the function *TryPartition* (line 1.17) because they have higher capability of providing a lower approximation error (and they can be implemented in the proposed IC sensor as well as partitions with lower resolution). The function *TryPartition*($A$, $B$, $C$) evaluates if the partition, $A$, meets or not the condition to be included in the partition list, *Pool*. The condition is that the data set, $B$, should belong to no less than the $C$ in percentage of the hyper-rectangles of the partition, $A$. Otherwise, such high resolution model has too many undefined affine functions. Finally, the partition list is explored (with the function *Optimization*) to determine the partition that provides the lowest approximation error. The function *Optimization*(i,B) (line 1.34) applies Levenberg-Marquardt algorithm to find the parameters $F$ that achieve the minimum MSE (and, hence, the Root Mean Square Error, RMSE) for each partition candidate, $i$, considering the data set $B$ [26]. Since the partition candidate (hyper-rectangle) is fixed, this optimization finds the best linear (affine) regression for the data. The $F$ parameters of hyper-rectangles not covered by data are fixed as the arithmetic mean of the $F$ in the neighbourhood. Once the algorithm finishes, if all the coefficients, $f_{ij}$, associated with the input $x_j$ ($i = 1, \ldots, P$), are zero, that input can be removed, thus resulting $x = \{x_1, \ldots, x_m\}$ with $m \leq n$.

The proposed sensor can be configured to provide several PWAR functions depending on the application. Therefore, it can be used to sense different variables. The configuration data are the number of input variables, and the number of intervals in which each dimension is divided that is, the hyper-rectangular partition is configurable. The sensor is also configurable in the affine functions of each hyper-rectangle by changing the value of the parameters associated with each hyper-rectangle.

The input variables can be a same variable measured at several sampling times as well as the output measured at previous instants, for example, $x = \{\alpha[k] \quad \alpha[k-1] \quad \beta[k] \quad y[k-1]\}$.

---

**Algorithm 1** Pseudo-code of PWAR virtual sensor algorithm.

---

**Require:** $B = [x_{11}, \cdots, x_{m1}, y_1, \cdots, x_{1M}, \cdots, x_{mM}, y_M], p, q, C$

$\quad p_1 = q; p_2 = \cdots = p_m = 0;$
$\quad part = \neg SUCCESS; Pool = [\ ]; Emax = 1;$
$\quad$ **while** $p_1 \geq 0$ **do**
$\quad\quad p_2 = p - \sum_{j=1 \ \& \ j \neq 2}^{m} p_j;$
5: $\quad\quad$ **if** $p_2 > q$ **then**
$\quad\quad\quad p_2 = q;$
$\quad\quad$ **end if**
$\quad\quad$ **while** $p_2 \geq 0$ **do**
$\quad\quad\quad \cdots$
10: $\quad\quad\quad$ **while** $p_{m-1} \geq 0$ **do**
$\quad\quad\quad\quad p_m = p - \sum_{j=1 \ \& \ j \neq m}^{m} p_j;$
$\quad\quad\quad\quad$ **if** $p_m > q$ **then**
$\quad\quad\quad\quad\quad p_m = q;$
$\quad\quad\quad\quad$ **end if**
15: $\quad\quad\quad\quad$ **while** $p_m \geq 0 \ \& \ part = \neg SUCCESS$ **do**
$\quad\quad\quad\quad\quad A = \{p_1, \cdots, p_m\};$
$\quad\quad\quad\quad\quad part = \text{TryPartition}(A, B, C);$
$\quad\quad\quad\quad\quad$ **if** part==SUCCESS **then**
$\quad\quad\quad\quad\quad\quad Pool = [Pool \ || \ A];$
20: $\quad\quad\quad\quad\quad$ **end if**
$\quad\quad\quad\quad\quad p_m = p_m - 1;$
$\quad\quad\quad\quad$ **end while**
$\quad\quad\quad\quad p_m = 0;$
$\quad\quad\quad\quad p_{m-1} = p_{m-1} - 1;$
25: $\quad\quad\quad$ **end while**
$\quad\quad\quad \cdots$
$\quad\quad\quad p_3 = 0;$
$\quad\quad\quad p_2 = p_2 - 1;$
$\quad\quad$ **end while**
30: $\quad\quad p_2 = 0;$
$\quad\quad p_1 = p_1 - 1;$
$\quad$ **end while**
$\quad$ **for** $i \in Pool$ **do**
$\quad\quad [RMSE, F] = \text{Optimization}(i, B);$
35: $\quad\quad$ **if** RMSE $< Emax$ **then**
$\quad\quad\quad Emax = RMSE;$
$\quad\quad\quad L = i;$
$\quad\quad$ **end if**
$\quad$ **end for**
40: **return** $L, F$

---

### 2.2. Trusted Sensing Based on AEGIS and PUF

The virtual measurement, $y$, is encrypted and authenticated by using the AEGIS algorithm, which is a dedicated authentication encryption algorithm. AEGIS is based on the AES encryption round function, providing the advantage of a computational cost about half that of AES [23]. If the nonce is not reused (which should be the case for a true nonce), AEGIS provides a high security since state and key can only be recovered by exhaustive search.

AEGIS algorithm takes the cryptographic key, *key*, a number used only once, *nonce*, and a plaintex, in this case the virtual measurement, *y*, and provides a ciphertext, $C_t$, and an authentication tag, *tag*.

$$[C_t, tag] = \text{AEGIS}(key, nonce, y) \tag{2}$$

An interesting feature of the proposed sensor is that the secret key, is not stored in the IC but it is recovered whenever needed. From a security point of view, a secret that is not stored is much more difficult to discover. Among the wide number of PUFs proposed in the literature, SRAM-based PUFs [27] are selected in this work since the proposed IC sensor requires SRAMs for its virtual sensing functionality.

Since an SRAM cell is composed of two cross-coupled inverters, as shown in Figure 2, the start-up value is imposed by the inverter which begins to conduct. The conditions that make one inverter be the winner can be intrinsic or external. Intrinsic conditions are related to mismatching between the inverters, while external conditions are aging, ambient temperature, power supply voltage value ($V_{dd}$) or ramp-up time (i.e., the time to reach $V_{dd}$ after power-on) [27–29]. As discussed in [27], there are SRAM cells whose intrinsic conditions dominate over the external conditions. Hence, although the external conditions change, their start-up values are mostly the same. This type of cells will be named herein as ID cells. There are also SRAM cells whose external conditions dominate over the intrinsic conditions so that they are able to extract the noise of the external conditions as a source of entropy. They will be named herein as RND cells.

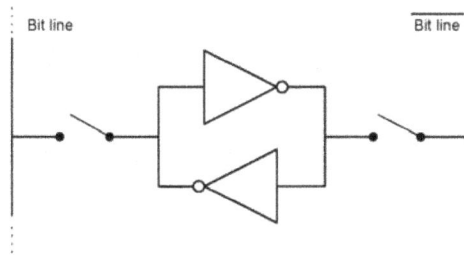

**Figure 2.** Schematic of a SRAM cell.

SRAM ID cells are used in the proposed IC sensor to recover the cryptographic key. The reliability of these stable cells is measured by the maximum fractional Hamming distance, $max_{\text{ID,HD}}$, between pairs of responses, *R*, generated by the same *n* cells at *m* different times:

$$max_{\text{ID,HD}} = \max_{\substack{i = 1, ..., m-1 \\ j = i+1, ..., m}} \left\{ \frac{\text{HD}(R_i, R_j)}{n} \right\} \tag{3}$$

Ideally, $max_{\text{ID,HD}}$ is zero but some bit flipping of the start-up values is unavoidable.

The SRAM cells are classified by using Algorithm 2 to obtain an ID_mask and a RND_mask. The SRAM is powered-up and down several times under different operating conditions.

HDA is used to obfuscate and to recover the key [25]. It is based on an error correcting repetition code whose correction capability should to be able to correct the maximum bit flipping probability of $max_{\text{ID,HD}}$ in (3). Algorithm 3 describes how Helper Data, *H*, are generated from a secret key and the start-up values of the SRAM.

---

**Algorithm 2** Pseudo-code of Masks Extraction algorithm.

---

**Require:** Number, J, of conditions to evaluate. Number of measurements, K, per condition
  **for** i=1 to J **do**
    **for** j=1 to K **do**
      power down and up the SRAM
      **if** j=1 **then**
        save the start-up values
      **else**
        compare the start-up values with the stored ones
      **end if**
      **for** all the cells of SRAM **do**
        **if** cell value does not change **then**
          add cell to ID_mask$_i$
        **else**
          add cell to RND_mask$_i$
        **end if**
      **end for**
    **end for**
  **end for**
  **for** all the cells of SRAM **do**
    **if** cell belongs to all $\{\text{ID\_mask}_1, \dots, \text{ID\_mask}_J\}$ **then**
      add cell to ID_mask
    **else if** cell belongs to all $\{\text{RND\_mask}_1, \dots, \text{RND\_mask}_J\}$ **then**
      add cell to RND_mask
    **end if**
  **end for**
  **return** ID_mask, RND_mask

---

**Algorithm 3** Pseudo-code of Helper Data Generation algorithm.

---

**Require:** $key = [k_1, \dots, k_a]$, start-up values of SRAM
  Each bit of the key is repeated $r$ times $\Longrightarrow key_r = [k_{11}, \dots, k_{1r}, \dots, k_{a1}, \dots, k_{ar}]$.
  $R =$ concatenation of the $a \cdot r$ SRAM ID cells start-up values at Helper Data generation step.
  $H = key_r \oplus R$.
  **return** $H$

---

The secret key is recovered as described in Algorithm 4. A response, $R'$, slightly different to $R$, since even ID cells may provide some bit flipping, is obtained by concatenating the new start-up values of the $a \cdot r$ SRAM ID cells used to generate the Helper Data. The key is recovered by using an ECC with codeword length $r$, employed to correct up to $\lfloor \frac{r}{2} \rfloor$ errors.

---

**Algorithm 4** Pseudo-code of Key Recovering algorithm.

---

**Require:** $H$, start-up values of SRAM
  $R' =$ concatenation of the $a \cdot r$ SRAM ID cells start-up values at key recovering step.
  $key' = H \oplus R' = key_r \oplus R \oplus R'$
  $key = ECC(key')$.
  **return** $key$

---

Helper Data do not reveal anything about the cryptographic key because the start-up values of SRAM ID cells obfuscate it. Similarly, Helper Data do not reveal anything about the intrinsic nature of the IC sensor because the cryptographic key obfuscates it. A way to measure the non-sensitiveness of Helper Data is to evaluate the average fractional Hamming distance, $avg_{\text{InterHD}}$, between all the possible pairs of Helper Data of $p$ different sensors that use the same key, as follows:

$$avg_{\text{InterHD}} = \frac{2}{p\,(p-1)} \sum_{i=1}^{p-1} \sum_{j=i+1}^{p} \frac{\text{HD}\left\{R_i, R_j\right\}}{n} \tag{4}$$

This way, instead of storing the key, only Helper Data are stored and the integrity of the sensor is ensured. If the sensor is copied, its SRAM ID cells will have other intrinsic features and they will not be able to recover the key.

The start-up values of the SRAM cells are also exploited to generate the nonce required by AEGIS. The nonce is initialized by the start-up values of SRAM RND cells and then updated by a counter. A way to measure the change of the nonce seed is to evaluate the minimum fractional Hamming distance, $min_{\text{RND,HD}}$, between pairs of responses, $R$, generated by the same $n$ RND cells at $m$ different times:

$$min_{\text{RND,HD}} = \min_{\substack{i = 1, ..., m-1 \\ j = i+1, ..., m}} \left\{ \frac{\text{HD}\left(R_i, R_j\right)}{n} \right\} \tag{5}$$

A value of $min_{\text{RND,HD}}$ strictly greater than zero ensures that the nonce seed is time variant, as required.

AEGIS is based on the AES round function [23]. There are three variations. AEGIS-128 processes a 16-byte message block with five AES round functions in parallel. It consumes the least resources, but it is the slowest for large messages. AEGIS-128L processes a 32-byte message block with eight AES round functions in parallel. In terms of resource consumption, it is the biggest version, but it is also the fastest. Finally, AEGIS-256 processes a 32-byte message block with six AES round functions. Then, it offers an intermediate performance in terms of resource consumption and time response. Since the virtual measurement is smaller or equal to 128-bits, only one 16-byte state has to be processed. Hence, AEGIS-128 is more adequate in this application, since it is as fast as AEGIS-128L, using less resources.

AEGIS-128 uses a 128-bit key and a 128-bit nonce. The algorithm is based on the function $S_{i+1} = StateUpdate(S_i, m_i)$ described in Algorithm 5 where the function $AESRound(A)$ is the AES encryption round function being $A$ the state. The main operations in the AES round are SubBytes operation, ShiftRows and AES-MixColumns.

---

**Algorithm 5** State Update function.

1: **function** STATEUPDATE($S_i, m_i$)
2:     $S_{i+1,0} = \text{AESRound}(S_{i,4}) \oplus S_{i,0} \oplus m_i$
3:     $S_{i+1,1} = \text{AESRound}(S_{i,0}) \oplus S_{i,1}$
4:     $S_{i+1,2} = \text{AESRound}(S_{i,1}) \oplus S_{i,2}$
5:     $S_{i+1,3} = \text{AESRound}(S_{i,2}) \oplus S_{i,3}$
6:     $S_{i+1,4} = \text{AESRound}(S_{i,3}) \oplus S_{i,4}$
7:     **return** $S_{i+1}$
8: **end function**

---

The pseudo-code of the AEGIS-128 applied to the virtual measurement is described in Algorithm 6. The main steps of the algorithm are to initialize the state (line 6.1), process the nonce (line 6.6), process the measurement and generate the ciphertext (line 6.14), and generate the authentication tag (line 6.20). *constant1* and *constant2* are two 128-bit constants fixed by the algorithm.

---

**Algorithm 6** Pseudo-code of AEGIS algorithm.

---

**Require:** key, nonce, $y$
1: $S_{-10,0} = \text{key} \oplus \text{nonce}$
2: $S_{-10,1} = \text{constant1}$
3: $S_{-10,2} = \text{constant2}$
4: $S_{-10,3} = \text{key} \oplus \text{constant1}$
5: $S_{-10,4} = \text{key} \oplus \text{constant2}$
6: **for** $i = -10 : -1$ **do**
7:     **if** $i$ is odd **then**
8:         $m_i = \text{key} \oplus \text{nonce}$
9:     **else**
10:         $m_i = \text{key}$
11:     **end if**
12:     $S_{i+1} = StateUpdate(S_i, m_i)$
13: **end for**
14: $C_t = y \oplus S_{0,1} \oplus S_{0,4} \oplus (S_{0,2} \& S_{0,3})$
15: $S_1 = StateUpdate(S_0, y)$
16: $tmp = S_{1,3} \oplus 128$
17: **for** $i = 1 : 6$ **do**
18:     $S_{i+1} = StateUpdate(S_i, tmp)$
19: **end for**
20: $tag = S_{7,0} \oplus S_{7,1} \oplus S_{7,2} \oplus S_{7,3} \oplus S_{7,3}$
21: **return** $C_t, tag$

---

## 3. Architectural Description of the Sensor

The proposed sensor has two main behavioural modes. The configuration mode is carried out whenever the sensor is powered up. It recovers the key, generates a seed for the nonces, and establishes the relation between the input data and the data to estimate, that is, the PWAR model. The trusted sensing mode is set once the configuration mode is finished. It generates a PWAR virtual measurement which is encrypted and authenticated.

The proposed sensor is composed of three main units: the PWAR, the Cryptographic, and the Control units. Figure 3 illustrates the block diagram of the sensor architecture, including the main signals and buses that interconnect the blocks. Note that the SRAM is shared by two units, thus saving resources since it is not used simultaneously by both units.

The non-volatile memory (NVM) should be programmed by the sensor manufacturers or their authorized distributed channels prior to use the sensor. The NVM stores the information required by the sensor: (a) the partition $L$ and the parameters $F$ that fix the PWAR model to implement (obtained by applying Algorithm 1); (b) the RND and ID masks (obtained by classifying the SRAM cells as exposed in Algorithm 2); and (c) the Helper Data needed to recover the cryptographic key (generated as in Algorithm 3).

The whole design prioritizes a short response time, using parallel processing, and a low power consumption, enabling the blocks only whenever needed.

### 3.1. PWAR Unit

The architecture of this unit, based on the one exposed in [30], is composed of the SRAM, the Address Generator, and the Arithmetic blocks.

Each word in the SRAM stores the coefficients of the input and the offset of the affine function associated with an hyper-rectangle, $f_i \in F$. Therefore, the width of the memory is $(n + 1) \cdot n_{in}$ bits and the depth is the maximum number of hyper-rectangles, $P$, being $n_{in}$ the number of bits used to represent each coefficient of the input and the offset. During the trusted sensing mode, one word of the SRAM is read providing in parallel the parameters needed to compute the affine function related to an hyper-rectangle. In the configuration mode, all the PWAR parameters, *F*, are read from the NVM and stored in the SRAM.

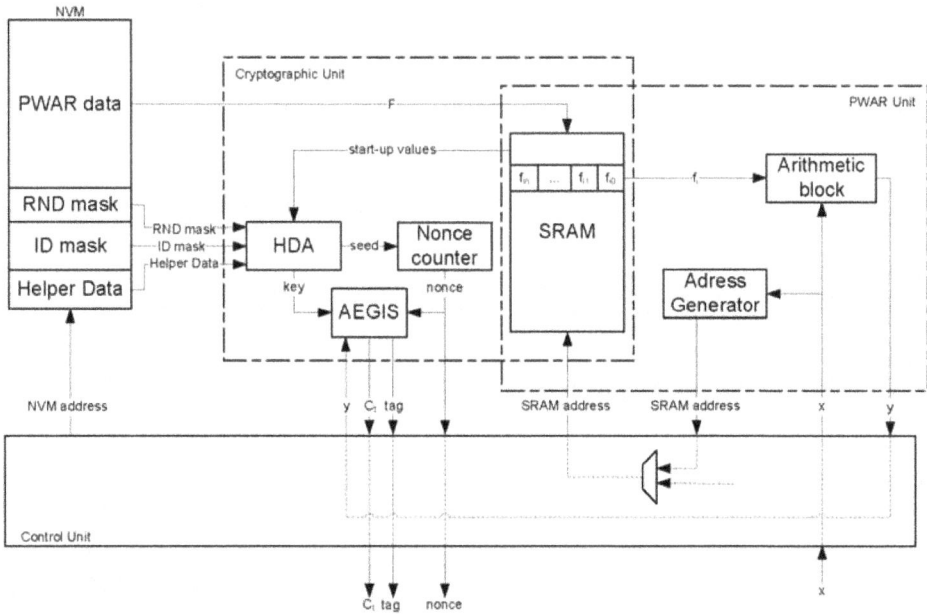

**Figure 3.** Architectural scheme of the proposed CMOS sensor.

The Address Generator block is only enabled during the trusted sensing mode. It determines the address of the SRAM where the parameters associated with the input are stored. For that purpose, it concatenates the $p_k$ most significant bits (MSBs) of each input $x_k$, as shown in Figure 4. The MSBs determine which of the $2^{p_k}$ intervals the input belongs to, and, hence, they determine the hyper-rectangle.

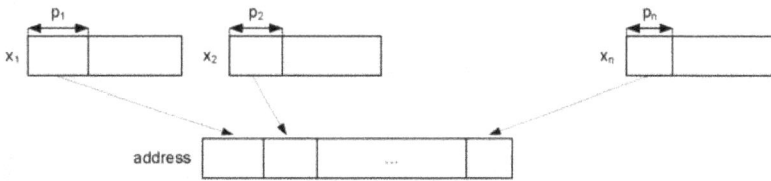

**Figure 4.** Generation of the SRAM address.

As the previous block, the Arithmetic block is only enabled during the trusted sensing mode. The Arithmetic block computes an affine function for a given input $x$, and for given parameters, $f_i$. The operation $y = \sum_{j=1}^{n} f_{ij} \cdot x_j + f_{i0}$ is carried out in parallel with $n$ multipliers and 1 adder. It uses fixed-point logic. The length of the fixed part is configured externally (it is one of the configuration parameters stored in the NVM).

*3.2. Cryptographic Unit*

The ID_mask, RND_mask and the Helper Data stored in the NVM are non-sensitive data. The cryptographic key, which is the sensitive information, cannot be recovered without the start-up values of the SRAM ID cells. Hence, hardware-based security is added to cryptographic-based security. During the configuration mode, both masks and the Helper Data are loaded from the NVM. During the configuration mode, the start-up values of the words in the lower part of the SRAM are read by the Cryptographic Unit. Afterwards, these cells are written with the PWAR parameters. The Cryptographic Unit is composed of the HDA block, the Nonce Counter, and the AEGIS block.

The HDA block is only enabled during the configuration mode. It regenerates the key by using the Helper Data, the ID_mask and the start-up values of the SRAM, as described in Algorithm 2. It also provides the seed for the nonces by taking the start-up values of the SRAM cells indicated by the RND_mask. The seed is the input of a counter that is enabled whenever a nonce is required in the trusted sensing mode. During the trusted sensing mode, the AEGIS block initializes the state serially, processes the nonce, and generates both the ciphertext and the tag, as explained in Algorithm 3. The State Update operation is implemented in parallel as well as the AES round included on it. Figure 5 illustrates the implementation of the State Update.

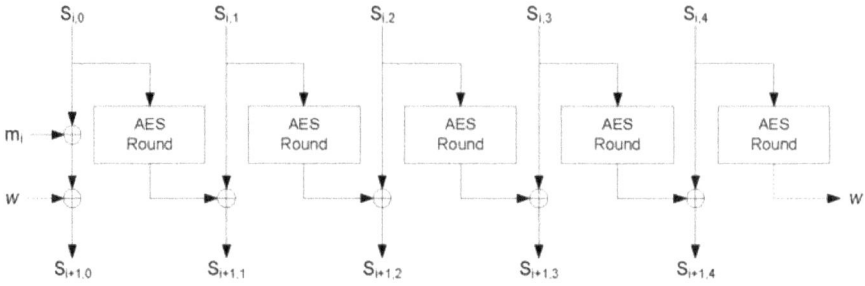

**Figure 5.** State Update.

### 3.3. Control Unit

The Control Unit copes with the two possible behavioural modes. It is implemented as a finite state machine (FSM). Figure 6 illustrates the states and the main operations. When the FSM is set in one state, only the required blocks are enabled, thus saving power consumption. The gray states correspond to the configuration mode, while the white states corresponds to the trusted sensing mode. Whenever the sensor is powered up, the configuration mode starts.

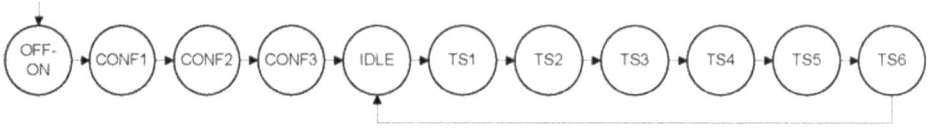

**Figure 6.** State diagram of the Control Unit.

The main operations in the configuration mode are as follows. The mask_ID, mask_RND, and Helper Data are read from the NVM (CONF1). The ID and the seed are generated (CONF2). The key is recovered, the nonce is initialized with the seed, configuration data of the PWAR are read from the NVM, and the PWAR Unit is configured (CONF3).

Once the configuration mode is finished, the FSM is set to state IDLE, where it remains until the sensor receives an input data to generate a new trusted PWAR virtual measurement.

In the trusted sensing mode, the main operations are the following. Input data ($x$) are read and a new nonce is generated (TS1). The Address Generator block generates the address and the AEGIS block processes the nonce (TS2). The SRAM provides the parameters stored in the address previously generated (TS3). The PWAR Unit generates the virtual measurement (TS4). The AEGIS block processes the virtual measurement and provides the encrypted virtual measurement (TS5), and finally the AEGIS block generates the authentication tag (TS6). Once the trusted virtual sensor output is provided, the FSM goes to the state IDLE and waits for a new input data.

If the sensor is powered down and then it is powered up, the FSM starts from the state OFF-ON and all the configuration states are carried out again. Consequently, a correct performance is ensured after (expected or unexpected) suspension of power supply since the sensor is always configured before entering the trusted sensing mode.

## 4. Implementation Results

The trusted virtual sensor was synthesized in a 90-nm CMOS technology provided by Taiwan Semiconductor Manufacturing Company (TSMC). A standard low-power dual-port SRAM IP module was employed for the SRAM. The register-transfer level specifications of the other blocks were synthesized using Design Vision tool from Synopsys. The Place and Route tool used was SoC Encounter from Cadence. ModelSim simulator from Mentor Graphics was used for timing simulations.

### 4.1. Features of the Design

The non-volatile memory (NVM) has an output bus with 12 bits, so that a 12-bit word can be read in a clock cycle. It stores: (a) 20,483 12-bit words (20,480 $\times$ 12 = 4096 $\times$ 5 $\times$ 12 bits defining the affine functions and 3 $\times$ 12 bits defining the PWAR Unit configuration); (b) 155 12-bit words associated with the RND_mask; (c) 360 12-bit words associated with the ID_mask; and (d) 310 12-bit words (3720 bits) associated with the Helper Data.

Concerning the PWAR Unit, the maximum number of input variables, $n$, is 4. The bits of both the input variables $\{x_1, \ldots, x_4\}$ and the parameters of the affine functions are 12. The bits of the output variable, $y$, are 26. The maximum number of hyper-rectangles, $P = 2^p$, is 4096 ($p = 12$), and the maximum number of intervals per input, $Q = 2^q$, is 128 ($q = 7$). Hence, the SRAM has 4096 $\times$ 60 bits.

Concerning the Cryptographic Unit, the key and the nonce have 128 bits. To recover the key, a binary repetition code with codeword length $r = 29$ is employed. Therefore, the ID size and the H size is 128 $\times$ 29 = 3712 bits. The seed of the nonce has the same size as the nonce, that is, 128 bits.

The timing responses of the main operations in both behavioural modes are detailed in the following.

The RND_mask (1860 bits) is read from the NVM in 155 clock cycles. To generate the seed of the nonce, 31 words of the SRAM (31 $\times$ 60 = 1860 bits) are read and they are selected or not for the 128-bit seed as indicated by the RND_mask. The seed generation takes 1860 clock cycles. The initialization of the Nonce Counter takes 1 clock cycle.

The ID_mask (4320 bits) is read from the NVM in 360 clock cycles. The Helper Data (3720 bits) are read in 310 clock cycles. To generate the response of 3720 ID cells, 72 words of the SRAM (72 $\times$ 60 = 4320 bits) are read and they are selected or not as indicated by the ID_mask. This takes 4320 clock cycles. The decoding to recover the secret key takes 5 clock cycles. The AEGIS needs 4 clock cycles to process the key. The PWAR configuration data and parameters are read in 3 + 4096 $\times$ 5 = 20,483 clock cycles.

To generate the PWAR virtual measurement, the input data are read in four clock cycles, the address of the SRAM is generated in one clock cycle, the read of a word from the SRAM takes one clock cycle, and the computation of the PWA function takes another one. Therefore, seven clock cycles are needed. Then, the AEGIS processes the nonce in 10 clock cycles. It processes the PWAR virtual measurement and provides the ciphertext in one clock cycle. Finally, the AEGIS block needs nine clock cycles to provide the authentication code. Hence, the AEGIS block takes 20 clock cycles to process the PWAR virtual measurement.

The configuration mode lasts the time that the NVM is read, since the other operations can be executed in parallel. Figure 7 shows how the parallelization is carried out. Then, it lasts at most $155 + 360 + 310 + 20,483 = 21,308$ clock cycles.

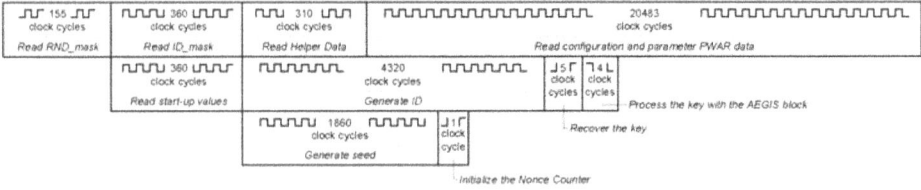

**Figure 7.** Timing of configuration mode.

Figure 8 shows the timing of the trusted sensing mode, it lasts $1 + 10 + 1 + 9 = 21$ clock cycles.

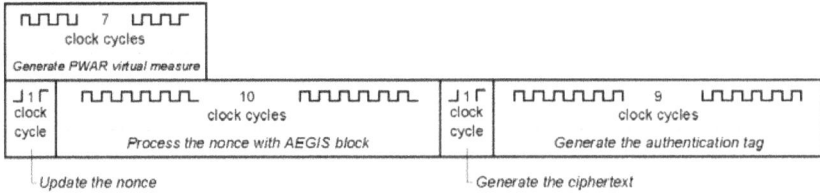

**Figure 8.** Timing of trusted virtual sensing mode.

Since the maximum operation of the trusted virtual sensor is 85 MHz, it can achieve response times lower than 0.25 microseconds.

Table 1 shows the power consumption during trusted sensing mode and the area of the main blocks provided by Design Vision tool from Synopsys.

**Table 1.** Area and power consumption during trusted sensing mode.

|  | Area | Power@50MHz |
| --- | --- | --- |
| HDA block | 0.0028 mm$^2$ | 0.13 mW |
| Nonce Counter | 0.0041 mm$^2$ | 0.15 mW |
| AEGIS | 0.14 mm$^2$ | 0.77 mW |
| Arithmetic block | 0.015 mm$^2$ | 0.24 mW |
| Address Generator block | 0.0007 mm$^2$ | 0.15 mW |
| SRAM | 0.67 mm$^2$ | 4.49 mW |
| Control Unit | 0.025 mm$^2$ | 1.19 mW |

The trusted virtual sensor occupies 0.86 mm$^2$ and consumes 7.12 mW at 50 MHz. The area occupied by the Cryptographic Unit is 17.1% and its power consumption is 14.8% (considering that the SRAM and the Control are needed by virtual sensing and, hence, re-used by Cryptographic Unit).

Figure 9 shows the layout of the trusted virtual sensor extracted with the SoC Encounter tool from Cadence. The largest gray rectangular area in the upper part of the layout is the SRAM.

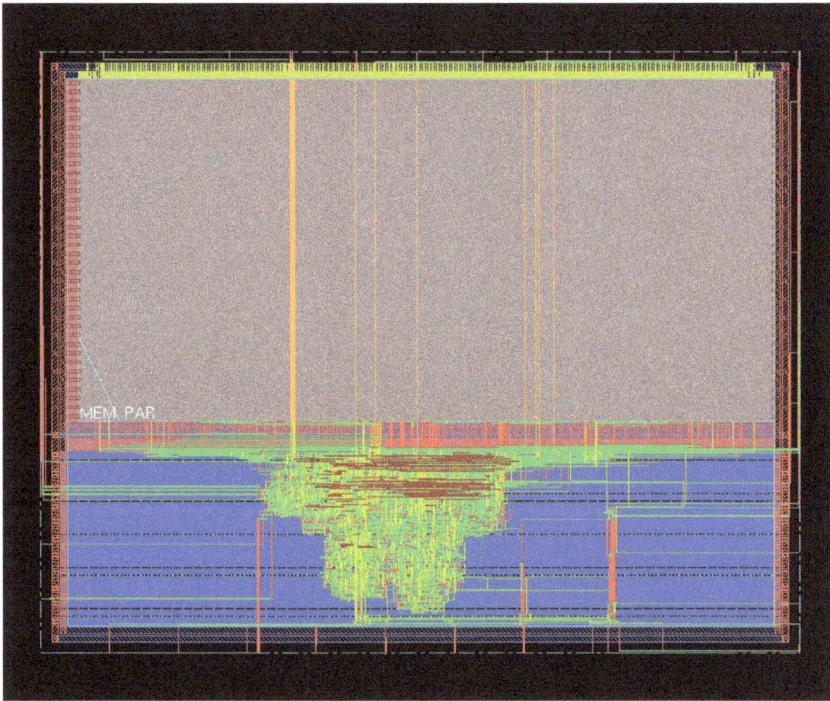

**Figure 9.** Layout of the trusted virtual sensor.

*4.2. Implementation Results Concerning Virtual Sensing*

To evaluate the capability of the IC prototype as virtual sensor, the PWAR Unit was fabricated in TSMC 90 nm technology and verified experimentally. As illustrative example, a problem from the automotive domain, the estimation of vehicle yaw rate, was selected to be solved. In literature, this problem was firstly addressed with a silicon micromachining (MEMS)-based sensor in [31]. Since MEMS-based sensors have inaccuracy problems including severe DC-offset, an additional module based on fuzzy logic was proposed in [32]. Lately, neural-network-based direct virtual sensors implemented in software were proposed in [6]. A PWAR-based model was found for the proposed IC prototype by using Algorithm 1. The set of input-output data needed to apply the algorithm (training and test subsets) was obtained from simulations of the vehicle modelled by the differential equations described in [6], discretized with a zero-order-hold approach with a sampling time of 10 ms, and including noise in the simulated outputs (lateral acceleration and yaw rate) as in [6]. The input variables are the longitudinal vehicle speed, $v_x(t)$, the steering angle, $\alpha_s(t)$, and the lateral acceleration, $a_y(t)$. After applying PWAR virtual sensor algorithm, Algorithm 1, (with $C = 80\%$), the best partition found divided the longitudinal speed, the steering angle, and the lateral acceleration into 4, 16 and 2 intervals, respectively. Figure 1 shows the yaw rate provided by the PWAR-based model for a fixed value of the longitudinal speed. Therefore, the sensor estimates the yaw rate as $\dot{\Psi} = f_{i0} + f_{i1} \cdot v_x + f_{i2} \cdot \alpha_s + f_{i3} \cdot a_y$ if $(v_x, \alpha_s, a_y) \in P_i$, with $i = 1, \ldots, 128$. The RMSE values of the PWAR-based sensor and the direct virtual sensor based on neural networks, DVS1 in [6], using simulated data, are shown in Table 2. The RMSE is lower for the PWAR-based virtual sensor.

A hardware-in-the-loop simulation was carried out in ModelSim. The evolution of the longitudinal vehicle speed, the steering angle, and the lateral acceleration scaled to their real values, are shown at the upper part of Figure 10. The yaw rate estimated by the sensor (also scaled to the real value) is shown in blue at the bottom of Figure 10. The yaw rate obtained by the model is shown in red. Noise can be observed in the lateral acceleration and the yaw rate of the model that makes it more realistic. To achieve a response time lower than the 10 ms sampling time, the sensor should work at more than 2.1 KHz.

**Table 2.** Error comparison between virtual sensors.

|  | Direct Virtual Sensor in [6] | | PWAR Virtual Sensor | |
|--|--|--|--|--|
|  | Training Data | Testing Data | Training Data | Testing Data |
| RMSE | 3.9% | 6.5% | 0.25% | 1.08% |

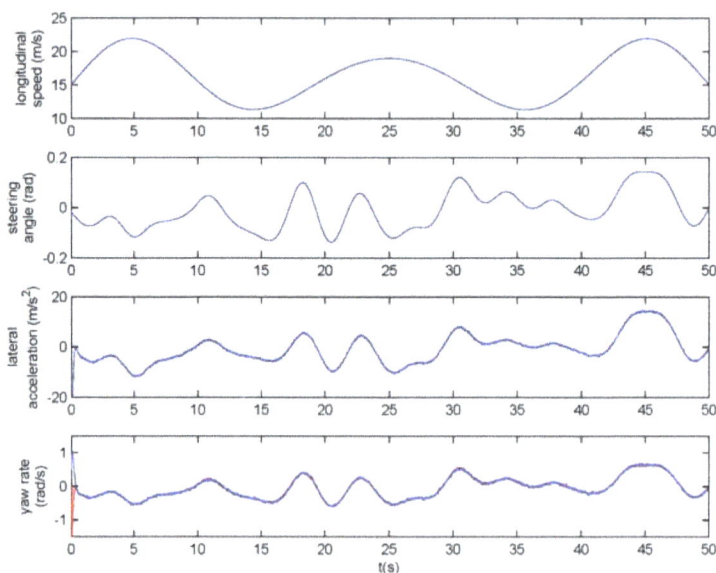

**Figure 10.** Hardware-in-the-loop simulation of the PWAR Unit.

*4.3. Implementation Results Concerning Trusted Sensing*

Since the behaviour of the SRAM is fundamental to ensure the trusted sensing mode of the prototype, the SRAM was fabricated in TSMC 90-nm technology and analyzed experimentally. Ten samples of $4096 \times 60$ SRAMs were measured as follows. The cells of the SRAM were classified three times after 20 measurements each time by using Algorithm 2 with $J = 3$ and $K = 20$. Nominal operation conditions were considered three times: power supply $V_{dd} = 1.2V$ and temperature $T = 25\,°C$. The $V_{dd}$ was supplied to the SRAM with a measured ramp-up time of 18.2 ms. The average of ID cells found in percentage was $87.27 \pm 0.23\%$ and that of RND cells was $7.15 \pm 0.16\%$. To obtain an ID of 3760 bits, as required, 72 words of 60 bits are enough since 3770 bits (ID cells) are obtained in average from 4320 SRAM cells. To obtain a seed for the nonces of 128 bits, 31 words of 60 bits are enough. On average, the 7.15% of $31 \times 60 = 1860$ cells are 132 RND cells.

To evaluate the robustness of the key recovering in the designed prototype, Algorithm 2 was applied considering different operating conditions (with $J = 2$ and $K = 20$): (a) changing the value of $V_{dd}$ to 1.08V and to 1.32V ($\pm 10\%$ of the nominal value); (b) changing the value of T to 5 °C and 75 °C;

(c) after accelerated aging the ICs with the SRAM continuously during 171 h at high temperature of 75 °C (and nominal $V_{dd}$); and (d) decreasing the $V_{dd}$ ramp-up time to 2.32 ms.

The first row of Table 3 shows the average of maximum intra HD (3) between responses of SRAM ID cells taken at operation conditions with changes in $V_{dd}$, $T$, ramp-up time, and aging (before and after aging). Reliability is mostly affected by changes in ramp-up time and temperature. Since the start-up values of different SRAM ID cells of the same SRAM were proved to be independent, the probability that a string of 29 bits contains more than 14 errors (and, hence, it is not correctly decoded to extract the original key bit) is calculated with the following formula:

$$P_{\text{bit-error}} = 1 - \sum_{i=0}^{14} \binom{29}{i} p^i \cdot (1 - p)^{29-i} \tag{6}$$

Table 3. Features of the SRAM PUF in the IC sensor.

|  | Changes in $V_{dd}$ | Changes in $T$ | Changes in Ramp-Up Time | Changes in Aging |
|---|---|---|---|---|
| $max_{\text{ID,HD}}$ **(ID)** | $0.66 \pm 0.04\%$ | $2.75 \pm 0.07\%$ | $9.64 \pm 0.24\%$ | $0.29 \pm 0.01\%$ |
| $min_{\text{RND,HD}}$ **(RND)** | $34.31 \pm 0.29\%$ | $41.49 \pm 0.51\%$ | $39.02 \pm 0.49\%$ | $33.74 \pm 0.31\%$ |

Assuming the worst-case bit flipping probability in the first row of Table 3, $p = 9.64\%$, the $P_{\text{bit-error}}$ is $1.19 \times 10^{-8}$. Similarly, assuming $1.19 \times 10^{-8}$ as the worst-case bit-error probability in the key, the probability that the 128-bit key is not correctly decoded by the authentic sensor is $1.53 \times 10^{-6}$, which is quite low. The average inter HD (4) between responses of ID cells of 10 different SRAMs (corresponding to different sensors) was measured as $0.50 \pm 0.01$ in nominal conditions. As commented in Section 2.2, this result shows that Helper Data do not reveal information about the cryptographic key. Taking such average inter HD as bit flipping probability between a false sensor and a genuine one, the probability of correctly decoding one bit of the cryptographic key is 0.5, and, hence, the probability of regenerating the correct key by a false sensor is $2 \times 10^{-128}$, which is completely infeasible. The second row of Table 3 shows the average of minimum inter HD (5) between responses of RND cells taken at the different operation conditions. In any case, it is ensured that the nonce seed is always different, as required.

## 5. Conclusions

The VLSI design of trusted virtual sensors is very suitable for industrial applications where security is becoming increasingly important since it offers privacy, authenticity and integrity of the virtually sensed measurement and the circuit itself. The implementation of the design into a 90-nm CMOS technology occupies 0.86 mm$^2$ and consumes 7.12 mW when trusted sensing at 50 MHz. Working at maximum frequency, the trusted virtual sensor allows sampling times lower than 0.25 µs. The inclusion of security to the virtual sensor needs 17.1% of active area and 14.8% of power consumption.

**Acknowledgments:** This work was supported in part by TEC2014-57971-R project from Ministerio de Economía, Industria y Competitividad of the Spanish Government (with support from the P.O. FEDER of European Union) and 201750E010 (HW-SEEDS) project from CSIC. M.C. Martínez-Rodríguez is supported by FPI fellowship program for Students from Spanish Government. The work of M.A. Prada-Delgado and P. Brox was supported by V Plan Propio de Investigación through the University of Seville, Seville, Spain.

**Author Contributions:** Macarena C. Martínez-Rodríguez designed the trusted virtual sensor prototype. Macarena C. Martínez-Rodríguez and Piedad Brox designed, simulated and tested the fabricated PWAR Unit. Miguel A. Prada-Delgado tested and analyzed the fabricated SRAM PUF. Iluminada Baturone contributed to the design of the prototype, the selection of the algorithms and the experiments. All authors contributed to writing the paper.

**Conflicts of Interest:** The authors declare no conflict of interest.

## References

1. Li, H.; Yu, D.; Braun, J.E. A review of virtual sensing technology and application in building systems. *HVAC R Res.* **2011**, *17*, 619–645.
2. Heredia, G.; Ollero, A. Virtual Sensor for Failure Detection, Identification and Recovery in the Transition Phase of a Morphing Aircraft. *Sensors* **2010**, *10*, 2188–2201.
3. Sánchez, J.A.; Rodríguez, F.; Guzmán, J.L.; Arahal, M.R. Virtual Sensors for Designing Irrigation Controllers in Greenhouses. *Sensors* **2012**, *12*, 15244–15266.
4. Bustillo, A.; Correa, M.; Reñones, A. A Virtual Sensor for Online Fault Detection of Multitooth-Tools. *Sensors* **2011**, *11*, 2773–2795.
5. Stephant, J.; Charara, A.; Meizel, D. Virtual sensor: Application to vehicle sideslip angle and transversal forces. *IEEE Trans. Ind. Electron.* **2004**, *51*, 278–289.
6. Novara, C.; Ruiz, F.; Milanese, M. Direct Identification of Optimal SM-LPV Filters and Application to Vehicle Yaw Rate Estimation. *IEEE Trans. Control Syst. Technol.* **2011**, *19*, 5–17.
7. Zhang, B.; Du, H.; Lam, J.; Zhang, N.; Li, W. A Novel Observer Design for Simultaneous Estimation of Vehicle Steering Angle and Sideslip Angle. *IEEE Trans. Industr. Electron.* **2016**, *63*, 4357–4366.
8. Rubagotti, M.; Poggi, T.; Oliveri, A.; Pascucci, C.A.; Bemporad, A.; Storace, M. Low-complexity piecewise-affine virtual sensors: theory and design. *Int. J. Control* **2014**, *87*, 622–632.
9. Poggi, T.; Rubagotti, M.; Bemporad, A.; Storace, M. High-speed piecewise affine virtual sensors. *IEEE Trans. Ind. Electron.* **2012**, *59*, 1228–1237.
10. Oliveri, A.; Cassottana, L.; Laudani, A.; Fulginei, F.R.; Lozito, G.; Salvini, A.; Storace, M. Two FPGA-Oriented High Speed Irradiance Virtual Sensors for Photovoltaic Plants. *IEEE Trans. Ind. Inform.* **2017**, *13*, 157–165.
11. Comashi, F.; Genuit, B.A.G.; Oliveri, A.; Heemels, W.P.M.H.; Storace, M. FPGA implementations of piecewise affine functions based on multi-resolution hyperrectangular partitions. *IEEE Trans. Circuits Syst. I* **2012**, *59*, 2920–2933.
12. Pinelis, M. Automotive sensors and electronics: trends and developments in 2013. In Proceedings of the Automotive Sensors and Electronics Expo, Detroit, MI, USA, 9 October 2013.
13. Kim, J.; Lee, D.; Jeon, W.; Lee, Y.; Won, D. Security Analysis and Improvements of Two-Factor Mutual Authentication with Key Agreement in Wireless Sensor Networks. *Sensors* **2014**, *14*, 6443–6462.
14. Sampangi, R.V.; Sampalli, S. Butterfly Encryption Scheme for Resource-Constrained Wireless Networks. *Sensors* **2015**, *15*, 23145–23167.
15. Semiconductor Industry Association. Available online: https://www.semiconductors.org/issues/anticounterfeiting/anti_counterfeiting/ (accessed on 1 August 2013).
16. Meguerdichian, S.; Potkonjak, M. Security primitives and protocols for ultra low power sensor systems. In Proceedings of the 2011 IEEE SENSORS, Limerick, Ireland, 28–31 October 2011; pp. 1225–1227.
17. Kanuparthi, A.; Karri, R.; Addepalli, S. Hardware and Embedded Security in the Context of Internet of Things. In Proceedings of the 2013 ACM Workshop on Security, Privacy & Dependability for Cyber Vehicles, Berlin, Germany, 4 November 2013; pp. 61–64.
18. Pfeiffer, O. *Implementing Scalable CAN Security with CANcrypt: Authentication and Encryption for CANopen, J1939 and Other Controller Area Network or CAN FD Protocols*; Embedded Systems Academy Inc.: Hanover, Germany, 2017.
19. Liu, W.; Zhang, Z.; Li, M.; Liu, Z. A Trustworthy Key Generation Prototype Based on DDR3 PUF for Wireless Sensor Networks. *Sensors* **2014**, *14*, 11542–11556.
20. Martínez-Rodríguez, M.C.; Prada, M.; Brox, P.; Baturone, I. CMOS digital design of a trusted virtual sensor. In Proceedings of the 2017 IEEE Nordic Circuits and Systems Conference (NORCAS): NORCHIP and International Symposium of System-on-Chip (SoC), Linkoping, Sweden, 23–25 October 2017.
21. National Institute of Standards and Technology (NIST). The Keyed-Hash Message Authentication Code (HMAC). Available online: http://nvlpubs.nist.gov/nistpubs/fips/nist.fips.198-1.png (accessed on 20 April 2013).
22. Eiroa, S.; Baturone, I. FPGA implementation and DPA resistance analysis of a lightweight HMAC construction based on PHOTON hash family. In Proceedings of the 23rd International Conference on Field programmable Logic and Applications, Porto, Portugal, 2–4 September 2013; pp. 1–4.

23. Wu, H.; Preneel, B. AEGIS: A Fast Authenticated Encryption Algorithm. In Proceedings of the Selected Areas in Cryptography—SAC 2013: 20th International Conference, Burnaby, BC, Canada, 14–16 August 2013; pp. 185–201.

24. CAESAR: Competition for Authenticated Encryption: Security, Applicability, and Robustness. Available online: https://competitions.cr.yp.to/caesar.html (accessed on 26 April 2014).

25. Guajardo, J.; Kumar, S.S.; Schrijen, G.; Tuyls, P. FPGA Intrinsic PUFs and Their Use for IP Protection. In *Cryptographic Hardware and Embedded Systems-CHES 2007*; Paillier, P., Verbauwhede, I., Eds.; Springer: Berlin, Heidelberg, 2007; pp. 63–80.

26. Fletcher, R. *Practical Methods of Optimization*; John Wiley & Sons: Hoboken, NJ, USA, 2000.

27. Baturone, I.; Prada-Delgado, M.A.; Eiroa, S. Improved generation of identifiers, secret keys, and random numbers From SRAMs. *IEEE Trans. Inf. Forensics Secur.* **2015**, *10*, 2653–2668.

28. Gebara, F.; Kim, J.; Schaub, J.; Strumpen, V. Temperature-Profiled Device Fingerprint Generation and Authentication from Power-Up States of Static Cells. US Patent 8,495,431, 23 July 2013.

29. Cortez, M.; Hamdioui, S.; Kaichouhi, A.; van der Leest, V.; Maes, R.; Schrijen, G.J. Intelligent Voltage Ramp-Up Time Adaptation for Temperature Noise Reduction on Memory-Based PUF Systems. *IEEE Trans. Comput. Aided Des. Integr. Circuits Syst.* **2015**, *34*, 1162–1175.

30. Brox, P.; Martínez-Rodríguez, M.C.; Tena-Sánchez, E.; Baturone, I.; Acosta, A.J. Application specific integrated circuit solution for multi-input multi-output piecewise-affine functions. *Int. J. Circuit Theory Appl.* **2016**, *44*, 4–20.

31. Lutz, M.; Golderer, W.; Gerstenmeier, J.; Marek, J.; Maihofer, B.; Mahler, S.; Munzel, H.; Bischof, U. A precision yaw rate sensor in silicon micromachining. In Proceedings of the International Conference on Solid State Sensors and Actuators, TRANSDUCERS '97 Chicago, Chicago, IL, USA, 19 June 1997; Volume 2, pp. 847–850.

32. Kim, D.; Park, Y.; Lee, H. Sensor offset compensation for a vehicle yaw rate sensor using fuzzy logic. In Proceedings of the International Conference on Control, Automation and Systems, Seoul, Korea, 17–20 October 2007; pp. 362–366.

*Article*

# High Frequency Data Acquisition System for Modelling the Impact of Visitors on the Thermo-Hygrometric Conditions of Archaeological Sites: A Casa di Diana (Ostia Antica, Italy) Case Study

**Paloma Merello [1], Fernando-Juan García-Diego [2,\*], Pedro Beltrán [2] and Claudia Scatigno [3]**

[1]  Department of Accounting, University of Valencia, Av. de los Naranjos s/n, 46071 Valencia, Spain; paloma.merello@uv.es

[2]  Department of Applied Physics, Universitat Politècnica de València, Av. de los Naranjos s/n, 46022 Valencia, Spain; pbeltran@fis.upv.es

[3]  Nanoscience & Nanotechnology & Innovative Instrumentation (NAST Centre), University of Rome "Tor Vergata", Via della Ricerca Scientifica 1, 00133 Rome, Italy; claudia.scatigno@uniroma1.it

\*  Correspondence: fjgarcid@upv.es; Tel.: +34-610-633-671

Received: 2 January 2018; Accepted: 21 January 2018; Published: 25 January 2018

**Abstract:** The characterization of the microclimatic conditions is fundamental for the preventive conservation of archaeological sites. In this context, the identification of the factors that influence the thermo-hygrometric equilibrium is key to determine the causes of cultural heritage deterioration. In this work, a characterization of the thermo-hygrometric conditions of Casa di Diana (Ostia Antica, Italy) is carried out analyzing the data of temperature and relative humidity recorded by a system of sensors with high monitoring frequency. Sensors are installed in parallel, calibrated and synchronized with a microcontroller. A data set of 793,620 data, arranged in a matrix with 66,135 rows and 12 columns, was used. Furthermore, the influence of human impact (visitors) is evaluated through a multiple linear regression model and a logistic regression model. The visitors do not affect the environmental humidity as it is very high and constant all the year. The results show a significant influence of the visitors in the upset of the thermal balance. When a tourist guide takes place, the probability that the hourly temperature variation reaches values higher than its monthly average is 10.64 times higher than it remains equal or less to its monthly average. The analysis of the regression residuals shows the influence of outdoor climatic variables in the thermal balance, such as solar radiation or ventilation.

**Keywords:** visitors; temperature; relative humidity; thermo-hygrometric balance; continuous monitoring; preventive conservation

## 1. Introduction

For the last decades, preventive conservation of cultural and archaeological sites has been understood as the whole control process of the deterioration factors in order to prevent damage to the cultural heritage (CH) before it occurs and minimize future interventions [1]. It is acknowledged as important for safeguarding CH, both in terms of preservation and reducing the costs of future conservation measures [2].

The deterioration process is determined by factors such as the petrographic and chemical characteristics of the materials, the presence of mineral salts and organic substances on the surfaces, air pollution, sunlight, temperature, water content of the surface, etc. [3]. Therefore, preventive conservation requires knowledge of the specific characteristics of a specific CH site (materials, mineralogical conditions, natural ventilation, tourist campaigns, etc.), as well as the set of parameters connected to it, including microclimatic conditions (temperature, humidity, wind, air conditioning, etc.) [4], among others.

As thermo-hygrometric parameters affect chemical reactions, they appear as one of the most influential factors of the conservation of a CH site [5]. Microclimatic monitoring studies have been conducted in churches [6–8] museums [9–12] and archaeological sites [13–16].

CH has survived for many centuries in conditions that must be considered risky but stable (equilibrium), without avoiding damage to the materials, but resulting from a long adaptation process. Thus, abrupt changes of microclimate parameters interrupting this equilibrium conditions induce further damage to material until a new equilibrium is reached [17].

Therefore, it is important to characterize an archaeological site prior to carrying out comparative studies in the future for preventive conservation, either by regular studies to verify whether the conditions are constant or occasional ones, when the boundary conditions are altered.

In this vein, the presence of visitors could cause large deviations from the usual conditions [18] as humans can alter with their presence or behavior the aforementioned parameters, affecting the conservation of sites with interest in preservation. In addition, many of these variables are normally controlled for the comfort of visitors, starting then an interesting debate between the prevalence of preventive conservation and human comfort.

A big problem arises when the microclimate has been planned only for the wellbeing of visitors disregarding the needs of conservation that requires a constant climate [18]. Hopefully this has improved in recent decades and the conservation of heritage sites has been prioritized. Therefore, the management and planning for preserving historical buildings is strictly connected to their use, mainly for tourism.

There are many works that study the impact of visitors in tourist sites with protection interests such as Geoparks and other geological sites [19–22], museums [18] and other cultural and archaeological sites [16,17,23,24].

Casa di Diana (Ostia Antica, Italy) is a complex building, comprised of several rooms, including a *Mithraeum* (a place dedicated to the cult of the Persian god Mithra during Roman times). Its microclimate behaves like a *hypogeum* (an underground room simulating a cave, used for mithraic rites—as initiations), despite being structurally comparable to a semi-confined environment and being defined as such [15,16]. Casa di Diana is accessible to guided tours with advance booking.

In the case of hypogea environments, characterized by a great stability (constant humidity and temperature) [25] and the abundance of nutrients providing a suitable niche for phototrophic microorganisms when combined with artificial illumination [15,24] visitors can be considered as a possible factor breaking that equilibrium state.

The results of [24] in the *hypogeum* of San Callixtus Catacombs (Rome) indicate that a sharp temperature increase was due to the automatic ignition of lamps when people passed along the corridors, as well as other effects on air temperature due to human body heat and in $CO_2$ due to breathing.

A monitoring campaign of indoor $CO_2$ concentration was conducted in Casa di Diana with the aim of evaluate its effects on biological colonization [26,27], determining that the growth of vegetation is concentrated in the lowest area of the *podiae* (seats for the Romans that assisted the sacrificial ritual of killing the bull), which also corresponds to the path followed by visitors.

The CH is conserved in order to be disclosed, contemplated and admired by the visitors being these who maintain it but, nevertheless, they can contribute to its deterioration. It is necessary to reach a consensus of disclosure of our CH in a sustainable manner. This paper deals with a mathematical modeling in order to quantify the influence that visitors can produce in our CH. Considering the thermo-hygrometric conditions of Casa de Diana as a case study.

A monitoring system, able to record and store large amounts of data (monitoring frequency of 1 data point per minute) was installed [11,28]. The system was comprised of 29 probes, including temperature and relative humidity sensors installed in parallel, calibrated and synchronized with a microcontroller.

As far as we know, this is the first time that a quantitative study is carried out with the aim of modelling the influence of visitors on the upset of the thermo-hygrometric equilibrium of an archaeological site. The methodology proposed in this paper may serve as a reference for other CH sites.

The paper is organized as follows: after the Introduction, a description of the *Mithraeum* of Casa di Diana, the monitoring system, a description of the sample and the analysis procedure are presented in Section 2. Section 3 deals with the mathematical modelling of the impact of the visitors in the upset of the thermal balance. Finally, the Conclusions section summarizes the main remarks.

## 2. Materials and Methods

### 2.1. The Archaeological Site: "Casa di Diana" Mithraeum

The Casa di Diana Region I, Insula III, is a Roman building (130 CE) part of the famous archaeological site of Ostia Antica, located at 23 km from the center of Rome. The building, comprised of *tabernae* (a single room shop covered by a barrel vault within great indoor markets of ancient Rome; normally with overhead lighting openings) and *cenacula* (dining rooms), presents a very characteristic microclimate, especially inside two intercommunicating rooms (the *Mithraeum* and *pre-Mithraeum*). The principal building materials are bricks and pozzolanic mortar aligned with the *opus caementicium* (Roman concrete, based on a hydraulic-setting cement, was a material used in construction during the late Roman Republic until the fading of the Roman Empire) technique [25,26]. The two rooms are characterized by different heights due to the presence at the sides of *podium* [26].

The ventilation is natural and comes from several openings. No mechanical ventilation systems are installed. The rooms are covered by a roof and protected against the rain, but small areas are directly exposed to sunlight and outer air due to the openings [16].

The Casa di Diana is visited, one or two days a week, normally at 10:30 a.m., with a mean duration of about 30 min (only 10 min in the *Mithraeum*'s rooms).

### 2.2. Monitoring System

The sensors consist of probes that contain an 8-pin small-outline integrated circuit (SOIC), model DS2438 [29] (Maxim Integrated Products, Inc., Sunnyvale, CA, USA) that incorporates a temperature (T) sensor with an accuracy of $\pm 2$ °C as well as an analogue-to-digital voltage (A/D) converter which measures the output voltage of a relative humidity (RH) sensor (HIH-4000 [30], Honeywell International, Inc., Minneapolis, MN, USA). The sensor has repeatability between measures allowing to detect small variations [11,14]. Three electric wires come out from each probe: one wire for +5 V DC power supply, one for ground and another for data transfer.

Because each DS2438 [29] has a unique silicon serial number, multiple DS2438s [29] can exist on the same data bus. This allows multiple sensors that can be used in the system simultaneously with only one data line (1-wire communication protocol). The sensors, with this communication protocol, are installed in parallel between the data line and ground, allowing a simple and robust wiring.

The HIH-4000 [30] RH sensors, with an accuracy of $\pm 3.5\%$, were calibrated in the laboratory with a saturated solution of salt [11,14]. The voltage output of the HIH-4000 [30] sensors is measured by one of the four analogs to digital A/D converters of the DS2438 [29]. As specified by the manufacturer, the voltage output is proportional to voltage supply. The exact value of the voltage supply is measured for each probe by another A/D of the DS2438 [29] and applied in the calibration curves each time a humidity data is measured. Sensors are synchronized and their data recorded by an ATMega328 microcontroller (Microchip Technology Inc., Chandler, AZ, USA) as described in [11].

### 2.3. Sensors Location

A total of 29 probes were installed, 28 in the interior of the *Mithraeum* and *pre-Mithraeum* of Casa di Diana (Figure 1a), and one additional probe placed on the sill of a window as an outdoor climate control (Figure 1b). The coordinate locations (*xyz*) for sensors along the walls are described in Table 1, as it has been drawn on wall B (Figure 1, fuchsia). The origin started at the left corner of the wall and at the pedestrian height. The coordinate *x* indicates the horizontal distance to the origin, coordinate *y* indicates the vertical distance to the origin, coordinate *z* indicates the distance to the walls surface only for two sensors (#72, #73) that are not placed along the wall.

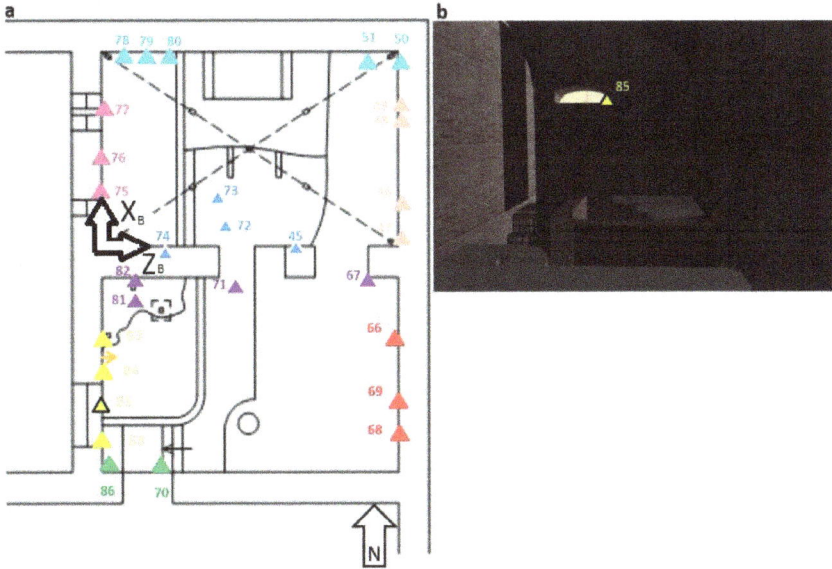

**Figure 1.** (**a**) Sensor locations (numbered triangles); (**b**) Sensor #85 (yellow triangle), outdoor climate control. Each sensor has a different color due to wall belonging. Each wall has a reference system (*xyz*) to individuate the sensor's position (coordinate allocation is expressed in Table 1). Legend: pink color corresponds to wall A, fuchsia color corresponds to wall B, azure corresponds to wall C, blue corresponds to wall D, yellow corresponds to wall E, red corresponds to wall F, green corresponds to wall G and violet corresponds to wall H.

**Table 1.** Sensors description and coordinated allocation.

| Reference System | Sensors N. | *x* (cm) | *y* (cm) | *z* (cm) |
|---|---|---|---|---|
| | 45 | 70 | 102 | 0 |
| BLUE | 72 | 424 | 0 | 94 |
| | 73 | 400 | 80 | 116 |
| | 74 | 512 | 140 | 0 |
| | 46 | 372 | 192 | 0 |
| PINK | 47 | 372 | 89 | 0 |
| | 48 | 197 | 142 | 0 |
| | 49 | 196 | 304 | 0 |
| | 75 | 126 | 163 | 0 |
| FUCHSIA | 76 | 176 | 302 | 0 |
| | 77 | 294 | 93 | 0 |

Table 1. *Cont.*

| Reference System | Sensors N. | $x$ (cm) | $y$ (cm) | $z$ (cm) |
|---|---|---|---|---|
| AZURE | 50 | 504 | 146 | 0 |
| | 51 | 425 | 201 | 0 |
| | 78 | 88 | 139 | 0 |
| | 79 | 174 | 148 | 0 |
| | 80 | 190 | 197 | 0 |
| GREEN | 70 | 226 | 217 | 0 |
| | 86 | 560 | 136 | 0 |
| YELLOW | 83 | 330 | 90 | 0 |
| | 84 | 245 | 120 | 0 |
| | 85 | 164 | 204 | 0 |
| | 88 | 32 | 169 | 0 |
| RED | 66 | 7 | 70 | 0 |
| | 68 | 242 | 154 | 0 |
| | 69 | 205 | 280 | 0 |
| VIOLET | 67 | 532 | 164 | 0 |
| | 71 | 290 | 106 | 0 |
| | 81 | 84 | −12 | 0 |
| | 82 | 86 | 80 | 0 |

*2.4. Database*

The continuous monitoring campaign system was started on 29 June 2014 and finished on 21 May 2015. The records of March 2015 were completely missing. In order to avoid possible abnormal measurements, data after 31 January 2015 were disregarded for the analysis.

The system records data with a frequency of 60 data points per hour (1 data per minute) being capable of recording 43.200 data points per month (30 days × 24 h/day × 60 data points/h). In this study we have a data base of 23,302,080 data, composing a data matrix of 401.760 rows (279 days × 24 h × 60 min) and 58 columns (29 sensors T + 29 sensors RH). Note that, sensors T#81-88 did not work during the month of August. On the other hand, sensor RH#77 was completely discarded due to instrument anomalies, during the entire data recording. Data were stored in Burrito software [31].

*2.5. Data Analysis*

The methodological procedure of the paper is as follows. First, an exploratory analysis of the thermo-hygrometric data is carried out to characterize the archaeological site. The starting hypothesis is coincident with [17], where events that contribute to upsetting the microclimatic balance considered the main source of damage. After, we select a subsample of sensors that present the ideal characteristics to model the hourly variation.

With this subsample, a multiple linear regression analysis and a logistic regression analysis are performed to assess the possible influence of the visitors in the upset of the microclimatic equilibrium.

Principal Component Analysis (PCA) was performed using Unscrambler version 9.7 (a chemometric software package from Camo, Woodbridge, NJ, USA), with a cross validation method. Other analyses were performed with the software packages OriginPro 8.5.1 (OriginLab, Northampton, MA, USA) and SPSS 22 (IBM, North Castle, NY, USA).

2.5.1. Dimension Reduction Techniques: PCA and Correspondence Analysis

PCA is a dimension reduction technique. It is an exploratory method whose objective is to summarize a large amount of data in a small number of dimensions, with the least possible loss of information. PCA looks for the projection according to which the data are best represented in terms of least squares, converting a set of observations of possibly correlated variables in a set of variable values without linear correlation called principal components.

This linear transformation is constructed through the covariance matrix or matrix of correlation coefficients. Due to the symmetry of this matrix, there is a complete set of eigenvectors and eigenvalues that lead to the transformation of the old coordinates into the coordinates of the new base.

The Correspondence Analysis is a statistical technique that is used to analyze, from a graphical point of view, the dependency relations of a set of categorical variables from data of a contingency table. Its objective is similar to that of PCA, but applied on categorical variables. Inertial matrices for columns and rows are calculated and diagonalized, obtaining their eigenvalues and eigenvectors, defining a space formed by the dimensions and the projection of the categories on it.

### 2.5.2. Multiple Linear Regression and Logistic Regression Models

The multiple linear regression considers that the values of the dependent variable ($Y$) have been generated by a linear combination of the values of $k$ explanatory (independent) variables ($x$) and a random term (error or residual, $u$):

$$Y = \beta_0 + \beta_1 x_1 + \beta_2 x_2 + \ldots + \beta_k x_k + u \tag{1}$$

The coefficients are estimated minimizing the residual variance. Balance is understood as the stability of microclimatic conditions. The hourly variation of temperature represents the deviation from the balance conditions and, therefore, possible source of deterioration [17]. Thus, in our study, the hourly variation of temperature is selected as the variable of interest.

The independent variables considered in the linear regression analysis are the continuous variable *visitors* (taking values from 0 to 60), and the dichotomous variable *visit*, which takes the value 1 on the day and time that a guided visit takes place in Casa di Diana. In addition, the control variable considered in the model are the sensor variable, the dummy variables month (*July, September, October, November, December*, taking January as a reference), the difference between the outdoors temperature and the monthly moving average of the outdoors temperature (*EXT85_MA*) and the dichotomous variable *EXT85d* which is equal to 1 when the hourly variation of the outdoors sensor exceeds the mean hourly variation of the outdoors sensors for that month, considering a moving average.

According to the characterization of Casa di Diana, a subsample of sensors of temperature is selected, since it is intended to observe variations in the temperature that in principle are not derived from the architectural characteristics of the place (see Sections 3.1 and 3.2). Finally, for the linear regression analysis a sample of 793,620 data, arranged in a matrix with 66,135 rows and 12 columns, is considered.

Note, that the models are replicated for the case in which the variable *visitors* are considered as quantitative variable. As worse adjustment results are obtained only the final model are presented in the section of results. The interaction of the variables *month* and *visit*, and the triple interaction of the variables *month, visit* and *cluster* are also considered. However, the adjustment of the models is not modified and, according to the parsimony principle, the simplest model is selected. On the other hand, logistic regression models the outcome of a categorical variable as a function of the independent variables. The logistic regression analysis is framed within the set of Generalized Linear Models (GLM) that uses the logit function as a link function. The model takes the following form:

$$p = \frac{1}{1 + e^{-(\beta_0 + \beta_1 x_1 + \beta_2 x_2 + \ldots + \beta_k x_k + u)}} \tag{2}$$

The odds ratio is a measure of how much higher/lower is the probability that an event will happen than the probability that it will not occur. It is calculated as:

$$\frac{p}{1 - p} = e^{\beta_0 + \beta_1 x_1 + \beta_2 x_2 + \ldots + \beta_k x_k + u}. \tag{3}$$

In Section 3.3, a logistic regression mathematical model is developed to determine the effect of visitors on the probability that the hourly variation represents a deviation from the equilibrium conditions; that is, when it presents a value higher than the monthly average value. The dummy variable *HV_largerMHV* (hourly variation larger than mean hourly variation) is defined. *HV_largerMHV* will take the value 1 when the hourly variation is larger than the monthly average of the hourly variation of the sensor, considering a monthly moving average, and the value 0 otherwise. It is understood that hourly variations larger than the monthly average hourly variation are likely to be considered upset of thermal balance.

## 3. Results and Discussion

### 3.1. Characterization of "Casa di Diana"

A PCA is performed on the original data in order to characterize the archaeological site. The dataset was previously auto scaled and a model with two principal components (T), explaining the 100% of the total variance (Figure 2a), and two principal components (RH), explaining the 89% of the total variance (Figure 2b), is selected to fit the data.

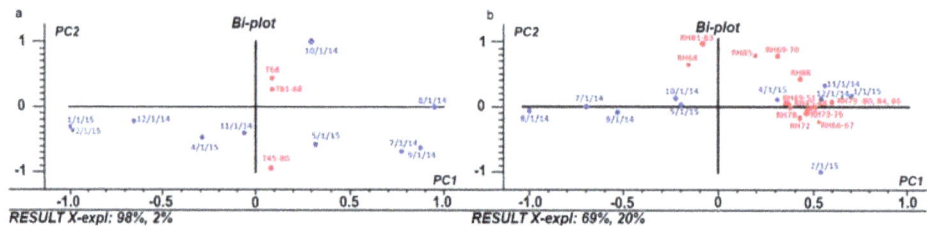

**Figure 2.** PCA Bi-plot (scores and loadings). (**a**) Tempera–ture PC1-PC2; (**b**) Relative Humidity PC1-PC2. Legend: blue color (scores), red color (loadings), *X-expl* refers to the percentage of variability of each sensor (*X*) explained by components 1 (PC1) and 2 (PC2), respectively.

Regarding temperature (Figure 2a), two clear groups are identified. One group is comprised by 8 sensors (#68, #81–#88) that were situated between the south west *pre-Mithraeum* wall and the north *pre-Mithraeum* wall (Figure 1a—yellow and orange triangles). This is the more ventilated area due to the window and main entrance. All the rest of the sensors (#45–#80) follow the same behavior. There is not found a particular relationship between wall and season (except for October, that seems to have an own trend). Regarding RH (Figure 2b), more data variability is found and two groups are identified. One group is composed by sensors with different behaviors and located in a more ventilated area. The other group comprises sensors located along the more protected walls, away from openings, windows and frontal walls, where the outdoor exchanges are frequent, and present very similar behavior between them.

In order to understand the thermo-hygrometric behavior during the year, Multi-Curve plots for the daily mean temperature (Tday$_m$) and RH (RHday$_m$) are represented for each sensor (Figure 3). In general, Tday$_m$ remains stable between July and at the end of October (26.6–20.7 °C). The higher amplitude is present during the winter season (17.7–3 °C, Figure 3a). Regarding RHday$_m$, it remains stable along the entire monitoring campaign with values between 99–84% (Figure 3b). Note that during the autumn-winter, the RH values are closed to saturation (97–99%) (Figure 3b).

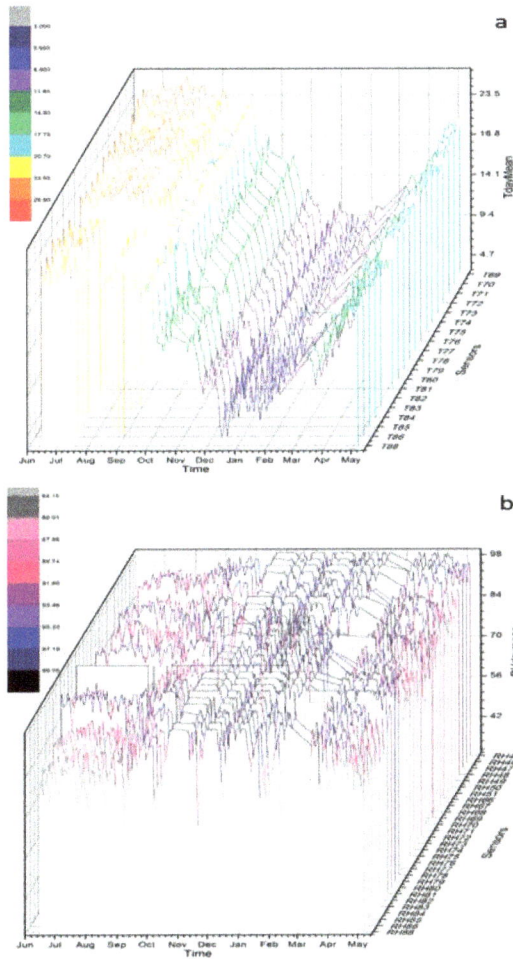

**Figure 3.** Multi-Curve Plot. (**a**) Tday$_m$ vs. time; (**b**) RHday$_m$ vs. time. Sensor locations are indicated in Figure 1.

Taking into account the walls behavior, the sensors are grouped by wall belonging and the maximum and minimum daily values for T and RH are calculated (Tmax/Tmin and RHmax/RHmin, respectively).

Regarding Tmax (Figure 4a), it remains stable along the walls. Sensors in wall "e" seem to have less variability than sensors on the other walls. RHmax has more variability; sensors in walls "a, c, d, f, h" have less data variability than sensors in walls "b, e, g". The difference between those two groups is in accordance with the presence or absence of air flow, as the first group (more data variability) was directly exposed to air flow (Figure 4c).

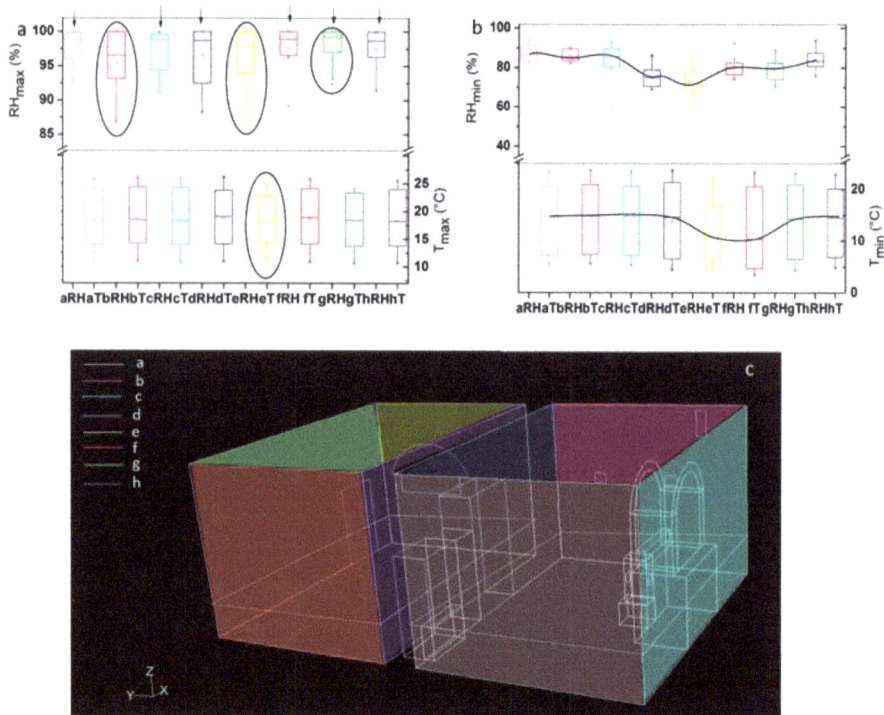

**Figure 4.** (**a**) Box & Whisker Tmax and RHmax; (**b**) Box & Whisker Tmin and RHmin; (**c**) 3D representation. The colors are according with the wall belonging.

The minimum values of T and RH corroborate the main results addressed by the maximum values. In addition to the previously considered factors, high levels of RH (rising damp, presence of probably seeping water) and wall temperature lower than air temperature induce surface condensation. Temperature sensors installed in previous monitoring campaigns and different materials indicated that the temperature of rock and mortars was always lower than air temperature, subsequently inducing water condensation on the surfaces of the material surfaces, as well as efflorescence and sub-efflorescence [15,16].

### 3.2. Modelling the Impact of Visitors on the Hourly Variation

According to the characterization, the data of the sensors #45–#80, excluding sensor #68, are taken, since it is intended to observe variations in the temperature that in principle are not derived from the architectural characteristics of the place. The HR data has been disregarded since it has been very constant during the whole monitoring campaign. Furthermore, the hours when there was no data for the outdoors sensor (#85) have been eliminated from the sample.

The hourly variation of temperature is selected as dependent variable. The independent variables considered in the analysis are variable visitors, visit, month (*July, September, October, November, December*), EXT85_MA and the dichotomous variable *EXT85d*.

A first stepwise regression analysis results in an adjusted $R^2$ of 39.2%. The model is significant with *p*-value < 0.001 and the coefficients of the significant variables appear in Table 2. The visits increase the hourly variation, with an aggravating effect when the outdoors hourly variation is also higher than its expected monthly average. Note that when a visit takes place the hourly variation increases by 0.106 °C; when the hourly variation is also higher than its expected monthly average and a visit takes place, it results in an increase of 0.17 °C (0.106 + 0.064).

**Table 2.** Estimated coefficients of the independent variables (first multiple linear regression model).

| Parameter | Coefficient | Significance | Parameter | Coefficient | Significance |
|---|---|---|---|---|---|
| *Constant* | −0.084 | 0.000 | *June* | −0.042 | 0.000 |
| *EXT85d* | 0.217 | 0.000 | *July* | −0.023 | 0.000 |
| *EXT85_MA* | 0.001 | 0.003 | *September* | −0.006 | 0.003 |
| *VISIT* | 0.106 | 0.000 | *December* | −0.017 | 0.000 |
| *EXT85d * VISIT* | 0.064 | 0.000 | | | |

It can be deduced from the analysis of the residuals that there are four clearly identifiable data clusters with a different regression line (Figure 5). A cluster analysis of the residuals is performed and the conglomerates (hereinafter, clusters of residuals) have the characteristics specified in Table 3.

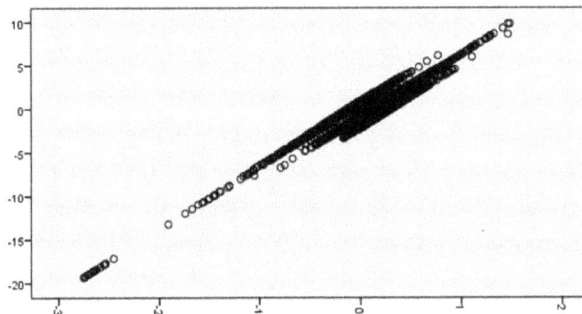

**Figure 5.** Scatter plot of the standardized residuals (vertical axis) according to the dependent variable (horizontal axis).

**Table 3.** Characteristics and compositions of the clusters of residuals.

| Cluster | Number of Cases | % Cases | Center |
|---|---|---|---|
| 1 | 19,349 | 29.26 | −0.13525 |
| 2 | 38 | 0.06 | −2.00502 |
| 3 | 5,863 | 8.87 | −0.2594 |
| 4 | 40,885 | 61.82 | 0.02867 |
| Total | 66,135 | 100.00 | |

A priori it seems reasonable that these groups of observations with a particular relationship with the dependent variable may correspond to sets of sensors with common temperature characteristics (average and variability) caused by their placement in the archaeological site, despite the fact that the sample consists of the sensors with more similar behavior. However, a correspondence analysis for the variable cluster of residuals and the categorical variable sensor does not seem to reveal any direct relationship between the categories of both variables. In addition, to analyze this possibility, a regression analysis is performed, including as categorical variables the groups of sensors that seem to have a similar behavior according to the perceptual map of correspondence analysis. None of these variables is significant, so the influence of the sensor variable is discarded.

Therefore, it can be deduced that the clusters of residuals are related to specific observations (independent of the sensor) with the influence of variables not considered in the model, such as ventilation or solar radiation, for example.

The aim of the work is to evaluate the influence of visits on the upset of the thermal equilibrium of the archaeological site, so it was decided to carry out a second regression model including the clusters of residuals that allows evaluating the differences between these subgroups. Future research will be necessary to characterize these clusters so that prevention measures can be designed in the most appropriate way. A Computational Fluid Dynamics study is being developed. Air sensors have been installed and, in the future, the recorded data will allow improving the evaluation of the model residuals [32].

With the introduction of three dummy variables that allow modeling the four clusters of residuals, and all their interactions with the rest of the independent variables, a model with an adjusted $R^2$ of 85.4% is obtained, which represents a great improvement in the predictive capacity with respect to the first model. The model is significant with $p$-value $< 0.001$ and the coefficients of the variables introduced in the final model appear in Table 4.

**Table 4.** Estimated coefficients of the independent variables (second multiple linear regression model).

| Parameter | Coefficient | Significance | Parameter | Coefficient | Significance |
|---|---|---|---|---|---|
| Constant | −0.214 | 0.000 | cl4 | 0.161 | 0.000 |
| EXT85-Mont_av | 0.002 | 0.000 | cl3 * VISIT | 0.075 | 0.000 |
| EXT85d | 0.217 | 0.000 | cl4 * VISIT | 0.066 | 0.000 |
| VISIT | 0.031 | 0.000 | cl3 * EXT85d | 0.031 | 0.000 |
| EXT85d * VISIT | 0.087 | 0.000 | cl2 * November | −1.155 | 0.000 |
| June | −0.045 | 0.000 | cl3 * June | −0.088 | 0.000 |
| July | −0.036 | 0.000 | cl3 * July | −0.044 | 0.000 |
| September | −0.011 | 0.000 | cl3 * September | −0.068 | 0.000 |
| October | −0.01 | 0.000 | cl3 * October | −0.036 | 0.000 |
| December | −0.016 | 0.000 | cl3 * December | −0.069 | 0.000 |
| cl2 | −1.232 | 0.000 | cl4 * July | 0.009 | 0.000 |
| cl3 | 0.407 | 0.000 | cl4 * September | −0.005 | 0.027 |

The hourly variation depends on the month as stated by the parameters of the variables *July, September, October, November, December*. As these parameters are negative, *January* is the month historically presenting the larger hourly variations in temperature. Depending on the cluster, the effect of the month is increased or reduced.

The visits always increase the hourly variation, with an aggravating effect when the outdoors hourly variation is also higher than its expected monthly average. Note that when a visit takes place the hourly variation increases by 0.031 °C for observations belonging to cluster 1; when the hourly variation is also higher than its expected monthly average and a visit takes place, it results in an increase of 0.118 °C for cluster 1. Furthermore, observations of clusters 3 and 4 present a greater sensitivity to visits. When a visit takes place, the hourly variation increases of 0.106 °C and 0.097 °C for observations belonging to cluster 3 and 4, respectively. This significant effect of visits on the thermal upset is in line with the findings of the study of $CO_2$ [26,27].

An analysis of the residuals of the model (Figure 6) shows that new clusters appear. Despite this, we have been able to identify the influence of visitors, which is the main objective of the paper. An individualized analysis per sensor has discarded the possibility of autocorrelation of residuals. In addition, abnormal residuals are present in all sensors, which surely must coincide with some of the identified clusters reinforcing the idea that these clusters are related to some external variables not considered in the model. Sensor #69 has the largest residuals variance (Figure 7).

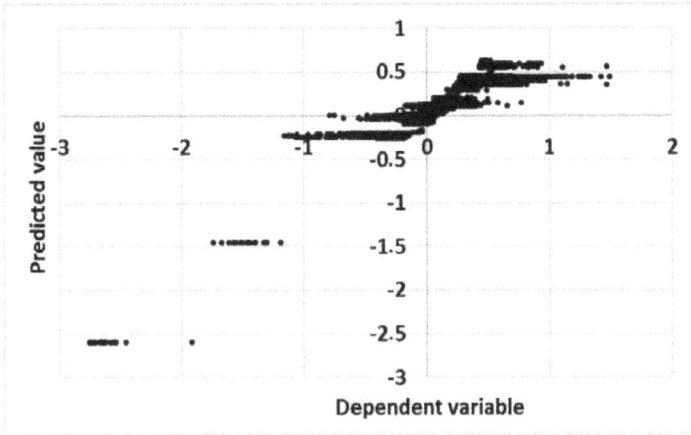

**Figure 6.** Predicted value versus dependent variable of the model.

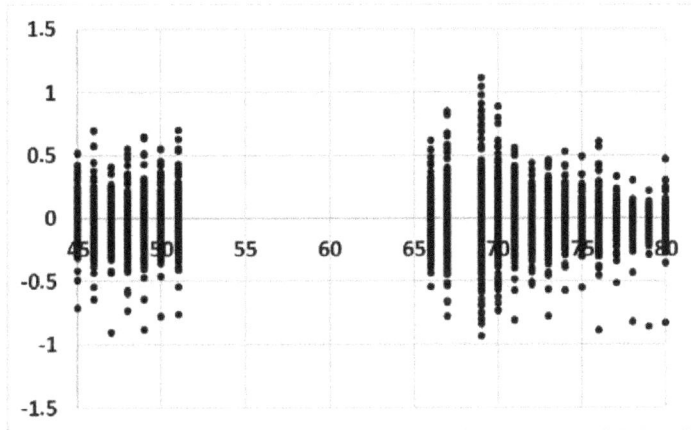

**Figure 7.** Residuals of the model (vertical axis) depending on the sensor (horizontal axis).

*3.3. Modelling the Impact of Visitors on the Upset of the Thermal Balance*

In this section, a logistic regression mathematical model is developed to determine the effect of visitors on the probability that the hourly variation represents a deviation from the equilibrium conditions. *HV_largerMHV* is the dependent variable; the independent variables considered in the analysis are the dichotomous variable visit, *EXT85_MA* and the dichotomous variable *EXT85d*.

The Hosmer Lemeshow test tests the null hypothesis that the fitted model is correct, therefore as *p*-value > 0.05 so there is lack of evidence against the null hypothesis. The model has a Cox and Snell's $R^2$ equal to 0.418 and a Nagelkerke's $R^2$ equal to 0.567. Variable *EXT85_MA* resulted non-significant. The model takes the following form:

$$odds = \frac{p}{1-p} = e^{-2.105+3.538EXT85d+2.635VISIT-0.569EXT85d\times VISIT} \tag{4}$$

Thus, when a visit takes place the probability that the hourly variation reaches values higher than its monthly average is 10.64 times the probability that the balance is maintained, when the rest of the variables remain constant.

The classification table (Table 5) shows the number of cases that have been classified (predicted) by the model as 0 or 1 and compares with the original classification (observed) for each case (observation). In the classification table (Table 5) we can verify that our model has a high specificity (87.4%) and a remarkable sensitivity (83.6%). T default limit point of the calculated $Y$ probability is set to 50% (0.5). It has been tested that, despite decreasing the limit point, the classification is maintained.

**Table 5.** Classification table of the logistic regression model.

| Observed | | Predicted | | |
|---|---|---|---|---|
| | | 0 | 1 | Percentage |
| HV_largerMHV | 0 | 35,078 | 5066 | 87.4 |
| | 1 | 4275 | 21,716 | 83.6 |
| total | | 39,353 | 26,782 | 85.9 |

Note that no improvement of the model is achieved when including the variable visitors as quantitative variable.

## 4. Conclusions

The innovative of the sensors technology—high memory of data storing (10 years and more) and high frequency of data recording (1 data/min)—allow us obtaining a big database and modelling, for the first time, the influence of visitors in the microclimate of an archaeological site.

A prior exploratory analysis of the data allowed to individuate the more humid walls, the walls more protected to air flow, as well as individuating differences into the single building (between the *Mithraeum* and *pre-Mithraeum*).

Exploratory analyzes have shown that Casa de Diana has a stable microclimate in terms of RH and that the largest variations occur in temperature and during autumn-winter. This paper is based on the idea that the main cause of damage in cultural heritage sites is the upset of the balance of microclimatic conditions. For the first time, the upset of equilibrium is mathematically modeled both in terms of hourly variations and in terms of the probability of significant changes.

The multiple linear regression model stated the influence of the visits on an increase in the hourly variation of the temperature. However, the residuals of the model determined the existence of observations with linearly differentiated behaviors. Therefore, more research is needed including other variables such as ventilation and solar radiation, with the purpose of identifying and characterizing the observations included in these clusters.

The logistic regression model assessed that when a visit takes place the probability of the upset of the thermal balance is 10.64 times larger than the probability that the balance is maintained.

Due to the high humidity of the archaeological site, the visitors did not influence this parameter; however, they affect the temperature despite having open windows to the outdoors and no mechanism of temperature control.

Note that the main goal of the study is proposing a monitoring system and a mathematical modelling methodology for the evaluation of the significant influence of visitors, more than the quantification of the damage, serving as a guide for other CH sites. In order to quantify the damages produced by the visitors, a second study without visitors should be carried out in the future.

**Acknowledgments:** The authors would like to thank the staff of the archaeological area of Ostia Antica for the permission to work in this house. This work is partially supported by the projects HAR2013-47895-C2-1-P and HAR2013-47895-C2-2-P from MINECO.

**Author Contributions:** C.S., P.M. and F.J.G.-D. conceived and designed the experiment; F.J.G.-D. and P.B. developed, calibrated and installed the monitoring system; C.S. and P.M. analyzed the data and wrote the paper.

**Conflicts of Interest:** The authors declare no conflict of interest.

# References

1. Camuffo, D. *Microclimate for Cultural Heritage: Conservation, Restoration, and Maintenance of Indoor and Outdoor Monuments*; Elsevier: Waltham, MA, USA, 2013; ISBN 9780444632968.
2. Kontozova-Deutsch, V.; Cardell, C.; Urosevic, M.; Ruiz-Agudo, E.; Deutsch, F.; Van Grieken, R. Characterization of indoor and outdoor atmospheric pollutants impacting architectural monuments: Th case of San Jerónimo Monastery (Granada, Spain). *Environ. Earth Sci.* **2011**, *63*, 1433–1445. [CrossRef]
3. Arnold, A.; Zehnder, K. Monitoring wall paintings affected by soluble salts. In *The Conservation of Wall Paintings*; Cather, S., Ed.; Getty Conservation Institute: London, UK, 1987; pp. 1–36. ISBN 0-89236-162-X.
4. La Gennusa, M.; Rizzo, G.; Scaccianoce, G.; Nicoletti, F. Control of indoor environments in heritage buildings: Experimental measurements in an old Italian museum and proposal of a methodology. *J. Cult. Herit.* **2005**, *6*, 147–155. [CrossRef]
5. Fernández-Navajas, Á.; Merello, P.; Beltrán, P.; García-Diego, F.J. Multivariate thermo-hygrometric characterisation of the archaeological site of plaza de l'Almoina (Valencia, Spain) for preventive conservation. *Sensors* **2013**, *13*, 9729–9746. [CrossRef] [PubMed]
6. Camuffo, D.; Bernardi, A.; Sturaro, G.; Valentino, A. The microclimate inside the Pollaiolo and Botticelli rooms in the Uffizi Gallery, Florence. *J. Cult. Herit.* **2002**, *3*, 155–161. [CrossRef]
7. Corgnati, S.P.; Filippi, M. Assessment of thermo-hygrometric quality in museums: Method and in-field application to the "Duccio di Buoninsegna" exhibition at Santa Maria della Scala (Siena, Italy). *J. Cult. Herit.* **2010**, *11*, 345–349. [CrossRef]
8. Legnér, M. On the Early History of Museum Environment Control—Nationalmuseum and Gripsholm Castle in Sweden, c.1866–1932. *Stud. Conserv.* **2011**, *56*, 125–137.
9. Camuffo, D.; Sturaro, G.; Valentino, A. Thermodynamic exchanges between the external boundary layer and the indoor microclimate at the Basilica of Santa Maria Maggiore, Rome, Italy: The problem of conservation of ancient works of art. *Bound.-Layer Meteorol.* **1999**, *92*, 243–262.
10. Tabunshikov, Y.; Brodatch, M. Indoor air climate requirements for Russian churches and cathedrals. *Indoor Air* **2004**, *14*, 168–174. [CrossRef] [PubMed]
11. García-Diego, F.J.; Zarzo, M. Microclimate monitoring by multivariate statistical control: The renaissance frescoes of the Cathedral of Valencia (Spain). *J. Cult. Herit.* **2010**, *11*, 339–344. [CrossRef]
12. Zarzo, M.; Fernández-Navajas, A.; García-Diego, F.J. Long-term monitoring of fresco paintings in the Cathedral of Valencia (Spain) through humidity and temperature sensors in various locations for preventive conservation. *Sensors* **2011**, *11*, 8685–8710. [CrossRef] [PubMed]
13. Maekawa, S.; Lambert, F.; Meyer, J. Environmental monitoring at Tiwanaku. *MRS Online Proc. Libr. Arch.* **1995**, *352*, 885–892. [CrossRef]
14. Merello, P.; García-Diego, F.J.; Zarzo, M. Microclimate monitoring of Ariadne's house (Pompeii, Italy) for preventive conservation of fresco paintings. *Chem. Cent. J.* **2012**, *6*, 145. [CrossRef] [PubMed]
15. Scatigno, C.; Gaudenzi, S.; Sammartino, M.P.; Visco, G. A microclimate study on hypogea environments of ancient roman building. *Sci. Total Environ.* **2016**, *566*, 298–305. [CrossRef] [PubMed]
16. Scatigno, C.; Prieto-Taboada, N.; Martinez, M.P.; Conte, A.M.; García-Diego, F.J.; Madariaga, J.M. Analitycal techniques for the characterisation of historical building materials: Case study "Casa di Diana" Mithraeum (Archeological site in Ostia Antica, Italy). In *Advances in Materials Science Research*; Nova Science Publishers, Inc.: New York, NY, USA, 2016; p. 31.
17. Visco, G.; Plattner, S.H.; Fortini, P.; Sammartino, M. A multivariate approach for a comparison of big data matrices. Case study: Thermo-hygrometric monitoring inside the Carcer Tullianum (Rome) in the absence and in the presence of visitors. *Environ. Sci. Pollut. Res.* **2017**, *24*, 13990–14004.
18. Camuffo, D.; Van Grieken, R.; Busse, H.J.; Sturaro, G.; Valentino, A.; Bernardi, A.; Shooter, D.; Gysels, K.; Deutsch, F.; Wieser, M.; et al. Environmental monitoring in four European museums. *Atmos. Environ.* **2001**, *35*, S127–S140. [CrossRef]
19. Walpole, M.J.; Goodwin, H.J.; Ward, K.G. Pricing policy for tourism in protected areas: Lessons from Komodo National Park, Indonesia. *Biol. Conserv.* **2001**, *15*, 218–227. [CrossRef]
20. Wang, W.; Ma, Y.; Ma, X.; Wu, F.; Ma, X.; An, L.; Feng, H. Seasonal variations of airborne bacteria in the Mogao Grottoes, Dunhuang, China. *Int. Biodeter. Biodegr.* **2010**, *64*, 309–315. [CrossRef]

21. Guo, W.; Chung, S. Using Tourism Carrying Capacity to Strengthen UNESCO Global Geopark Management in Hong Kong. *Geoheritage* **2017**, 1–13. [CrossRef]

22. Buckley, R.; Zhong, L.; Ma, X. Visitors to protected areas in China. *Biol. Conserv.* **2017**, *209*, 83–88. [CrossRef]

23. García-Hernández, M.; de la Calle-Vaquero, M.; Yubero, C. Cultural heritage and urban tourism: Historic city centres under pressure. *Sustainability* **2017**, *9*, 1346. [CrossRef]

24. Sanchez-Moral, S.; Luque, L.; Cuezva, S.; Soler, V.; Benavente, D.; Laiz, L.; Gonzalez, J.M.; Saiz-Jimenez, C. Deterioration of building materials in Roman catacombs: The influence of visitors. *Sci. Total Environ.* **2005**, *349*, 260–276. [CrossRef] [PubMed]

25. Scatigno, C.; Prieto-Taboada, N.; García-Florentino, C.; Fdez-Ortiz de Vallejuelo, S.; Maguregui, M.; Madariaga, J.M. Combination of in situ specyroscopy and chemometric techniques to discriminate different types of Roman bricks and the influence of microclimate environment. *Environ. Sci. Pollut. Res.* **2017**, 1–15.

26. Scatigno, C.; Moricca, C.; Tortolini, C.; Favero, G. The influence of environmental parameters in the biocolonization of the Mithraeum in the roman masonry of casa di Diana (Ostia Antica, Italy). *Environ. Sci. Pollut. Res.* **2016**, *23*, 13403–13412. [CrossRef] [PubMed]

27. Cardarelli, E.; De Donno, G.; Oliveti, I.; Scatigno, C. Three-dimensional reconstruction of a masonry building through electrical and seismic tomography validated by biological analyses. *Near Surf. Geophys.* **2018**, 16. [CrossRef]

28. García-Diego, F.J.; Fernández-Navajas, Á.; Beltrán, P.; Merello, P. Study of the Effect of the Strategy of Heating on the Mudejar Church of Santa Maria in Ateca (Spain) for Preventive Conservation of the Altarpiece Surroundings. *Sensors* **2013**, *13*, 11407–11423. [CrossRef] [PubMed]

29. Honeywell. HIH-4000 Series Integrated Circuit Humidity Sensor Datasheet. Available online: http://www.farnell.com/datasheets/1685535.pdf (accessed on 15 December 2017).

30. Maxim Integrated Products. DS2438 Smart Battery Monitor Datasheet. Available online: https://datasheets.maximintegrated.com/en/ds/DS2438.pdf (accessed on 15 December 2017).

31. Fernández-Navajas, Á.; Merello, P.; Beltrán, P.; García-Diego, F.-J. Software for Storage and Management of Microclimatic Data for Preventive Conservation of Cultural Heritage. *Sensors* **2013**, *13*, 2700–2718. [CrossRef] [PubMed]

32. García-Diego, F.J.; Scatigno, C.; Merello, P.; Bustamante, E. Preliminary data of CFD modeling to asses the ventilation in an archaeological building. In Proceedings of the ARQUEOLÓGICA 2.0—8th International Congress on Archaeology Computer Graphics, Cultural Heritage and Innovation, València, Spain, 5–7 September 2016; pp. 504–507.

*sensors*

MDPI

*Article*

# Implementation and Operational Analysis of an Interactive Intensive Care Unit within a Smart Health Context

**Peio Lopez-Iturri [1], Erik Aguirre [1], Jesús Daniel Trigo [1,2], José Javier Astrain [2,3], Leyre Azpilicueta [4], Luis Serrano [1,2], Jesús Villadangos [2,3] and Francisco Falcone [1,2,*]**

[1] Department of Electrical and Electronic Engineering, Public University of Navarre, 31006 Pamplona, Navarra, Spain; peio.lopez@unavarra.es (P.L.-I.); erik.aguirre@unavarra.es (E.A.); jesusdaniel.trigo@unavarra.es (J.D.T.); lserrano@unavarra.es (L.S.)

[2] Institute of Smart Cities, Public University of Navarre, 31006 Pamplona, Navarra, Spain; josej.astrain@unavarra.es (J.J.A.); jesusv@unavarra.es (J.V.)

[3] Department of Mathematical Engineering and Computer Science, Public University of Navarre, 31006 Pamplona, Navarra, Spain

[4] School of Engineering and Sciences, Tecnologico de Monterrey, 64849 Monterrey, Mexico; leyre.azpilicueta@itesm.mx

* Correspondence: francisco.falcone@unavarra.es; Tel.: +34-948-169-741

Received: 31 December 2017; Accepted: 26 January 2018; Published: 29 January 2018

**Abstract:** In the context of hospital management and operation, Intensive Care Units (ICU) are one of the most challenging in terms of time responsiveness and criticality, in which adequate resource management and signal processing play a key role in overall system performance. In this work, a context aware Intensive Care Unit is implemented and analyzed to provide scalable signal acquisition capabilities, as well as to provide tracking and access control. Wireless channel analysis is performed by means of hybrid optimized 3D Ray Launching deterministic simulation to assess potential interference impact as well as to provide required coverage/capacity thresholds for employed transceivers. Wireless system operation within the ICU scenario, considering conventional transceiver operation, is feasible in terms of quality of service for the complete scenario. Extensive measurements of overall interference levels have also been carried out, enabling subsequent adequate coverage/capacity estimations, for a set of Zigbee based nodes. Real system operation has been tested, with ad-hoc designed Zigbee wireless motes, employing lightweight communication protocols to minimize energy and bandwidth usage. An ICU information gathering application and software architecture for Visitor Access Control has been implemented, providing monitoring of the Boxes external doors and the identification of visitors via a RFID system. The results enable a solution to provide ICU access control and tracking capabilities previously not exploited, providing a step forward in the implementation of a Smart Health framework.

**Keywords:** Intensive Care Unit; visitor control application; hospital; 3D Ray Launching; Smart Health; radio planning

## 1. Introduction

Sustainability in the provision of health and social services has become one of the main challenges of public administrations in a wide sense. This interest is coupled to the advent of Smart Cities and Smart Regions, in which efficient resource management as well as different governance and data handling models are proposed. Within this context, multiple solutions and studies have been proposed to implement Ambient Assisted Living (AAL) environments to aid elderly persons, people with several degrees of disability or with neurodegenerative diseases [1]; the implementation of Social

Sensor networks to aid people in risk of social exclusion [1]; or the combination of e-Health/m-Health services within the context of Smart Cities to provide Smart Health scenarios [2]. In this sense, the introduction of communication systems coupled to medical and social service practice allows monitoring physiological signals which aid in the treatment of chronic diseases in a controlled household environment. In the case of social sensor systems, user behavior, which can also be combined with biomedical signals, provides monitoring capabilities, which enable users to reside in their households whilst guaranteeing normalized living conditions. Field trials have been performed in the case of users with neurological disorders, providing promising results in terms of inclusion [1]. Within the broad health context, hospitals and health centers are a specific use case, given by their size, user density and large variety of activities which take place. Multiple solutions have been adopted to enhance effective hospital management, such as the use and extension of medial health records, automated pharmacy control, patient location and tracking, among others.

ICUs are one of the most demanding scenarios, requiring precise time responsiveness and resource management. ICUs offer particular challenges, such as large number of sensitive equipment, signal acquisition and processing or false alarm handling, among others. Multiple specialized medical equipment can be simultaneously in operation, with different medical specialists requiring access to data, both real time as well as in relation with parameter trends. These data provide relevant information in relation with items such as treatment control, such as sedation, in which adequate treatment doses can be adjusted to the specific patient needs over time. One of the elements that give rise to operational issues is the presence of false alarms, due to inadequate signal interpretation. Not only technical challenges arise, but also logistical elements, such as adequate space usage, logistics of disposable material, accompanying person access and control or unauthorized user access also have to be dealt with. This holds especially true in the case of visitor access control, fundamental in patient recovery process, but usually overlooked on premises access. Studies have been performed to enhance patient control in sedation by using fuzzy logic control systems [3] or by using expert systems in the control of cardiopulmonary disease [4]. Signal processing and integrity is also a relevant issue, especially in order to control the appearance of false alarms [5], for example in the case of ECG signal detection in the case of arrhythmias [6,7], or arterial blood pressure signals [8]. Enhanced performance has also been achieved by employing video signal analysis, in the case of neonatal ICU [9], as well as by processing of audio alarm signals within a similar scenario [10]. Wireless sensor network integration has also been studied in ICU scenarios, with the aim of controlling patient location and position, employing bed weight sensors coupled to wireless transceivers [11].

In this work, an interactive ICU environment is implemented by integrating a Wireless Sensor Network (WSN), in which tracking capabilities and signal I/O acquisition and processing is enabled. One of the main issues in this context is information integrity, given by potential interference values as well by strict coverage/capacity relations (i.e., coverage distance between transceivers in which the required transmission rate is maintained). Therefore, it is compulsory to achieve the required signal levels above transceiver sensitivity thresholds in relation with interference background levels. Precise wireless channel modeling is performed for the specific ICU conditions with the aid of deterministic in-house 3D Ray Launching algorithm, as well as an extensive set of interference channel measurements. Wireless system analysis leads to system configuration, which is tested under real site conditions at the ICU of "Hospital Complex of Navarra" (HCN), in Spain. A set of wireless sensor motes have been employed, as well as specific ICU management application software, in order to provide real time monitoring applications. The rest of the paper is organized as follows: Section 2 presents the ICU context and employed simulation techniques. Section 3 presents simulation as well as measurement results for the channel analysis. Once wireless channel estimations, in terms of coverage values and interference characterization, have been performed, the proposed ICU's Visitor Access Application and its functionalities are presented in Section 4. Final comments and concluding remarks complete this work.

## 2. Materials and Methods

### 2.1. Scenario under Test: Intensive Care Units (ICU)

2.1.1. Brief Overview of ICUs

An ICU is a healthcare facility that provides intensive medicine. Patients admitted to ICUs suffer from life-threatening—albeit potentially recoverable—conditions. Thus, they require continuous observation, monitoring and support from multidisciplinary specialists as well as a variety of medications and state-of-the-art medical equipment. Since ICUs are highly specialized and technological facilities, they must meet several functional, structural and organizational requirements so that security, quality and efficiency terms are guaranteed.

In a typical ICU, patients are usually allocated in a single ward with direct visual monitoring by the staff by means of a piece of glass, although closed circuit television is also possible. ICUs typically enable voice intercommunication between the ward and the monitoring zone. In addition, specific alarm systems for critical situations (e.g., cardiorespiratory arrest) are usually installed.

ICU-admitted patients, given their critically illness, require the monitoring of a number of vital signs, such as heart rate, blood pressure, respiratory rate, pulse oximetry, hourly urine output, temperature, or blood gases [12]. Generally speaking, intensive care is suitable for patients requiring support of one or more organs, being the lung the most commonly treated. Besides respiratory support, patients in critical care usually require support in body systems such as circulatory, renal, neurological and metabolic systems [13]. To provide such support, the medical equipment normally included in an ICU comprises a variety of devices, such as ventilators (for respiratory support), hemofiltration or hemodialysis equipment (for renal support), multi-parameter monitors (which continuously monitor bodily parameters such as electrocardiogram, heart rate, temperature, blood pressure, pulse oximetry, end-tidal $CO_2$, oxygen saturation, cardiac output, etc.), infusion pumps (for infusing fluids, medication or nutrients into the circulatory system), among other equipment [14,15].

Besides the patient, there are two major groups involved that are worth discussing here: the staff and the visitors. Concerning the staff, a multidisciplinary team—with a high staffing-to-patient ratio (i.e., with a high number of staff people vs. the number of patients)—is usually in charge of providing critical care. The intensivist is the foremost specialized physician, who is an expert in the care of critically ill patients. Besides intensivists, other specialists may contribute to ICUs, such as anesthetists, radiologists, infectologists, pathologists, surgeons, or neurologists. Specialized nurses work alongside the specialists. Since ICU nurses provide continued care for the patients, their work is of utmost importance. They monitor and care about the wellbeing of the patients and must be competent with the technical procedures and devices used in the ICU. Indeed, nurses' attitudes and perceptions of new technologies are strategic to successful implementation. Lastly, other healthcare professionals are usually involved in ICUs, e.g., physiotherapists, pharmacists, dietitians, and microbiologists, as well as other professionals, such as technicians, hospital porters, or cleaning staff [13,15,16].

Visitors play a crucial part in the patient recovery process. Traditionally, visits were widely restricted under the premises of higher risk of infection, which is better controlled today; the patient's need for rest and quietness; and the presumption that visitors may interfere with medical staff work. However, ICUs can be vastly distressing and the presence of relatives can be regarded as positive, since relatives can help in encouraging physical and intellectual stimulation as well as in providing adequate psychological an emotional support [16]. Thus, current philosophy tends to open visit policies [17], but it is still limited in practice and should be implemented with caution [18].

2.1.2. ICU of the "Hospital Complex of Navarra" (HCN)

This paper is focused on the ICU of the HCN located in Pamplona/Iruña (Spain). According to Navarra's government data, 1200 patients are admitted yearly into the ICU. The unit was remodeled in 2015 and is further divided into two ICU facilities: ICU-A and ICU-B. The analysis provided in this paper is focused in ICU-A, which comprises 24 beds, one within each of the 24 Boxes. Figure 1 shows the scenario.

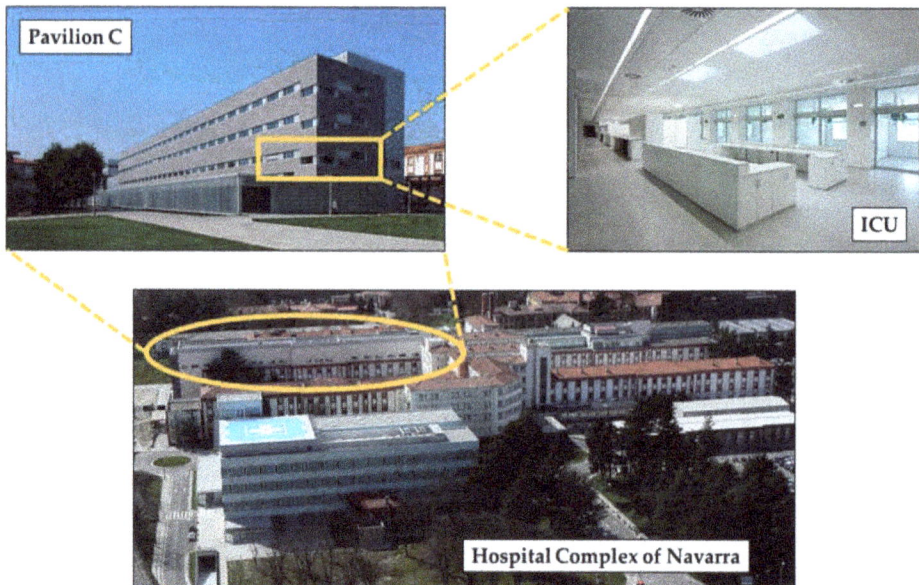

**Figure 1.** Scenario under analysis: the ICU-A of the HCN, located in Pamplona/Iruña, Navarre, Spain.

A wide range of medical equipment is allocated in each Box of the ICU-A, as described in Table 1. The devices with communication capabilities gather the medical data to the information system, so that every pre-configured parameter is stored and a graph is generated for the nursing staff. It is important to note that all communications are wired. All collected data are managed with the aid of specific software developed for the ICU-A. Figure 2 shows some of the most relevant screenshots of the management software, which can be accessed from all the computers within the ICU-A. In Figure 2b, the occupation of the Boxes is shown graphically. As can be seen, 20 out of the 24 Boxes are occupied (those in green and yellow), while four beds are free (Boxes 1, 2, 6 and 16). In Figure 2d, a screenshot of the parameters monitored and controlled by the medical staff can be seen. Additionally, nurses at the central zone control all the multi-parameter monitors that are sending the vital signs and handle the potential alarms (see Figure 2c). The names of patients and physicians have been redacted from all the screenshots.

Regarding the visiting policy, the ICUs of the HCN have embraced an open visiting policy, aligned with current trends. However, as an excessive number of visitors in a specific Box might be counter-productive, a recommendation of maximum two visiting persons at the same time for each Box is displayed on posters. At this time, there is no method for controlling or monitoring access, and nursing staff take over this purely managerial function. Thus, there is an opportunity to propose and provide new automatized wireless services for such purposes, so that nurses have more time to engage with their medical-related tasks.

**Table 1.** Medical equipment included in a ward of the ICU of the Hospital Complex of Navarra.

| Equipment | Parameters/Techniques |
|---|---|
| Multi parameter monitor | Electrocardiogram<br>Capnography<br>Arterial pressure<br>Pulse oximetry<br>Cardiac output<br>Electroencephalogram (EEG)<br>Bispectral Index (derived from EEG) |
| Renal monitor | Slow continuous ultrafiltration<br>Continuous venovenous hemodiafiltration<br>Continuous venovenous hemofiltration<br>Molecular Adsorbent Recirculating System (MARS)<br>Plasmapheresis |
| Extracorporeal membrane oxygenation (ECMO) machine | Venovenous ECMO<br>Venoarterial ECMO |
| Mechanical ventilator | Up to 60 different modes |
| High-flow oxygen therapy system | - |
| Hypothermia monitor | - |

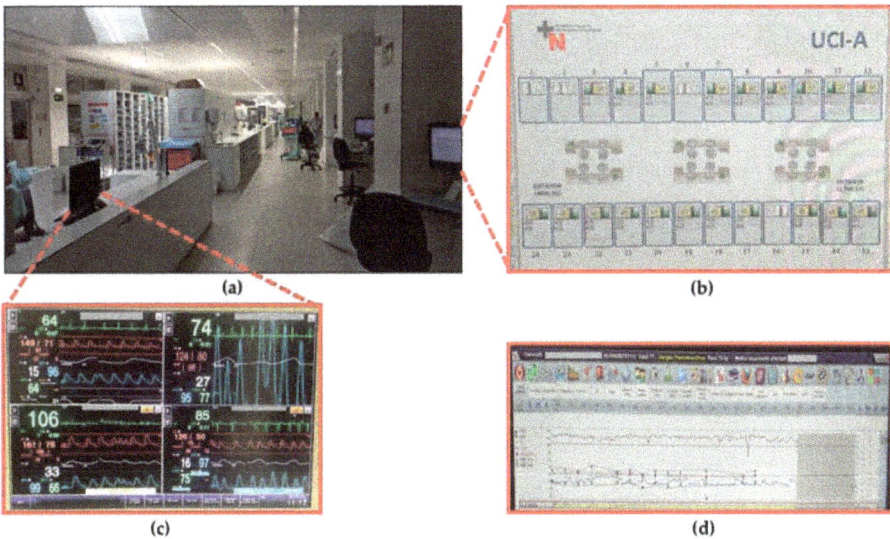

**Figure 2.** Screenshot of different tools of the management software of ICU-A: (**a**) ICU-A; (**b**) Boxes occupation tool; (**c**) Vital signs graphs; and (**d**) Monitored parameters screenshot.

## 2.2. Ray Launching Simulation Technique

### 2.2.1. Background

In this subsection, the simulation technique used for the radio planning tasks previous to the deployment of a WSN is presented. In general, hospital environments are very complex in terms of morphology, interference and variability of human being and equipment presence. It has been reported in the literature that wave propagation in hospitals and health environments is different from other buildings, due to the special construction of the walls. These walls contain metallic layers within their structure, e.g., operating rooms, X-ray rooms or magnetic resonance rooms, with lead or copper shielded walls. The radio channel characterization in such rooms is a key factor to assess and prevent EMC (Electromagnetic Compatibility) between different medical devices. In the same way, it is important to analyze the wave propagation to design an efficient use of wireless devices, even more when they are deployed around critical care medical equipment, where their immunity to electric field strength has to be taken into account.

There are some works in the literature that analyze wave propagation in hospital environments using different methodologies. The first approach comprehends empirical or statistical methods, which are based on measurements. The main advantage of these methods is that they finish quickly. However, they require adjustments based on in-situ measurements to give a good fit of the estimated results. In [19], the radio channel model for very high-speed radio systems (60 GHz) in hospital environments is presented, specifically with two possible applications: real time video streaming and angiography and ultrasonic imaging. The channel model parameters are described statistically and implementation guidelines in hospital environments are given in the work based on measurements. A novel statistical path loss model based on measurements taking into account the variability of height and the attenuation caused by the human beings at both frequencies, 6 and 8.5 GHz, under a hospital scenario is presented in [20]. It is shown that the proposed model is more flexible and accurate when compared with equivalent methodologies.

However, these methods do not consider all the elements in the environment. Thus, they can give erroneous results on the electromagnetic propagation estimation when the morphology of the scenario has a significant impact (due to its complexity), such as hospital environments, which usually have a complex structure of the walls, different type of scatterers, materials, mobile equipment, etc. Thus, the most widely used approach for propagation prediction in this type of scenarios corresponds to deterministic methods. The main principle of these methods is based on ray optics with advanced ray tracing techniques, or on solving Maxwell's equations. These methods are accurate, but are also time-consuming due to inherent computational complexity. Methods based on geometrical optics (GO), such as ray launching (RL) or ray tracing (RT), offer a reasonable trade-off between precision and required calculation time [21]. The difference between RL and RT is that, in the RL technique, rays are launched from a transmitter, and when they intersect with an obstacle, the new reflected, absorbed, diffracted, or scattered rays are created, while, in RT techniques, imaging techniques are usually employed, creating several possible paths that rays follow from the transmitter to receiver over the direct, reflect and diffracted rays. In [22], a ray tracing approach which considers information about intersection techniques and the hither neighbor cell is presented to characterize wave propagation between the control units (CUs) and Wireless Access point (AP) in a hospital scenario, showing the effects on received power in 60 GHz frequency. In reference [23], the simulation of 60 GHz radio channel in a hospital ultrasonic inspection room using ray tracing methodologies with a single reflection scattering is presented, showing a good match with measurements. In [24], the characterization of radio wave propagation in hospitals in the frequency range from 42.6 MHz to 5.2 GHz is presented. They use a simulation tool based on the finite-difference time-domain (FDTD) for the lower frequencies, and a ray optical tool for higher frequencies up to 5.2 GHz. They conclude that wall attenuation in hospital areas is not the usual because of the different metallic layers in their structure.

2.2.2. The Ray Launching Technique

Following the trend, in this work, an in-house developed three-dimensional (3D) RL tool has been used for the channel characterization of the Intensive Care Unit (ICU) area of the previously presented HCN. The proposed in-house 3D RL algorithm has been previously used and validated in large complex indoor environments [25–27], and also specifically in a hospital environment [28]. The algorithm is based on GO and the Uniform Theory of Diffraction (UTD). The principal basis of the algorithm is that rays are launched from the transmitter with a determined angular and spatial resolution. When a ray impacts an object, both reflected and refracted rays are created, and, when a ray hits an edge, a new family of diffracted rays is created. Electromagnetic phenomena such as reflection, refraction and diffraction are taken into account, as well as all the material properties of all the obstacles within the environment, considering its conductivity and relative permittivity at the frequency of the wireless system under analysis. Hybrid techniques can also be used with this in-house developed RL tool, depending on the dimensions and the complexity of the considered scenario. Different modules have been created in the RL tool, such as the Neural Network (NN) module [29], the Diffusion Equation (DE) approach [30], or Collaborative Filtering (CF) module [31]. These novel hybrid techniques can be used with the RL tool, leading to accurate results while computational burden decreases significantly.

The complete area of the UCI-A has been created for its simulation with the 3D Ray Launching tool. The schematic view of the created scenario can be seen in Figure 3, where the central zone (for the medical staff), Boxes for patients and the external aisles for the visitors are indicated. For its creation, all the dimensions and main furniture's size have been obtained from the CAD model of the building. The material properties of furniture, beds, walls, floor, ceiling and Boxes' equipment have been selected as close as possible to the real ones to obtain accurate simulation results. Conductivity and electric permittivity of all the obstacles within the environment have been considered at the frequency under analysis, as input parameters of the algorithm. The principle of the 3D RL algorithm is that rays are launched from the transmitter with an angular resolution of rays of $\theta$ and $\varphi$ as an input parameter. The scenario can be divided into a fixed number of cuboids with different dimensions in X, Y and Z axes. Thus, when rays are launched from the transmitter, each of them follows a different path and all the parameters associated with each ray are stored in each cuboid that it goes through, calculating also the distance traveled by each ray. When a ray hits an object, new reflected and refracted rays are created, following the principle of Snell's law, always taking into account the material properties of the object. In addition, when a ray hits an edge, a new family of diffracted rays are created, following diffraction laws (UTD).

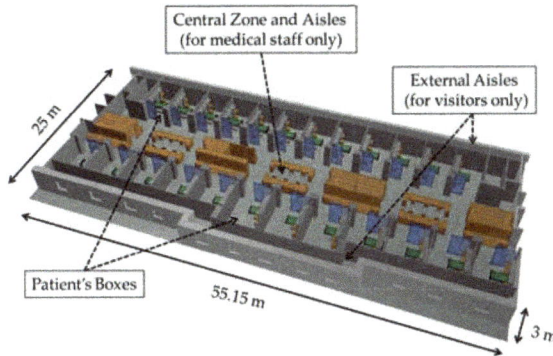

**Figure 3.** UCI-A scenario created for its simulation with the in-house 3D Ray Launching algorithm.

## 3. Results

This section has been divided in two subsections. First, with the aim of studying the availability of wireless communication frequency bands within the ICU, an assessment in terms of RF (Radio Frequency) noise is presented. Then, before the deployment of the real devices, radio planning tasks regarding the proposed WSN are carried out to obtain information about its adequacy.

### 3.1. RF Assessment of the Environment

Before choosing an adequate wireless communication technology for the deployment of the system, radio planning tasks have been carried out within the scenario. Specifically, background noise measurements have been performed at different typical communication frequencies where medical devices can operate, such as 433 MHz [32], 868 MHz [33], 2.4 GHz [33] and 5 GHz [34]. The aim of these measurements was to analyze the existence of potential interferences at the mentioned frequency bands, since the coexistence of wireless medical devices is an important factor for the deployment of new wireless networks [35]. For that purpose, different antennas have been used, connected to a FieldFox N9912A portable spectrum analyzer. The characteristics of the employed antennas are summarized in Table 2. Specifically, spectrograms have been measured on each point, as they provide information of the power level of the received signal within a specified frequency bandwidth during a time interval, which for these measurements was of 5 min. The measurements have been taken in different zones of the scenario: within boxes 2 and 6 (since there were no patients inside), on a table of the central zone and in the middle of the external aisles. All these measurement points are represented by blue dots in Figure 4a, whilst Figure 4b shows the employed spectrum analyzer within Box 2. Figure 5 shows all the measured spectrograms at 433 MHz band, Figure 6 at 868 MHz, Figure 7 at 2.4 GHz and Figure 8 at 5 GHz band. The shown bandwidths and central frequencies correspond to the data in Table 2.

**Table 2.** Characteristics of the antennas used for measurements.

| Antenna | Central Frequency | Bandwidth | Maximum Gain (dB) |
|---------|-------------------|-----------|-------------------|
| FLEXI-SMA90-433 | 433 MHz | 20 MHz | Unknown |
| ANT-868-CW-HWR | 868 MHz | 30 MHz | 2 |
| ACA-4HSRPP-2458 | 2.45 GHz | 100 MHz | 1 |
| ACA-4HSRPP-2458 | 5.5 GHz | 600 MHz | 1 |

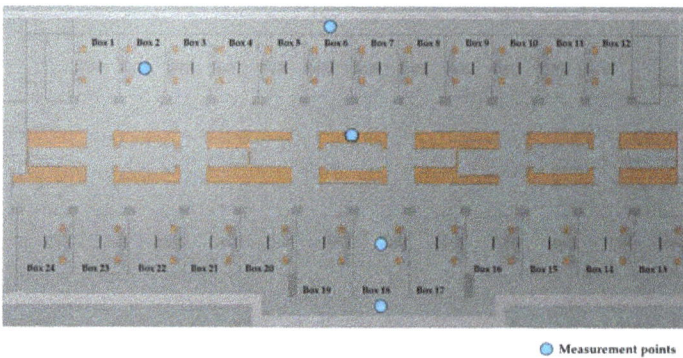

(a)

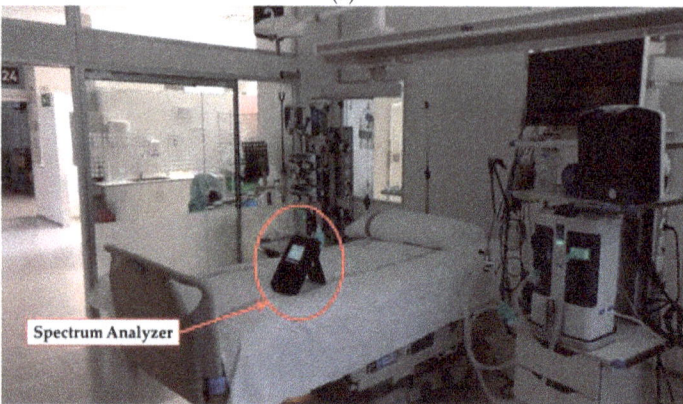

(b)

**Figure 4.** (a) Points within the scenario where spectrograms have been measured; and (b) the employed FieldFox N9912A spectrum analyzer of brand Agilent (Las Rozas, Spain) within Box 2.

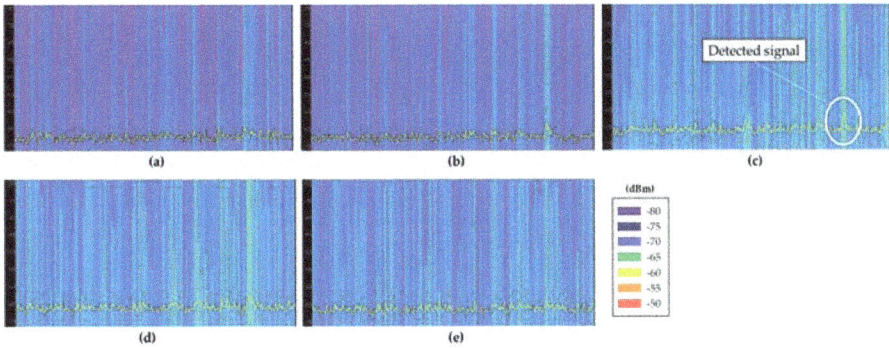

**Figure 5.** Measured spectrograms at 433 MHz central frequency with 20 MHz bandwidth: (**a**) within Box 2; (**b**) within Box 18; (**c**) aisle in front of Box 6; (**d**) aisle in front of Box 18; and (**e**) in the middle of the scenario.

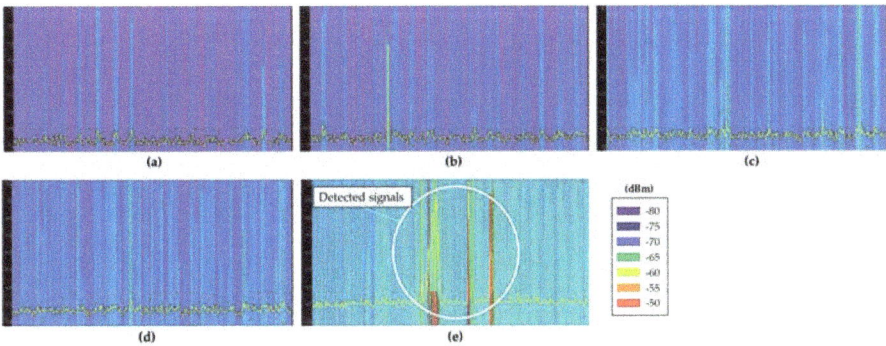

**Figure 6.** Measured spectrograms at 868 MHz central frequency with 30 MHz bandwidth: (**a**) within Box 2; (**b**) within Box 18; (**c**) aisle in front of Box 6; (**d**) aisle in front of Box 18; and (**e**) in the middle of the scenario (bandwidth of 300 MHz).

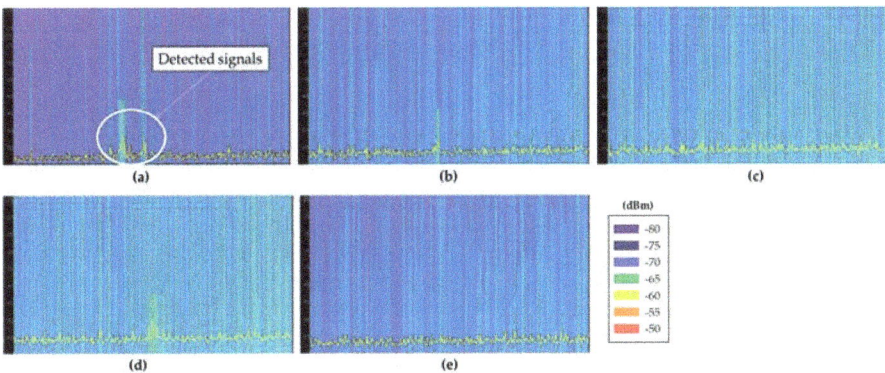

**Figure 7.** Measured spectrograms at 2.45 GHz central frequency with 100 MHz bandwidth: (**a**) within Box 2; (**b**) within Box 18; (**c**) aisle in front of Box 6; (**d**) aisle in front of Box 18; and (**e**) in the middle of the scenario.

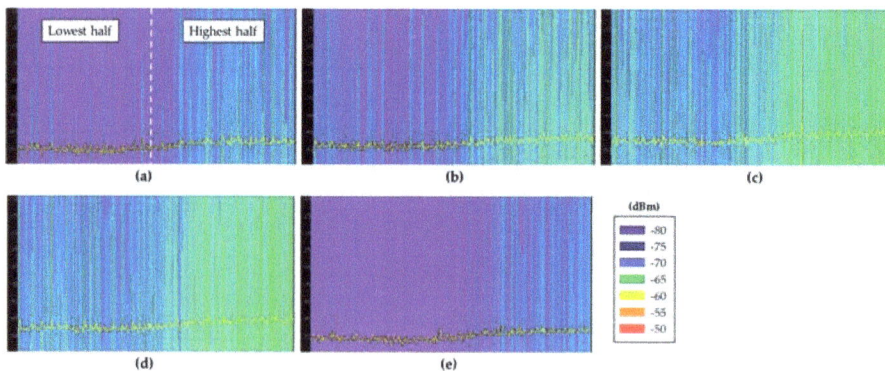

**Figure 8.** Measured spectrograms at 5.5 GHz center frequency with 600 MHz bandwidth: (a) within Box 2; (b) within Box 18; (c) aisle in front of Box 6; (d) aisle in front of Box 18; and (e) in the middle of the scenario.

As can be seen in the spectrograms, the analyzed communication bands are almost interference free except for the 868 MHz band. The 433 MHz band presents a signal around 440 MHz in all the measured spectrograms, but the rest of the band is free. On the other hand, the 868 MHz band is the most interfered band measured in this scenario. Figure 6e shows different signals with a high power level (i.e., similar to radio communication signal levels), which could greatly interfere a wireless communication. Note that the shown bandwidth of Figure 6e has been increased to 300 MHz to capture the whole interference. Due to its pattern, these detected signals are probably produced by an operating medical device. The 2.4 GHz band is almost signal free mainly because there is no WiFi access point on the whole floor of the building. Regardless, some signals can be seen at the center of the band, which probably are WiFi signals coming from WiFi access points outside this floor of the building or from the mobile devices of the medical staff present within the scenario when the measurements were carried out. Last, the 5 GHz band seems to be interference free. The background noise level difference that can be seen between the lowest half of the spectrum and the highest half is due to the employed antenna, which has not the same gain for all the measured bandwidth. Finally, it is worth noting how the detected background noise levels for all the frequency bands are higher for the measurements taken within the external aisles. This effect can be seen in all (c) and (d) spectrograms of Figures 5–8, which presents higher background noise level than the rest of measurement points. In summary, the closer to the street, the higher is the background noise level.

Taking into account these results, the 868 MHz band was discarded immediately as operation frequency band due to the high level of interference that can be found within the scenario. The rest of the measured bands are almost completely available for the deployment of wireless devices. However, among 433 MHz, 2.4 GHz and 5 GHz bands, the band of 2.4 GHz band has been chosen due to the advantages that provides and the great amount of devices that are available in the market. Specifically, ZigBee-based devices have been chosen in order to deploy the WSN within the ICU due to its low energy consumption (mainly due to the low transmission data rate of 250 kbps), high reliability in terms of the transmitting packets routing (auto-configurable mesh topology, up to 3 retransmissions by the use of ACKs and CSMA-Collision Avoidance) and the relatively low cost in terms of device cost and license free frequency band (ISM 2.4 GHz). Besides, as ZigBee's physical layer is based on the standard IEEE 802.15.4, it provides immunity to in-band interference and multipath propagation [36]. In the same way, there are 16 frequency channels in the 2.4 GHz band available, which give the possibility to avoid unexpected interferences by changing the transmission channel. In fact, as mentioned previously, Figure 7 shows some signals detected around 2.45 GHz. Therefore, ZigBee channel C, which corresponds to 2.41 GHz central frequency channel, has been selected to operate within the scenario, avoiding this potential interference at 2.45 GHz. In fact, most of the channels could be used, as they do not either overlap the interference. To show this graphically, Figure 9 shows a spectrogram measured in the middle of the scenario with a ZigBee-based XBee mote transmitting 10 dBm at ZigBee channel C. As can be seen, the frequency channel does not overlap with the 2.45 GHz potential interference. It is important to note that the measurement of Figure 9 has been made with the ZigBee mote just beside the spectrum analyzer, which is the reason of why the detected power is high.

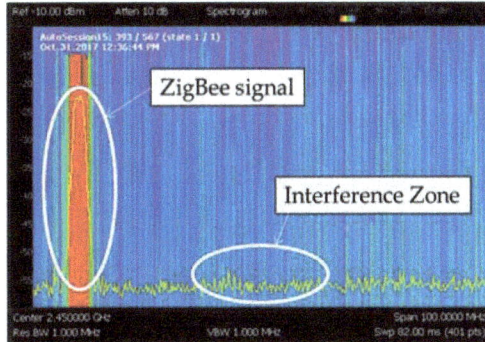

**Figure 9.** Measured spectrogram at 2.4 GHz band in the middle of the scenario with the XBee mote transmitting 10 dBm at ZigBee channel C (2.41 GHz).

### 3.2. Ray Launching Simulation Results

Once the wireless technology and the operating frequency channel for the deployment of the WSN have been selected, a new measurement campaign has been carried out in order to validate the presented 3D Ray Launching simulation tool for its use in this scenario. For that purpose, a Zigbee-based mote (XBee-Pro mote) connected to a laptop via USB cable has been placed on one of the tables used by the medical staff, and several received power level measurements have been taken along the internal aisles of the scenario. Figure 10a shows the position of the transmitter (represented by a red dot) and the measurements points (represented by green dots). Figure 10b,c shows how the transmitter has been deployed. The mote has been configured to operate at 2.41 GHz channel with its maximum transmitted power level of 18 dBm. For the received power level measurements, the corresponding antenna (see the 2.45 GHz antenna of Table 2) connected to the Agilent FieldFox N9912A spectrum analyzer has been used, the same way that can be seen in Figure 4b. All measurements have been taken at 1.5 m height.

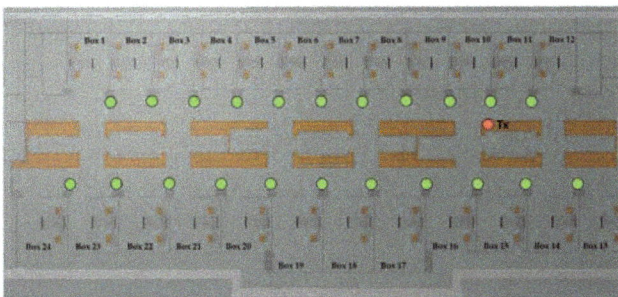

(a)

**Figure 10.** *Cont.*

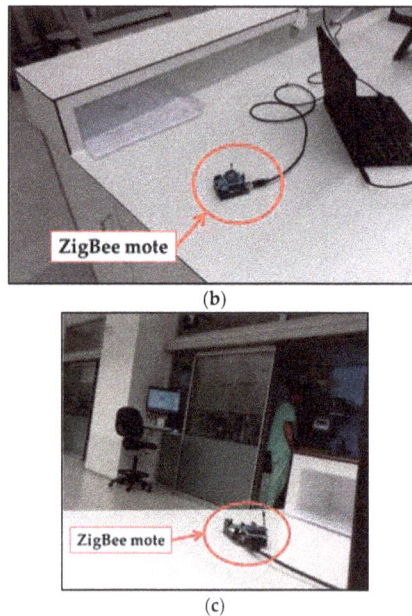

**Figure 10.** (**a**) Schematic view of the scenario with the position of the transmitter (red dot) and the measurement points (green dots); and (**b**,**c**) the detail of how the XBee-Pro mote has been deployed.

Once the measurements have been taken, the 3D Ray Launching simulation has been launched in order to obtain the estimated RF power distribution for the whole volume of the scenario. Table 3 shows the main parameters used for the 3D Ray Launching simulations, which have been selected as close as possible to the real parameters of the XBee-Pro module and the antenna of the spectrum analyzer. Figure 11 shows a bi-dimensional plane of the RF power distribution at height 1.5 m. Generally, in indoor complex environments, the most important radio propagation phenomenon is the multipath propagation. Due to the complex morphology of this scenario, the multipath propagation has a great impact on the radio propagation. This phenomenon causes rapid RF power variations along the distance. To see this effect, in Figure 12 three linear paths vs. received power level graphs are shown. These curves correspond to the three white dashed lines in Figure 11. The typical rapid RF power variations due to the multipath propagation can be clearly noticed in Figure 12. Now, the estimated received power level at the same positions of the measurement points in Figure 10a can be obtained to validate the simulation tool. Figure 13 shows the comparison between the measurements and the simulation results. The measurement points are numbered from 1 to 23, and each number refers to the point between two Boxes, e.g., measurement point number 1 refers to the green point in Figure 10a placed between Box 1 and Box 2, and so on. As can be seen in the comparison graph, for most of the points, the estimated values are very close to the measured ones, but there are a few points where the difference is quite high (more than 10 dB in some cases). These high error points are due to the different conditions between the simulated scenario and the real scenario: On the one hand, when the measurements were taken, the medical staff was working and constantly moving throughout the ICU (as can be seen in Figure 14), and there was some mobile equipment, such as X-ray equipment, which was moved throughout the scenario. On the other hand, simulations have been performed without this extra equipment and without the presence of human beings, which could have a great impact in terms of received power level [37]. Nevertheless, the 3D Ray Launching simulation tool can be considered validated as the mean error taking into account the measurements at the 22 points shown in Figure 10a and their corresponding simulation estimations (see Figure 13) is 3.49 dB with a standard deviation of 5.86 dB.

**Table 3.** 3D Ray Launching parameters.

| Parameter | Value |
|---|---|
| Operation Frequency | 2.41 GHz |
| Data rate | 250 kbps |
| Transmitted power level | 18 dBm |
| Antenna type | Monopole |
| Antenna gain (Transmitter) | 1.2 dB |
| Antenna gain (Receiver) | 1 dB |
| Permitted reflections | 6 |
| Launched rays angular resolution | 1° |
| Cuboids size | 0.5 m × 0.5 m × 0.5 m |

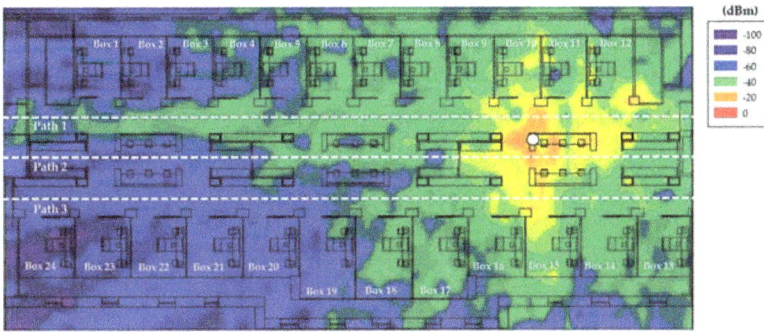

**Figure 11.** RF power distribution at height 1.5 m obtained by the 3D Ray Launching software. Transmitter is represented by a white dot.

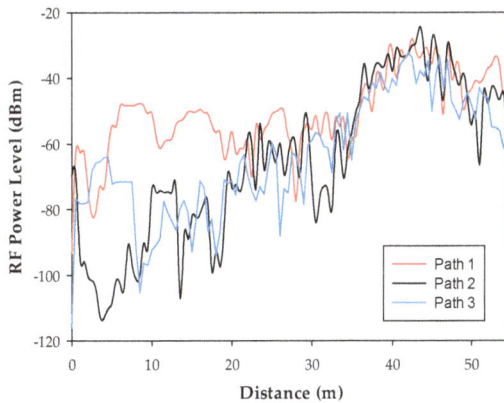

**Figure 12.** Three paths for linear RF power level distribution. The results correspond to the white dashed lines in Figure 11.

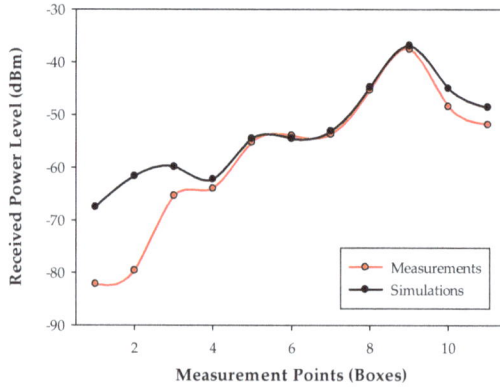

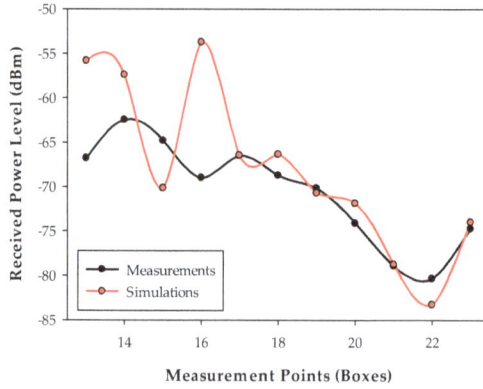

**Figure 13.** Measurements vs. 3D Ray Launching simulation results: (**a**) for Boxes 1–12; and (**b**) for boxes 13–24.

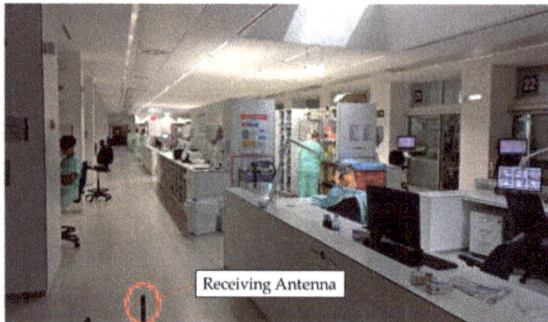

**Figure 14.** Picture of the UCI, taken during the measurements.

After the validation of the 3D Ray Launching tool, the assessment of a proposed ZigBee-based WSN deployment for the proposed application for the ICU is discussed next. As the WSN devices are going to be deployed on each of the external doors of all the boxes (as can be seen in Section 4), there is no need to find the best location for the transceivers. However, it is important to estimate if the proposed WSN will perform satisfactorily and optimally before its deployment, as the ICU is a scenario where less time and fewer tests are better for the patients and the medical staff. Therefore, new simulations have been performed to analyze the real positions of the transmitted elements within the scenario, i.e., placed on the external doors of the Boxes.

In Figure 15, the obtained estimations by means of the 3D Ray Launching algorithm can be seen. As an example, two specific positions of ZigBee transmitters are shown. The simulation parameters are the same of those shown in Table 3 but the transmitted power level, which has been set to 10 dBm (i.e., the maximum power level for the International version of ZigBee XBee-Pro modules). A key condition that must be fulfilled at receiver locations is to exceed the sensitivity value of the wireless modules, which in this case is −102 dBm. This threshold must be fulfilled at the receiver in order to be able to communicate with the transmitter. With the aim of showing this graphically, Figure 16a shows the sensitivity fulfillment planes corresponding to the results of Figure 15. As can be seen, the sensitivity is fulfilled for almost the whole scenario (at the presented height).

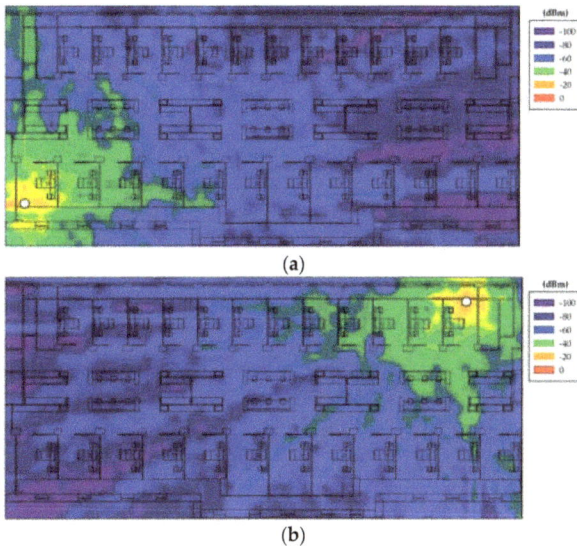

(a)

(b)

**Figure 15.** RF power distribution obtained by the 3D Ray Launching tool for two different locations of the proposed ZigBee-based WSN. The white dots represent the ZigBee devices on the external doors of Boxes 12 and 24.

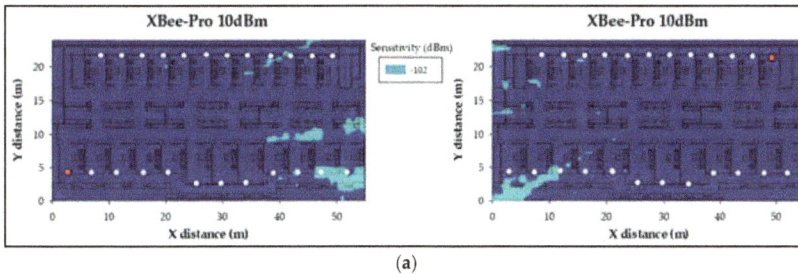

(a)

**Figure 16.** *Cont.*

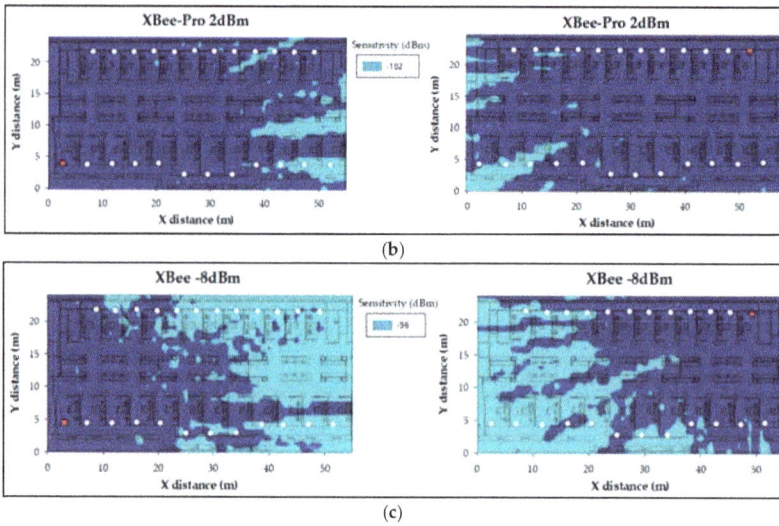

**Figure 16.** Sensitivity fulfillment planes corresponding to: (**a**) XBee-Pro motes transmitting 10 dBm; (**b**) XBee-Pro motes transmitting 2 dBm; and (**c**) XBee motes transmitting −8 dBm. Red dots represent the transmitter mote position.

However, there are other important aspects that must be taken into account, such as the power consumption of the wireless motes. As said, the transmitted power level has been firstly set to the maximum (10 dBm). However, since the sensitivity criteria are exceedingly fulfilled, new simulations have been performed setting the transmitted power to the minimum of the motes, i.e., 2 dBm. As can be seen in Figure 16b, the sensitivity fulfillment zone (dark blue) diminished significantly since the overall RF power within the scenario decreases. Nevertheless, there are still many ZigBee motes' locations that comply with the sensitivity threshold. Taking into account that a ZigBee network (operating in mesh topology) could route automatically the received packets to reach the net coordinator, it could be considered to transmit less power, even more taking into account that the used frequency channels are interference free.

Therefore, alternative motes have been analyzed and used for the deployment of the WSN, since the XBee-Pro modules' transmitted power cannot be less than 2 dBm. Specifically, the XBee motes have been chosen. These motes are a variant of the XBee-Pro modules which the main difference with the previous ones is that they are smaller, cheaper, they have a slightly worse sensitivity level (−96 dBm) and their transmitted power range goes between 2 dBm and −8 dBm. Now, simulations with the lowest value of −8 dBm have been performed, and the results in terms of sensitivity can be seen in Figure 16c. The sensitivity fulfillment has been drastically reduced, but the RF power distribution is still enough to communicate with other XBee motes of the WSN. Therefore, the ZigBee XBee motes have been chosen to deploy the wireless network as they will consume less energy (lower transmitted power) and the cost of the transceiver is lower.

A summary of all these performed simulations is shown in Table 4, which corresponds to the results shown in Figure 16.

**Table 4.** 3D Ray Launching parameters.

| ZigBee Module | Sensitivity | Transmitted Power | Corresponding Figure |
|---|---|---|---|
| XBee-Pro | −102 dBm | 10 dBm | Figure 16a |
| XBee-Pro | −102 dBm | 2 dBm | Figure 16b |
| XBee | −96 dBm | −8 dBm | Figure 16c |

## 4. Visitor Access Control Application

As previously mentioned, ICU visitors play a crucial part in the recovery of patients, but, at the same time, the current trend of open visits should be implemented with caution. In the ICU under analysis in this study, the access of visitors is made from the core of the hospital's external circulation. The visitors are allowed only at the preset visiting hours (except in specific cases), and the access to patients is via the Box doors of the external aisles. When visiting hours are finished, the access to external aisles is closed, and, therefore, the access to the Boxes. However, in this ICU, the access to the external aisles is out of reach of direct visual control from the reception area of the unit. Besides, the external doors of the Boxes have a simple manual lock, which can be easily leave open unintentionally by medical staff or cleaning staff. Thus, the patient's health could be put at risk (e.g., for isolated patients), and the security of the ICU itself could be compromised.

On the other hand, the current "open visit" policy recommends that only one or two companions could remain with the patient for as many hours as possible in order to provide psychological and emotional support to the patient. In any case, the hospital must ensure that the privacy and safety of patients is always preserved. For such reasons, ICUs need to define mobility limitation measures, and also to implement access control and monitoring mechanisms. Therefore, a Visitor Access Control Application has been developed in this study.

The application consists on the monitoring of the Boxes external doors and the identification of visitors via a RFID system. For the monitoring of the Boxes external doors, a ZigBee-based implementation has been selected, as seen previously. Figure 17 shows the working scenario as well as the external aisles, the remote control device (the ZigBee coordinator) and the used open/close sensors. Each external door of the ICU care Boxes will have an open/close sensor that notifies to the remote control device the change of the Box door status (opened/closed) via a ZigBee communication. Besides, a RFID reader will permit (or not) the access to the visitor (who should have received a RFID tag) to the specific Box, thus, each visitor can access only the Box corresponding to his/her relative.

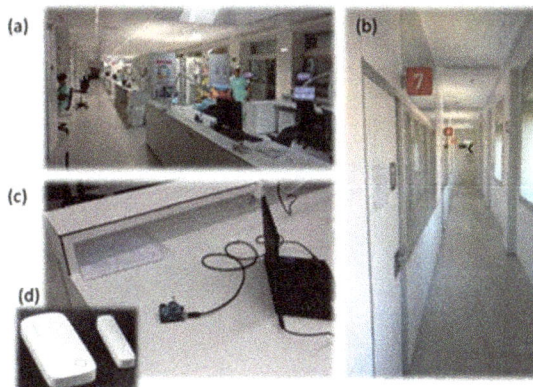

**Figure 17.** Working scenario: (**a**) central Zone; and (**b**) external aisles; and used wireless devices: (**c**) remote control device; and (**d**) open/close sensors.

The system uses the lightweight messaging protocol MQTT (Message Queue Telemetry Transport) to minimize data packet size and power consumption. MQTT allows implementing passive (push) and active (pull) monitoring of the doors status as well as efficient distribution of the gathered information. Passive monitoring implies that sensors are themselves responsible for notifying the change of status to the remote central service (i.e., the ZigBee coordinator of the network). Each time a door is opened/closed, the sensor notifies it to the central service without requiring any kind of user intervention. Sensors push the information to the service. On the other hand, active monitoring implies a pull mechanism since it is done on demand. Each time a user (medical staff) needs to know the status of a given door, or a set of them, the central service interrogates the different sensors involved and refreshes the situational awareness. Figure 18 shows the architecture of the proposed monitoring system. The application and the system provide the possibility to extend the access control to medical staff and medical equipment such as portable RX equipment or portable ultrasound scanners, but at this estate of development they have not been implemented. Figure 19 illustrates the application interface.

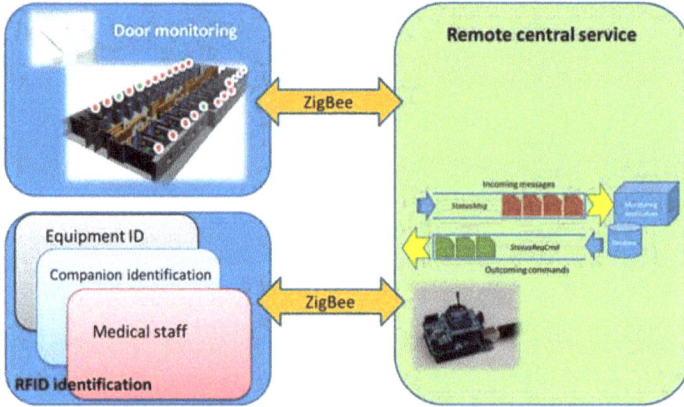

**Figure 18.** System architecture schema.

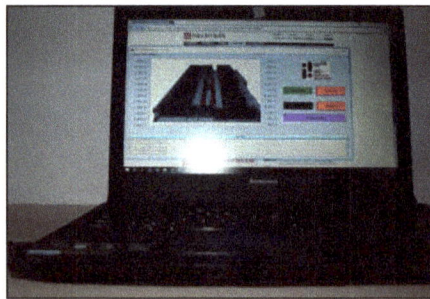

(a)

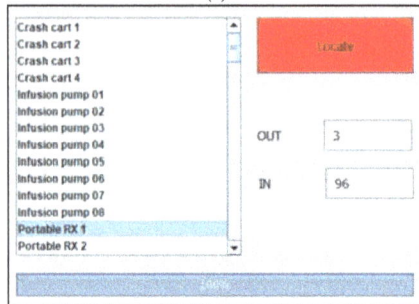

(b)

**Figure 19.** *Cont.*

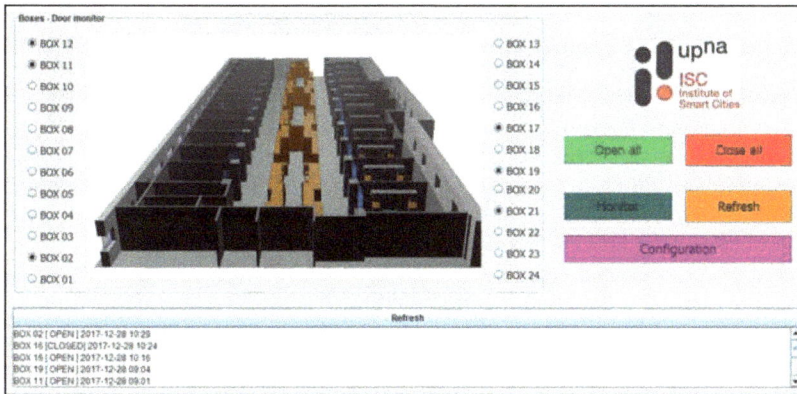

(c)

**Figure 19.** Application interface: (**a**) application running on a laptop; (**b**) equipment location control; and (**c**) boxes door monitoring.

Finally, it is important to note that the system could be modified (if required) to supply devices with energy harvesting methods (those devices with low energy consumption). In the same way, a greater level of confidentiality can be provided to the wireless communications if messages are encrypted. This can be easily done as the used ZigBee motes have the possibility to activate the cryptography option (in exchange of higher energy consumption and latency increment of the system).

## 5. Discussion

In this work, a context aware ICU environment has been developed, by combining a WSN formed by adapted motes with ad-hoc application software to enable user interaction. Due to the specific nature of ICUs, an extensive wireless channel analysis has been performed in order to gain insight in limitations given by interference, as well as to guarantee coverage/capacity relations, which are given by interference values as well as by signal fading. Precise wireless channel characterization is compulsory in order to consider large signal variations present in complex indoor scenarios, given mainly by multipath propagation components. Deterministic 3D RL simulations provide channel information for the complete volumetric scenario, providing information in relation with WSN transceiver mote location, applicable to a wide range of elements within the ICU, such as instrumentation, material/waste logistics or perimeter tracking and user location and visitors access control. The employed 3D RL code implemented in-house has been optimized by means of convergence analysis in order to optimize simulation parameters (given by angular resolution and number of rays until extinction). Moreover, hybrid simulation based on the combination of 3D RL with Neural Network interpolators, or Collaborative Filtering techniques allow analyzing large volumetric scenarios in a precise manner. In this way, full coverage estimation for the complete scenario volume are obtained, which enable deployment strategies in any potentially required location. On site measurements have been conducted at typical WSN operational frequencies (433 MHz, 868 MHz, 2.4 GHz and 5 GHz), in which interference analysis indicates that their operation is feasible. Time/frequency spectrograms have been obtained from the measurement campaign done within the ICU, in which interference conditions, which can be included with the deterministic simulation estimations, have been obtained. In relation with path loss estimations, signal levels are within sensitivity range for almost all the locations within ICU, enabling constant communication capabilities. An ICU specific Visitor Access Control Application has been developed, based on Xbee wireless sensor nodes and a lightweight architecture implemented for the application framework, providing visitor identification and the monitoring of Boxes doors. More functionalities could be added if necessary, such as user tracking, material/waste logistics and potential extension to I/O data acquisition, such as biomedical signals or environmental signals.

The proposed ICU WSN system and application is currently being tested, with the aim of integrating the solution within a broader set of Smart Health systems at Hospital Complex of Navarra. In this sense, future work is envisaged, related with system interoperability, analysis of user dynamics or application extension with new functionalities, among others. Additionally, besides technological enhancements, further effort is required for the validation of the service, for example through usage assessment performed mainly by ICU nurses, given that the proposed system primarily alters the workload of this group of professionals—although including the other actors involved would also be of interest.

**Acknowledgments:** The team of the ICU of "Hospital Complex of Navarra/Complejo Hospitalario de Navarra (CHN)", and particularly, Juan Pedro Tirapu and Asier García de Iturrospe, Telecommunication Engineer from Government of Navarra, for providing the access and the technical information (network layout, biomedical instrumentation, software, etc.) of the ICU of Complejo Hospitalario de Navarra.

**Author Contributions:** Peio Lopez-Iturri and Luis Serrano conceived and designed the experiments. Peio Lopez-Iturri, Erik Aguirre and Jesús Daniel Trigo made the measurements. Peio Lopez-Iturri and Erik Aguirre conducted the simulations. Peio Lopez-Iturri and Francisco Falcone made the radio planning analysis. José Javier Astrain and Jesús Villadangos designed and developed the application. Peio Lopez-Iturri, Leyre Azpilicueta, Jesús Daniel Trigo and Francisco Falcone wrote the paper.

**Conflicts of Interest:** The authors declare no conflict of interest.

## References

1. Aguirre, E.; Led, S.; Lopez-Iturri, P.; Azpilicueta, L.; Serrano, L.; Falcone, F. Implementation of Context Aware e-Health Environments Based on Social Sensor Networks. *Sensors* **2016**, *16*, 310. [CrossRef] [PubMed]

2. Solanas, A.; Patsakis, C.; Conti, M.; Vlachos, I.S.; Ramos, V.; Falcone, F.; Postolache, O.; Perez-Martinez, P.A.; Di Pietro, R.; Perrea, D.N.; et al. Smart Health: A Context-Aware Health Paradigm within Smart Cities. *IEEE Commun. Mag.* **2014**, *52*, 74–81. [CrossRef]

3. Shieh, J.; Fu, M.; Huang, S.; Kao, M. Comparison of the Applicability of Rule-Based and Self-Organizing Fuzzy Logic Controllers for Sedation Control of Intracranial Pressure Pattern in a Neurosurgical Intensive Care Unit. *IEEE Trans. Biomed. Eng.* **2006**, *53*, 1700–1705. [CrossRef] [PubMed]

4. Gholami, B.; Bailey, J.M.; Haddad, W.M.; Tannenbaum, A.R. Clinical Decision Support and Closed-Loop Control for Cardiopulmonary Management and Intensive Care Unit Sedation Using Expert Systems. *IEEE Trans. Control Syst. Technol.* **2012**, *20*, 1343–1350. [CrossRef] [PubMed]

5. Clifford, G.; Silva, I.; Moody, B.; Li, Q.; Kella, D.; Chahin, A.; Kooistra, T.; Perry, D.; Mark, R. False alarm reduction in critical care. *Physiol. Meas.* **2016**, *37*, 5–23. [CrossRef] [PubMed]

6. Behar, J.; Oster, J.; Li, Q.; Clifford, G.D. ECG Signal Quality during Arrhythmia and Its Application to False Alarm Reduction. *IEEE Trans. Biomed. Eng.* **2013**, *60*, 1660–1666. [CrossRef] [PubMed]

7. Krasteva, V.; Jekova, I.; Leber, R.; Schmid, R.; Abächerli, R. Real-time arrhythmia detection with supplementary ECG quality and pulse wave monitoring for the reduction of false alarms in ICUs. *Physiol. Meas.* **2015**, *37*, 1273–1297. [CrossRef] [PubMed]

8. Lameski, P.; Zdravevski, E.; Koceski, S.; Kulakov, A.; Trajkovik, V. Suppression of Intensive Care Unit False Alarms Based on the Arterial Blood Pressure Signal. *IEEE Access* **2017**, *5*, 5829–5836. [CrossRef]

9. Villarroel, M.; Guazzi, A.; Jorge, J.; Davis, S.; Watkinson, P.; Green, G.; Shenvi, A.; McCormick, K.; Tarassenko, L. Continuous non-contact vital sign monitoring in neonatal intensive care unit. *Healthc. Technol. Lett.* **2014**, *1*, 87–91. [CrossRef] [PubMed]

10. Raboshchuk, G.; Nadeu, C.; Jancovic, P.; Peiro Lilja, A.; Kokuer, M.; Muñoz Mahamud, B.; de Veciana, A.R. A Knowledge-Based Approach to Automatic Detection of Equipment Alarm Sounds in a Neonatal Intensive Care Unit Environment. *J. Transl. Eng. Health Med.* **2017**. [CrossRef]

11. Silva, R.; Silva, J.; Silva, A.; Pinto, F.; Simek, M.; Boavida, F. Wireless Sensor Networks in Intensive Care Units. In Proceedings of the IEEE International Conference on Communications Workshops, Dresden, Germany, 14–18 June 2009; pp. 1–5.

12. Smith, G.; Nielsen, M. Criteria for admission. *BMJ* **1999**, *318*, 1544–1547. [CrossRef] [PubMed]

13. Palanca, I.; Sánchez, A.; Elola, J. *Unidad de Cuidados Intensivos: Estándares y Recomendaciones*; Ministerio de Sanidad y Política Social: Madrid, Spain, 2010.

14. Haslam, J.; Ball, J.; Rhodes, A.; MacNaughton, P. Monitoring. In *ABC of Intensive Care*; Nimmo, G.R., Singer, M., Eds.; Wiley-Blackwell: Hoboken, NJ, USA, 2011; ISBN 978-1-4051-7803-7.

15. Batchelor, A.; Nightingale, N. General principles of intensive management. In *ABC of Intensive Care*; Nimmo, G.R., Singer, M., Eds.; Wiley-Blackwell: Hoboken, NJ, USA, 2011; ISBN 978-1-4051-7803-7.

16. Bennett, D.; Bion, J. Organisation of intensive care. *BMJ* **1999**, *318*, 1468–1470. [CrossRef] [PubMed]

17. Giannini, A.; Garrouste-Orgeas, M.; Latour, J.M. What's new in ICU visiting policies: Can we continue to keep the doors closed? *Intensiv. Care Med.* **2014**, *40*, 730–733. [CrossRef] [PubMed]

18. McAdam, J.L.; Puntillo, K.A. Open visitation policies and practices in US ICUs: Can we ever get there? *Crit. Care* **2013**, *17*, 171. [CrossRef] [PubMed]

19. Kyrö, M.; Haneda, K.; Simola, J.; Takizawa, K.; Hagiwara, H.; Vainikainen, P. Statistical Channel Models for 60 GHz Radio Propagation in Hospital Environments. *IEEE Trans. Antennas Propag.* **2012**, *60*, 1569–1577. [CrossRef]

20. Cui, P.-F.; Yu, Y.; Lu, W.-J.; Liu, Y.; Zhu, H.-B. Measurement and Modeling of Wireless Off-Body Propagation Characteristics under Hospital Environment at 6–8.5 GHz. *IEEE Access* **2017**, *5*, 10915–10923. [CrossRef]

21. Iskander, M.F.; Yun, Z. Propagation prediction models for wireless communication systems. *IEEE Trans. Microw. Theory Tech.* **2002**, *50*, 662–673. [CrossRef]

22. Islam, M.J.; Reza, A.W.; Kausar, A.S.M.Z.; Ramiah, H. New Ray Tracing Method to Investigate the Various Effects on Wave Propagation in Medical Scenario: An Application of Wireless Body Area Network. *Sci. World J.* **2014**, *2014*, 306270. [CrossRef] [PubMed]

23. Järveläinen, J.; Haneda, K.; Kyro, M.; Kolmonen, V.-M.; Takada, J.; Hagiwara, H. 60 GHz Radio Wave Propagation Prediction in a Hospital Environment Using an Accurate Room Structural Model. In Proceedings of the Lourghborough Antennas & Propagation Conference, Loughborough, UK, 12–13 November 2012.

24. Schäfer, T.M.; Wiesbeck, W. Simulation of Radiowave Propagation in Hospitals Based on FDTD and Ray-Optical Methods. *IEEE Trans. Antennas Propag.* **2005**, *53*, 2181–2388. [CrossRef]

25. Granda, F.; Azpilicueta, L.; Vargas-Rosales, C.; Lopez-Iturri, P.; Aguirre, E.; Astrain, J.J.; Villadangos, J.; Falcone, F. Spatial Characterization of Radio Propagation Channel in Urban Vehicle-to-Infrastructure Environments to Support WSNs Deployment. *Sensors* **2017**, *17*, 1313. [CrossRef] [PubMed]

26. Azpilicueta, L.; López-Iturri, P.; Aguirre, E.; Martínez, C.; Astráin, J.J.; Villadangos, J.; Falcone, F. Evaluation of Deployment Challenges of Wireless Sensor Networks at Signalized Intersections. *Sensors* **2016**, *16*, 1140. [CrossRef] [PubMed]

27. Azpilicueta, L.; Astrain, J.J.; Lopez-Iturri, P.; Granda, F.; Vargas-Rosales, C.; Villadangos, J.; Perallos, A.; Bahillo, A.; Falcone, F. Optimization and Design of Wireless Systems for the Implementation of Context Aware Scenarios in Railway Passenger Vehicles. *IEEE Trans. Intell. Transp. Syst.* **2017**, *18*, 2838–2850. [CrossRef]

28. Aguirre, A.; López-Iturri, P.; Azpilicueta, L.; Rivares, C.; Astráin, J.J.; Villadangos, J.; Falcone, F. Design and Performance Analysis of Wireless Body Area Networks in Complex Indoor e-Health Hospital Environments for Patient Remote Monitoring. *Int. J. Distrib. Sens. Netw.* **2016**, *12*. [CrossRef]

29. Azpilicueta, L.; Rawat, M.; Rawat, K.; Ghannouchi, F.; Falcone, F. A Ray Launching-Neural Network Approach for Radio Wave Propagation Analysis in Complex Indoor Environments. *IEEE Trans. Antennas Propag.* **2014**, *62*, 2777–2786. [CrossRef]

30. Azpilicueta, L.; Falcone, F.; Janaswamy, R. A Hybrid Ray Launching-Diffusion Equation Approach for Propagation Prediction in Complex Indoor Environments. *IEEE Antennas Wirel. Propag. Lett.* **2017**, *16*, 214–217. [CrossRef]

31. Casino, F.; Azpilicueta, L.; López-Iturri, P.; Aguirre, E.; Falcone, F.; Solanas, A. Optimised Wireless Channel Characterisation in Large Complex Environments by Hybrid Ray Launching-Collaborative Filtering Approach. *IEEE Antennas Wirel. Propag. Lett.* **2017**, *16*, 780–783. [CrossRef]

32. Kim, S.; Brendle, C.; Lee, H.-Y.; Walter, M.; Gloeggler, S.; Krueger, S.; Leonhardt, S. Evaluation of a 433 MHz Band Body Sensor Network for Biomedical Applications. *Sensors* **2013**, *13*, 898–917. [CrossRef] [PubMed]

33. Conway, G.A.; Scanlon, W.G. Wearable Antennas for Medical Monitoring Systems. In Proceedings of the International Workshop on Antenna Technology (iWAT), Seoul, Korea, 4–6 March 2015.

34. Hemapriya, D.; Viswanath, P.; Mithra, V.M.; Nagalakshmi, S.; Umarani, G. Wearable Medical Devices—Design Challenges and Issues. In Proceedings of the IEEE International Conference on Innovations in Green Energy and Healthcare Technologies (ICIGEHT'17), Coimbatore, India, 16–18 March 2017.

35. Balid, W.; Al Kalaa, M.O.; Rajab, S.; Tafish, H.; Refai, H.H. Development of Measurement Techniques and Tools for Coexistence Testing of Wireless Medical Devices. In Proceedings of the IEEE Wireless Communications and Networking Conference Workshops (WCNCW), Doha, Qatar, 3–6 April 2016.

36. Staniec, K. IEEE 802.15.4 (ZigBee) immunity to in-band interference and multipath propagation. In Proceedings of the Radio and Antenna Days of the Indian Ocean (RADIO), Belle Mare, Mauritius, 21–24 September 2015.

37. Lopez-Iturri, P.; Aguirre, E.; Azpilicueta, L.; Astrain, J.J.; Villadangos, J.; Falcone, F. Implementation and Analysis of ISM 2.4 GHz Wireless Sensor Network Systems in Judo Training Venues. *Sensors* **2016**, *16*, 1247. [CrossRef] [PubMed]

![sensors logo] *sensors*

MDPI

*Article*

# Analysis of 3D Scan Measurement Distribution with Application to a Multi-Beam Lidar on a Rotating Platform

Jesús Morales , Victoria Plaza-Leiva , Anthony Mandow, Jose Antonio Gomez-Ruiz *, Javier Serón and Alfonso García-Cerezo

Robotics and Mechatronics Lab, Andalucía Tech, Universidad de Málaga, 29071 Málaga, Spain; jesus.morales@uma.es (J.M.); victoriaplaza@uma.es (V.P.-L.); amandow@uma.es (A.M.); jseron@uma.es (J.S.); ajgarcia@uma.es (A.G.-C.)
* Correspondence: janto@uma.es

Received: 4 December 2017; Accepted: 25 January 2018; Published: 30 January 2018

**Abstract:** Multi-beam lidar (MBL) rangefinders are becoming increasingly compact, light, and accessible 3D sensors, but they offer limited vertical resolution and field of view. The addition of a degree-of-freedom to build a rotating multi-beam lidar (RMBL) has the potential to become a common solution for affordable rapid full-3D high resolution scans. However, the overlapping of multiple-beams caused by rotation yields scanning patterns that are more complex than in rotating single beam lidar (RSBL). In this paper, we propose a simulation-based methodology to analyze 3D scanning patterns which is applied to investigate the scan measurement distribution produced by the RMBL configuration. With this purpose, novel contributions include: (i) the adaption of a recent spherical reformulation of Ripley's $K$ function to assess 3D sensor data distribution on a hollow sphere simulation; (ii) a comparison, both qualitative and quantitative, between scan patterns produced by an ideal RMBL based on a Velodyne VLP-16 (Puck) and those of other 3D scan alternatives (i.e., rotating 2D lidar and MBL); and (iii) a new RMBL implementation consisting of a portable tilting platform for VLP-16 scanners, which is presented as a case study for measurement distribution analysis as well as for the discussion of actual scans from representative environments. Results indicate that despite the particular sampling patterns given by a RMBL, its homogeneity even improves that of an equivalent RSBL.

**Keywords:** 3D laser scanner; multi-beam lidar; spatial data analysis; tilting mechanism; 3D scan pattern analysis

## 1. Introduction

There is a growing interest in portable and affordable three-dimensional (3D) lidar systems for new applications that can benefit from accurate and speedy range measurements, such as progress tracking in construction sites [1], precision agriculture [2], medical imaging [3], intelligent surveillance [4], or textile tailoring [5]. An alternative to high-end terrestrial scanners, such as those used for digital terrain modeling or forest inventory [6], is to obtain dense 3D data by aggregating multiple views from a less expensive sensor. At present, the most common approach to build compact 3D devices from low-cost sensors is tilting or spinning a 2D rangefinder by mounting it onto a one degree-of-freedom (DOF) rotation mechanism. Many examples of this type of customized rotating single-beam lidar (RSBL), mainly from the robotics community, can be found in the literature (e.g., [7–10]).

In the last few years, automotive applications have fostered an active market for new compact and cost-effective multi-beam lidar (MBL) rangefinders, such as those developed by Velodyne (Morgan Hill, CA, USA) [11]. As opposed to single-beam 2D sensors, in multi-beam rangefinders the rotating mirror is replaced by a spinning structure that holds a number of independent laser transceivers to scan different elevation angles within a given vertical field of view (FOV). This arrangement favors high data rates when compared to other 3D lidar configurations, but commercial MBLs offer limited vertical resolution and have to be rotated in order to produce a complete spherical FOV [12].

As MBLs are becoming increasingly popular and affordable, customized rotating multi-beam lidars (RMBL) built by adding one DOF to a commercial MBL may arguably become a common solution to obtain affordable rapid full-3D high resolution scans in the near future. This idea is supported by the introduction of customized RMBLs in recent research works [12–14].

The analysis of the scan measurement distribution, which can be appreciated qualitatively by simulating measured points on a hollow sphere [15], is crucial to exploit the potential capabilities of a 3D scanner. One major difference of RMBLs with respect to other 3D lidar configurations is the complexity of the resulting scanning pattern. Contrary to RSBLs, the additional DOF in RMBLs causes multiple scan beams to overlap, so measurements are not distributed with constant horizontal and vertical angular resolutions. Thus, the adoption of spatial data distribution indicators, such as Ripley's $K$ function [16], can be useful to analyze complex 3D scanning patterns from new rangefinder configurations and devices, not only qualitatively but also quantitatively.

In this paper, we investigate the promising RMBL configuration by proposing a simulation-based methodology to analyze the particular scanning patterns resulting from the addition of a DOF to a MBL. With this purpose, novel contributions include: (i) the adaption of a spherical formulation of the $K$ function [17] to assess 3D sensor data distribution on a simulated hollow sphere; (ii) a comparison, both qualitative and quantitative, of general scan patterns produced by an ideal VLP-16-based RMBL with other 3D scan alternatives (i.e., RSBL and MBL); and (iii) a new RMBL implementation consisting of a portable tilting platform for VLP-16 scanners, which is presented as a case study for measurement distribution analysis as well as for the discussion of actual scans from representative environments.

The rest of the paper is organized as follows. Section 2 reviews related work. Section 3 defines an ideal RMBL based on Velodyne MBL specifications. Section 4 presents the analysis methodology. Section 5 analyzes RMBL scan patterns and offers comparisons with alternative 3D configurations. Section 6 describes a new VLP-16-based tilting platform, which is used to scan three representative environments that are discussed in Section 7. Finally, Section 8 offers the conclusions.

## 2. Related Work

This section reviews related work. First, it offers an overview of affordable 3D lidar sensors. Then, it discusses methods used to evaluate 3D scan data.

### 2.1. Affordable Solutions for 3D Lidar Sensors

For more than fifteen years, the most common alternative to high-end 3D lidar sensors has been to build customized devices where a less expensive off-the-shelf rangefinder is rotated by a servo-drive mechanism. Table 1 offers a chronology of representative works in which customized 3D lidar sensors have been proposed, analyzed or applied to particular problems. Most of these works correspond to RSBLs, whose aim is to achieve 3D point clouds by rotating a 2D sensor. The first works were based mainly on the 180° 2D Sick LMS200 (Waldkirch, Germany) scanner, which was later substituted by lighter and more compact devices such as the 270° Hokuyo UTM-30LX (Osaka, Japan).

More recently, MBL rangefinders commercialized by Velodyne are becoming increasingly popular and affordable. MBLs can be considered as a hybrid between 2D and 3D scanners, as they consist on a spinning structure that holds a number of independent laser transceivers to scan planes with different fixed elevation angles within a limited vertical FOV [11].

**Table 1.** Representative examples of customized lidar systems based on a commercial device with an extra DOF.

|  | Type | Device | Major Application |
|---|---|---|---|
| Batavia 2002 [7] | RSBL | Sick | Obstacle detection |
| Wulf 2003 [15] | RSBL | Sick LMS200 | Density analysis |
| Weingarten 2006 [18] | RSBL | (2) Sick LMS200 | Indoor scenario reconstruction |
| Dias 2006 [19] | RSBL | Sick LMS200 | Device comparison |
| Sheh 2006 [8] | RSBL | Hokuyo URG-04LX | Sensor configuration analysis |
| Ueda 2006 [20] | RSBL | Hokuyo URG-04LX | Mapping |
| Yoshida 2010 [9] | RSBL | Hokuyo UTM-30LX | Mapping |
| Morales 2011 [21] | RSBL | Hokuyo UTM-30LX | Mapping and environment modeling |
| Xiao 2013 [22] | RSBL | Hokuyo UTM-30LX | Indoor mobile robot |
| Neumann 2014 [13] | RMBL | Velodyne HDL-64E | Underground mapping |
| Morales 2014 [23] | RSBL | Hokuyo UTM-30LX | Boresight calibration |
| Alismail 2015 [24] | RSBL | Hokuyo UTM-30LX-EX | Calibration for 3D mapping |
| An 2015 [25] | RSBL | Hokuyo URG-30LX | Plane extraction from indoor robot |
| Martinez 2015 [26] | RSBL | Hokuyo UTM-30LX-EX | UGV and UAV environment modeling |
| Özbay 2015 [27] | RSBL | Hokuyo UTM-30LX | UGV obstacle modeling |
| Moon 2015 [28] | RSBL | SICK LMS511-pro | Cargo ship modeling |
| Shaukat 2016 [29] | RSBL | Hokuyo UTM-30LX | RGB-D terrain modelling |
| Schubert 2016 [30] | RSBL | Hokuyo UTM-30LX | Robot mapping |
| Leingartner 2016 [31] | RMBL | Velodyne HDL-64E | Mapping |
| Neumann 2016 [12] | RSBL RMBL | Hokuyo UTM-30LX-EW and Velodyne VLP-16 | RMBL and MBL comparison |
| Kang 2016 [32] | RSBL | Hokuyo UTM-30LX | 6 DOF calibration |
| Droeschel 2017 [10] | RSBL | Hokuyo UTM-30LX-EW | Robot mapping |
| Klamt 2017 [14] | RMBL | Velodyne VLP-16 | Robot mapping |

The Velodyne VLP-16 and HDL-32 rangefinders are representative examples of the most affordable end of commercial multi-beam sensors. Their major specifications are summarized in Table 2. The VLP-16 (or Puck) [33] is a more compact and lightweight device. From an operational standpoint, the major differences between these two sensors lie in the number of laser transceivers and the vertical FOV: The VLP-16 scanner has 16 individual laser/detectors arranged in a $30°$ FOV, which yields a vertical resolution of $2.0°$ with a data rate of 300,000 points/s, whereas the HDL-32 has 32 transducers within a FOV of $41.3°$ with a vertical resolution of $1.33°$ and a correspondingly higher data rate. In contrast with the VLP-16, whose FOV is symmetrical with respect to the horizontal plane, the FOV of the HDL-32 has a downward shift. Furthermore, the increased vertical resolution of the HDL-32 has a significant impact in the cost of the sensor, which is substantially more expensive than the VLP-16. These features have favored the adoption of MBLs in mobile applications, where dynamic point cloud registration along the vehicle's path compensates for device limitations in vertical resolution and FOV [34]. Nevertheless, MBLs have to be rotated to yield a complete spherical FOV [12].

Given the compact size, high data rate, and decreasing cost of MBLs, customized 3D sensors built as rotating multi-beam lidars (RMBL) have the potential to become a common solution to obtain affordable rapid full-3D high resolution scans in the near future. The emergence of the RMBL configuration is indicated by recent examples shown in Table 1. For instance, the rotating 2D Hokuyo UTM-30LX-EW used by the Momaro robot in the DARPA Robotics Challenge of 2015 [10] has recently been replaced by a rotating Velodyne VLP-16 [14]. Moreover, Neumann et al. [13] built an RMBL based on a high-end 15 kg Velodyne HDL-64E to map underground mines from a wheeled robot. In a later work, these authors have developed a rotating platform based on a Velodyne VLP-16 that also includes a 2D Hokuyo and other sensors [12]. Moreover, a tilting HDL-64E has been used in [31] for robotic tunnel mapping.

**Table 2.** Manufacturer specifications for the VLP-16 and HDL-32 sensors [33].

|  | VLP-16 | HDL-32 |
|---|---|---|
| Laser/detector pairs | 16 | 32 |
| Range | 1 m to 100m | 1 m to 70 m |
| Accuracy | ±3 cm | ± 2 cm |
| Data | Distance/Calibrated reflectivities | Distance/Calibrated reflectivities |
| Data Rate | 300,000 points/s | 700,000 points/s |
| Vertical FOV | $30° : [-15°, +15°]$ | $41.3° : [-30.67°, +10.67°]$ |
| Vertical Resolution | $2.0°$ | $1.33°$ |
| Horizontal FOV | $360°$ | $360°$ |
| Horizontal Resolution | $0.1°$ to $0.4°$ (programmable) | $0.08°$ to $0.35°$ (programmable) |
| Size | 103 mm × 72 mm | 85.3 mm × 149.9 mm |
| Weight | 0.83 Kg | 1.3 Kg |

### 2.2. Evaluation of Scan Data

Most of the works addressing 3D scan data quality have focused on calibration methods to compensate for inaccurate intrinsic parameters with evaluation of the accuracy, repeatability and stability of 3D measurements. This is particularly necessary for customized devices, which are prone to construction misalignments [23]. In particular, intrinsic calibration of RSBLs has been extensively treated in the literature (e.g., [23,24,32]). Besides, some recent works have proposed calibration methods to improve factory parameters in MBLs [35,36]. Temporal instability of MBL measurements poses another relevant calibration problem [37]. In this sense, temporal variability of calibration parameters and performance deviations between individual beams have been evaluated for the Velodyne VLP-16 [38]. As for the RMBL configuration, to our knowledge, no specific calibration methods have been proposed yet.

Not so many works have explicitly addressed the analysis of 3D lidar data in terms of the resulting measurement distribution, which is fundamental to exploit the potential capabilities of a particular sensor/DOF combination for a given application [12,15]. Wulf and Wagner [15] analyzed the scanning patterns resulting from different arrangements of scan directions and rotation axes for a 180° 2D Sick LMS200. This influential work studied the non-homogeneous distribution of range measurements of RSBLs by proposing a qualitative illustration of measured points on both a simulated hollow sphere surrounding the 3D scanner as well as on actual scans from representative environments. In [39], a subsampling method for RSBLs aimed at improving the homogeneity of measurements on the hollow sphere. Alismail and Browning [24] used a synthetic hollow cuboid with the RSBL at its center for quantitative assessment of the scanning pattern for calibration purposes. Furthermore, Schubert et al. [30] aimed at optimizing the alignment of a 2D rangefinder with respect to the additional DOF. With this purpose, they claim that a cost function can be computed from the density distribution of points on the hollow sphere. Regarding RMBL sensors, Neumann et al. [12] offered a comparison between several high-end MBLs and customized RMBLs by using both a qualitative analysis of example scenes and quantitative performance indices that are representative of particular device specifications, such as scanning time, data rate, and average point density on the sphere.

### 2.3. Our Approach

In this work, we focus on the analysis of measurement distribution of scan data, so calibration aspects fall outside of the scope of the paper. The analysis of spatial data distribution is especially interesting for the emergent RMBL configuration, since the vertical and horizontal resolutions of the resulting measurements is uneven due to the overlapping of multiple beams during rotation.

The review of published works indicates that commonly used indicators of spatial data distribution, such as Ripley's $K$ function [16], have not been considered in the analysis of 3D scanning patterns. The $K$ function is useful to investigate the homogeneity of points for different ranges of distances, which has been applied to identify clusters from actual 3D scans of natural terrain [40]. Interestingly, a recent definition of the $K$ function for spherical point-pattern analysis on planetary-scale distributions [17] allows that the use of this indicator can be extended to analyze scan patterns on the hollow sphere.

Thus, we propose to adapt spherical formulation of the $K$ function [17] within an simulation-based analysis approach, both qualitative and quantitative, of the scan patterns produced by 3D lidar data. This analysis is applied to study general scan patterns produced by an ideal full-sphere RMBL based on the VLP-16 characteristics [33]. The proposed analysis approach allows comparing the ideal RMBL sensor with other 3D scan alternatives (i.e., SMBL and MBL). Besides, it can also be used to analyze a particular sensor implementation and to contrast it with the ideal RMBL. The latter point is illustrated through a case study with the *Velomotion-16* RMBL. This new sensor is an addition to the few RMBLs that have been reported recently [12–14,31].

## 3. Rotation of a Multi-Beam Lidar Sensor

This section provides a general definition for an RMBL whose rotation axis is parallel to one of the MBL frame axes and presents the computation of Cartesian point clouds. Without loss of generality, the VLP-16 [33] will be considered in this work for the addition of a rotation mechanism. This MBL sensor is especially suitable to build an RMBL on account of its more accessible cost, lighter weight, symmetric FOV, and compact size. Nevertheless, the following definitions could be extended to any MBL.

The local frame $X_v Y_v Z_v$ of the VLP-16 is shown in Figure 1a. This frame has its origin in the optical center, with the $Y_v$ axis in the forward direction and $Z_v$ pointing upwards. The VLP-16 scans points in spherical coordinates $(R, \omega, \alpha)$. With this information, Cartesian coordinates $(x_v, y_v, z_v)$ can be obtained for each measured point:

$$x_v = R\cos(\omega)\sin(\alpha), \tag{1}$$

$$y_v = R\cos(\omega)\cos(\alpha), \tag{2}$$

$$z_v = R\sin(\omega). \tag{3}$$

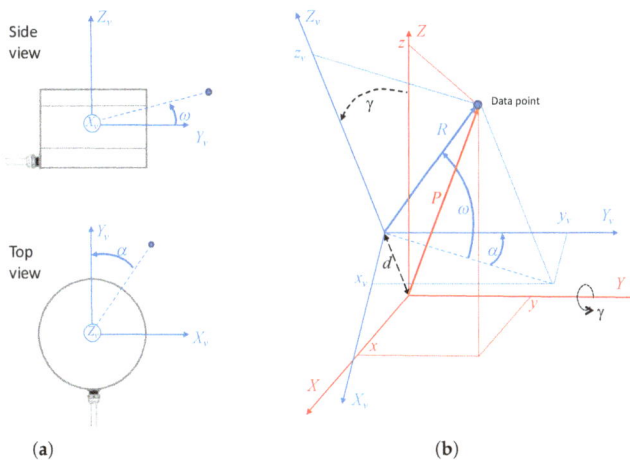

**Figure 1.** Reference frames and data point coordinates: (**a**) Velodyne VLP-16 sensor; and (**b**) RMBL based on a VLP-16. The VLP-16 local frame is represented in blue; the RMBL local frame is represented in red.

The local frame $XYZ$ of the RMBL resulting from the addition of a rotating mechanism (i.e., spinning or tilting) to the VLP-16 is illustrated in Figure 1b. When the rotation angle $\gamma$ is null, $Z_v$ is aligned with $Z$, and $X_v$ and $Y_v$ are parallel to $X$ and $Y$, respectively. Let us consider that the rotation axis is parallel to one of the VLP-16 axes; in this case, the rotation axis is $Y$. It should be noted that rotation about the $Y$ and $X$ axes would be similar, as the VLP-16 has a 360° horizontal FOV, whereas rotation about $Z$ would be pointless as this is redundant with the spinning motion of the multi-beam transceivers. Furthermore, in practice, the rotation axis should be at some small distance $d$ below the VLP-16 for the sake of compactness and shadow avoidance.

Then, Cartesian coordinates $(x, y, z)$ of data points in the RMBL frame can be computed as:

$$x = R(\cos(\omega)\sin(\alpha)\cos(\gamma) + \sin(\omega)\sin(\gamma)) + d\sin(\gamma), \tag{4}$$

$$y = R\cos(\omega)\cos(\alpha), \tag{5}$$

$$z = R(\sin(\omega)\cos(\gamma) - \cos(\omega)\sin(\alpha)\sin(\gamma)) + d\cos(\gamma). \tag{6}$$

## 4. Analysis Methodology

This section proposes a simulation-based methodology to analyze, both qualitatively and quantitatively, the spatial distribution of laser beams in 3D lidars. First, the simulation of sensor points on a hollow sphere is defined. Qualitative analysis is done from a visualization of point patterns projected on the sphere and also on orthogonal planes. For a quantitative analysis, two spatial descriptive statistics are considered: sampling density and homogeneity. We propose adapting a spherical extension of Ripley's $K$ function to evaluate scanning homogeneity.

### 4.1. Numerical Simulation

The proposed methodology considers the set of points computed by simulating a scan from a sensor that is placed at the center of a virtual hollow sphere [15,30]. This structure allows analyzing the homogeneity and the beam density in all directions around the sensor. By using a sphere, the analysis depends exclusively on sensor characteristics and is independent of the orientation and distance of the target surface (e.g., as in planar targets). Moreover, the azimuth and elevation of the points in the sphere are independent of the sphere radius.

To analyze the general patterns produced by different scan configurations, Section 5 will consider ideal sensors that are independent of particular device considerations. To generate a complete sphere, an ideal RMBL sensor based on a VLP-16 lidar can be simulated by considering a constant angular velocity of the tilt motion from $\gamma = 0°$ to $\gamma = 180°$. Different angular velocities correspond to different scan resolutions for the additional DOF. For the sake of simplicity, it will also be assumed that the sphere radius is large enough to make the deviation between the center of the sphere (i.e., the RMBL's origin) and its optical center negligible (i.e., $d \approx 0$); thus, all ranges $R$ coincide with the sphere radius and Equations (1)–(3) can be applied. Moreover, as points are intended to represent beam directions, no noise is considered in the simulations. Besides, for generalization, no shadows or other FOV limitations due to a particular mechanism are considered.

For a qualitative analysis, it is also interesting to consider how the angular distribution of points on the sphere would be translated onto planar surfaces in a synthetic environment. In the proposed analysis, the sensors is placed at a height of 1.5 m over a ground plane that is parallel to the local $XY$ plane and at 10 m from planes (representing walls) that are parallel to $YZ$ and $XZ$ planes. We have preferred this synthetic configuration rather than placing the sensor at the center of a 10 m × 10 m cube [24] because it is more representative of ground-based lidar applications.

### 4.2. Sampling Density

Sampling density for different scanning angles [15,24] can be represented as a 2D histogram on the sphere. With this aim, the sphere surface is partitioned with a triangular mesh obtained by recursive icosahedron sphere tessellation [41]. In this work, the sphere surface has been partitioned into 5120 bins.

### 4.3. Spatial Distribution Analysis of Scan Data with the K Function

A set of points in the plane is considered homogeneous if the same number of points occurs in any circular region of a given area. A common approach to analyze data homogeneity is Ripley's K function [16]. The comparison of the K functions for complete spatial randomness (CSR) and for a given point set allows determining whether points have a random, dispersed or clustered distribution over a range of distances.

In this work, we need to evaluate homogeneity of points on a sphere. With this purpose, we adopt the K function proposed in [17], which modifies the planar K function and the corresponding CSR reference for spherical surfaces.

The K function for spherical CSR is given by:

$$K_{csr}(r/R) = 2\pi R^2 (1 - \cos(r/R)) \tag{7}$$

where $R$ is the radius of the sphere and $r$ is the great-circle distance, with $r/R \in [0, \pi]$.

Given a set of $n$ sphere points $p_1, \ldots, p_n$, the estimation of the K function can be expressed as:

$$\hat{K}(r/R) = \frac{8\pi R^2}{n(n-1)} \sum_{i=1}^{n} \sum_{j=i+1}^{n} I(\theta(p_i, p_j) \leq r/R), \tag{8}$$

where $\theta(p_i, p_j))$ represents the angle corresponding to the great-circle distance between $p_i$ and $p_j$, and $I(\cdot)$ is the indicator function.

Equation (8) is a modification with respect to [17] that accounts for the reciprocity of great circle distances between pairs of points, so the number of required computations is halved. Then, in order to apply this concept to 3D sensor homogeneity assessment, this equation is applied to the set of simulated scan points for discretized angle increments $\Delta r/R$ in the interval $[0, \pi]$. For an RMBL based on a VLP-16, a value of $\Delta r/R = \pi/151$ is appropriate so that all bins contain a representative number of samples.

## 5. Analysis of Ideal RMBL Scanning Patterns and Comparison with Other 3D Configurations

The methodology defined in Section 4 has been applied to an ideal full-sphere VLP-16-based RMBL, which is compared with alternative 3D lidar configurations.

### 5.1. Qualitative Analysis

A visualization of sphere points given by the 3D scanners is offered in Figure 2. These points are shown in local sensor coordinates for a sphere of radius $R = 10$ m. Figure 2a,b illustrates the inhomogeneous beam pattern of the Velodyne sensors as well as their differences in vertical resolution and FOV. Moreover, Figure 2c,d shows points from RMBLs with different tilt speeds (i.e., different vertical resolutions). Points are distributed over the complete sphere but patterns due to the combination of the VLP-16 beams and the additional rotation are visible, especially in the lower resolution case. These patterns can also be appreciated on the lateral views of the spheres shown in Figure 3. Furthermore, some pattern distortion is appreciable in the central vertical strip in Figure 3a,b, which corresponds to a range of $[-15°, +15°]$ around the extremes of the tilting motion. This can be explained because the eight VLP-16 transducers with positive $\omega$ elevation values when $\gamma = 0°$ are overlapped with the eight transducers with negative $\omega$ when $\gamma = 180°$, and vice versa.

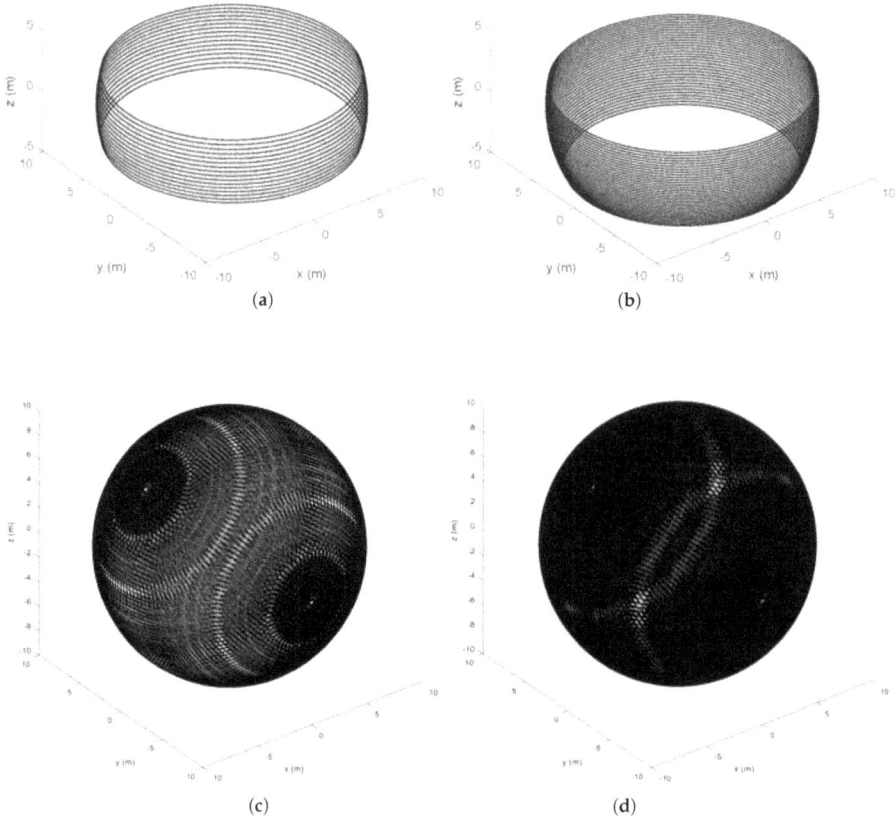

**Figure 2.** Representation of lidar beams as sphere points: (**a**) VLP-16; (**b**) HDL-32; (**c**) RMBL with tilting speed of 120°/s (29 frames); and (**d**) RMBL with tilting speed of 50°/s (71 frames).

The translation of scan points onto planar surfaces in a synthetic environment is illustrated in Figure 4. The figure shows how the VLP-16 provides scarce information about the ground. This is improved in the HDL-32, but the maximum height of wall points is reduced. The RMBL offers denser data and a wider FOV for both ground and walls. Besides, it can be observed that RMBL point patterns on the target planes depend on the sensor orientation. Thus, the wall perpendicular to the $X$ axis is scanned with a higher density on the sides (i.e., similar to a good peripheral vision), whereas the wall perpendicular to the $X$ axis is sampled with a higher density in the area that is close to the tilting axis. These differences should be considered when deciding on the sensor orientation with respect to the target surfaces in a particular application. Moreover, Figure 4d shows that a pattern of blind spots remains despite increasing the vertical scan resolution.

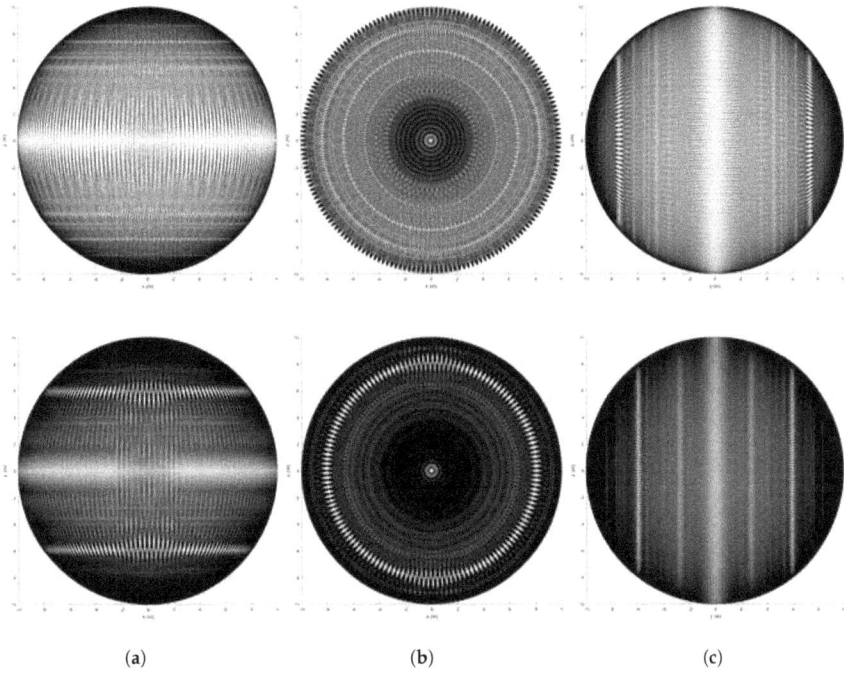

(a)        (b)        (c)

**Figure 3.** Different views of the sphere points given by the RMBL with tilting speed of 120°/s (top) and 50°/s (bottom): XY plane (**a**); XZ plane (**b**); and YZ plane (**c**).

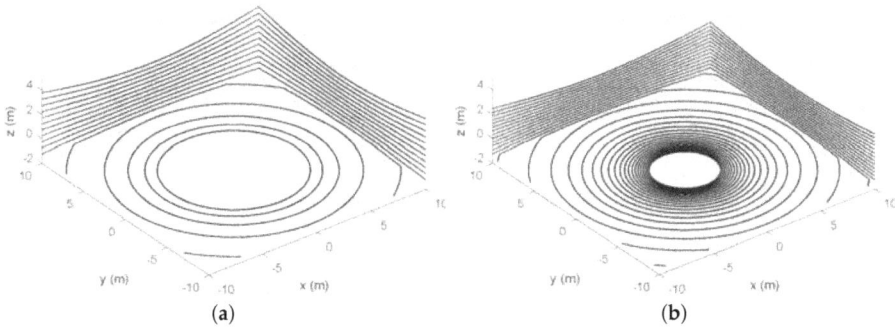

(a)        (b)

**Figure 4.** *Cont.*

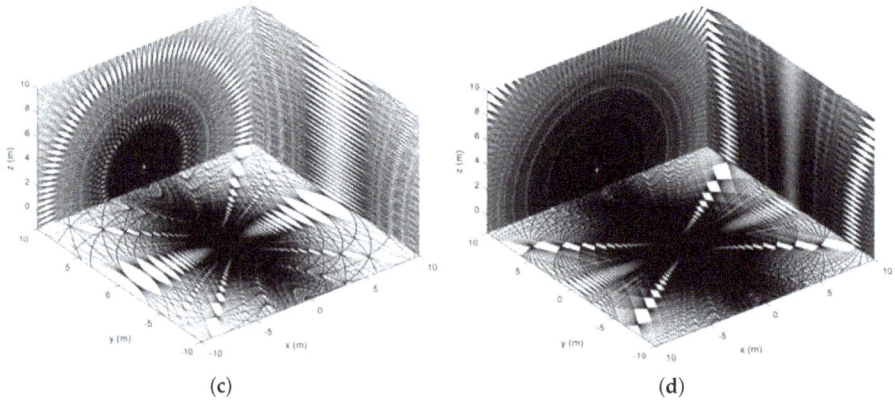

(c)                                                    (d)

**Figure 4.** Orthogonal plane points of lidar sensors: (**a**) VLP-16; (**b**) HDL-32; (**c**) RMBL with tilting speed of 120°/s (29 frames); and (**d**) RMBL with tilting speed of 50°/s (71 frames).

## 5.2. Sampling Density

Beam density histograms for the RMBL with tilting speeds of 120°/s and 50°/s are presented in Figure 5. For each case, the color scale has been normalized with respect to the maximum and minimum number of points per bin. The density increases notably in the polar regions (i.e., around the intersection with the rotation axis), which is a common trait with rotating 2D scanners [39]. The figure reflects that the slower scan offers more density, but also that the density distributions are similar regardless of tilting speed. However, a slightly darker and wider equator band in Figure 5b would indicate a lower density in relation to the poles for the slower scan, which can be attributed to greater polar oversampling. The triangles with the maximum number of beams are in the polar regions, with 365 points for the 120°/s scan and 881 for 50°/s. In the lower resolution scan, the triangles with the minimum number of points (i.e., 25) lie around the equator band. When the tilt resolution is increased, the bins with fewer points (i.e., 95) are in two parallel bands closer to each polar region. These parallel dead zones are not completely eliminated by increasing tilt resolution, as can be seen in Figure 3. The dead zones can be explained by the elevation gap between the VLP-16 transducers.

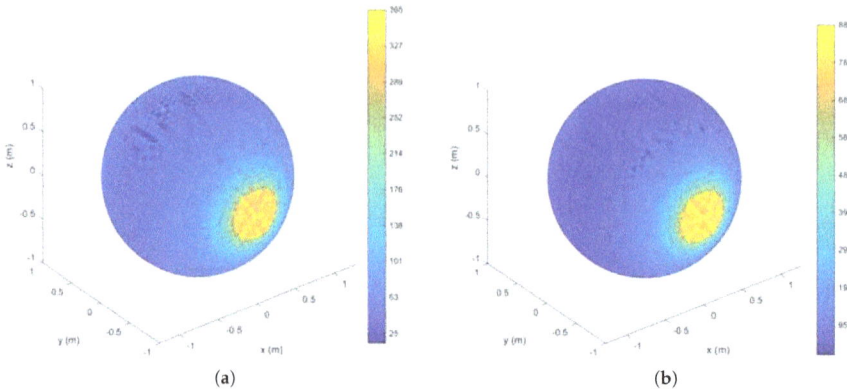

(a)                                                    (b)

**Figure 5.** 2D histogram of beam density on the sphere: (**a**) RMBL with tilting speed of 120°/s (29 frames); and (**b**) RMBL with tilting speed of 50°/s (71 frames). Color bars indicate points/bin.

The same data is represented in Figure 6 in a classical histogram graph. This figure indicates that the number of points is quite homogeneous with the exception of two groups of peaks that correspond to the polar regions. Again, the shapes of the histograms for both tilting speeds (i.e., elevation resolutions) are very similar. For the 120°/s speed, the mean is 10.48 points/deg$^2$ (84.47 points/bin) with a standard deviation of 6.98 points/deg$^2$ (56.24 points/bin); for 50°/s, the mean value is 25.16 points/deg$^2$ (202.73 points/bin) with a standard deviation of 16.62 points/deg$^2$ (133.92 points/bin).

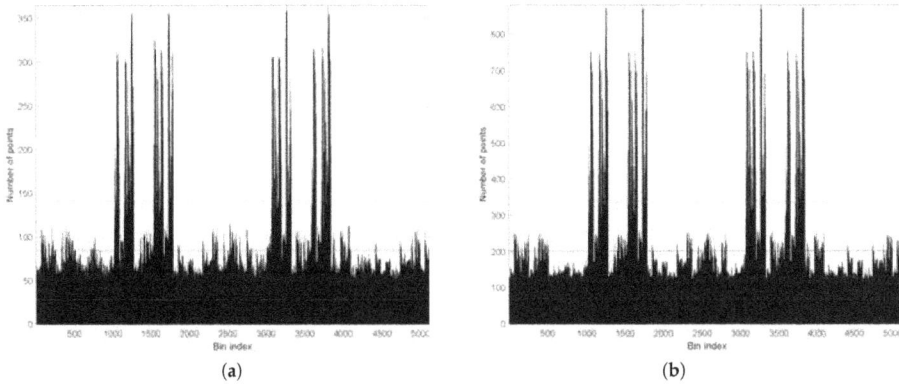

(a)                                                  (b)

**Figure 6.** Beam density histogram: (**a**) RMBL with tilting speed of 120°/s (29 frames); and (**b**) RMBL with tilting speed of 50°/s (71 frames). The red solid lines indicate the mean value and the blue dotted lines are the standard deviation.

*5.3. Spatial Distribution Analysis of Scan Data with the K Function*

The spherical *K* function has been computed for the hollow sphere data patterns that correspond to different MBL, RSBL, and RMBL sensor configurations. Two different representations of the resulting functions are given in Figure 7: the deviation of $\hat{K}$ with respect to the reference $K_{csr}$ and the first derivative of $\hat{K}$.

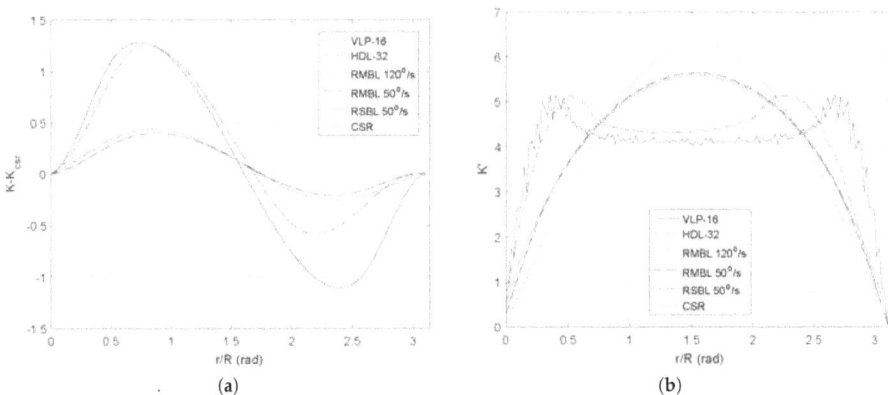

(a)                                                  (b)

**Figure 7.** Homogeneity analysis of point patterns on the sphere: (**a**) deviation of $\hat{K}$ with respect to the reference $K_{csr}$; and (**b**) first derivative of $K_{csr}$ and $\hat{K}$.

The deviation of $\hat{K}$ with respect to the CSR reference value $K_{csr}$ in Equation (7) is represented in Figure 7a for different great-circle distances *r* normalized by the sphere radius *R*. In particular,

Equation (8) has been used to compute $\hat{K}$ for the VLP-16, the HDL-32, two different tilting speeds of a VLP-16 based RMBL, and a generic RSBL with the same horizontal resolution as the VLP-16 and a tilt speed of 50°/s. It can be appreciated that the spherical point pattern homogeneity for the RMBL is independent of the tilting speed, as the curves for 120°/s and 50°/s are overlapped.

Positive values in Figure 7a indicate clustering, i.e., that the average number of neighbor points for that particular range of evaluation distances is higher than the average for the whole distribution, whereas negative values indicate dispersion. The graph indicates that clustering is similar for the VLP-16 and HDL-32 sensors, while dispersion is larger for the VLP-16. In the VLP-16, the maximum values for both clustering and dispersion are similar, as opposed to the HDL-32, where clustering is significantly larger than dispersion. This difference can be explained by the asymmetrical FOV of the HDL-32 (see Figure 2a,b), as samples from the lower beams (i.e., the lower values of the $Z$ coordinate) are denser. Correspondingly, the higher sampling density of the poles in the RMBL sensor (as shown in Figure 5) provokes that clustering is larger than dispersion, which also happens in the RSBL. All in all, the curves for the RMBL are clearly closer to $K_{csr}$ than the Velodyne sensors and offer some improvement over the RSBL.

Figure 7b presents the derivative of $\hat{K}$ with respect to the normalized great-circle distance $r/R$. This graph is interesting because its values for a given distance represent observed frequencies of points separated by that particular distance. The discretization of the $r/R$ axis in 151 sections explains the noise-like aspect of the VLP-16 curve (i.e., the one with the smallest number of samples). The reference curve $K_{csr}$ is symmetrical with respect to the great-circle distance and reaches its maximum at the central value (i.e., for points at about a distance of 90°). This central peak value does not appear in the VLP-16 and HDL-32 curves, where large portions of the sphere are not sampled. Furthermore, some asymmetry is evident for the HDL-32, which is consistent with its asymmetric FOV. The tilting sensors give results that are much closer to the $K_{csr}$, with a slight advantage of the RMBL with respect to the RSBL.

## 6. Implementation of a Portable Tilting Mechanism for a Velodyne VLP-16

The analysis methodology presented in Section 4 can be used to assess the scan measurement distribution of a real MBRL device and to establish a comparison with the ideal full-sphere values obtained in Section 5. An implementation of an RMBL consisting on a tilting multi-beam laser scanner has been developed in this work. This new device, named *Velomotion-16*, has been designed as a light portable platform based on the Velodyne VLP-16 scanner (see Figure 8).

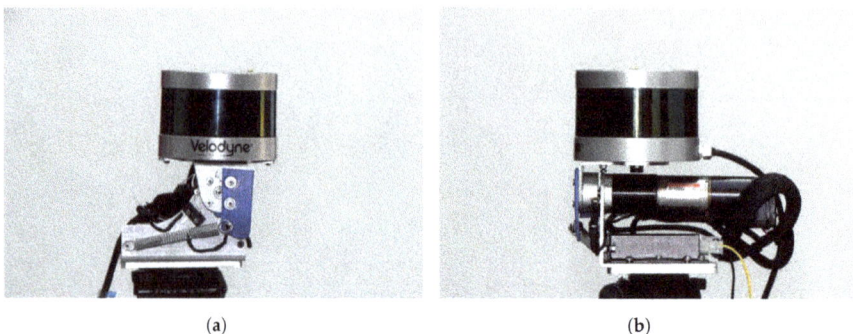

(a)                                                    (b)

**Figure 8.** Views of the Velomotion-16 RMBL sensor: (**a**) side; and (**b**) front.

### 6.1. Velomotion-16 System Description

Several views of the *Velomotion-16* sensor design are presented in Figure 9. This figure shows the frames for VLP-16 frame and the *Velomotion-16* using the same notation as in Figure 1, including the tilting angle $\gamma$ and the relative distance between the rotation mechanism and the optical center, which is $d = 6$ cm.

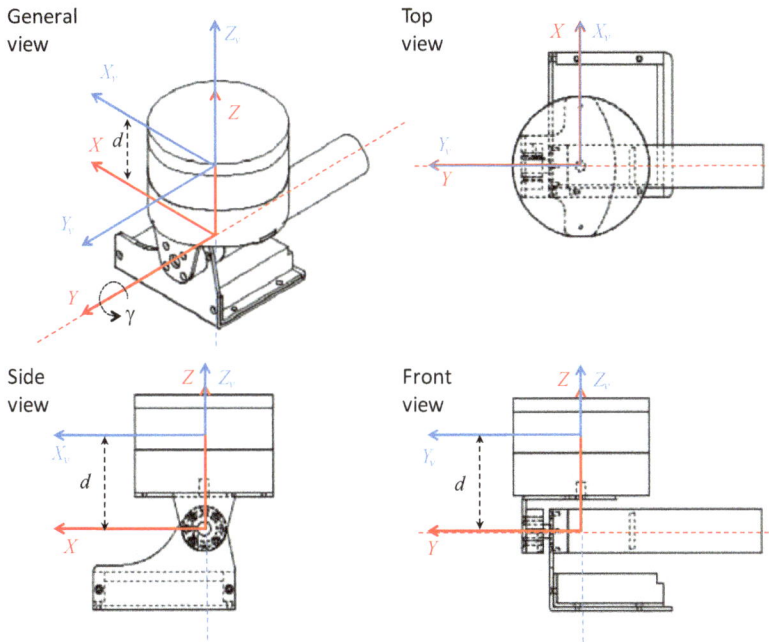

**Figure 9.** *Velomotion-16* reference frames and tilting parameters. The VLP-16 local frame is represented in blue; the *Velomotion-16* local frame is represented in red.

The main specifications of the tilting platform are presented in Table 3, where some parameters are inherited from the constituent VLP-16 sensor. In this device, there is a mechanical limitation regarding the additional DOF, which is in the range $[-45°, 0°]$. This means that the vertical FOV is asymmetrical with respect to the horizontal plane, as it is $[-60°, 15°]$ in the forward direction and $[-15°, 60°]$ backwards.

**Table 3.** Specifications of Velomotion-16, as used in the case study (VLP-16 device included). When values are inherited from the VLP-16, this is indicated.

| | |
|---|---|
| Range | 1 m to 100 m (VLP-16) + offset $\leq d$ |
| Accuracy | $\pm 3$ cm (VLP-16) |
| Data Rate | 300,000 points/s (VLP-16) |
| $d$ | 6 cm |
| Tilting range | $[-45°, 0°]$ |
| Tilting speed | $0.05°/s$ to $56.25°/s$ (programmable) |
| Vertical FOV | $75°: [-60°, +15°]$ (forwards), $[-15°, 60°]$ (backwards) |
| Vertical Resolution | uneven |
| Mean vertical resolution | $5.2°\cdot 10^{-3}$ to $0.59°$ (programmable) |
| Horizontal FOV | $360°$ (VLP-16) |
| Horizontal Resolution | $0.1°$ to $0.4°$ (programmable) (VLP-16) |
| Size | 105 mm width $\times$ 95 mm height $\times$ 165 mm depth |
| Weight | 1.9 kg (+0.7 kg wires) |

The range of Velomotion-16 is inherited from that of the VLP-16, but it is affected by a positive offset not greater than $d$. The actual limits of the sensing range depend on $(\alpha, \omega, \gamma)$. Given that a range measurement $P$ (see Figure 1) is:

$$P = \sqrt{x^2 + y^2 + z^2},$$ (9)

then, the actual minimum and maximum range values for $P$ can be computed using Equations (4)–(6) with $R = 1$ m and $R = 100$ m, respectively.

From a mechanical standpoint, the portable tilting platform consists of two L-shaped links with a rotational joint. The base link has been designed to accommodate the controller card and the motor-gear-break set and includes two switches to restrict the displacement. The VLP-16 support link has been designed to be lightweight, to achieve a small $d$ value, and to make the $Y$ and the $Y_v$ axes parallel, as in Figure 1. A cylindrical coupling piece joins the output axis from the reduction gear with the VLP-16 support. Furthermore, a spring avoids the clearance between the base link and the VLP-16 support.

A general overview of the system architecture is shown in Figure 10. The tilting motion is achieved by an EC brushless motor with encoder and brake and an EPOS2 controller, both by Maxon (Sachseln, Switzerland). Two 12 V batteries are used to provide 12 V power to the system (including the VLP-16) and 24 V to the brake. Motion control is performed through a trapezoidal profile in which speed, acceleration and deceleration can be specified by the user to produce a particular scanning resolution. The fastest scan is achieved by setting the tilting speed to $56.25°/s$, which corresponds to eight VLP-16 scans. Conversely, high density scans can be obtained by programming slower tilt speeds, which can be as low as $0.05°/s$.

As illustrated in Figure 10, the PC host sends capture commands and receives 3D data as Robot Operating System (ROS) messages from the VLP-16 via Ethernet. Moreover, the PC sends the goal position of the motion profile and receives the current angle from the tilting platform through a USB connection. A ROS driver has been developed to synchronize consecutive VLP-16 scans with the corresponding tilt angles in order to generate a dense point cloud.

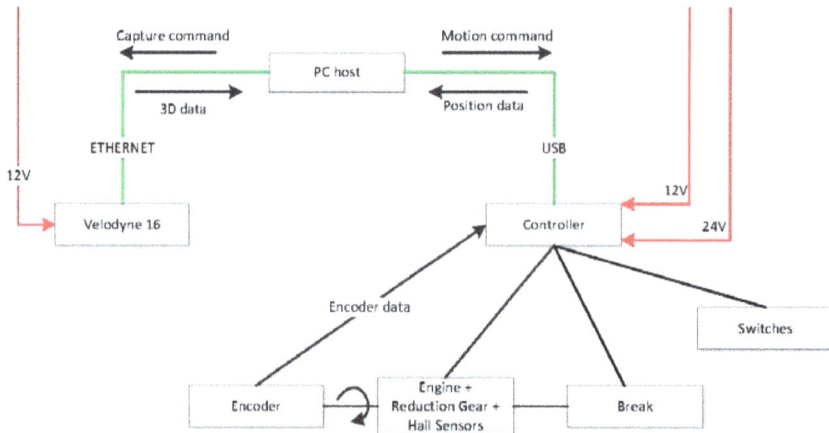

**Figure 10.** Velomotion-16 system architecture.

*6.2. Analysis of the Scan Measurement Distribution for Velomotion-16*

The effects on the scan measurement distribution produced by a particular sensor construction can be identified by applying the proposed analysis methodology. This is illustrated in Figure 11 for

Velomotion-16, which has a limited tilting range. The results in the figure correspond to a tilting speed of $50°/s$.

The complete FOV originated by the tilting range, which can be clearly appreciated in Figure 11a, has a downwards orientation in the negative direction of the $X$ axis (i.e., backwards). The hollow sphere patterns are similar to those of the ideal sensor with the exception of the scan lines in the extremes of the tilt range, which are sparser because they are not overlapped. This sparseness can be appreciated in the forward floor plane points of Figure 11b. Orthogonal plane points also reveal the efficacy of Velomotion-16 to scan ground points in the backward direction, whereas the forward direction is more appropriate to capture higher vertical structures.

The point density histograms in Figure 11c,d also indicate unscanned regions as well as a slighter scan density in the extremes of the tilt range. The mean point density has been reduced from $25.16$ points/deg$^2$ ($202.73$ points/bin) in the ideal full-sphere to $6.29$ points/deg$^2$ ($50.68$ points/bin) in the Velomotion-16 sensor. Similarly, the standard deviation has changed from $16.62$ points/deg$^2$ ($133.92$ points/bin) to $7.85$ points/deg$^2$ ($63.22$ points/bin).

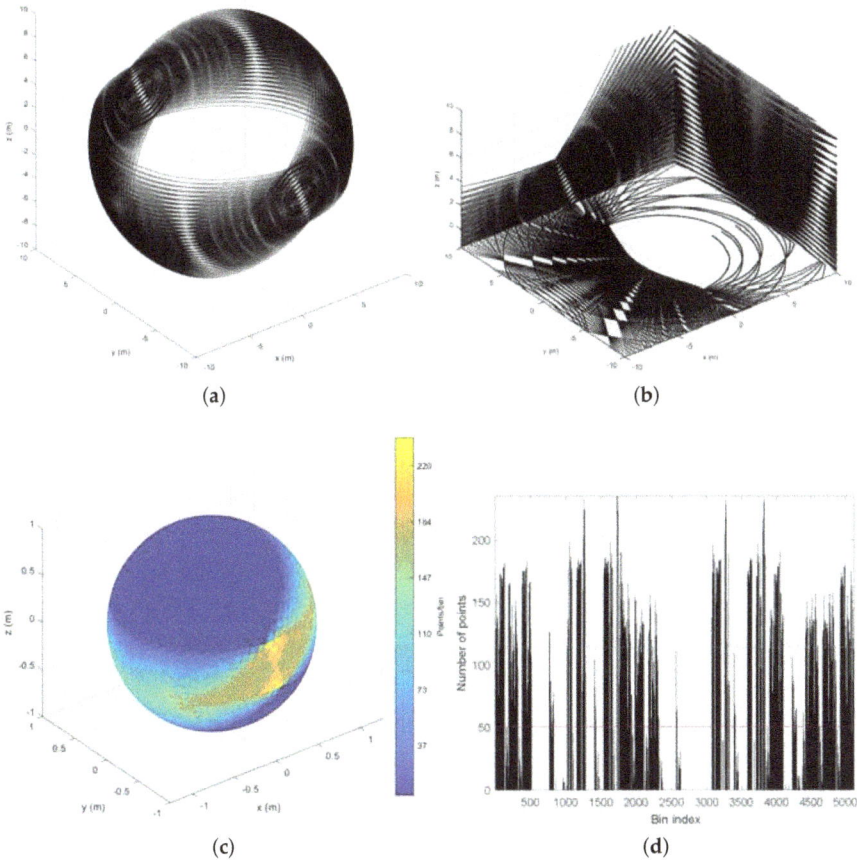

(a)

(b)

(c)

(d)

**Figure 11.** *Cont.*

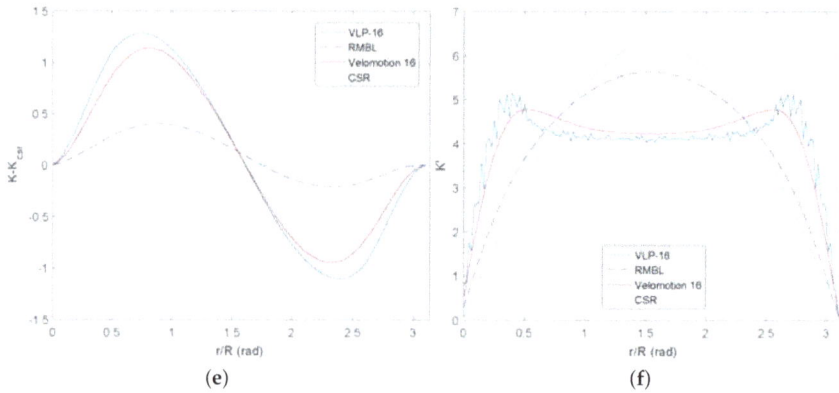

(e)                (f)

**Figure 11.** Analysis of 3D scan measurement distribution for *Velomotion-16* with tilting speed of 50°/s (17 frames): (**a**) hollow sphere points; (**b**) orthogonal plane points; (**c**) point density on the sphere (dark blue means no measurements); (**d**) point density as a histogram; (**e**) homogeneity analysis as $\hat{K}$ - $K_{csr}$; and (**f**) homogeneity analysis as the first derivative of $\hat{K}$.

As for the $K$ function, the difference of the Velomotion-16 with respect the CSR value (shown in Figure 11e) clearly improves homogeneity with respect to the original VLP-16 sensor, in both clustering and dispersion. However, the limited tilt range of the Velomotion-16 causes that the $K$ function is closer to the VLP-16 than to the ideal RMBL. This homogeneity results can also be appreciated in the first derivative representation of the estimated $K$ function seen in Figure 11f.

## 7. Discussion of Example Scans

The purpose of this section is to illustrate 3D lidar data from actual scans obtained in three representative scenes of indoor and outdoor environments: a building hall, an outdoor parking area in a urban setting, and a quarry with irregular terrain, respectively (see Figure 12). Actual scans have been obtained by Velodyne's VLP-16 and HDL-32 MBLs as well as for the *Velomotion-16* RMBL with two different tilting speeds (i.e., vertical resolutions): the fastest scan speed given by the sensor (i.e., 56.25°/s) and a slow high resolution speed of 1.07°/s. The sensors have been placed on a tripod at a height of about 1.2 m.

(a)                (b)

**Figure 12.** *Cont.*

(c)

**Figure 12.** Photos of experimental scenes: (**a**) building hall (indoor); (**b**) parking area (urban); and (**c**) quarry (terrain).

Table 4 summarizes sensor performance by presenting the scan time for each case, as well as the resulting number of points for each scene. The total number of points given by the *Velomotion-16* is substantially larger in the indoor environment due to out-of-range measurements in the urban environment. This difference is not so important in the Velodyne sensors, as their FOV is very limited in the upwards direction. The fast *Velomotion-16* requires 0.8 s (i.e., eight times the Velodyne scan time) to capture eight VLP-16 scans, which produces a greater number of points (i.e., 7.21 times and 7.86 times, for the indoor and urban scenes, respectively) than a single VLP-16 scan as well as a wider FOV. Furthermore, the fast *Velomotion-16* also improves the number of points and the FOV of the HDL-32, which is a considerably more expensive sensor. By adjusting the *Velomotion-16* $\gamma$ speed to a slower value, the resulting number of points, and the subsequent data density, can be greatly improved, as indicated by the numbers given by the slow case in the table.

**Table 4.** Sensor performance in example scans.

|  | Scan Time (s) | $\gamma$ Speed (°/s) | Indoor: Points | Urban: Points | Terrain: Points |
|---|---|---|---|---|---|
| *Velomotion-16* (slow) | 42.05 | 1.07 | 11,348,103 | 6,833,873 | 6,269,304 |
| *Velomotion-16* (fast) | 0.80 | 56.25 | 201,851 | 158,351 | 125,237 |
| VLP-16 | 0.10 | - | 27,998 | 20,141 | 16,954 |
| HDL-32 | 0.10 | - | 68,080 | 58,875 | 51,362 |

The results for these experiments are presented in Figures 13–15, where the color grading represents elevation. These illustrations confirm the improvement in data density and FOV provided by the *Velomotion-16* with respect to both Velodyne lidars. The particular mechanical rotation limits of the *Velomotion-16* favor a denser resolution of the floor in the forward direction and a better measurement of higher objects in the backwards direction. Interestingly, even the fast *Velomotion-16* offers a high point density in the immediate floor area. In the quarry scene, the forward scan direction provides a detailed scan of the ground terrain and the excavated wall, especially in the slow scan.

As for the particular scan patterns presented in Figure 4, these can also be appreciated in the floor and vertical walls of these actual scans. In general, these patterns become difficult to identify when the resolution is increased in the fast *Velomotion-16*, but the pattern of small blind spots can be seen on the wall at the center-left part of Figure 13d, the floor at the bottom-right side of Figure 14d, and the bottom-left diagonal and the top-right area of Figure 15d.

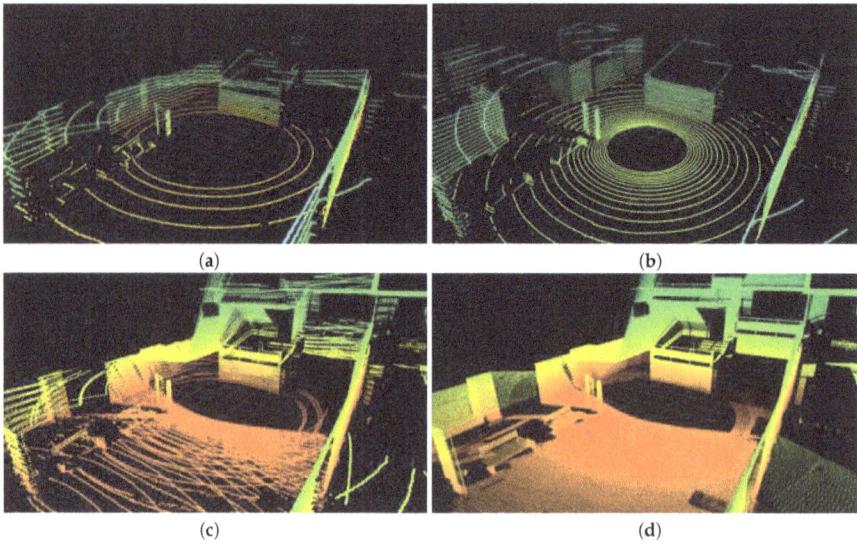

**Figure 13.** Scans from an indoor scene: (**a**) VLP-16; (**b**) HDL-32; (**c**) Fast *Velomotion-16*; and (**d**) Slow *Velomotion-16*.

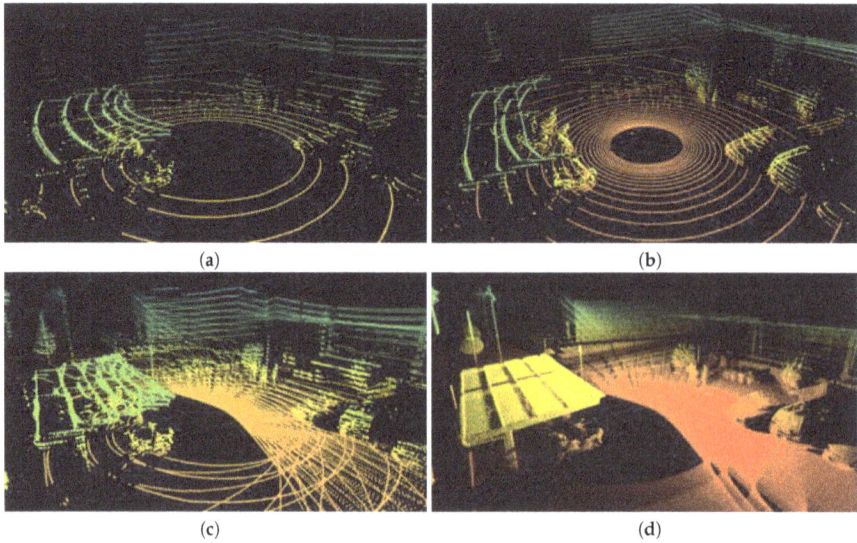

**Figure 14.** Scans from the urban scene: (**a**) VLP-16; (**b**) HDL-32; (**c**) Fast *Velomotion-16*; and (**d**) Slow *Velomotion-16*.

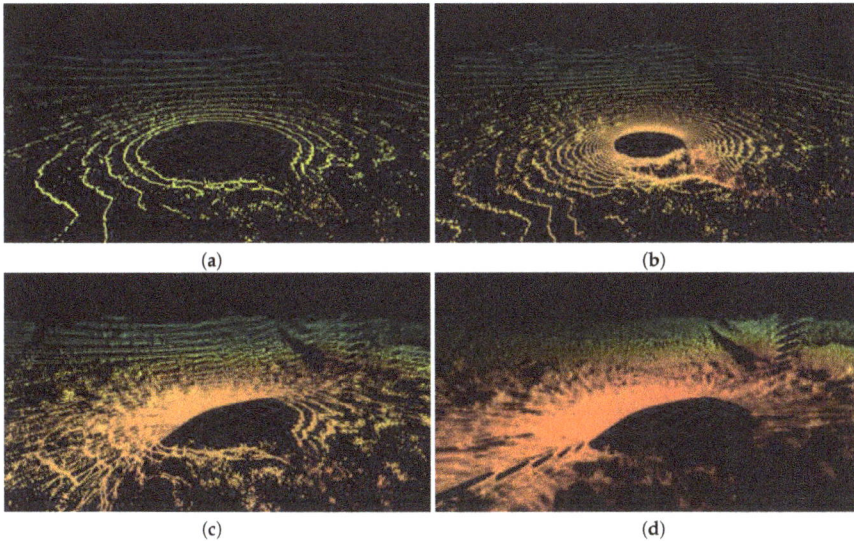

**Figure 15.** Scans from an outdoor terrain scene: (**a**) VLP-16; (**b**) HDL-32; (**c**) Fast *Velomotion-16*; and (**d**) Slow *Velomotion-16*.

## 8. Conclusions

In this work, we have addressed rotating multi-beam lidar (RMBL) sensors, a type of customized 3D rangefinders built by adding a rotation mechanism to a commercial multi-beam lidar (MBL). Recent published examples using VLP-16 (Puck), the most affordable and lightest MBL by Velodyne, indicate that the RMBL configuration has the potential to become a common solution to get low-cost, rapid, and full-3D high resolution scans, as has happened with customized rotating single-beam lidars (RSML) during the last fifteen years. However, contrary to RSBLs, the additional DOF in RMBLs causes multiple scan beams to overlap creating complex scanning patterns.

Particularly, we have proposed a simulation-based methodology to analyze 3D scanning patterns, which has been applied to investigate the complex scan measurement distribution produced by the RMBL configuration. With this major purpose, novel contributions offered in the paper include the following: (i) the adaption of a recent spherical reformulation of Ripley's $K$ function to assess 3D sensor data distribution on hollow sphere simulations; (ii) a comparison, both qualitative and quantitative, of scan patterns produced by an ideal RMBL based on a Velodyne VLP-16 (Puck) and those of other 3D scan alternatives (i.e., rotating 2D lidar and MBL); and (iii) a discussion of experimental scans from three representative environments obtained from a new RMBL implementation consisting of a portable tilting platform for VLP-16 scanners, which have been compared with the VLP-16 and HDL-32E MBLs.

Qualitative analysis evidences particular sampling patterns provoked by the RMBL configuration. Most of these patterns are difficult to appreciate when the resolution of the additional rotation is increased with slower scans. Nevertheless, rows of characteristic small blind spots remain visible even with high resolution rotation. Apart from that, similarly to RSBLs, the measurement density has focal points in the rotation axis, which has to be considered when placing the sensor for a particular application.

Besides, the analysis of the spatial distribution of scan measurements with the spherical $K$ function indicates that homogeneity is independent of the rotation speed. However, it has been observed that the $K$ function deviates from complete spatial randomness due to unsampled regions, such as those resulting from the limited field of view (FOV) of MBLs, as well as by poles (or focal points) in RMBLs and RSBLs. The comparison of *K* function estimations between ideal full-sphere RMBLs and an equivalent RSBL yields similar results, with a slight advantage of the multi-beam based case.

The *Velomotion-16* sensor has been presented as case study to discuss actual scans from representative scenes as well as an illustration of the use of the proposed methodology to analyze the scan measurement distribution of an actual sensor with a limited tilt range. Thus, experimental example scans obtained by a VLP-16 on a tilting mechanism have illustrated a practical implementation of the RMBL configuration. These scans have shown that even a fast tilt (in less than one second) of the VLP-16 provides an environment description that can be considerably richer (both in FOV and number of points) than that of the HDL-32. A much higher level of detail can be appreciated in scans taken in less than a one-minute span.

Given the significant difference in price between the VLP-16 and the HDL-32 (not to mention other high-end 3D lidars), these results support the feasibility of customized RMBLs based on the least expensive MBLs (like the VLP-16) in applications demanding affordable and compact high-resolution point clouds without the FOV and vertical resolution limitations of commercial MBLs. The use of this type of sensor for robotic mapping can be done with stop-and-go scans or with continuous tilting, but the latter requires a more complicated registration process [13]. Furthermore, other future applications could benefit from the availability of a portable and affordable sensor producing high-resolution point clouds with no FOV limitations in a stop-and-go fashion. These potential applications could include: intelligent surveillance of public and private spaces; modeling and progress tracking in construction sites; modeling of caves, tunnels or narrow spaces in speleology, archeology, mining, and search and rescue; and body imaging for medicine, prosthetics, textile tailoring and other applications.

Further potentially interesting aspects include an analysis of accuracy and precision of the *Velomotion-16* device. We are investigating these aspects in ongoing studies. Moreover, it will be interesting to test the applicability of calibration methods devised for 2D lidar with a rotation mechanism (e.g., [23]) when applied to the rotation of multiple beams.

In the future, the proposed simulation methodology can be useful to assess the effect of other particular rotating mechanisms with respect to ideal measure distributions. Furthermore, it will also be interesting to analyze the effect of rotating other MBLs, since there is an increasingly active MBL market for new compact and cost-effective devices (e.g., the 32-beam Ultra Puck VLP-32C by Velodyne [11]).

**Acknowledgments:** This work was partially supported by the Spanish project DPI2015-65186-R. The authors are grateful to anonymous reviewers for their valuable comments.

**Author Contributions:** The simulations and spherical *K* function analysis were performed by J.M. The *Velomotion-16* sensor was conceived and implemented by V.P.-L., J.S., and A.G.-C. The manuscript was written by A.M. and J.A.G.-R. The design and analysis of experiments were done by J.A.G.-R., A.M., and V.P.-L. The work presented in the paper was coordinated by A.M. and was conceived within research projects led by A.G.-C. and A.M.

**Conflicts of Interest:** The authors declare no conflict of interest.

## References

1. Omar, T.; Nehdi, M.L. Data acquisition technologies for construction progress tracking. *Autom. Constr.* **2016**, *70*, 143–155.
2. Yandun, F.; Reina, G.; Torres-Torriti, M.; Kantor, G.; Auat Cheein, F. A Survey of Ranging and Imaging Techniques for Precision Agriculture Phenotyping. *IEEE/ASME Trans. Mechatron.* **2017**, doi:10.1109/TMECH.2017.2760866.
3. Chromy, A.; Zalud, L. Robotic 3D scanner as an alternative to standard modalities of medical imaging. *SpringerPlus* **2014**, *3*, doi:10.1186/2193-1801-3-13.
4. Benedek, C. 3D people surveillance on range data sequences of a rotating Lidar. *Pattern Recognit. Lett.* **2014**, *50*, 149–158.
5. Vitali, A.; Rizzi, C. A virtual environment to emulate tailor's work. *Comput.-Aided Des. Appl.* **2017**, *14*, 671–679.

6.   Brede, B.; Lau, A.; Bartholomeus, H.M.; Kooistra, L. Comparing RIEGL RiCOPTER UAV LiDAR derived canopy height and DBH with terrestrial LiDAR. *Sensors* **2017**, *17*, doi:10.3390/s17102371.

7.   Batavia, P.H.; Roth, S.A.; Singh, S. Autonomous coverage operations in semi-structured outdoor environments. In Proceedings of the IEEE/RSJ International Conference on Intelligent Robots and Systems, Lausanne, Switzerland, 30 September–4 October 2002; Volume 1, pp. 743–749.

8.   Sheh, R.; Jamali, N.; Kadous, M.W.; Sammut, C. A low-cost, compact, lightweight 3D range sensor. In Proceedings of the Australasian Conference on Robotics and Automation, ACRA 2006, Auckland, New Zealand, 6–8 December 2006.

9.   Yoshida, T.; Irie, K.; Koyanagi, E.; Tomono, M. A sensor platform for outdoor navigation using gyro-assisted odometry and roundly-swinging 3D laser scanner. In Proceedings of the IEEE/RSJ International Conference on Intelligent Robots and Systems, Taipei, Taiwan, 18–22 October 2010; pp. 1414–1420.

10.  Droeschel, D.; Schwarz, M.; Behnke, S. Continuous mapping and localization for autonomous navigation in rough terrain using a 3D laser scanner. *Robot. Auton. Syst.* **2017**, *88*, 104–115.

11.  Velodyne LIDAR, Inc. Products. Available online: http://velodynelidar.com/products.html (accessed on 27 January 2018).

12.  Neumann, T.; Dülberg, E.; Schiffer, S.; Ferrein, A. A rotating platform for swift acquisition of dense 3D point clouds. In Proceedings of the International Conference on Intelligent Robotics and Applications, Wuhan, China, 16–18 August 2016; Volume 9834, pp. 257–268.

13.  Neumann, T.; Ferrein, A.; Kallweit, S.; Scholl, I. Towards a mobile mapping robot for underground mines. In Proceedings of the 2014 PRASA, RobMech and AfLaT International Joint Symposium, Cape Town, South Africa, 27–28 November 2014; pp. 279–284.

14.  Klamt, T.; Behnke, S. Anytime Hybrid Driving-Stepping Locomotion Planning. In Proceedings of the IEEE/RSJ International Conference on Intelligent Robots and Systems, Vancouver, BC, Canada, 24–28 September 2017; pp. 1–8.

15.  Wulf, O.; Wagner, B. Fast 3D Scanning Methods for Laser Measurement Systems. In Proceedings of the International Conference on Control Systems and Computer Science, Bucharest, Romania, 2–5 July 2003; Volume 1, pp. 312–317.

16.  Ripley, B.D. Modelling Spatial Patterns. *J. R. Stat. Soc. Ser. B Methodol.* **1977**, *39*, 172–212.

17.  Robeson, S.M.; Li, A.; Huang, C. Point-pattern analysis on the sphere. *Spatial Stat.* **2014**, *10*, 76–86.

18.  Weingarten, J.; Siegwart, R. 3D SLAM using planar segments. In Proceedings of the 2006 IEEE/RSJ International Conference on Intelligent Robots and Systems, Beijing, China, 9–13 October 2006; pp. 3062–3067.

19.  Dias, P.; Matos, M.; Santos, V. 3D reconstruction of real world scenes using a low-cost 3D range scanner. *Comput.-Aided Civ. Infrastruct. Eng.* **2006**, *21*, 486–497.

20.  Ueda, T.; Kawata, H.; Tomizawa, T.; Ohya, A. Mobile SOKUIKI Sensor System-Accurate Range Data Mapping System with Sensor Motion. In Proceedings of the International Conference on Autonomous Robots and Agents, Palmerston North, New Zealand, 12–14 December 2006.

21.  Morales, J.; Martínez, J.L.; Mandow, A.; Pequeño-Boter, A.; García-Cerezo, A. Design and development of a fast and precise low-cost 3D laser rangefinder. In Proceedings of the IEEE International Conference on Mechatronics, Istanbul, Turkey, 13–15 April 2011; pp. 621–626.

22.  Xiao, J.; Zhang, J.; Adler, B.; Zhang, H.; Zhang, J. Three-dimensional point cloud plane segmentation in both structured and unstructured environments. *Robot. Auton. Syst.* **2013**, *61*, 1641–1652.

23.  Morales, J.; Martínez, J.L.; Mandow, A.; Reina, A.J.; Pequeño Boter, A.; García-Cerezo, A. Boresight Calibration of Construction Misalignments for 3D Scanners Built with a 2D Laser Rangefinder Rotating on Its Optical Center. *Sensors* **2014**, *14*, 20025–20040.

24.  Alismail, H.; Browning, B. Automatic Calibration of Spinning Actuated Lidar Internal Parameters. *J. Field Robot.* **2015**, *32*, 723–747.

25.  An, S.Y.; Lee, L.K.; Oh, S.Y. Line segment-based fast 3D plane extraction using nodding 2D laser rangefinder. *Robotica* **2015**, *33*, 1751–1774.

26.  Martínez, J.L.; Morales, J.; Reina, A.J.; Mandow, A.; Pequeño-Boter, A.; García-Cerezo, A. Construction and calibration of a low-cost 3D laser scanner with 360 degrees field of view for mobile robots. In Proceedings of the IEEE International Conference on Industrial Technology, Seville, Spain, 17–19 March 2015; pp. 149–154.

27. Özbay, B.; Kuzucu, E.; Gül, M.; Öztürk, D.; Taşci, M.; Arisoy, A.M.; Şirin, H.O.; Uyanik, I. A high frequency 3D LiDAR with enhanced measurement density via Papoulis-Gerchberg. In Proceedings of the International Conference on Advanced Robotics, ICAR 2015, Istanbul, Turkey, 27–31 July 2015; pp. 543–548.

28. Moon, Y.G.; Go, S.J.; Yu, K.H.; Lee, M.C. Development of 3D laser range finder system for object recognition. In Proceedings of the IEEE/ASME International Conference on Advanced Intelligent Mechatronics, Busan, Korea, 7–11 July 2015; pp. 1402–1405.

29. Shaukat, A.; Blacker, P.C.; Spiteri, C.; Gao, Y. Towards camera-LIDAR fusion-based terrain modelling for planetary surfaces: Review and analysis. *Sensors* **2016**, *16*, doi:10.3390/s16111952.

30. Schubert, S.; Neubert, P.; Protzel, P. How to build and customize a high-resolution 3D laserscanner using off-the-shelf components. In Proceedings of the Conference Towards Autonomous Robotic Systems, Guildford, UK, 19–21 July 2016; Volume 9716, pp. 314–326.

31. Leingartner, M.; Maurer, J.; Ferrein, A.; Steinbauer, G. Evaluation of Sensors and Mapping Approaches for Disasters in Tunnels. *J. Field Robot.* **2016**, *33*, 1037–1057.

32. Kang, J.; Doh, N.L. Full-DOF Calibration of a Rotating 2-D LIDAR with a Simple Plane Measurement. *IEEE Trans. Robot.* **2016**, *32*, 1245–1263.

33. Velodyne LIDAR, Inc. Datasheets. Available online: http://velodynelidar.com/docs/datasheet/ (accessed on 27 January 2018).

34. Vlaminck, M.; Luong, H.; Goeman, W.; Philips, W. 3D scene reconstruction using Omnidirectional vision and LiDAR: A hybrid approach. *Sensors* **2016**, *16*, doi:10.3390/s16111923.

35. Atanacio-Jiménez, G.; González-Barbosa, J.J.; Hurtado-Ramos, J.B.; Ornelas-Rodríguez, F.J.; Jiménez-Hernández, H.; García-Ramirez, T.; González-Barbosa, R. LIDAR velodyne HDL-64E calibration using pattern planes. *Int. J. Adv. Robot. Syst.* **2011**, *8*, 70–82.

36. Chen, C.Y.; Chien, H.J. On-site sensor recalibration of a spinning multi-beam LiDAR system using automatically-detected planar targets. *Sensors* **2012**, *12*, 13736–13752.

37. Chan, T.O.; Lichti, D.D. Automatic In Situ calibration of a spinning beam LiDAR system in static and kinematic modes. *Remote Sens.* **2015**, *7*, 10480–10500.

38. Glennie, C.L.; Kusari, A.; Facchin, A. Calibration and stability analysis of the VLP-16 laser scanner. In Proceedings of the International Archives of the Photogrammetry, Remote Sensing and Spatial Information Sciences—ISPRS Archives, Prague, Czech Republic, 12–19 July 2016; Volume 40, pp. 55–60.

39. Mandow, A.; Martínez, J.; Reina, A.; Morales, J. Fast range-independent spherical subsampling of 3D laser scanner points and data reduction performance evaluation for scene registration. *Pattern Recognit. Lett.* **2010**, *31*, 1239–1250.

40. Tonini, M.; Abellan, A. Rockfall detection from terrestrial lidar point clouds: A clustering approach using R. *J. Spatial Inf. Sci.* **2014**, *8*, 95–110.

41. Baumgardner, J.R.; Frederickson, P.O. Icosahedral discretization of the two-sphere. *SIAM J. Numer. Anal.* **1985**, *22*, 1107–1115.

MDPI

*Article*

# An Intraoperative Visualization System Using Hyperspectral Imaging to Aid in Brain Tumor Delineation

**Himar Fabelo** [1,*], **Samuel Ortega** [1], **Raquel Lazcano** [2], **Daniel Madroñal** [2], **Gustavo M. Callicó** [1], **Eduardo Juárez** [2], **Rubén Salvador** [2], **Diederik Bulters** [3], **Harry Bulstrode** [4], **Adam Szolna** [5], **Juan F. Piñeiro** [5], **Coralia Sosa** [5], **Aruma J. O'Shanahan** [5], **Sara Bisshopp** [5], **María Hernández** [5], **Jesús Morera** [5], **Daniele Ravi** [6], **B. Ravi Kiran** [7], **Aurelio Vega** [1], **Abelardo Báez-Quevedo** [1], **Guang-Zhong Yang** [6], **Bogdan Stanciulescu** [8] and **Roberto Sarmiento** [1]

[1] Institute for Applied Microelectronics (IUMA), University of Las Palmas de Gran Canaria (ULPGC), Las Palmas de Gran Canaria 35017, Spain; sortega@iuma.ulpgc.es (S.O.); gustavo@iuma.ulpgc.es (G.M.C.); avega@iuma.ulpgc.es (A.V.); abaez@iuma.ulpgc.es (A.B.-O.); roberto@iuma.ulpgc.es (R.S.)

[2] Centre of Software Technologies and Multimedia Systems (CITSEM), Technical University of Madrid (UPM), Madrid 28031, Spain; raquel.lazcano@upm.es (R.L.); daniel.madronal@upm.es (D.M.); ejuarez@sec.upm.es (E.J.); ruben.salvador@upm.es (R.S.)

[3] Wessex Neurological Centre, University Hospital Southampton, Tremona Road, Southampton SO16 6YD, UK; dbulters@nhs.net

[4] Department of Neurosurgery, Addenbrookes Hospital and University of Cambridge, Cambridge CB2 0QQ, UK; hb252@cam.ac.uk

[5] Department of Neurosurgery, University Hospital Doctor Negrin, Las Palmas de Gran Canaria 35010, Spain; adamszolna@wp.pl (A.S.); pinerbrains1@yahoo.es (J.F.P.); coralia.sosa@gmail.com (C.S.); aruosha@gmail.com (A.J.O.); sarabisshop@hotmail.com (S.B.); hhdez.maria@gmail.com (M.H.); jmormol@gobiernodecanarias.org (J.M.)

[6] The Hamlyn Centre, Imperial College London (ICL), London SW7 2AZ, UK; d.ravi@imperial.ac.uk (D.R.); g.z.yang@imperial.ac.uk (G.-Z.Y.)

[7] Laboratoire CRISTAL, Université Lille 3, Villeneuve-d'Ascq 59653, France; ravi.kiran@esiee.fr

[8] Ecole Nationale Supérieure des Mines de Paris (ENSMP), MINES ParisTech, Paris 75006, France; bogdan.stanciulescu@mines-paristech.fr

* Correspondence: hfabelo@iuma.ulpgc.es; Tel.: +34-928-451-220

Received: 15 December 2017; Accepted: 30 January 2018; Published: 1 February 2018

**Abstract:** Hyperspectral imaging (HSI) allows for the acquisition of large numbers of spectral bands throughout the electromagnetic spectrum (within and beyond the visual range) with respect to the surface of scenes captured by sensors. Using this information and a set of complex classification algorithms, it is possible to determine which material or substance is located in each pixel. The work presented in this paper aims to exploit the characteristics of HSI to develop a demonstrator capable of delineating tumor tissue from brain tissue during neurosurgical operations. Improved delineation of tumor boundaries is expected to improve the results of surgery. The developed demonstrator is composed of two hyperspectral cameras covering a spectral range of 400–1700 nm. Furthermore, a hardware accelerator connected to a control unit is used to speed up the hyperspectral brain cancer detection algorithm to achieve processing during the time of surgery. A labeled dataset comprised of more than 300,000 spectral signatures is used as the training dataset for the supervised stage of the classification algorithm. In this preliminary study, thematic maps obtained from a validation database of seven hyperspectral images of in vivo brain tissue captured and processed during neurosurgical operations demonstrate that the system is able to discriminate between normal and tumor tissue in the brain. The results can be provided during the surgical procedure (~1 min), making it a practical system for neurosurgeons to use in the near future to improve excision and potentially improve patient outcomes.

**Keywords:** hyperspectral imaging instrumentation; brain cancer detection; image processing

## 1. Introduction

Currently, patients with brain cancer continue to have very poor survival rates. Surgery is one of the mainstays of treatment, together with radiotherapy and chemotherapy [1]. Brain tumors are classified based on their histology and molecular parameters [2]. Malignant gliomas are the most common form of primary brain tumors in adults and cause between 2 and 3% of cancer deaths worldwide [3]. Since brain tumors diffusely infiltrate into the surrounding normal brain tissue (especially gliomas), it is extremely difficult for the surgeon to accurately differentiate between tumor and normal brain tissue with the naked eye. In some cases, unintentionally leaving behind tumor tissue after the resection is unavoidable, and in other cases, too much normal brain tissue is resected in an effort to ensure complete excision. Over-resection can produce permanent neurological deficits that affect patient quality of life [4]. In contrast, several studies have demonstrated that tumor tissue left behind during surgery is a major cause of morbidity and mortality and represents the most common cause of tumor progression [5–7].

Several image guidance tools, such as intra-operative neuro-navigation, intra-operative magnetic resonance imaging (iMRI), and fluorescent tumor markers (for example 5-aminolevulinic acid, 5-ALA), have been commonly used to assist surgeons in the identification of brain tumor boundaries. However, these technologies have several limitations. One limitation is related to the brain shift phenomenon [8]. During craniotomy, the opening of the skull and dura inevitably leads to movement of the brain. This typically manifests as herniation of the brain into the craniotomy defect under pressure from the underlying tumor, or the slump of the brain due to drainage of cerebrospinal fluid and the administration of mannitol. Similarly, following resection of the tumor, the residual brain tissue may slump towards the surgical cavity. This brain deformation invalidates the patient-to-image mapping and reduces the effectiveness of using pre-operative images for intra-operative surgical guidance. Thus, neuronavigation systems relying on preoperative image data have decreasing accuracy as the surgical procedure progresses [9–11]. iMRI solves the problem of brain shift, mapping the tumor margins intra-operatively, but this method has poor spatial resolution and significantly extends the duration of the surgery, with a limited number of images that can be obtained [12]. Finally, although 5-ALA can identify the tumor boundaries, it produces relevant knock-on effects for the patient and can only be used for high-grade tumors [13,14]. Thus, there is no current device that helps in the accurate definition of brain tumor boundaries during surgical procedures. A label-free and non-ionizing imaging modality would be an ideal solution to this problem.

Hyperspectral imaging (HSI) is a non-contact, non-ionizing, and minimally invasive sensing technique that has been used in medical applications for more than two decades [15,16]. Unlike standard red, green, and blue (RGB) or multispectral images (which have a few more bands than the RGB image), hyperspectral (HS) images cover a wide range of the electromagnetic spectrum, and are able to capture a large number of contiguous and narrow spectral bands. This high amount of information conforms the spectral signature, which offers the possibility of distinguishing between each type of material or substance presented in the captured scene. HSI is an emerging imaging modality, and promising results have been shown with respect to cancer detection. Akbari et al. performed a study to identify gastric tumors in human ex vivo tissues, employing an HS system capable of capturing images ranging in size between 1000 and 250 nm [17]. From their experiments, they determined that the spectral regions between 1226 and 1251 nm and 1288 and 1370 nm are the most suitable ranges for distinguishing between non-cancerous and cancerous gastric tissue. Laryngeal cancer has been investigated by Regeling et al. using a flexible endoscopy coupled to an HSI system that is able to obtain HS cubes in the region between 390 and 680 nm [18]. Additionally, in this area Kester et al. developed a real-time snapshot HSI endoscope system based on an image mapping technique that is capable of operating at frames rates of 5.2 fps (frames per second), obtaining HS cubes in the range between 450 and 650 nm, with a spatial resolution of 100 μm [19]. In prostate cancer, Akbari et al. employed an HSI system to capture in vivo images (in the range between 450 and 950 nm) of mice affected by human prostate tumors [20]. Their results showed a maximum sensitivity of 92.8% and a specificity of 96.9% in the classification of malignant and non-malignant regions. Several studies have been carried out employing HSI for breast cancer diagnosis. Hou et al. developed a laser diode-induced hyperspectral system especially designed for breast cancer diagnosis, achieving higher accuracy and resolution as well as faster processing than other brain cancer diagnosis systems [21]. In addition, ex vivo breast cancer tissues were studied by Kim et al. to extract their regions of interest and thus differentiate between cancerous and non-cancerous tissues, employing a hyperspectral system that covered the region between 380 and 780 nm [22]. The same group also worked in the classification of these ex vivo breast cancer tissues using HSI, obtaining sensitivity and specificity of 98% and 99%, respectively [23]. In vivo colorectal tumors were also studied by Han et al. using a flexible hyperspectral colonoscopy system to discriminate between malignant colorectal tumors and normal mucosa in human patients [24]. Moreover, in vitro colon biopsy samples were analyzed by Masood et al. using a HSI system based on a tuned light source and a charge-coupled device (CCD) camera coupled to a microscope with 40× magnification (covering the range between 440 and 700 nm), obtaining accuracy results of 90% in the differentiation of benign and malignant patterns [25]. In vitro hyperspectral colon tissue images were also classified and segmented using morphological analysis and wavelet-based segmentation in [26,27]. HSI has also been applied to analyze skin cancer using visible-to-near-infrared (VNIR) information, obtaining promising results in the discrimination between melanoma and normal skin [28,29]. Other types of tumors have been also studied and analyzed using HSI, such as those of the head and neck [30], oral tissue [31], and tongue [32–34]. Nevertheless, HSI systems are not standardized, as different technologies were used in these studies. HS cameras generally use CCD sensors for VNIR applications (covering the range between 400 and 1000 nm) while indium gallium arsenide (InGaAs) sensors are used for near-infrared (NIR) applications (covering the range between 1000 and 1700 nm), since the quantum efficiency of the CCD sensors is quite low above 1000 nm. As a result, in some applications, more than a single camera is required to cover a broadband spectral range to study the suitable spectral range of the application, as is done in the creation of a spectral signature library for abdominal organs, arteries, and veins [35], or in the study of detection and analysis of intestinal ischemia during surgery [36]. The illumination systems used in HSI applications are mainly based on halogen or xenon lamps, and sometimes, optical fibers are used for light transmission, like in the diffuse reflectance spectroscopy used for early detection of malignant changes in the oral cavity [37].

The work presented in this paper was done as part of the HypErspectraL Imaging Cancer Detection (HELICoiD) project [38–40]. HELICoiD is a European Future and Emerging Technologies (FET) project with the goal of developing a demonstrator capable of discriminating between tumor and normal brain tissue, which can be used during neurosurgical operations. This demonstrator is designed to help surgeons with brain tumor resection, avoiding the excessive extraction of normal tissue and preventing small remnants of tumors from being left behind. Such precise delimitation of the tumors boundaries will improve the results of the surgery and is expected to improve patient outcomes. Although some parts of the system have been already described in previous works [41–45], in this paper we present, for the first time, a comprehensive description of the full system, including parts not previously addressed like the integration with hardware acceleration. We also present the measurements of the total times (for acquisition and processing), and the results using the complete training database and data from five new patients (which were not employed to train the classifier) to validate the overall system.

## 2. Materials and Methods

This section describes the HSI instrumentation developed for the detection of brain cancer intraoperatively. Figure 1 shows the block diagram of the demonstrator where all the parts of the system and their interconnections are presented. The acquisition platform is formed by two pushbroom HS cameras, covering the spectral range from 400 to 1700 nm, and the illumination system, mounted on a scanning platform guided by a high-precision stepper motor. The control unit is in charge of managing all the components of the system, while the hardware accelerator has the goal of speeding up the HS brain cancer detection algorithm in order to perform intraoperatively. The electromechanical elements allow the demonstrator's operator to focus and obtain the image in optimal conditions. Finally, the user interface was developed in a user-friendly way, facilitating the use of the system by non-expert users. Each of these parts will be described in detail in the following sections.

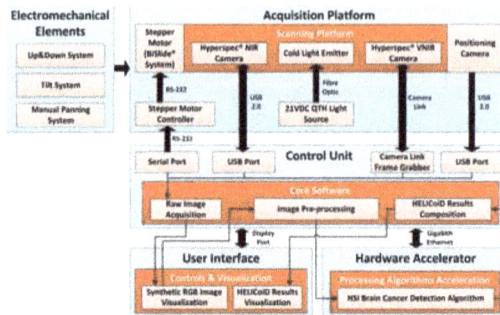

**Figure 1.** The HELICoiD demonstrator block diagram. QTH: Quartz Tungsten Halogen.

### 2.1. Acquisition Platform

The acquisition platform locates all the elements required to capture the HS images (also called HS cubes). Two HS cameras that cover the spectral range from 400 to 1700 nm are employed. Using these two cameras, two different HS cubes are generated: one in the VNIR spectral range (from 400 to 1000 nm) and another one in the NIR spectral range (from 900 to 1700 nm). Four different elements compose the acquisition platform: the HS cameras, the scanning platform, the illumination system, and the positioning camera. Figure 2 summarizes all the elements that are placed in the acquisition platform of the demonstrator.

### 2.1.1. Hyperspectral Cameras

HS cameras are mainly classified into four different types depending on the method employed to obtain the HS cube: whiskbroom (point-scanning) cameras, pushbroom (line-scanning) cameras, cameras based on spectral scanning (area-scanning or plane-scanning), and snapshot (single shot) cameras [46]. The HS cameras selected for the acquisition platform of the system are the Hyperspec® VNIR A-Series (Figure 2a) and the Hyperspec® NIR 100/U (Figure 2b) cameras, manufactured by Headwall Photonics Inc. (Fitchburg, MA, USA). These HS cameras are based on a line-scanning technique. The camera sensor is a two-dimensional detector array in which one of the spatial dimensions and the complete spectral dimension of the scene are captured in one single shot (called a frame). The second spatial dimension is obtained by shifting the camera's field of view (FOV) relative to the scene by means of a linear motion system. These cameras offer the best compromise between spectral and spatial resolution and acquisition time. The spectral range covered by both cameras is between 400 and 1700 nm (VNIR and NIR). This range has been selected with the aim of finding the most relevant spectral regions where the tumor and normal brain tissues can be distinguished using machine learning algorithms. The main characteristics of the selected cameras are as follows:

- The Hyperspec® VNIR A-Series model covers spectral range from 400 to 1000 nm. It has a dispersion per pixel of 0.74 nm and a spectral resolution of 2–3 nm (with a 25-μm slit), and is able to capture 826 spectral bands and 1004 spatial pixels. This device integrates a silicon CCD detector array (Adimec 1000-m, Adimec Electronic Imaging, Inc., Woburn, MA, USA) with a minimum frame rate of 90 fps. This sensor is a monochromatic camera connected to the control unit using a PIXCI® Camera Link Interface (EPIX, Inc., Buffalo Grove, IL), which provides a data transmission rate up to 255 MB/s. The lens used in this camera is a Xenoplan 1.4 (Schneider Optics, Hauppauge, NY, USA) with a focal length of 22.5 mm and a broadband coating for the spectral range of 400 to 1000 nm.
- The Hyperspec® NIR 100/U model covers the spectral range from 900 to 1700 nm. It has a dispersion per pixel of 4.8 nm and a spectral resolution of 5 nm (with a 25-μm slit), being able to capture 172 spectral channels and 320 spatial pixels. This system incorporates an indium gallium arsenide (InGaAs) detector array (Xeneth XEVA 5052, Xenics nv, Leuven, Belgium), which provides a fast response, high quantum efficiency, and low dark current for the sensor area. This system has a frame rate of up to 100 fps. This camera is connected to the control unit by a USB 2.0 interface with a transfer rate up to 60 MB/s. The lens used with this camera is a Kowa LM25HC-SW 1.4 (Kowa Optimed Deutschland GmbH, Düsseldorf, Germany) with 25 mm of focal length and a broadband coating for the spectral range of 800–2000 nm.

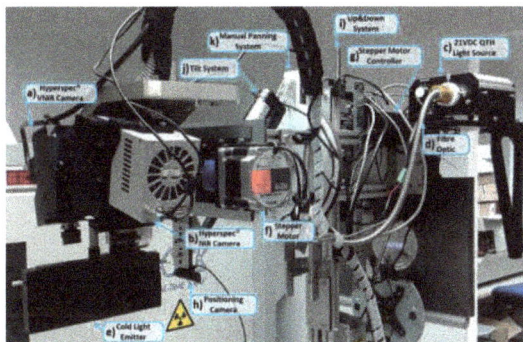

**Figure 2.** The HELICoiD demonstrator acquisition platform. (**a,b**) VNIR and NIR HS cameras mounted on the scanning platform; (**c–e**) QTH light source connected to the fiber optic system for the light transmission to obtain cold light emission in the scanning platform; (**f,g**) Stepper motor coupled to the spindle and connected to the stepper-motor controller to perform the linear movement of the cameras; (**h**) Positioning of the RGB camera used to identify the position of the cameras' field of view (FOV); (**i**) The Up&Down system used to focus the HS cameras; (**j**) and (**k**) Tilt and manual panning systems employed to correctly orientate the scanning platform.

### 2.1.2. Illumination System

HS cameras require strong and precise illumination of the scene to be captured in order to avoid external interferences produced by the environmental illumination where the capture is being performed. The illumination system used in this demonstrator is based on a quartz tungsten halogen (QTH) lamp of 150 W with a broadband emission between 400 and 2200 nm. This type of lamp is suitable for HS applications due to the high homogeneity of its spectrum across the entire spectral range [47]. The light source where the lamp is installed is a TechniQuip's Model 21 DC source light (TechniQuip, Pleasanton, CA, USA) connected to an optical fiber that transmits the light to a cold light emitter, ending in double glass isolation with an air chamber in the middle. Using this cold light system, the high temperature produced by the QTH lamp is isolated from the brain surface, since a high temperature irradiating over the brain surface can cause damage and even premature cell death [48]. Figure 2c shows the light source placed in the back of the system connected to the optical fiber (Figure 2d) that transmits the light to the cold light emitter located in the scanning platform (Figure 2e).

Although the illumination system employed in this demonstrator is able to avoid the interference of environmental illumination, HSI requires calibration of the raw images to be performed for correct processing of the data. In the calibration process, the significant signal variations caused by the non-uniform illumination over the surface of the captured scene are corrected. The acquired raw image is calibrated using white and dark reference images. These reference images are acquired by the system with the VNIR and NIR cameras separately, but in the same illumination conditions inside the operating theatre before the start of the operation. A white reference image is acquired from a Spectralon® tile (SphereOptics GmbH, Herrsching, Germany), a type of material that reflects the 99% of the incoming radiation in the full spectral range considered in this work. This white reference is placed at the same location where the patient's head will be placed during the surgery, thus taking into account all the real light contributions. The dark reference image is obtained by keeping the camera shutter closed and is used to avoid the dark currents produced by the camera sensor. The HS-calibrated image is calculated by Equation (1), where $\beta$ is the calibrated image, $\alpha$ is the raw image, and $\gamma$ and $\delta$ are the white and dark reference images, respectively:

$$\beta = 100 \cdot \frac{\alpha - \delta}{\gamma - \delta} \tag{1}$$

Figure 3a shows the white reference tile spectrum obtained with the VNIR camera, while Figure 3b,c respectively present raw and calibrated spectrum examples of normal brain tissue pixels. In Figure 3d the representation of the white reference tile spectrum obtained with the NIR camera can be seen, and in Figure 3e,f, the raw and calibrated spectra of a normal brain tissue pixel are shown. Based on the repeatability experiments performed with the system and taking into account that the white reference tile is used only a few minutes for the calibration, through measurements it is confirmed that the spectrum of the certified white reference tile does not show perceptible changes over time.

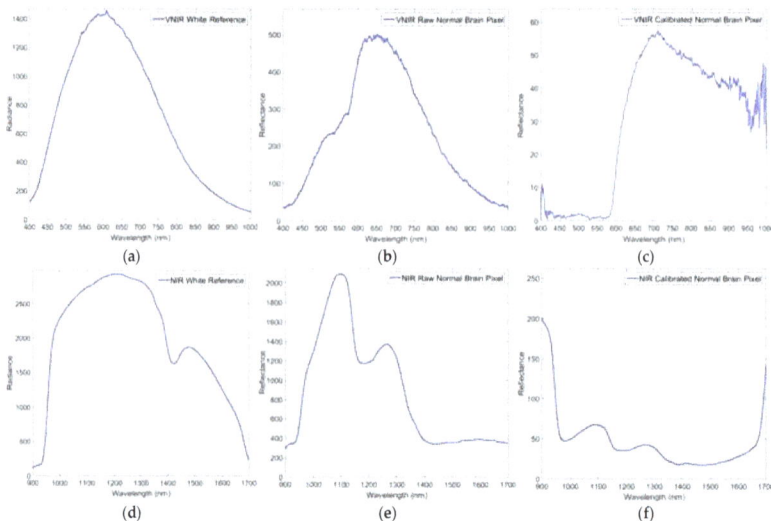

**Figure 3.** Calibration process of a certain pixel of the VNIR and NIR cameras. (**a**) The VNIR white reference spectrum; (**b**,**c**) The VNIR raw and calibrated spectra of a pixel of normal brain tissue; (**d**) The NIR white reference spectrum; (**e**,**f**) The NIR raw and calibrated spectra of a pixel of normal brain tissue.

2.1.3. Scanning Platform

Commonly, in the HS found in laboratories based on pushbroom cameras, the camera is usually fixed and the sample to be captured is moved, although some few examples can be found of moving cameras [49]. In brain tumor applications, it is not possible to move the brain of the patient to perform the capture; instead, the HS cameras (Figure 2a,b) are installed in a scanning platform together with a cold light emitter (Figure 2e). The scanning platform provides the necessary movement for the pushbroom scanning. This scanning platform is composed of a spindle and a stepper motor, called the BiSlide® motor-driven assembly (Velmex, Inc. Bloomfield, NY, USA, Figure 2f). The spindle has a size of 1 m and allows the cameras to capture a scene of a maximum size of 230 mm in the X-axis. The step resolution of the scanning platform is 6.17 μm. The stepper motor is managed by a Velmex VXM® stepping motor controller (Velmex, Inc. Bloomfield, NY, USA, Figure 2g). This motor controller is connected to the control unit via a serial protocol and its programming is accomplished through a Recommended Standard 232 (RS-232) protocol.

2.1.4. Positioning Camera

The positioning camera is installed in the acquisition platform to visualize the area that will be captured by the HS cameras. Since every HS camera sensor captures only one spatial line of the scene, it is not possible to determine the exact position of the current pushbroom frame over the brain. For this reason, the inclusion of an additional standard RGB camera in the acquisition platform was required, correctly aligned with the FOV of the HS cameras, in order to identify the area of the brain surface to be captured. However, unlike the HS cameras, this positioning camera is placed in a fixed position. This camera permits the user to visualize the complete area that is going to be captured by the cameras, allowing the system to be easily positioned in the correct place. Figure 2h shows the positioning camera placed in the acquisition platform below the scanning platform.

2.1.5. Electromechanical Elements

Three different electromechanical elements were installed in the HS acquisition system. These elements provide several degrees of freedom to the system, which are required to focus and orientate the cameras in a convenient way for obtaining high quality images. The Up&Down system (Figure 2i) allows the movement of the acquisition platform in the Y-axis to focus the camera. Keeping the HS images well focused is fundamental for obtaining good quality spectral signatures. Effectively, the spectral signature of each pixel is distorted in the case they are unfocused. The focus of the system is performed by looking an X-Lambda image (all the bands of the captured line in a spatial 2D image) captured by the sensor, where the lambda is the wavelength. The focusing distance between the exposed brain tissue and the lens of the cameras is 40 cm. This distance is determined by the distribution of the HS cameras in the scanning platform. The FOV of both cameras is oriented and aligned to the beam of the cold light emitter to obtain the highest reflectance value in the sensor. Furthermore, this distance is determined by the minimum security distance (30 cm) that must exist between the exposed brain and the nearest element of the demonstrator (in this case, the cold light emitter). The Up&Down system is composed of a 24-VDC motor coupled to a spindle, allowing a displacement of ±7.75 cm. On the other hand, the tilt system (Figure 2j) is composed of a 12-VDC linear actuator that permits the rotation of the scanning platform 40° forward and backward. Finally, the manual panning system (Figure 2k) is employed to manually rotate (up to 45° to the left and 45° to the right) the scanning platform, using an aluminum plate.

*2.2. Control Unit*

The control unit (CU) is responsible for managing all the subsystems that comprise the demonstrator. This CU is a computer based on an Intel® Core™ i7-4770k 3.5 GHz quad-core processor, with 8 GB of Random Access Memory (RAM) and a high-capacity 512 GB solid-state drive with write speeds exceeding 500 MB/s. Specific software was developed to manage and integrate the different elements that conform the acquisition platform, allowing the user to perform the HS image acquisition in an easy and effective way. Furthermore, the CU is in charge of executing the HS brain cancer detection algorithm together with the hardware accelerator in order to finally present the tumor boundary prediction.

HS Image Acquisition Software

Customized software for image acquisition was developed due to the need to automate and accelerate the capture of both HS cameras of the system. The simplification of the acquisition procedure ensures easy interaction of the user with the system as well as reduced time needed to capture the HS images during neurosurgical procedures.

To develop this software, three different software development kits (SDKs) were integrated, belonging to the two HS cameras and the stepper motor controller. Figure 4a shows the HS image acquisition software flow diagram for the capturing procedure. Firstly, after running the program, the scanning platform is initialized, detecting and establishing the absolute zero of the motor position. Then, the platform is positioned at the center of the scanning area. Taking into account the x-size value of the capturing area established by the user through the graphical user interface (GUI), the scanning platform is moved to the initial position. The VNIR capturing process is performed starting from the right to the left of the platform with the stepper motor speed fixed to 3 mm/s. This speed is calculated according to the pixel size (0.1287 mm and 0.48 mm for the VNIR and NIR cameras, respectively) and the frame rate of the camera (90 fps and 100 fps for the VNIR and NIR cameras, respectively). When the VNIR capture is done, the stepper motor stops at the final position, waits a few milliseconds to stabilize the system structure, and fixes the speed to 5 mm/s. Then, the NIR capturing process begins. This capture is performed starting from the left to the right of the platform. After that, the stepper motor moves the scanning platform to the central position. Then, the synthetic RGB images of both HS cubes are generated by selecting three bands that correspond with red (708.97 nm), green (539.44 nm), and blue (479.06 nm) colors for the VNIR image, and three bands of the NIR cube to generate a false color RGB image (red: 1094.89 nm, green: 1247.44 nm and blue: 1595.45 nm). These bands are selected to maintain the compatibility with the original software (Hyperspec® III software, Headwall Photonics Inc., Fitchburg, MA, USA) provided by the camera manufacturer. Using this technique for the acquisition process, a speedup of 3× with respect to the original software is achieved. The maximum image size provided by the system is 1004 × 1787 pixels (129 × 230 mm) for the VNIR image, and 320 × 479 pixels (153 × 230 mm) for the NIR image, with spatial resolutions of 128.7 μm and 480 μm, respectively.

Figure 4b shows the acquisition system being used during a neurosurgical operation and the RGB synthetic images of the captured HS cubes (VNIR and NIR) where their image sizes and relative spatial resolutions can be seen. The time employed by the system to obtain the maximum size image using the manufacturer's software is ~240 s for the VNIR image and ~140 s for the NIR image. However, employing the acquisition software developed in this work, the acquisition time for the maximum image size is reduced to ~80 s and ~40 s for the VNIR and NIR cameras, respectively.

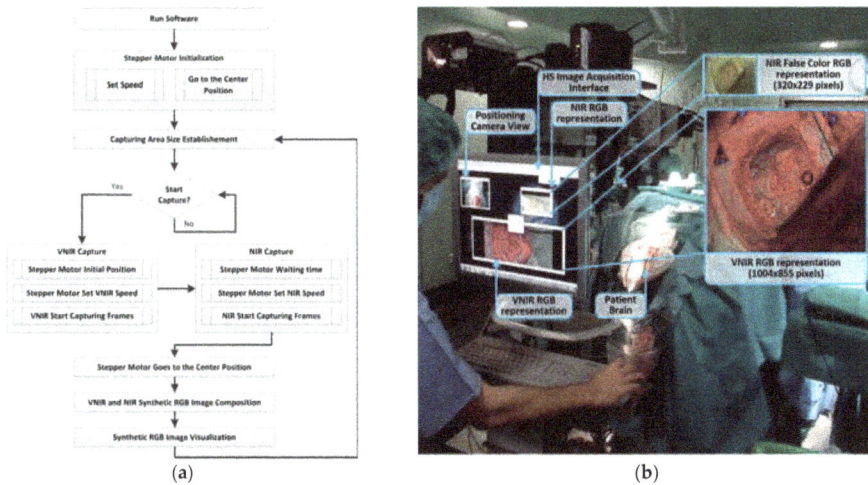

**Figure 4.** (**a**) HS image acquisition software flow diagram; (**b**) HS image acquisition user interface (and the RGB representations of each HS cube) being used during a neurosurgical intervention at the University Hospital Doctor Negrin of Las Palmas de Gran Canaria (Spain).

### 2.3. Hardware Accelerator

Due to the high computational cost of the developed HS brain cancer detection algorithm and the large amount of data generated by the HS cameras, it is necessary to use a hardware accelerator (HA) where the most time-consuming parts of the algorithm are implemented. Therefore, the algorithm must be highly parallelized for processing to be completed during neurosurgical operations.

The HA selected for this purpose is the Kalray Massively Parallel Processor Array (MPPA®) EMB01 board (Kalray S.A., Montbonnot Saint Martin, France) with a multiple instruction, multiple data (MIMD) many-core processor [50]. This accelerator is focused on computationally-intensive low-power embedded applications. The MPPA® EMB01 processing performance reaches 230 GFlops, which, for the 5-W power consumption reported, turns into 46 GFlops/W, a much higher figure compared to other kinds of high-performance platforms.

The MPPA® EMB01 board contains a standard host ×86 ComExpress module working as an embedded computer, and a carrier board containing the MPPA-256 many-core chip. Figure 5a shows the MPPA® board (in the center of the image) connected to a preliminary environment developed to execute the hardware accelerated part of the algorithm. The host module side of the board (Figure 5b) is composed of an AMD G-T40E Dual Core Processor with an integrated graphics processor unit (GPU) running a CentOS 7 GNU/Linux operative system (OS) instance with 4 GB of RAM, 1 Peripheral Component Interconnect Express (PCIe) Gen2×2 for communication with the MPPA®-256 many-core chip, and a 16-GB solid-state drive (SSD) as a system disk. The carrier board can be seen in Figure 5c. It features an MPPA®-256 many-core processor (under the fan). It also contains 4 GB of RAM and 64 MB of flash memory plus the host PCIe Gen2×2 port to communicate with the dual core processor.

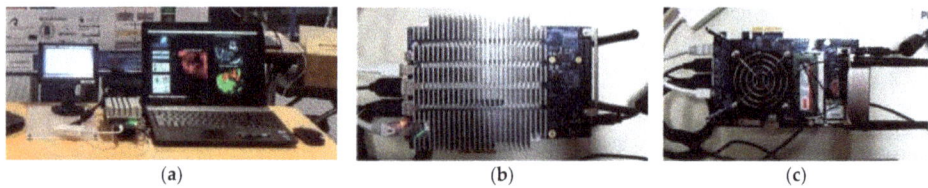

**Figure 5.** The Kalray MPPA® EMB01 Platform: (**a**) Developer environment; (**b**) EMB01 top view where the host module is located; (**c**) EMB01 bottom view where the carrier board is placed.

The Kalray MPPA-256 is a single-chip many-core processor that assembles 256 user cores distributed in 16 clusters running at 400 MHz. This chip comprises 256 user cores—32-bit very long instruction word (VLIW) processors with floating point units—distributed in several computing clusters. Additionally, this platform contains quad-core input/output (I/O) subsystems to manage the communications with the clusters. A network-on-a-chip (NoC) manages the synchronization and communications among the compute clusters and the I/O subsystem. Each cluster gathers 2 MB of memory—which is shared among the 16 cores—as well as a resource management (RM) core aimed at running the cluster operating system (NodeOS) and managing events and interrupts, and a direct memory access (DMA) module to transfer data from the shared memory to the NoC and vice versa. This architecture presents two main advantages: first, the system parallelization complexity is maintained within reasonable limits as the MPPA® includes mechanisms such as POSIX (Portable Operating System Interface), OpenMP, and OpenCL; and secondly, in comparison with other architectures like GPUs or field programmable gate arrays (FPGAs), the MPPA® platform leads in terms of energy efficiency [51].

*2.4. HS Training Database*

Employing the HELICoiD demonstrator, a total of 36 HS cubes of in vivo brain tissue belonging to 22 different patients were acquired from two different hospitals (the University Hospital Doctor Negrin at Las Palmas de Gran Canaria, Spain, and the University Hospital of Southampton, Hampshire, UK) in two data acquisition campaigns. The study protocol and consent procedures were approved by the Comité Ético de Investigación Clínica-Comité de Ética en la Investigación (CEIC/CEI) for the University Hospital Doctor Negrin, and the National Research Ethics Service (NRES) Committee South Central–Oxford C for the University Hospital of Southampton. Written informed consent was obtained from all subjects.

The creation of the training dataset (the gold standard employed to train the HS brain cancer detection classifier) was performed in the following way. Firstly, after performing the craniotomy and durotomy, the operating surgeons placed some sterilized rubber ring markers over the brain surface areas that they considered with relative certainty to be made up of tumor or normal tissue, using the information provided by an image-guided navigation system based on preoperative computed tomography (CT) or magnetic resonance imaging (MRI), as well as macroscopic appearance. In the cases where the tumor area was superficial, markers were placed on the brain surface before the resection started. Figure 6a shows an example of the synthetic RGB representation of a captured HS cube where the markers were used to identify the normal tissue (top marker) and the tumor tissue (bottom marker) affected by metastatic breast carcinoma. In the cases where the tumor was in a deeper layer with respect to the normal tissue and it was clearly identified, no markers were used and the operating surgeon identified the tumor and healthy area immediately after the operation using the synthetic RGB image. After marker placement, the operator of the system captured an HS image. Depending on the location of the tumor, the images were acquired immediately after the dura removal (Figure 6a) when the tumor was superficially located, or in an advanced stage of the tumor resection (Figure 6c) when the tumor was deep-seated. Glioblastoma (GBM) heterogeneity is one of the main problems in establishing a gold standard for a training and validation dataset. For this reason, when possible, several images were captured at different stages of the operation of both of the necrotic core and the enhanced rim of the tumor tissue. Once the HS image was obtained, the operating surgeon performed a biopsy of the tissue located within the tumor tissue marker or within the clearly identified tumor area. The resected tissue was sent to the pathologist to confirm the presence or absence of tumor, and to specify its histopathological diagnosis (grade and type of tumor). The average size of the resected tumor sample obtained for pathological analysis was $0.5 \times 0.5$ mm, with a 0.2-mm depth, since HSI technique cannot practically penetrate into the tissue (in the case of NIR, the depth was of 1 mm at most). Normal tissue markers were only used as a reference for the labeling process carried out after the completion of the operation. It is not ethical to biopsy what is known to be normal brain tissue, as this can result in damage to the patient. In this preliminary study, the spectral differences between grey matter and white matter in normal brain sample were not taken into account. These differences were not relevant in this study as the intention was to only resect tumor tissue. The labeling of the HS cubes was performed using histopathological information (from the tumor tissue samples) and the knowledge of the operating surgeon (from the normal tissue samples) to create a training dataset for the supervised classifier of the HS brain cancer detection algorithm.

In order to increase the training dataset, a methodology for extracting the gold standard information from the HS cubes, based on the spectral angle mapper (SAM) algorithm [52], was developed and designed using Matlab® GUIDE application. This SAM algorithm is an automated method for comparing the spectra of the pixels of a HS image with a well-known spectrum obtained from a reference pixel. The tool was employed by the corresponding operating surgeon after the completion of the operation to create the gold standard map for each captured HS image. Four different classes were established in this study: normal tissue, tumor tissue, blood vessel/hypervascularized tissue, and background (i.e., other materials or substances that can be presented in the surgical scenario but are not relevant for the tumor resection procedure). Therefore, normal class involves both grey matter and white matter tissue. The procedure to generate the neurosurgeon's gold standard map is as follows. The user (usually the operating surgeon) loads the HS cube and selects a reference pixel, looking the synthetic RGB image at the location where a biopsy is done (where the tumor marker is placed) or at a location far enough from the tumor margins where the surgeon can be quite confident that the tissue is abnormal (in the case of tumor labeling). In the case of normal tissue, blood vessel/hypervascularized tissue, and background classes, the labeling is performed by selecting a reference pixel by the naked eye based on the surgeon's knowledge and experience. Then, the most similar pixels to the selected reference pixel are highlighted, computed by using the SAM measurement, and the user configures the threshold that varies the tolerances on the selected pixels. Once the user considers that only the pixels belonging to one class have been highlighted, the selected pixels are assigned to that class. Neurosurgeons are instructed to select only a few sets of very reliable pixels instead of a wider set of uncertain pixels. Figure 6b,d shows an example of a gold standard map, where the labeled pixels that belong to tumor tissue, normal tissue, blood vessels/hypervascularized tissue, and background are identified with red, green, blue, and black colors, respectively.

In the end, the reliability of the training dataset is guaranteed by the use of (a) intraoperative MRI neuronavigation for locating tumor tissue; (b) the operating surgeon's knowledge and experience in the labeling of normal tissues, blood vessels/hypervascularized tissues, and background samples; and (c) the pathological analysis of the resected tissues for the tumor labeling.

After a preliminary analysis of the spectral signatures of both HS cubes (VNIR and NIR), only the VNIR images were labeled and used to generate the training dataset for the brain cancer detection algorithm. This was because of the practical impossibility of performing reliable labeling of the NIR images due to the low spatial resolution of these HS cubes (Figure 4b). Figure 6e,f show the mean and standard deviations of the VNIR spectral signatures of normal brain tissue (green color), blood vessels/hypervascularized tissue (blue color), and tumor tissue (red color) affected by GBM. In Figure 6e, the intra-patient variability (of one patient affected by GBM) of the spectral signatures can be seen, while in Figure 6f, the inter-patient variability (of 13 patients affected by GBM) is shown. In these cases, the tumor samples were obtained from the center of the tumor in the brain surface identified using the intraoperative neuronavigation system. These spectral signatures were extracted from the VNIR HS cube after the application of the pre-processing chain of the HS brain cancer detection algorithm (described in the next section). Figure 6g shows the average spectral signatures of each tumor type comprising the training database. As can be seen in this figure, there are significant spectral differences between these types of tumors. However, this study has mainly addressed the discrimination between tumor tissue (involving all the types of tumors) and normal tissue.

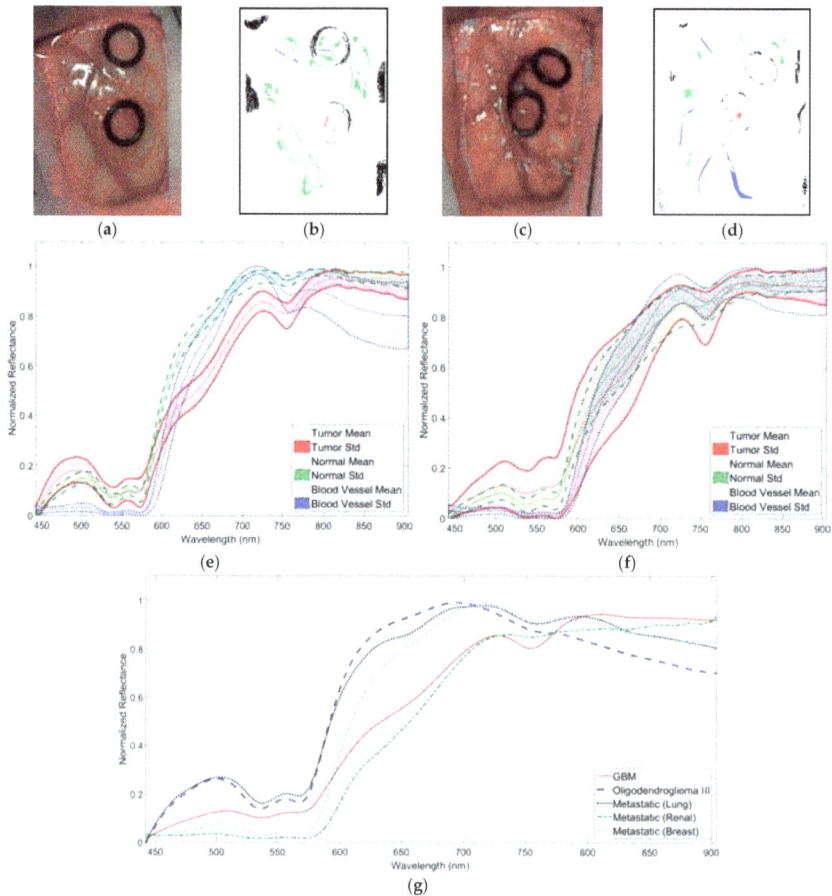

**Figure 6.** (**a**) Synthetic RGB representation of a VNIR HS cube of the a patient's brain surface affected by a metastatic breast carcinoma (bottom marker) identified before resection started and (**b**) the training map where normal tissue, tumor tissue, blood vessels/hypervascularized tissue, and background were labeled using green, red, blue, and black colors, respectively; (**c**,**d**) Synthetic RGB representation and training map of the same patient but in an advanced stage of the tumor resection; (**e**) Mean and standard deviation of the pre-processed labeled spectral signatures of one patient affected by a GBM tumor (red), with labeled normal tissue pixels (green), and labeled blood vessels/hypervascularized tissue pixels (blue); (**f**) Mean and standard deviation of the pre-processed labeled spectral signatures of 13 patients affected by GBM tumors (with the same color identification); (**g**) Mean values of the pre-processed labeled spectral signatures of each type of tumor available in the training dataset.

Table 1 details the total number of pixels labeled per each class and type of tissue. The tumor class involves two different primary tumors (GBM and grade III anaplastic oligodendroglioma) and three different secondary tumors, also called metastatic tumors (lung, renal, and breast carcinomas). After labeling all the available data, a total of 377,556 spectral signatures were obtained from the training database. Using this training dataset, the supervised classification stage of the HS brain cancer detection algorithm was trained in order to generate the classification maps from a new patient during the surgical procedure. Although different types of tumors were included in the training database, for this preliminary study only one multiclass supervised classifier was generated to differentiate primarily between tumor and normal tissue. Only one classifier was used instead of a different one per each type of tissue due to the reduced number of samples obtained for each type of tumor.

**Table 1.** Training dataset summary.

| Class | | | #Patients | #Labelled Pixels |
|---|---|---|---|---|
| Normal | | | 21 | 117,242 |
| | Primary (G-IV) | GBM | 8 | 12,641 |
| | Primary (G-III) | Anaplastic Oligodendroglioma | 2 | 1844 |
| Tumor | | Lung | 2 | 1936 |
| | Secondary | Renal | 1 | 21 |
| | | Breast | 1 | 325 |
| Blood Vessel/Hypervascularized Tissue | | | 22 | 57,429 |
| Background | | | 21 | 186,118 |
| Total (22 Patients, 36 Captures): | | | | 377,556 |

In order to determine the suitable percentage of samples of the training database that should be used to train the supervised algorithm, several experiments to generate and evaluate the supervised model were carried out employing different number of training samples. Each experiment was performed following a 10-fold cross-validation method to calculate the average overall accuracy result. Figure 7 shows the overall accuracy results varying the percentage of training samples with increments of 2%, starting at 2% and finishing at 100%. The evolution of the overall accuracy shows that when more than 75% of the training samples are used, the results stabilize, with overall accuracy of around 97.5%. With this experiment, it can be seen that there is no overfitting effect and the use of all the training samples will provide the best classification map.

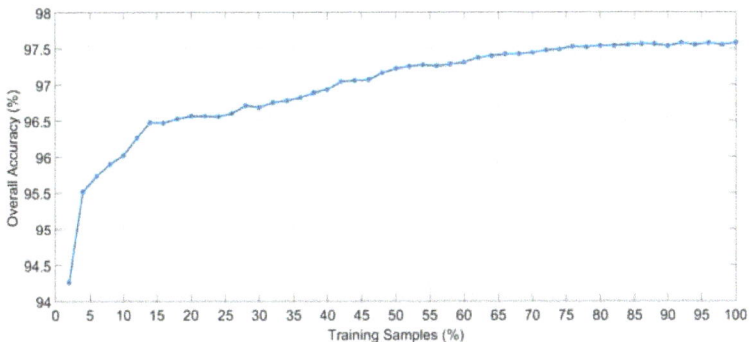

**Figure 7.** Overall accuracy evolution depending on the percentage of training samples employed to generate the supervised classification model of the HS brain cancer detection algorithm.

*2.5. Brain Cancer Detection Algorithm Implementation*

The HS brain cancer detection algorithm developed in this research work aims to exploit both the spatial and spectral features of the HS images. The whole algorithm can be divided into two main steps: the off-line process and the in situ process. The off-line process is the part of the algorithm in which the information previously provided by the experts in labeled samples is employed to train the supervised stage of the algorithm. On the other hand, the in situ process is carried out during surgery inside the operating theatre when a new HS image is acquired from the undergoing patient. This part of the algorithm is implemented and accelerated using the HELICoiD demonstrator.

In summary, the in situ process is based on five main steps. Firstly, a new hypercube is acquired during a surgical procedure. Secondly, a pre-processing chain is applied to homogenize the spectral signatures of the HS cube. Thirdly, a supervised pixel-wise classification is performed in order to obtain a classification map, where different types of tissues are identified according to the information previously provided by medical doctors. The supervised classifier employed is the support vector machine (SVM) algorithm [53], previously trained in the off-line process with the HS training dataset. Fourthly, once the supervised classification map is obtained, a spatial–spectral homogenization is accomplished using k-nearest neighbors (k-NN) filtering, where a one-band representation of the hypercube is employed. The dimensionality reduction algorithm used to obtain the one-band representation of the hypercube is the principal component analysis (PCA) algorithm [45]. Finally, in order to obtain the definitive classification map (also called the HELICoiD three maximum density (TMD) map), the spatial–spectral supervised classification map is fused with a segmentation map, obtained via unsupervised learning, employing hierarchical K-means (HKM) clustering. The algorithm used to fuse both images is based on a majority voting (MV) approach.

Figure 8a depicts the different blocks of the HS brain cancer detection algorithm, as well as their distribution in the implementation onto both platforms and the execution scheduling. Furthermore, the RGB representation of the outputs obtained at each step of the algorithm is also shown. The blue block represents the steps of the algorithm that were mapped to the CU, while the green block represents the steps mapped to the HA. As can be observed, the pre-processing stage, the HKM clustering and the MV algorithm are executed on the CU. In contrast, the spatial–spectral supervised classification stage, where the PCA, the SVM classification, and the KNN filtering are performed, is executed on the HA due to its high computational load.

The data flow sequence of the implementation follows the next steps. Firstly, the raw image is pre-processed on the CU and the resulting HS cube is sent to the HA through the Gigabit Ethernet interface, to be employed as the input of the PCA and SVM classification algorithms. The same HS cube is used in the CU as the input of the HKM clustering algorithm. Secondly, HKM clustering is executed on the CU, while the spatial–spectral supervised classification—PCA, SVM classification and KNN filtering—is executed on the HA. Both the unsupervised and the supervised stages are executed simultaneously. In addition, the PCA algorithm and the SVM classification are executed in parallel in the HA. Finally, once the previous stages are finalized, the MV algorithm is executed on the CU to compute the final HELICoiD TMD map. This TMD map is a RGB representation of the first three major probabilities per cluster obtained from the HKM clustering algorithm, where the brain tumor is marked in red. This image is shown to the user (the neurosurgeon) through the HS processing interface. Figure 8b shows the different parts that comprise the HELICoiD demonstrator in relation to HS data processing.

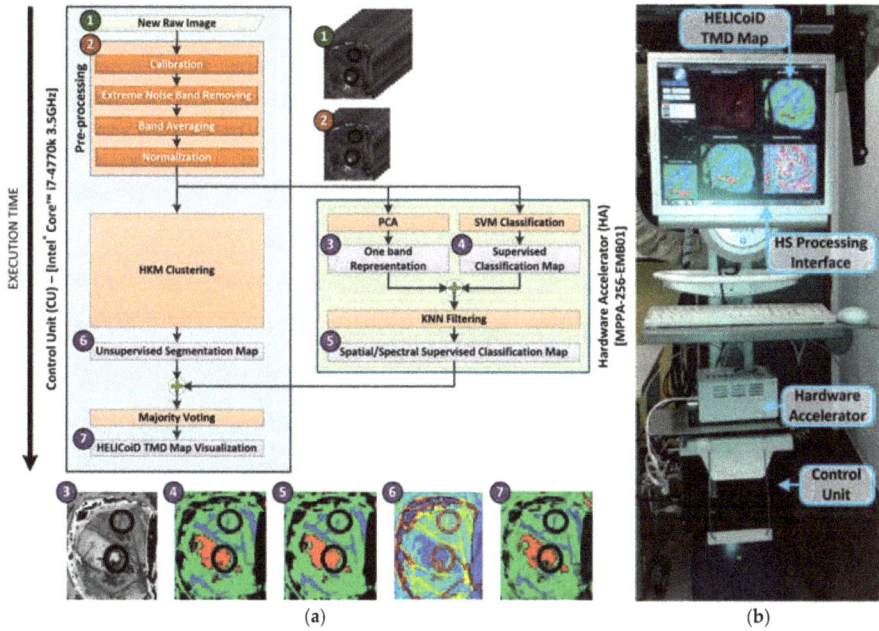

**Figure 8.** (a) HS brain cancer detection algorithm implementation flow diagram and the RGB representation of the output of each step; (b) Different parts related to the HS data processing of the HELICoiD demonstrator.

## 3. Experimental Results and Discussion

The validation of the HELICoiD demonstrator was performed during neurosurgical operations at the University Hospital Doctor Negrin of Las Palmas de Gran Canaria, employing the data of four different patients affected by different types of tumors. Table 2 details the characteristics of the validation database used to test the developed HS brain cancer detection system and the corresponding pathological diagnosis. Seven different images were included. These images involved normal brain tissue acquired during the first stage of the surgical operation, used to test if the system included false positives when no tumor as present in the image, and three different types of primary tumors.

**Table 2.** Validation HS image dataset characteristics.

| Image ID | Size (MB) | #Pixels | Dimension (Width × Height × Bands) | Pathological Diagnosis |
|---|---|---|---|---|
| P1C1 | 362.62 | 224,770 | 495 × 456 × 826 | Normal Brain |
| P1C2 | 197.90 | 122,670 | 471 × 262 × 826 | Primary Grade II Oligodendroglioma |
| P2C1 | 225.35 | 139,682 | 332 × 423 × 826 | Normal Brain |
| P2C2 | 276.99 | 171,699 | 364 × 474 × 826 | Primary GBM |
| P3C1 | 402.26 | 249,344 | 513 × 488 × 826 | Normal Brain |
| P3C2 | 230.34 | 143,560 | 485 × 296 × 826 | Primary GBM |
| P4C1 | 372.47 | 230,878 | 480 × 483 × 826 | Primary Grade I Meningioma |

The TMD maps of the validation database obtained by the HELICoiD demonstrator during the surgical operations and their respective synthetic RGB images are shown in Figures 9 and 10. The TMD maps are represented in four colors that can be mixed depending on the density of each class presented in the image. Figure 9 shows the results obtained from the normal brain images. In these results, it can be seen that the system does not present any false positives in the parenchymal area, and normal tissue and blood vessels are clearly identified. Furthermore, bright pixels, which can be found in the images due to the light reflections over the arachnoid of the brain or due to the presence of surgical serum in the surface, are identified as background pixels. On the other hand, Figure 10 shows the results obtained from the HS images of the brain surface affected by a tumor, where the tumor areas are surrounded with a yellow line in the synthetic RGB representations. These results offer a clear indication that the HELICoiD demonstrator is able to identify the tumor tissue presented in the images. In Figure 10b, there are some false positives in the bottom corner of the TMD map, however, this false information is located outside the area of exposed brain parenchyma and thus, it does not affect the neurosurgeon decisions during the tumor resection. It is worth noting that two of the cases (Figure 10b,d) identify two tumor types (grade II oligodendroglioma and grade I meningioma) for which there are no spectral signatures within the training database. These results highlight the robustness and the generalization capabilities of the HELICoiD demonstrator to identify other types of tumor rather than only the ones available in the HS training database. Finally, it should be mentioned that the tumor identification becomes more difficult when the tumor is located deeper in the brain. Figure 10f,h show the TMD maps of GBM tumors at an advanced stage of the surgical procedure. It can be seen that, in case of Figure 10f, the tumor tissue is clearly identified although it is located in a deep layer. However, in Figure 10h, there is no correct identification of the tumor tissue due to problems with shadows and the presence of extravasated blood in the tumor area. Since HSI is not able to penetrate into the surface, extravasated blood present in the image is identified as a hypervascularized tissue class (blue color) in the TMD map.

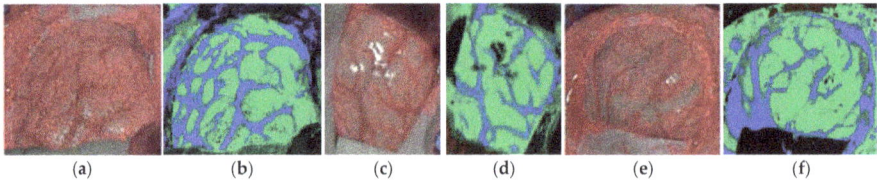

**Figure 9.** Normal brain image results obtained from the validation database employing the HELICoiD demonstrator: (**a**,**b**) synthetic RGB image and TMD map of the P1C1 HS image; (**c**,**d**) synthetic RGB image and TMD map of the P2C1 HS image; (**e**,**f**) synthetic RGB image and TMD map of the P3C1 HS image.

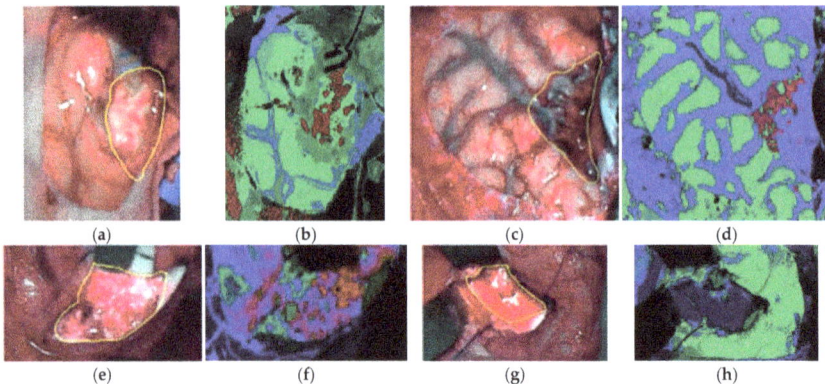

**Figure 10.** Tumor tissue identification results obtained from the validation database employing the HELICoiD demonstrator: (**a**,**b**) synthetic RGB image and TMD map of the P2C2 HS image; (**c**,**d**) synthetic RGB image and TMD map of the P4C1 HS image; (**e**,**f**) synthetic RGB image and TMD map of the P1C2 HS image; (**g**,**h**) synthetic RGB image and TMD map of the P3C2 HS image.

Table 3 shows the execution times obtained using the HELICoiD demonstrator to acquire and process the validation database during surgery. To assess the processing times obtained using the hardware acceleration in the spatial-spectral supervised classification stage, Table 3 also shows the processing times obtained when the whole algorithm is implemented in the CPU, i.e., sequential time results. The total processing time required in the accelerated version is computed taking into account the maximum time obtained between the spatial–spectral supervised classification (PCA + SVM + KNN) and the unsupervised clustering (HKM). In summary, when the hardware accelerator is not employed, the spatial–spectral supervised classification is the most time-consuming stage. In contrast, an average speedup factor of 24× is achieved in the spatial–spectral supervised classification stage when the hardware accelerator is employed, becoming the unsupervised clustering the limiting factor in this case. These results show that the proposed system provides a TMD map of the captured scene during the surgery in approximately 1 min, depending on the size of the captured image.

**Table 3.** Acquisition and processing time comparison between the sequential (Seq.) and accelerated (Acc.) implementations of the proposed algorithm for different size of images.

| Image ID | Processing Type | Acquisition Time (s) | Pre-Processing (s) | Transmission (s) | PCA + SVM (s) | KNN (s) | HKM (s) | MV (s) | Total Processing Time (s) |
|---|---|---|---|---|---|---|---|---|---|
| P1C1 | Seq. | 19.98 | 15.07 | 0.00 | 11.32 | 378.87 | 39.68 | 0.009 | 444.95 |
| | Acc. | | | 14.00 | 6.02 | 8.16 | | | 68.76 * |
| | Speedup | N/A ¥ | N/A ¥ | 0.00 | 1.88 | 46.45 | N/A ¥ | N/A ¥ | 6.47 |
| P1C2 | Seq. | 19.02 | 6.50 | 0.00 | 5.90 | 196.64 | 21.87 | 0.004 | 230.92 |
| | Acc. | | | 7.15 | 4.35 | 4.23 | | | 35.53 * |
| | Speedup | N/A ¥ | N/A ¥ | 0.00 | 1.36 | 46.44 | N/A ¥ | N/A ¥ | 6.50 |
| P2C1 | Seq. | 13.40 | 9.35 | 0.00 | 6.72 | 158.66 | 24.96 | 0.005 | 199.70 |
| | Acc. | | | 8.07 | 4.48 | 3.48 | | | 42.38 * |
| | Speedup | N/A ¥ | N/A ¥ | 0.00 | 1.50 | 45.62 | N/A ¥ | N/A ¥ | 4.71 |
| P2C2 | Seq. | 14.70 | 12.59 | 0.00 | 8.96 | 212.96 | 30.45 | 0.006 | 264.97 |
| | Acc. | | | 9.56 | 5.02 | 4.66 | | | 52.61 * |
| | Speedup | N/A ¥ | N/A ¥ | 0.00 | 1.78 | 45.74 | N/A ¥ | N/A ¥ | 5.04 |
| P3C1 | Seq. | 20.71 | 19.72 | 0.00 | 13.68 | 434.96 | 44.57 | 0.008 | 512.93 |
| | Acc. | | | 13.34 | 6.72 | 9.44 | | | 77.63 * |
| | Speedup | N/A ¥ | N/A ¥ | 0.00 | 2.03 | 46.10 | N/A ¥ | N/A ¥ | 6.61 |
| P3C2 | Seq. | 19.58 | 8.94 | 0.00 | 7.73 | 234.90 | 25.75 | 0.005 | 277.33 |
| | Acc. | | | 9.45 | 4.66 | 5.08 | | | 44.15 * |
| | Speedup | N/A ¥ | N/A ¥ | 0.00 | 1.66 | 46.27 | N/A ¥ | N/A ¥ | 6.28 |
| P4C1 | Seq. | 19.38 | 13.84 | 0.00 | 11.49 | 377.60 | 41.59 | 0.007 | 444.52 |
| | Acc. | | | 12.36 | 6.29 | 8.15 | | | 67.79 * |
| | Speedup | N/A ¥ | N/A ¥ | 0.00 | 1.83 | 46.34 | N/A ¥ | N/A ¥ | 6.56 |

\* The total time obtained in the accelerated version is computed taking into account the maximum time obtained between the spatial-spectral supervised classification and the unsupervised clustering; ¥ Measurement not available.

## 4. Conclusions

In this study, a novel visualization system based on HSI was developed to aid surgeons in the difficult task of identifying brain tumor boundaries during neurosurgical procedures. The identification of tumor boundaries and tumor infiltration into normal brain tissue is extremely important in order to avoid excessive resection of normal brain tissue and to avoid unintentionally leaving behind residual tumor. Using only RGB information, the naked eye cannot be used to accurately determine the boundaries of the tumor, especially in gliomas where tumor heterogeneity is extremely high. In addition, intraoperative neuronavigation based on CT and MRI is problematic due to brain shift, producing a significant error between the real position of the tumor boundaries respect to the CT or MRI information. As a proof-of-concept, the demonstrator developed in this study was able to generate thematic maps of the exposed brain surface using spectral information of the VNIR range (between

400 and 1000 nm). These thematic maps differentiate between four different classes: normal tissue, tumor tissue, blood vessels/hypervascularized tissue, and background. In these maps, the tumor boundaries can be easily identifiable. Only the information obtained from the VNIR camera has been employed to generate the gold standard for the training of the classification algorithm and validate its results. Due to the low spatial resolution of the NIR camera, it is not possible to perform reliable labeling of the NIR HS cubes. Although some preliminary analysis of the NIR images performed by the research team reveal that the use of the NIR spectral range could help in the identification of blood vessels and extravasated blood, NIR images alone are not relevant for the goal of this study. An HS brain cancer detection algorithm, based on unsupervised and supervised machine learning approaches, was developed and implemented in the system. The supervised algorithm was trained by employing a labeled dataset composed of more than 300,000 spectral signatures, extracted by medical doctors from 36 different HS cubes captured with the acquisition system from 22 different patients from Spain and UK. In this preliminary study, only one multiclass classifier was generated for the supervised part of the algorithm, employing all the types of tumors available in the training database to distinguish mainly between tumor and normal tissue, without identifying the different types of tumors. The implementation of the algorithm was partitioned between the control unit and a hardware accelerator, where the higher computational tasks were implemented in a many-core platform to achieve intraoperative processing (~1 min). The demonstrator was validated using seven HS images obtained in four neurosurgical operations. The TMD maps demonstrate that the system did not introduce false positives in the parenchymal area when no tumor was present and it was able to identify different types of tumor that were not present in the training database. Currently, further investigations are being carried out by the research team in order to enlarge the training database and the validation database with more patients and types of tumors. Additionally, the fusion of both types of HS images (VNIR and NIR) is being investigated in order to investigate if the NIR information could help to more accurately distinguish the boundaries between the tumor tissue and the surrounding hypervascularized normal tissue. Furthermore, an extensive clinical validation of the system must be carried out. In this clinical validation, a comprehensive pathological analysis of the entire tumor area outlined by the TMD map (especially in the boundaries between tumor and the surrounding normal tissue) must be performed as well as to correlate the results with the MRI information in order to know if the tumor infiltration into normal brain tissue can be properly identified by the system. Additionally, through clinical validation, the relation between the improvement of the patient outcomes and the use of the system during the surgery will be studied.

**Acknowledgments:** This work has been supported in part by the European Commission through the FP7 FET Open Programme ICT–2011.9.2 European Project HELICoiD "HypErspectral Imaging Cancer Detection" under Grant Agreement 618080.

**Author Contributions:** H.F. and S.O. supervised the project, developed the demonstrator and the algorithm, designed and performed experiments, and wrote the manuscript. R.L. and D.M. implemented and accelerated the algorithm, performed experiments, and edited the manuscript. G.M.C. supervised the project, designed experiments, and edited the manuscript. R.Sal. and E.J. supervised the algorithm implementation and acceleration and edited the manuscript. D.B. acquired and analyzed data and edited the manuscript. H.B., A.S., J.F.P., C.S., A.J.O., S.B., M.H. and J.M. acquired and analyzed data. D.R. and B.R.K. developed the algorithm and performed experiments. A.V. developed the demonstrator. A.B. implemented the algorithm and edited the manuscript. G.-Z.Y., B.S. and R.Sar. supervised the algorithm development.

**Conflicts of Interest:** The authors declare no conflict of interest. The founding sponsors had no role in the design of the study; in the collection, analyses, or interpretation of data; in the writing of the manuscript, and in the decision to publish the results.

**Ethical Statements:** All subjects gave their informed consent for inclusion before they participated in the study. The study was conducted in accordance with the Declaration of Helsinki, and the protocol was approved by the Ethics Committee of the University Hospital Doctor Negrin and the University Hospital of Southampton (Acta-09/2013-CEIC-130069 and REC-14/SC/0108 respectively).

## References

1. Improving Outcomes for People with Brain and Other CNS Tumours. Available online: https://www.nice.org.uk/guidance/csg10 (accessed on 30 January 2018).
2. Louis, D.N.; Perry, A.; Reifenberger, G.; von Deimling, A.; Figarella-Branger, D.; Cavenee, W.K.; Ohgaki, H.; Wiestler, O.D.; Kleihues, P.; Ellison, D.W. The 2016 World Health Organization Classification of Tumors of the Central Nervous System: A summary. *Acta Neuropathol.* **2016**, *131*, 803–820. [CrossRef] [PubMed]
3. Robson, D.K. Pathology & Genetics. Tumours of the Nervous System. World Health Organisation ClassificationofTumours. *J. Pathol.* **2001**, *193*, 276. [CrossRef]
4. Stummer, W.; Tonn, J.C.; Mehdorn, H.M.; Nestler, U.; Franz, K.; Goetz, C.; Bink, A.; Pichlmeier, U. Counterbalancing risks and gains from extended resections in malignant glioma surgery: A supplemental analysis from the randomized 5-aminolevulinic acid glioma resection study: Clinical article. *J. Neurosurg.* **2011**, *114*, 613–623. [CrossRef] [PubMed]
5. Sanai, N.; Berger, M.S. Glioma extent of resection and its impact on patient outcome. *Neurosurgery* **2008**, *62*, 753–764. [CrossRef] [PubMed]
6. Sanai, N.; Berger, M.S. Operative Techniques for Gliomas and the Value of Extent of Resection. *Neurotherapeutics* **2009**, *6*, 478–486. [CrossRef] [PubMed]
7. Petrecca, K.; Guiot, M.; Panet-Raymond, V.; Souhami, L. Failure pattern following complete resection plus radiotherapy and temozolomide is at the resection margin in patients with glioblastoma. *J. Neurooncol.* **2013**, *111*, 19–23. [CrossRef] [PubMed]
8. Gerard, I.J.; Kersten-Oertel, M.; Petrecca, K.; Sirhan, D.; Hall, J.A.; Collins, D.L. Brain shift in neuronavigation of brain tumors: A review. *Med. Image Anal.* **2017**, *35*, 403–420. [CrossRef] [PubMed]
9. Kast, R.E.; Auner, G.W.; Rosenblum, M.L.; Mikkelsen, T.; Yurgelevic, S.M.; Raghunathan, A.; Poisson, L.M.; Kalkanis, S.N. Raman molecular imaging of brain frozen tissue sections. *J. Neurooncol.* **2014**, *120*, 55–62. [CrossRef] [PubMed]
10. Reinges, M.H.T.; Nguyen, H.H.; Krings, T.; Hütter, B.O.; Rohde, V.; Gilsbach, J.M.; Black, P.M.; Takakura, K.; Roberts, D.W. Course of brain shift during microsurgical resection of supratentorial cerebral lesions: Limits of conventional neuronavigation. *Acta Neurochir. (Wien)* **2004**, *146*, 369–377. [CrossRef] [PubMed]
11. Nimsky, C.; Ganslandt, O.; Hastreiter, P.; Fahlbusch, R. Intraoperative compensation for brain shift. *Surg. Neurol.* **2001**, *56*, 357–364. [CrossRef]
12. Ganser, K.A.; Dickhaus, H.; Staubert, A.; Bonsanto, M.M.; Wirtz, C.R.; Tronnier, V.M.; Kunze, S. Quantification of brain shift effects in MRI images. *Biomed. Tech. (Berl.)* **1997**, *42*, 247–248. [CrossRef] [PubMed]
13. Stummer, W.; Pichlmeier, U.; Meinel, T.; Wiestler, O.D.; Zanella, F.; Reulen, H.J. Fluorescence-guided surgery with 5-aminolevulinic acid for resection of malignant glioma: a randomised controlled multicentre phase III trial. *Lancet Oncol.* **2006**, *7*, 392–401. [CrossRef]
14. Floeth, F.W.; Sabel, M.; Ewelt, C.; Stummer, W.; Felsberg, J.; Reifenberger, G.; Steiger, H.J.; Stoffels, G.; Coenen, H.H.; Langen, K.J. Comparison of 18F-FET PET and 5-ALA fluorescence in cerebral gliomas. *Eur. J. Nucl. Med. Mol. Imaging* **2011**, *38*, 731–741. [CrossRef] [PubMed]
15. Lu, G.; Fei, B. Medical hyperspectral imaging: A review. *J. Biomed. Opt.* **2014**, *19*, 10901. [CrossRef] [PubMed]
16. Calin, M.A.; Parasca, S.V.; Savastru, D.; Manea, D. Hyperspectral imaging in the medical field: Present and future. *Appl. Spectrosc. Rev.* **2014**, *49*, 435–447. [CrossRef]
17. Akbari, H.; Uto, K.; Kosugi, Y.; Kojima, K.; Tanaka, N. Cancer detection using infrared hyperspectral imaging. *Cancer Sci.* **2011**, *102*, 852–857. [CrossRef] [PubMed]
18. Regeling, B.; Thies, B.; Gerstner, A.O.H.; Westermann, S.; Müller, N.A.; Bendix, J.; Laffers, W. Hyperspectral Imaging Using Flexible Endoscopy for Laryngeal Cancer Detection. *Sensors* **2016**, *16*, 1288. [CrossRef] [PubMed]
19. Kester, R.T.; Bedard, N.; Gao, L.; Tkaczyk, T.S. Real-time snapshot hyperspectral imaging endoscope. *J. Biomed. Opt.* **2011**, *16*, 56005. [CrossRef] [PubMed]
20. Akbari, H.; Halig, L.V.; Schuster, D.M.; Osunkoya, A.; Master, V.; Nieh, P.T.; Chen, G.Z.; Fei, B. Hyperspectral imaging and quantitative analysis for prostate cancer detection. *J. Biomed. Opt.* **2012**, *17*, 760051. [CrossRef] [PubMed]

21. Hou, Y.; Ren, Z.; Liu, G.; Zeng, L.; Huang, Z. Design of a novel LD-induced hyper-spectral imager for breast cancer diagnosis based on VHT grating. In Proceedings of the 2011 Symposium on Photonics and Optoelectronics, Wuhan, China, 16–18 May 2011.

22. Kim, B.; Kehtarnavaz, N.; LeBoulluec, P.; Liu, H.; Peng, Y.; Euhus, D. Automation of ROI extraction in hyperspectral breast images. In Proceedings of the 2013 35th Annual International Conference of the IEEE Engineering in Medicine and Biology Society (EMBC), Osaka, Japan, 3–7 July 2013; Volume 2013, pp. 3658–3661. [CrossRef]

23. Pourreza-Shahri, R.; Saki, F.; Kehtarnavaz, N.; Leboulluec, P.; Liu, H. Classification of ex-vivo breast cancer positive margins measured by hyperspectral imaging. In Proceedings of the 2013 IEEE International Conference on Image Processing, ICIP 2013—Proceedings, Melbourne, VIC, Australia, 15–18 September 2013; pp. 1408–1412.

24. Han, Z.; Zhang, A.; Wang, X.; Sun, Z.; Wang, M.D.; Xie, T. In vivo use of hyperspectral imaging to develop a noncontact endoscopic diagnosis support system for malignant colorectal tumors. *J. Biomed. Opt.* **2016**, *21*, 16001. [CrossRef] [PubMed]

25. Masood, K.; Rajpoot, N.M. Texture based classification of hyperspectral colon biopsy samples using CLBP. In Proceedings of the IEEE International Symposium on Biomedical Imaging: From Nano to Macro, 2009, ISBI 09, Boston, MA, USA, 28 June–1 July 2009; pp. 1011–1014. [CrossRef]

26. Masood, K.; Rajpoot, N.; Rajpoot, K.; Qureshi, H. Hyperspectral Colon Tissue Classification using Morphological Analysis. In Proceedings of the International Conference on Emerging Technologies, Peshawar, Pakistan, 13–14 Novrmber 2006; pp. 735–741.

27. Rajpoot, K.M.; Rajpoot, N.M. Wavelet based segmentation of hyperspectral colon tissue imagery. In Proceedings of the INMIC 2003: IEEE 7th International Multi Topic Conference, Islamabad, Pakistan, 8–9 December 2003; pp. 38–43.

28. Dicker, D.T.; Lerner, J.; Van Belle, P.; Barth, S.F.; Guerry, D., 4th; Herlyn, M.; Elder, D.E.; El-Deiry, W.S. Differentiation of normal skin and melanoma using high resolution hyperspectral imaging. *Cancer Biol. Ther.* **2006**, *5*, 1033–1038. [CrossRef] [PubMed]

29. Gaudi, S.; Meyer, R.; Ranka, J.; Granahan, J.C.; Israel, S.A.; Yachik, T.R.; Jukic, D.M. Hyperspectral Imaging of Melanocytic Lesions. *Am. J. Dermatopathol.* **2014**, *36*, 131–136. [CrossRef] [PubMed]

30. Akbari, H.; Halig, L.V.; Zhang, H.; Wang, D.; Chen, Z.G.; Fei, B. Detection of Cancer Metastasis Using a Novel Macroscopic Hyperspectral Method. *Proc. SPIE* **2012**, *8317*, 831711. [CrossRef]

31. Laffers, W.; Westermann, S.; Regeling, B.; Martin, R.; Thies, B.; Gerstner, A.O.H.; Bootz, F.; Müller, N.A. Early recognition of cancerous lesions in the mouth and oropharynx: Automated evaluation of hyperspectral image stacks. *HNO* **2015**. [CrossRef]

32. Zhi, L.; Zhang, D.; Yan, J.Q.; Li, Q.L.; Tang, Q.L. Classification of hyperspectral medical tongue images for tongue diagnosis. *Comput. Med. Imaging Graph.* **2007**, *31*, 672–678.

33. Liu, Z.; Wang, H.; Li, Q. Tongue tumor detection in medical hyperspectral images. *Sensors* **2012**, *12*, 162–174. [CrossRef] [PubMed]

34. Liu, Z.; Yan, J.; Zhang, D.; Li, Q.-L. Automated tongue segmentation in hyperspectral images for medicine. *Appl. Opt.* **2007**, *46*, 8328–8334. [CrossRef] [PubMed]

35. Akbari, H.; Kosugi, Y.; Kojima, K.; Tanaka, N. Blood vessel detection and artery-vein differentiation using hyperspectral imaging. In Proceedings of the 31st Annual International Conference of the IEEE Engineering in Medicine and Biology Society, Minneapolis, MN, USA, 3–6 September 2009; pp. 1461–1464. [CrossRef]

36. Akbari, H.; Kosugi, Y.; Kojima, K.; Tanaka, N. Detection and Analysis of the Intestinal Ischemia Using Visible and Invisible Hyperspectral Imaging. *IEEE Trans. Biomed. Eng.* **2010**, *57*, 2011–2017. [CrossRef] [PubMed]

37. Jayanthi, J.L.; Nisha, G.U.; Manju, S.; Philip, E.K.; Jeemon, P.; Baiju, K.V.; Beena, V.T.; Subhash, N. Diffuse reflectance spectroscopy: diagnostic accuracy of a non-invasive screening technique for early detection of malignant changes in the oral cavity. *BMJ Open* **2011**, *1*, e000071. [CrossRef] [PubMed]

38. Kabwama, S.; Bulters, D.; Bulstrode, H.; Fabelo, H.; Ortega, S.; Callico, G.M.; Stanciulescu, B.; Kiran, R.; Ravi, D.; Szolna, A.; et al. Intra-operative hyperspectral imaging for brain tumour detection and delineation: Current progress on the HELICoid project. *Int. J. Surg.* **2016**, *36*, S140. [CrossRef]

39. Salvador, R.; Ortega, S.; Madroñal, D.; Fabelo, H.; Lazcano, R.; Marrero, G.; Juárez, E.; Sarmiento, R.; Sanz, C. HELICoiD: Interdisciplinary and collaborative project for real-time brain cancer detection. In Proceedings of the ACM International Conference on Computing Frontiers 2017, CF 2017, Siena, Italy, 15–17 May 2017.

40. Szolna, A.; Morera, J.; Piñeiro, J.F.; Callicó, G.M.; Fabelo, H.; Ortega, S. Hyperspectral Imaging as A Novel Instrument for Intraoperative Brain Tumor Detection. *Neurocirugia* **2016**, *27*, 166.

41. Fabelo, H.; Ortega, S.; Kabwama, S.; Callico, G.M.; Bulters, D.; Szolna, A.; Pineiro, J.F.; Sarmiento, R. HELICoiD project: A new use of hyperspectral imaging for brain cancer detection in real-time during neurosurgical operations. In Proceedings of the SPIE 9860, Hyperspectral Imaging Sensors: Innovative Applications and Sensor Standards 2016; SPIE- Commercial + Scientific Sensing and Imaging, Baltimore, MD, USA, 17–21 April 2016.

42. Salvador, R.; Fabelo, H.; Lazcano, R.; Ortega, S.; Madroñal, D.; Callicó, G.M.; Juárez, E.; Sanz, C. Demo: HELICoiD tool demonstrator for real-time brain cancer detection. In Proceedings of the Conference on Design and Architectures for Signal and Image Processing, DASIP, Rennes, France, 12–14 October 2016.

43. Fabelo, H.; Ortega, S.; Guerra, R.; Callicó, G.; Szolna, A.; Piñeiro, J.F.; Tejedor, M.; López, S.; Sarmiento, R. A novel use of hyperspectral images for human brain cancer detection using in-vivo samples. In Proceedings of the BIOSIGNALS 2016–9th International Conference on Bio-Inspired Systems and Signal Processing; Part of 9th International Joint Conference on Biomedical Engineering Systems and Technologies, BIOSTEC 2016, Rome, Italy, 21–23 February 2016.

44. Madroñal, D.; Fabelo, H.; Lazcano, R.; Callicó, G.M.; Juárez, E.; Sanz, C. Parallel implementation of a hyperspectral image linear SVM classifier using RVC-CAL. In Proceedings of the SPIE–The International Society for Optical Engineering, Edinburgh, UK, 26–29 September 2016; Volume 10007.

45. Lazcano, R.; Madroñal, D.; Salvador, R.; Desnos, K.; Pelcat, M.; Guerra, R.; Fabelo, H.; Ortega, S.; Lopez, S.; Callico, G.M.; Juarez, E.; Sanz, C. Porting a PCA-based hyperspectral image dimensionality reduction algorithm for brain cancer detection on a manycore architecture. *J. Syst. Archit.* **2017**, *77*, 101–111. [CrossRef]

46. Wu, D.; Sun, D.-W. Advanced applications of hyperspectral imaging technology for food quality and safety analysis and assessment: A review—Part I: Fundamentals. *Innov. Food Sci. Emerg. Technol.* **2013**, *19*, 1–14. [CrossRef]

47. Elvidge, C.D.; Keith, D.M.; Tuttle, B.T.; Baugh, K.E. Spectral identification of lighting type and character. *Sensors* **2010**, *10*, 3961–3988. [CrossRef] [PubMed]

48. Quinn, P.J. Effects of temperature on cell membranes. *Symp. Soc. Exp. Biol.* **1988**, *42*, 237–258. [PubMed]

49. Akbari, H.; Kosugi, Y. Hyperspectral imaging: A new modality in surgery. In *Recent Advances in Biomedical Engineering*; InTech: London, UK, 2009.

50. De Dinechin, B.D.; Ayrignac, R.; Beaucamps, P.E.; Couvert, P.; Ganne, B.; De Massas, P.G.; Jacquet, F.; Jones, S.; Chaisemartin, N.M.; Riss, F.; et al. A clustered manycore processor architecture for embedded and accelerated applications. In Proceedings of the 2013 IEEE High Performance Extreme Computing Conference, HPEC 2013, Waltham, MA, USA, 10–12 September 2013.

51. Madroñal, D.; Lazcano, R.; Fabelo, H.; Ortega, S.; Salvador, R.; Callicó, G.M.; Juarez, E.; Sanz, C. Energy consumption characterization of a Massively Parallel Processor Array (MPPA) platform running a hyperspectral SVM classifier. In Proceedings of the 2017 Conference on Design and Architectures for Signal and Image Processing, DASIP, Dresden, Germany, 27–29 September 2017.

52. Chang, C.-I. New hyperspectral discrimination measure for spectral characterization. *Opt. Eng.* **2004**, *43*, 1777. [CrossRef]

53. Madroñal, D.; Lazcano, R.; Salvador, R.; Fabelo, H.; Ortega, S.; Callico, G.M.; Juarez, E.; Sanz, C. SVM-based real-time hyperspectral image classifier on a manycore architecture. *J. Syst. Archit.* **2017**, *80*. [CrossRef]

*sensors*

MDPI

*Review*

# Crowdsensing in Smart Cities: Overview, Platforms, and Environment Sensing Issues

Oscar Alvear [1,2,*], Carlos T. Calafate [1], Juan-Carlos Cano [1] and Pietro Manzoni [1]

[1] Department of Computer Engineering, Universitat Politecnica de Valencia, 46022 Valencia, Spain; calafate@disca.upv.es (C.T.C.); jucano@disca.upv.es (J.-C.C.); pmanzoni@disca.upv.es (P.M.)
[2] Department of Electrical Engineering, Electronics and Telecommunications, Universidad de Cuenca, Cuenca 010150, Ecuador
* Correspondence: oscar.alvear@alttics.com

Received: 3 December 2017; Accepted: 31 January 2018; Published: 4 Febuary 2018

**Abstract:** Evidence shows that Smart Cities are starting to materialise in our lives through the gradual introduction of the Internet of Things (IoT) paradigm. In this scope, crowdsensing emerges as a powerful solution to address environmental monitoring, allowing to control air pollution levels in crowded urban areas in a distributed, collaborative, inexpensive and accurate manner. However, even though technology is already available, such environmental sensing devices have not yet reached consumers. In this paper, we present an analysis of candidate technologies for crowdsensing architectures, along with the requirements for empowering users with air monitoring capabilities. Specifically, we start by providing an overview of the most relevant IoT architectures and protocols. Then, we present the general design of an off-the-shelf mobile environmental sensor able to cope with air quality monitoring requirements; we explore different hardware options to develop the desired sensing unit using readily available devices, discussing the main technical issues associated with each option, thereby opening new opportunities in terms of environmental monitoring programs.

**Keywords:** Smart city; Internet of Things; crowdsensing; sensor design

---

## 1. Introduction

Smart cities are revolutionising our view of the world, and their functioning achieves a very high level of integration, coordination, and cooperation between ordinary objects, providing them with some degree of intelligence. This novel paradigm provides a plethora of systems and technological tools aimed at increasing our life quality, minimising the environmental impact of everyday activities, and optimising resource usage. Such effects are more noticeable in urban areas with millions of citizens, the so called *mega cities*, which in the near future will be more and more common [1]. The main concept behind a Smart City is the integration of the physical world with the virtual world [2]. This is achieved by providing additional capabilities such as environmental sensing and automatic behaviour to common objects, allowing to capture and to analyse the data from the real world to ensure a better operation of the virtual one.

There are several technologies/perspectives that simplify the process of creating a Smart City. Figure 1 provides a global overview of available technologies from diverse perspectives, which cover different aspects that allow creating a Smart City. Thus, in a Smart City, all daily objects, called things, are equipped with extra capabilities, usually sensing and/or acting capabilities, along with communication capabilities, to share information and to optimise their functional operation. This way, and from a communications perspective, the Internet of Things (IoT) [3] focuses on the intercommunication between all things, as well as on the communication between things and data servers (Cloud or Fog/Edge). On the other hand, from an operational perspective, Cyber Physical Systems (CPS) focus on the integration of these physical things with the computational process [4–6] to improve its functionally. Finally, from a service perspective, Cloud Computing [7,8] and Edge/Fog [9–11] Computing focus on the data processing, and on the structure of the Central servers or Local devices. In the remainder of the paper, we will focus on the Internet of Things perspective, analysing available technologies and their adequacy in terms of implementation.

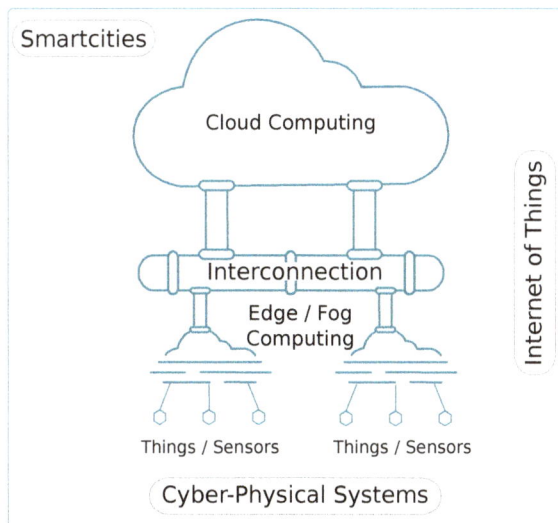

**Figure 1.** Smart city structure.

Any city has several areas of concern to the authorities. In a smart city, all of these areas must have some level of intelligence to minimise management efforts. Thus, there are various subareas of interest including Smart Governance, Smart Mobility, Smart Utilities, Smart Buildings, and Smart Environment, where the adoption of this paradigm can have a clear impact, being highly beneficial [12,13].

Sensing processes are one of the most important tasks in a smart city because they allow retrieving the different parameters involved in different control processes. Examples of such processes include transportation, energy management, air conditioning, etc. However, controlling air pollution in smart cities stands out as a key issue, as it has severe consequences on human health, thereby making environment sensing a critical task and a prominent service.

Currently, controlling pollution levels is an on-going effort undertaken by most European cities, which invest a considerable amount of money in controlling the different hazards produced by poor air quality. Considering these aspects, an index known as AQI (Air Quality Index [14]) was created to classify the air quality, and it specifies different healthy risks, from Low to Very High.

Air quality assessment mainly relies on static monitoring stations, meaning that most cities are endowed with at least one of these stations. Overall, there are about 1500 air quality monitoring stations in Europe. However, the installation and maintenance of these stations is quite expensive. In addition, installing new air monitoring stations is tough since, in crowded places, there is no room for allocating them, meaning that they are typically installed in remote locations, such as parks or sparsely populated areas, which is prone to produce data with little representativeness for the overall population.

There are several air pollutant types, and various techniques can be adopted to measure them. In general, air pollutants can be of two types: (i) primary air pollutants, which are gases or particles emitted directly into the atmosphere, including carbon monoxide (CO), carbon dioxide ($CO_2$), particulate matter smaller than 10 microns ($PM_{10}$), or particulate matter smaller than 2.5 microns ($PM_{2.5}$); and (ii) secondary air pollutants, which are gases produced by a chemical reaction between primary pollutants and some environment element, including ozone ($O_3$), which is produced by the combination of nitrogen oxides (NOx), Oxygen ($O_2$), Volatile Organic Compounds (VOC), and sunlight [15]. In addition, some pollutants are gases (i.e., $CO_2$, CO, and Ozone), while others are suspended particles in the air ($PM_{10}$, and $PM_{2.5}$). Usually, the process of gas monitoring relies on chemical elements that react to the presence of these gases, while the monitoring of suspended particles mostly relies on optical detection procedures.

By embracing the Smart City paradigm, crowdsensing becomes a solution able to cope with air pollution monitoring since this novel paradigm assumes that a significant number of users perform collaborative sensing tasks, thereby collecting data from different populated locations while doing their daily activities. The collected data is periodically transmitted to a central server (Cloud) for data storage and processing. Overall, this strategy implies that the sensors used must be cheap and tiny enough for comfortable and easy transportation. Otherwise, it becomes hard to achieve a widespread distribution and adoption. Besides, there must be a communications link for transmitting the acquired data to a cloud-based server, where data are constantly being stored and processed.

To fulfil the first two requirements concerning price and size, we get advantage by using emerging prototyping platforms such as Raspberry Pi or Arduino, which allow achieving the aforementioned goals when combined with low-cost sensors available on the market. Such solutions, besides being inexpensive, have the advantage of using compact, powerful, and easy-to-use hardware that is widely adopted nowadays both in research and industry. These compact platforms can also be integrated in vehicles, i.e., cars, bikes, or public transportation, making it easy to monitor different air pollutants while the vehicle travels around the city.

On the other hand, smartphones are nowadays widely adopted devices that have become ubiquitous. Thus, it becomes interesting to exploit their communication interfaces, such as Wi-Fi or Cellular, for transmitting the acquired data from mobile sensors. With these issues in mind, mobiles sensors must be able to create a communication link with smartphones to make such interaction possible.

Recent literature suggests that, although there are several hardware options for creating different types of sensors, there are none specific for air pollution monitoring from a crowdsensing perspective. Hence, it becomes necessary to analyse the different options for creating a small, low-cost, mobile sensor able to communicate with off-the-shelf smartphones in an Internet of Things type of environment.

This paper is organised as follows: Section 2 presents a literature overview of crowdsensing projects. Section 3 provides an overview of the current IoT platforms and protocols that are able to support environment-sensing solutions under the crowdsensing paradigm. Section 4 presents an analysis of available architectures and technologies for creating a crowdsensing system while defining the basic requirements for it. Section 5 presents a comparison of possible hardware configurations for creating a mobile sensor able to perform air monitoring, transmitting the collected data via some mobile sink (e.g., a smartphone) towards a datacenter for storage and processing. Section 6 studies the main requirements of this type of sensors, thoroughly assessing some developed configurations that fulfil the aforementioned requirements. Finally, Section 7 presents the conclusions of the present work.

## 2. Crowdsensing Architectures in the IoT Context: Literature Overview

Following the Smart City paradigm, and focusing on the data collection domain, the concept of crowdsourcing has been introduced to refer to scenarios where a large group of people, through different devices and technologies, actively participate in the data acquisition process [16]. Once data are collected, they are sent to a central server for analysis, and feedback will eventually be returned to citizens through actions and services aimed at improving their life quality.

Crowdsensing is a subtype of crowdsourcing where sensors are the actual sources of the data gathered [17]. If air quality sensors are used, crowdsensing becomes a good alternative to traditional stationary air quality stations whereby small sensors are distributed to a large group of people that seamlessly contribute to the system while doing their everyday tasks [18].

In the literature, we can find several works related to crowdsensing systems applied to air monitoring. Air-cloud [19] is a system to monitor the concentrations of $PM_{2.5}$ using crowdsensing in China. It focuses on the analysis of the mechanical sensor design to optimise air reception, as well as on data fusion techniques. The sensor calibration process relied on neural network tuning using laboratory measurements as input.

U-Air [20] shows how to analyse the data obtained from different sources, such as traffic levels, weather conditions, and pollution using various Big Data techniques, evidencing how these techniques allow inferring environmental pollution levels with better granularity.

Researchers from Zurich University [21] have also designed a prototype to monitor ozone levels using a sensor that is connected to the smartphone via an USB cable. The Smart City project in Serbia [22] seeks similar goals, relying on a commercial sensor by Libelium [23] for measuring various air pollutants using the public transport system.

Devarakonda et al. [24] propose a system with two types of sensors: (i) a sensor based on Arduino Mega128 to install in vehicles that costs about $700; and (ii) a personal sensor that is smaller than previous ones, and that costs about $400 for end-user applications. In this work, Google technology is used for data processing.

Mead et al. [25] analyse the behaviour of electrochemical sensors to monitor air pollution in urban scenarios. They design two types of sensors (static and mobile) based on the PIC18F67J10 microchip, and they show how to deal with electrochemical sensors.

Manna et al. [26] propose a sensor to monitor air pollution on roads, and to track vehicles which cause pollution, by using a sensor based on Arduino, an electrochemical CO gas sensor, and RFID technology.

In addition, we can find several technological approaches for fabricating small sensors. For instance, Hjiri et al. [27] use Al-doped ZnO (AZO) nanoparticles to create a highly sensitive CO gas sensor. Instead, Borini et al. [28] use graphene oxide to create humidity and temperature sensors with ultrafast response times (30 ms). Chen et al. [29] analyse the use of nanowires to fabricate gas sensors due to their characteristics: ultrasensitive, higher selectivity, low power consumption, and fast response. In addition, Zaatar et al. [30] show how to use the fiber optic evanescent wave to monitor air pollution.

Even though there are different studies that provide a wide set of approaches for air pollution monitoring through crowdsensing, or that provide an isolate analysis of sensing technologies, there is no specific hardware solution that is widely available for regular users, and different alternatives have not yet been compared in a detailed manner. Thereby, an exhaustive analysis of the different solutions available to create a small, low-cost device endowed with both air quality sensors and communication capabilities, would represent a step forward in this challenging and fascinating area.

## 3. Internet of Things Protocols Overview

In recent years, the Internet of Things (IoT) has become one of the most challenging research topics, offering a wide range of novel solutions for Smart Cities [3,31]. These proposals mainly analyse the intercommunication between devices, and involve a large variety of domains like home-based solutions [32], intelligent transportation systems [33], healthcare [34], safety and security [35,36], industrial control [37], and environmental monitoring. Thus, the analysis of the sensor design must be able to cope with IoT protocols, which will be described in this section.

The main principle underlying the IoT paradigm is that all "things" are, or must be, connected to the Internet, and interact with each other to create developed areas that promote sustainability and a high life quality in multiple key areas [13,38].

The main characteristic of these things is that they are constrained devices such as small sensors, meaning they have restricted processing/storage capacity, restricted battery, restricted communication characteristics, i.e., Low Bandwidth, Low Data Rate, Low Coverage, etc. With this in mind, the Internet of Things has created a subset of protocols divided into various layers, similar to the traditional Internet stack, but taking into account the restrictions of the Internet of Things in terms of processing, battery capacity, and communication capabilities of embedded devices. Figure 2 summarises the different layers defined for IoT, and the differences towards the traditional Internet: (i) the Infrastructure Layer typically relies on wireless technologies, like ZigBee, LoRa, or Bluetooth Low Energy (BLE); (ii) the Addressing Layer focuses on the analysis of the addressing issues to achieve compatibility with Internet protocols; (iii) the Transport Layer is the same than for the TCP/IP (Internet) protocol stack, so either TCP or UDP are available, although UDP is typically used; (iv) the Messaging Layer defines protocols to transmit data towards the servers; (v) the Message Format Layer defines encoding types to store and transmit data; and (vi) the Semantic Layer defines the structure of the data.

Below, we will describe the most important protocols involved in the Internet of Things at these different layers.

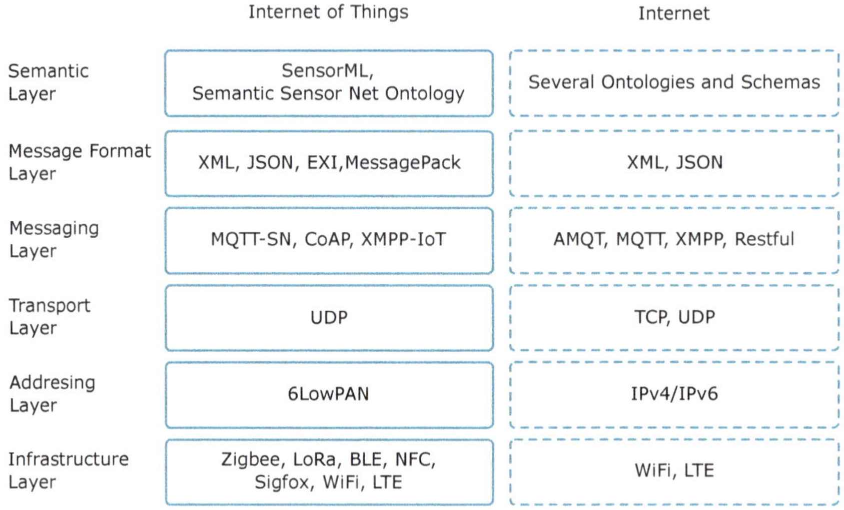

**Figure 2.** IoT protocols.

*3.1. Infrastructure Layer*

Currently, there are several communication technologies for the Internet of Things. Notice that, while any communications technology would allow us to create a network for IoT, not all of them take device restrictions into account. Theoretically, 5G allows a lot of possibilities to be offered in the context of the Internet of Things [39]. Similarly, new technologies such as LoRA or SIGFOX allow network sensors to remain connected due to their large coverage.

Below, we make a brief analysis of the possible wireless technologies for IoT.

**5G Network** [40] is the fifth generation cellular network architecture, designed to support great amounts of data, high speed, configurability, etc. from new emerging technologies such as Internet of Things. Currently, it is in the first phase, where new standards and services will be defined, but soon it shall become the default cellular network technology.

**ZigBee** [41] is based on the **IEEE 802.15.4** standard, and it was designed for Wireless Sensor Networks. Its main characteristics are its small size and low power consumption. Usually, the transmission range can vary from 10 to 100 m, depending on the output power. The main drawback of this technology, though, is that current smartphones are not equipped with ZigBee interfaces.

**Wi-Fi** [42] is based on the IEEE 802.11 standard, and it was designed for Wireless Local Area Networks. The evolution of this technology provides several variants operating in the 2.4 GHz or in the 5 GHz band, being currently the 802.11n version the most widespread option. The transmission range for standard interfaces is about 100 m.

**LoRa** [43] is a LPWAN (Low-Power Wide-Area Network) technology designed to optimise different aspects such as communication range, battery lifetime, and costs, supporting thousands of devices headed for the Internet of Things in several domains including sensing, metering, and machine-to-machine (M2M) communications.

Theoretically, LoRa achieves a transmission range of more than 15 km in rural environments, and of more than 2 km in dense urban areas. Its bandwidth ranges between 250 bps and 50 Kbps in different frequencies: 169 MHz, 433 MHz, and 868 MHz in Europe, and 915 MHz in North America.

**SIGFOX** [44] is an emerging technology that offers a proprietary telecommunications network to support Internet of Things solutions. It was designed for LPWA (Low-Power Wide-Area) networks operating in the ISM 868 MHz band, reaching distances greater than 1 km. Since the selected ISM band is restricted, the communications could be of up to 12 bytes per message, and up to 140 messages per day.

**Near Field Communications (NFC)** [45] was designed for communications between two nearby devices (closer than 4 cm). It main target applications are smartphone-based payments and IoT solutions such as access control, or inventory systems. However, its distance requirements and intermittent connectivity features make it a poor option for our purposes.

**Bluetooth** [46] was designed for Personal Area Networks, purposely having a maximum coverage range of 10 meters by default. Currently, it is used for transmitting information between personal devices, such as smartphones, smartwatches, and headsets.

**Bluetooth Low Energy** [47], or **Bluetooth Smart**, is the name under which Bluetooth version 4 is known. Its main advantage, when compared to previous versions, is using very low power, being nowadays one of the best options for IoT applications. Similarly to previous Bluetooth specifications, the coverage range is of 10 m.

### 3.2. Addressing Layer

The addressing layer defines the logic address of the packets by assigning a specific address to all possible nodes. These protocols deal with the packet forwarding problem, which in the TCP/IP model is handled by IPv4 and/or IPv6.

**6lowPAN** [48] (IPv6 over Low power Wireless Personal Area Networks) allows using IPv6 over networks based mainly in the IEEE 802.15.4 standard.

It is designed for resource-constrained devices by reducing the size of the address to 64 or 16 bits, depending on whether it is for a Local Network or a Personal Area Network, respectively; it uses default values for specifying the network.

### 3.3. Messaging Layer

The messaging layer defines protocols for data transmission systems considering IoT restrictions.

**RESTful** [49] (Representational state transfer) is a Web-based architecture to exchange or to manipulate Web resources through a textual representation using preset stateless operations; it means that each message must have all the required information to complete the request. It follows a client/server model based on the HTTP protocol, and relies on its functions: (i) GET, to retrieve a resource; (ii) POST, to create a resource; (iii) PUT, to change the state of a resource; and (iv) DELETE, to delete a resource. The data representation typically adopts the XML or JSON formats.

Figure 3 presents a basic overview of a Restful Architecture under the client/server model.

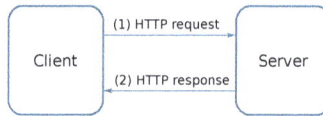

**Figure 3.** Basic RESTful architecture.

There is a specification [50] for constrained nodes and networks called Constrained RESTful Environments (CoRE) Link Format. It specifies a set of links to discover resources, and to access these resources in a Machine-to-Machine (M2M) environment.

**MQTT** [51] (Message Queue Telemetry Transport) is a lightweight messaging protocol based on the publisher/subscriber scheme that runs on top of the TCP/IP protocol. It is also designed for constrained networks with limited bandwidth.

The MQTT is composed by three elements: publishers, subscribers, and a message broker. A subscriber, which wants to receive a message related to a specific topic, must subscribe to the message broker; next, when a publisher sends/publishes a message related a certain topic, it is transmitted to all subscribers subscribed to this topic.

MQTT has three transmission Quality of Service (QoS) levels: (i) QoS 0: At most once. The message is sent once, but it does not check for ACKs to confirm message reception. (ii) QoS 1: At least once. The message could be sent more than once to each subscriber. (iii) QoS 2: Exactly once. The message is sent exactly once using four-way handshaking.

MQTT-SN [52] (MQTT Sensor Network) has been designed to be similar to MQTT, but considering the restrictions of wireless communication environments, such as limited bandwidth, short message length, etc., running over UDP or on Non-IP environments. For interoperating with standard MQTT environments, it needs an MQTT-SN Gateway which connects MQTT-SN nodes, such as constrained sensors, to the MQTT network. Figure 4 shows a basic MQTT architecture.

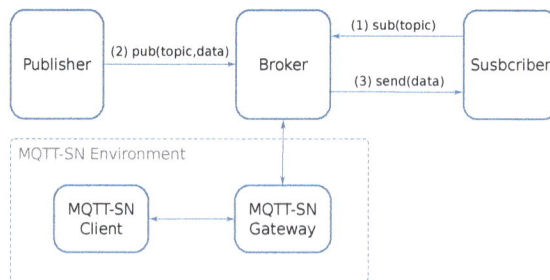

**Figure 4.** Basic MQTT (Message Queue Telemetry Transport) architecture.

**CoAP** [53] (Constrained Application Protocol) is a generic web protocol designed for constrained environments with restricted network capacities and restricted devices, allowing these devices to communicate with the Internet or other constrained devices. It implements a compressed subset of the REST model implementing GET, POST, PUT, and DELETE operations over UDP.

CoAP reduces the message header and restricts message exchange, reducing the network overhead. It is very useful for Machine-to-Machine (M2M) communication, and for the Internet of Things (IoT).

CoAP can be easily translated to HTTP for seamless integration with existing Web systems, while reducing network requirements.

**XMPP** [54] (Extensible Messaging and Presence Protocol) is a message-oriented communications protocol based on XML. Initially, it was called Jabber, and it was designed for instant messaging (IM). It allows federations among various XMPP servers, and even communication with different technologies using XMPP gateways. Currently, it is also used for VoIP, video, gaming, or even for IoT applications. Figure 5 shows the basic architecture of an XMPP-based system.

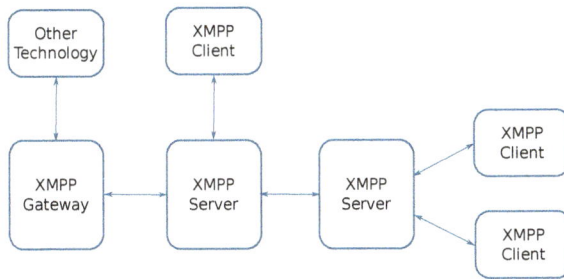

**Figure 5.** Basic XMPP (eXtensible Messaging and Presence Protocol) architecture.

The specification for IoT is XEP-0323: Internet of Things—Sensor Data [55], which provides the architecture, basic operations, and data structures for sensor data communication, including a hardware abstraction model for the interconnection of constrained devices.

**sMAP** [56] (Simple Measuring and Actuation Profile) is an example of how RESTful web services can be simplified, while still allowing instruments and other producers of physical information to directly publish their data in a central server.

*3.4. Message Format Layer*

The Message Format layer presents all data encoding types to store and transmit structured data for IoT applications.

**XML** [57] (eXtensible Markup Language) is a markup language for encoding documents in a text format that is understandable by both human and machines. XML is designed to store data units called entities, where all data structures and document descriptions are achieved through markups. Using these markups, it is possible to create any logical data structure in an easy way.

**JSON** [58] (JavaScript Object Notation) is a format notation to encode structured data (attribute-value pair or array) using human-readable text. It was designed to replace XML by reducing its complexity. It is very common in web systems, especially in AJAX-style ones. It uses pairs (object_id:object_value) and brackets to provide complex object structuring for fitting data in a text document.

**EXI** [59] (Efficient XML Interchange) is a binary and compact representation of XML or JSON documents. It aims at resource-constrained devices and networks, attempting to reduce the size of the data and the computational requirements when compared to other compressors such as gzip. The EXI coder is based on events, and it follows a simplified Huffman coding to create a binary document.

**MessagePack** [60] is a binary serialization format that encodes messages faster and in a more compact manner than traditional methods such as JSON or XML. This is possible since small integer values are encoded in a single byte, while strings require only an extra byte to identify them. This simplifies the encoding process, but it has some limitations, such as the size of strings or numbers, the number of the key/value association map, etc.

Table 1 shows a representation of sensor data using XML and JSON encoding, and Table 2 shows a representation of sensor data using EXI and MessagePack encoding, respectively. We can observe that the EXI encoding and the MessagePack are much smaller (163 and 186 bytes) than the encoding achieved using XML (474 bytes) or JSON (357 bytes).

**Table 1.** Example of data representation using typical encoding types.

| XML Encoding (474 bytes) | JSON Encoding (357 bytes) |
|---|---|
| ```<br><?xml version="1.0" encoding="UTF-8" ?><br><trace><br>  <id>trace1</id><br>  <values><br>    <captures><br>      <latitude>39.470577</latitude><br>      <longitude>-0.3336604</longitude><br>      <ozone>56</ozone><br>    </captures><br>    <captures><br>      <latitude>39.470652</latitude><br>      <longitude>-0.3343365</longitude><br>      <ozone>68</ozone><br>    </captures><br>    <captures><br>      <latitude>39.470892</latitude><br>      <longitude>-0.3359987</longitude><br>      <ozone>59</ozone><br>    </captures><br>  </values><br></trace><br>``` | ```<br>{<br>  "trace": {<br>    "id": "trace1",<br>    "values": {<br>      "captures": [<br>        {<br>          "latitude": "39.470577",<br>          "longitude": "-0.3336604",<br>          "ozone": "56"<br>        },<br>        {<br>          "latitude": "39.470652",<br>          "longitude": "-0.3343365",<br>          "ozone": "68"<br>        },<br>        {<br>          "latitude": "39.470892",<br>          "longitude": "-0.3359987",<br>          "ozone": "59"<br>        }<br>      ]<br>    }<br>  }<br>}<br>``` |

**Table 2.** Example of data representation using compressed (binary) encoding types.

| EXI Encoding (163 bytes) | MessagePack Encoding (186 bytes) |
|---|---|
| 80 40 67 47 26 16 36 5a 80 24 06 d2 c9 50 08 84<br>3a 39 30 b1 b2 98 80 ee cc 2d 8e ac ae 75 00 48<br>25 8d 85 c1 d1 d5 c9 95 ce a0 00 00 96 c6 17 46<br>97 47 56 46 5a 80 44 2c cc e4 b8 d0 dc c0 d4 dc<br>dc 0a 6c 6f 6e 67 69 74 75 64 65 a8 04 43 0b 4c<br>0b 8c cc cc cd 8d 8c 0d 00 66 f7 a6 f6 e6 5a 80<br>44 10 d4 d9 00 0c 02 cc ce 4b 8d 0d cc 0d 8d 4c<br>80 0c 83 0b 4c 0b 8c cc cd 0c cc cd 8d 40 0d 01<br>0d 8e 10 00 c0 2c cc e4 b8 d0 dc c0 e0 e4 c8 00<br>c8 30 b4 c0 b8 cc cc d4 e4 e4 e0 dc 00 d0 10 d4<br>e5 ea 80 | 81 a5 74 72 61 63 65 82 a2 69 64 a6 74 72 61 63<br>65 31 a6 76 61 6c 75 65 73 81 a8 63 61 70 74 75<br>72 65 73 93 83 a8 6c 61 74 69 74 75 64 65 a9 33<br>39 2e 34 37 30 35 37 37 a9 6c 6f 6e 67 69 74 75<br>64 65 aa 2d 30 2e 33 33 33 36 36 30 34 a5 6f 7a<br>6f 6e 65 a2 35 36 83 a8 6c 61 74 69 74 75 64 65<br>a9 33 39 2e 34 37 30 36 35 32 a9 6c 6f 6e 67 69<br>74 75 64 65 aa 2d 30 2e 33 33 34 33 33 36 35 a5<br>6f 7a 6f 6e 65 a2 36 38 83 a8 6c 61 74 69 74 75<br>64 65 a9 33 39 2e 34 37 30 38 39 32 a9 6c 6f 6e<br>67 69 74 75 64 65 aa 2d 30 2e 33 33 35 39 39 38<br>37 a5 6f 7a 6f 6e 65 a2 35 39 |

## 3.5. Semantic Layer

The Semantic layer presents all approaches that describe a logical representation of things in the IoT context. Therefore, in sensing approaches, we can find several representative examples:

**SensorML** [61] is an standard model based on XML encoding for describing sensors and measurement processes. It is developed by the Open Geospatial Consortium, describing a wide range of sensors for different types of architectures, including remote sensors, in-situ sensors and dynamic sensors, among others.

**Semantic Sensor Net Ontology** [62] describes sensors and observations, avoiding to describe domain concepts, location, time, etc. It is developed by the W3C Semantic Sensor Networks Incubator Group (SSN-XG).

**Web of Things** [63] specifies a data model to describe physical devices connected to the Web (Internet) using JSON encoding. It was created for the Mozilla project, and was formally submitted to W3C for discussion.

Taking into account that the previously described protocols, a sensor device must be able to cope with a subset of them to allow the exchange of data between the sensors and a central server.

## 4. Mobile Sensing Requirements

Normally, the sensing process is made through a Wireless Sensor Network (WSN) [64,65], which is composed by a set of nodes or sensors that collect data and send it towards a central sink or gateway. The latter carries out processing tasks, or merely resends collected data to a server for storage and processing. Usually, all sensors are resource-constrained devices, and the central sink or gateway has fewer restrictions, often being connected to the power network. Currently, most Wireless Sensor Networks are based on the IoT architecture since it allows us to work with constrained devices and restricted networks.

Figure 6 shows the basic structure of a Wireless Sensor Network. We can see that the communications link between the sensors and the sink/gateway is wireless, typically relying on ZigBee, and that the communications link between the sink/gateway and the central server is typically a more robust link, either wired (e.g., Ethernet) or wireless (e.g., Wi-Fi or Cellular).

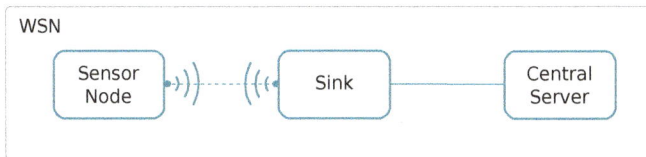

**Figure 6.** Wireless Sensor Network structure.

The support for mobility in a Wireless Sensor Network [66,67] can be achieved through different strategies, including sensor mobility, as shown in Figure 7a, or by having mobility on both sensors and gateway, as shown the Figure 7b. Finally, we have crowdsensing architectures where the gateway and the sensor are the same, or are packed together. Commonly, the best way to implement crowdsensing is through smartphones, since nearly all people carry one with them nowadays, and they are endowed with several sensors and communication interfaces. Figure 8 shows an example of a crowdsensing architecture where a smartphone is used as the gateway between the sensor and a Central Server.

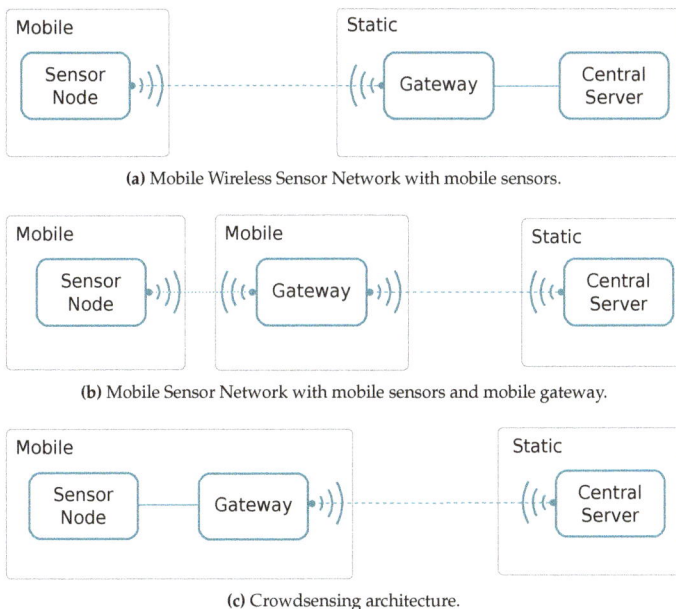

**(a)** Mobile Wireless Sensor Network with mobile sensors.

**(b)** Mobile Sensor Network with mobile sensors and mobile gateway.

**(c)** Crowdsensing architecture.

**Figure 7.** Different types of Mobile Sensor Networks' structures.

Crowdsensing solutions need to be widely disseminated and adopted by users to be successful. In addition, to achieve such widespread acceptance, the impact on the users' everyday activities must be low. This means that any deployed application must operate in the background and avoid consuming excessive resources, while requiring only a minimal user participation. Concerning the sensor itself, if external to the smartphone, it should be cheap, small, easy to use, and comfortable to carry.

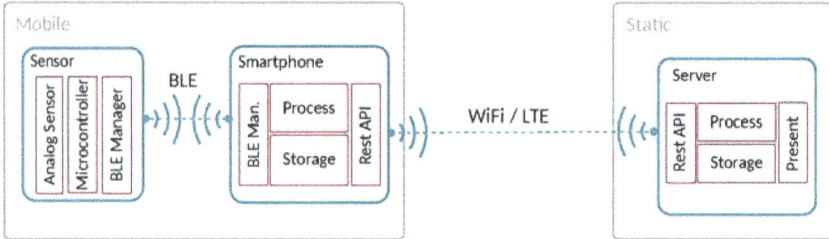

**Figure 8.** Crowdsensing architecture overview.

Crowdsensing approaches have two basic architectural components [68]: a mobile component for the data acquisition process, and a central server for data storage and processing.

The mobile component must be able to collect environmental parameters, transmitting them towards the central server. The data acquisition process is based on smartphone sensors, or on small external sensors accessible via smartphone, and the transmission process usually relies on smartphone connectivity towards the Internet. Despite delegating transmission tasks on smartphones, external sensing devices must still be endowed with communication capabilities to transfer the collected data to the smartphone. Thus, the sensing device should be equipped with a wireless communications interface, being technologies such as Wi-Fi, Bluetooth, RFID, NFC, and ZigBee good candidate solutions.

The central processing server must be able to receive the transmitted data from the sensors, store and process the data, as well as properly present the obtained results to system managers. In addition, in some cases, they perform remote communication with the mobile devices for configuration tasks, thereby allowing to dynamically change the sensing behaviour.

Taking the aforementioned considerations into account, Figure 8 shows a basic hardware architecture applicable to air quality sensing applications that should include: (i) a mobile sensor; (ii) a smartphone; and (iii) a central server. The proposed architecture resembles various approaches from different authors [19–21,68]. Moreover, as shown in Figure 9, the crowdsensing process basically includes five different tasks: (i) sampling process; (ii) filtering process; (iii) data transfer; (iv) data processing; and (v) results presentation. Notice that all these tasks could be done by different hardware components; for instance, the filtering task could be done by the sensor, the smartphone, or even the central server, depending on the system characteristics. Moreover, characteristics associated to sensing, filtering, and transmission tasks could be defined based on parameters obtained from the processing step.

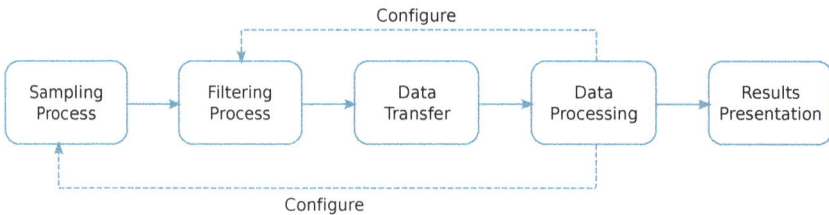

**Figure 9.** Crowdsensing steps.

**Sampling process** refers to the process of capturing pollutant measurements, including the calibration process, where electrical signals are translated to pollution units, filters, fault detection and diagnosis, etc.

Sensor calibration, in Commercial Off-The-Shelf (COTS) sensors such as electrochemical ones, is a process that depends on the physical sensor characteristics, temperature, etc. Basically, electrical outputs must be translated to pollution units, and often there is no lineal relation. The calibration is commonly made in advanced laboratories, taking into account samples taken with different pollution levels and for different temperatures and humidity conditions [69,70]. However, in urban scenarios, the auto calibration procedure is too complicated because all sensors are distributed among different users. Alvear et al. [71] proposed a method to calibrate off-the-shell sensors using mathematical regressions based on high-accuracy samples obtained through the fixed stations deployed in a city. Once a mobile sensor is near to these stations, these samples are used to adjust the translation equation (electrical signal to pollution values).

Using COTS, the sampling error and its diagnostics can be a problem [71]. Nevertheless, when focusing on a Crowdsensing solution using a large number of mobile sensors (smart city scenario), this problem could be solved by accounting for redundant data and statistical analysis (i.e., Kriging allows us to deal with sampling error).

**Filtering process** refers to deleting redundant and/or wrong measurements caused by the sensors reading oscillations. By using mobile sensors, the filtering process also has to deal with temporal variations, as describe in [68,71], adjusting samples to a same temporal fragment.

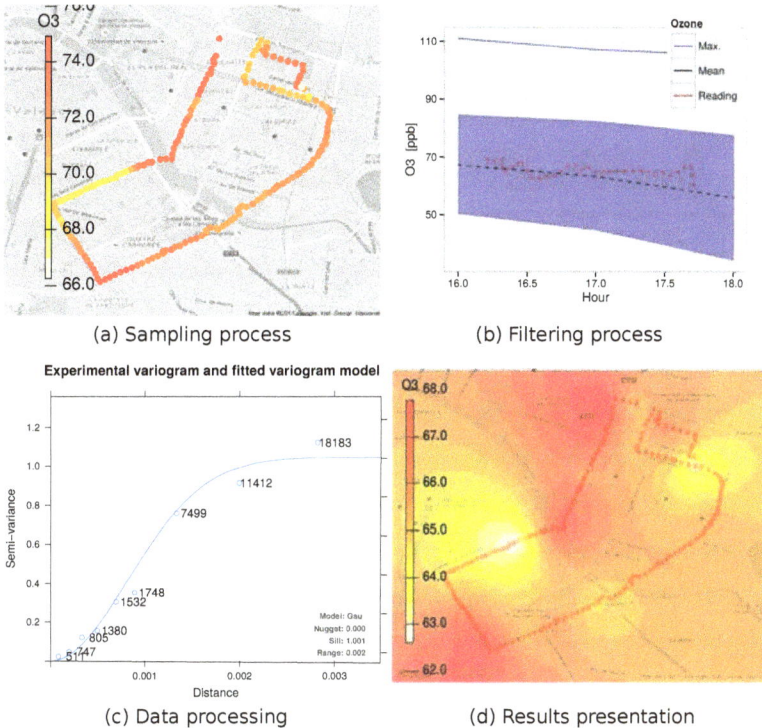

(a) Sampling process

(b) Filtering process

(c) Data processing

(d) Results presentation

**Figure 10.** Data handling process as described in [68]: (**a**) sampling process; (**b**) filtering process after adjusting the temporal variations; (**c**) data analysis using a semivariogram of the captured data used for interpolating the entire area using the Kriging technique; and (**d**) pollution distribution map.

**Data transfer process** refers to the upload of data from the sensor to the cloud (Central servers), including sensor-smartphone and smartphone-server communications. In is achieved through the previously described IoT protocols.

**Data processing** refers to the interpolation technique used to recreate a pollution distribution map. It could be made by different methods (Kriging, IDW, and Nearest neighbour Spatial Averaging) as described in [72]. Currently, the most used method is the Kriging interpolation technique, where a semivariogram is calculated for create a complete pollution map.

**Results presentation** refers to the way results are presented to the system administrator. The most useful representation is a graphical map for the target region.

Figure 10 presents the data handling process, as the authors described in [68], showing the four processes for handling pollution information in order to recreate a complete pollution map for a certain target area.

*4.1. Basic Smartphone Software Architecture*

Concerning smartphones, they are devices widely used nowadays for any task, and characterised by powerful computing capabilities, large amounts of memory, and several embedded sensors and communication interfaces [73]. We consider smartphones as the best gateway option for connecting mobile sensors with a central server. In addition, they can perform CPU-intensive tasks such as data filtering or data fusion, simplifying sensor requirements and design to mere data acquisition and data relaying towards the smartphone.

Since the smartphone must act as a gateway between sensor and cloud server, it must manage at least two network interfaces: one to collect data from the sensor (Sensing middleware), and another one to upload data to a central server (Cloud middleware). Although both tasks must run independently, the data uploading process is often not made in real time, contrarily to the sensor data collection process, which is a task that should be done very frequently, especially if we aim at a simplified sensor design, as shown in Figure 11. Moreover, modules to process and store the collected data are also needed. By assuming that the smartphone becomes responsible for all the computationally intensive tasks, it becomes necessary to analyse which are the actual basic requirements when designing a mobile sensor for crowdsensing solutions.

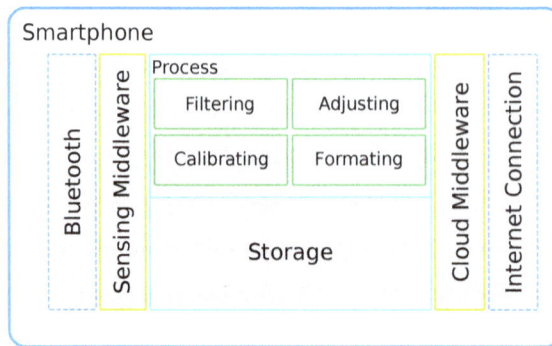

**Figure 11.** Smartphone software architecture overview.

*4.2. Functional Requirements of the Mobile Sensor*

Despite relying on smartphones for providing system intelligence, basic mobile sensor requirements still involve:

**Processing**: The sensor must be able to process the measured data, perform basic filtering tasks, and transfer data to an external device. Anyway, in terms of processing power, requirements are low.

**Storage**: By assuming that a links towards a smartphone or a similar device is available, the sensor does not need to actually store large amounts of collected data. In fact, since data can be seamlessly relayed to the smartphone in real time, the sensor can limit its internal storage to only a few samples.

**Communication**: Sensors do not need to have a direct connection to the cloud server via Internet, but they still need to transfer the collected data to the smartphone. Thus, sensors require a communications link compatible with current smartphone technologies like Wi-Fi, Bluetooth, or NFC.

**Autonomy**: Sensors must be able to operate for long periods using a small power supply. Thus, energy optimisation becomes a key requirement to take advantage of small batteries.

**Size**: Sensors need to be transported by users, or to be quickly installed in vehicles (e.g., bicycle, motorcycles, and cars). Thus, they must be small enough for the sake of aesthetics and comfort.

**Price**: To be attractive to users, sensors must be as cheap as possible. Otherwise, it becomes difficult to meet the broad dissemination requirements of crowdsensing approaches.

*4.3. Basic Mobile Sensor Design*

To fulfil the technical requirements, a basic mobile sensor should be composed of a sensor device able to monitor the differences between different pollutant levels, a communications module for transferring the data collected, and a microcontroller/microcomputer acting as a central element for managing all tasks.

Figure 12 shows a basic mobile sensor design, and the main characteristics to consider. As shown, the sensor hardware module must be able to connect to a microcontroller/microcomputer, a connection that typically relies on an analog/digital port. Similarly, the communications module must also be connected to the microcontroller via an UART or USB port. Thus, the microcontroller becomes a central element in the sensing module, being responsible for managing the interactions between all the elements.

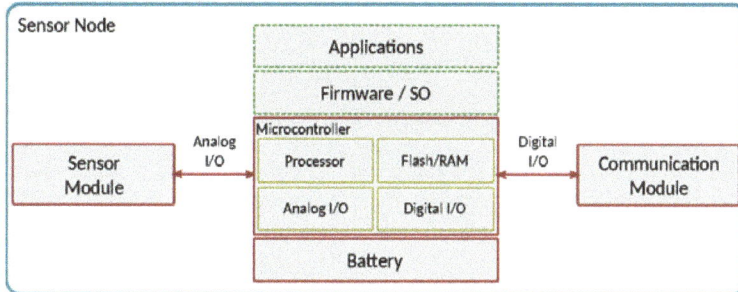

**Figure 12.** Mobile sensor design.

Overall, the sensing module must be equipped with different analog ports, UART/USB ports, a processor, and flash memory (ROM or RAM).

**5. Overview of Available Hardware and Software**

In recent years, the appearance of different embedded prototyping platforms, such as Raspberry Pi or Arduino, which are complemented by a large variety of compatible electronic components, paved the way for the creation of diverse applications related to IoT. When specifically focusing on environmental monitoring requirements, we find that there are different development options, including different types of sensor, and various communication interfaces.

Commercially, several companies offer small and yet powerful boards, along with a large variety of electronic components for personalising them according to user needs. In addition, there are various companies offering extra modules or add-ons such as Seeedstudio [74], which offers their own sensors for developing personalised frameworks based on Grove technologies. Similarly, Adafruit [75] provides different embedded platforms, as well as all kinds of electronic components, including personalised add-ons for batteries, communication modules, and sensor boards compatible with the most widely extended platforms: Raspberry Pi, Arduino, BeagleBone, Intel Edison, or Intel Galileo.

Focusing on final solutions, TST [76] offers products in the field of Smart Cities (Waste Management, Industrial Control, Light Control, etc.), basing their solutions on their own hardware platform. Likewise, Libelium [23] is a company providing various products in the field of monitoring (Environment monitoring, Agriculture, Water monitoring, etc.); most of the components, and the programming tool used, are based on the Arduino platform. However, from a crowdsensing perspective, the solutions offered are inadequate due to the relatively large sizes of the devices, being mostly oriented for public infrastructure deployment.

Based on the state of the art, we now provide an in-depth analysis of the different hardware and software components applicable to our mobile air quality sensing context.

To enable air quality data acquisition, specialised pollution sensors must be connected to a microcontroller or microcomputer via an analog or digital port. Moreover, for communication tasks, the microcontroller/microcomputer must be connected to the communications module via an USB port or UART interface. In this sense, available options will depend on the microcontroller/microcomputer characteristics.

### 5.1. Microcontroller/Microcomputer-Based Embedded Systems

Despite the lightweight processing constraints, there are several options available for embedded systems acting as central elements in the sensor design. In fact, it is possible to use microcomputers such as Raspberry Pi, BeagleBone, or Intel Edison, which use a standard operating system, and that allow developing applications for sensing tasks in a straightforward manner. Alternatively, it is possible to use a microcontroller board such as Arduino, and develop application-specific firmware instead.

**Raspberry Pi** is one of the most popular microcomputers worldwide. It is a low-cost and small-sized computer that allows connecting standard PC peripherals including a monitor, a keyboard, and a mouse. It was designed to explore computing, and it supports different Operating Systems: Raspbian, which is based on Debian, and also Ubuntu Mate or Windows 10 IoT Core, thereby allowing to use several programming languages. In addition, all Raspberry Pi versions benefit from several input/output ports operating at 5 Volts, thus being ideal for all sorts of IoT projects.

There are different versions of the Raspberry Pi, as shown Figure 13, being model A and type 2 the most commonly used. They have different characteristics, e.g., Type B offers better performance in terms of memory and processing, but Model A consumes much less power. Recently, Raspberry Pi 3 has appeared, offering better features than previous versions, being the major difference the integration of a Bluetooth and a Wi-Fi module, thereby facilitating communication tasks in the scope of IoT projects.

Raspberry Mod. A                    Raspberry Pi 2                    Raspberry Pi 3

**Figure 13.** Overview of different Raspberry Pi models.

**BeagleBone** is a small computer running a Linux Operating System called Angstrom, and supporting various software distributions such as Android or Ubuntu. It has an USB port for connecting distinct peripherals, along with an HDMI port for video connection, allowing to use it as a regular computer. It has two 46 pin headers which operate at 3.3 V, allowing to augment the available functionalities by connecting different digital or analog devices like sensors or actuators.

Commercially, we can find several Beaglebone versions, being Beaglebone Black the most commonly used (see Figure 14).

**Figure 14.** Beaglebone Black microcomputer overview.

**Intel Edison** is a tiny but powerful computer developed by Intel. It is designed for IoT applications, targeting at both prototypes and commercial solutions with performance constraints. It supports a modified Linux distribution (Yocto) as its Operating System, and it integrates both Wi-Fi and Bluetooth 4.0 interfaces. In addition, it can benefit from two types of expansion board: an Arduino Expansion board, and a mini breakout board (see Figure 15).

**Figure 15.** Intel Edison with two extension boards.

**Pycom** [77] is a microcontroller based on the ESP32 chip with 24 GPIO pins, 2 UARTs, 1 SPI and 1 I2C port, using a firmware based on micropython. It can be equipped with several communication interfaces such as Wi-Fi, Bluetooth Low Energy, LoRA, and Sigfox. Moreover, using an expansion board, it can integrate an SD Card, as well as different sensors such as Gyroscope, Accelerometer, or GPS (see Figure 16).

**Figure 16.** Pycom module.

**Arduino Uno** is an open source prototyping platform characterised by easy-to-use hardware and software. It has several analog and digital input/output pins to connect sensors, actuators or complementary boards, allowing to create a wide variety of IoT solutions. Arduino has its own programming language based on Wiring, and its own Arduino Software based on Processing [78]. It has a central microcontroller, and an USB port for programming and to supply power.

**Arduino nano** is a tiny prototyping platform that maintains the Arduino Uno concept, using the same programming languages and the same libraries. It was also developed for prototyping solutions, but it is smaller than the standard Arduino, and it has less available memory (see Figure 17).

**Figure 17.** Arduino Nano module (**right**) and Arduino Uno module (**left**).

Table 3 allows comparing these five embedded systems by providing a summary of the most significant technical details.

**Table 3.** Comparison of different processing module components.

| Board | CPU Speed | Memory/Storage | Power Comp. | Ports A/D | Size and Weight | USB Ports/Wireless Interfaces | Price | Operating Systems |
|---|---|---|---|---|---|---|---|---|
| Raspberry Pi Model A | 700 MHz | 256 MB/SD (4 GB) | 0.8 W | 0/30 | 6.5 × 5.5 cm/100 g | 1 USB port/- | $25 | Raspbian |
| Raspberry Pi 2 | 900 MHz | 1 GB/SD (4 GB) | 1.5 W | 0/30 | 8.4 × 5.5 cm/136 g | 4 USB ports/- | $35 | Raspbian Ubuntu Mate Windows 10 IoT |
| Raspberry Pi 3 | 1.2 GHz | 1 GB/SD (4 GB) | 1.8 W | 0/30 | 8.4 × 5.5 cm/136 g | 4 USB ports/Wi-Fi and Bluetooth | $40 | Raspbian Ubuntu Mate Windows 10 IoT |
| Beagle Bone | 720 MHz | 512 MB/SD (4 GB) | 1.0 W | 14/6 | 8.4 × 5.5 cm/100 g | 1 USB port/- | $75 | BeagleBone Linux |
| Intel Edison | 500 MHz | 1 GB/SD (4 GB) | 0.6 W | 14/6 | 5.9 × 2.8 cm/82 g | -/Wi-Fi and Bluetooth | $90 | Yocto Project |
| Pycom | 32 MHz | 1 kB/32 kB | 0.2 W | 14/6 | 6.8 × 5.4 cm/100 g | Wi-Fi, Bluetooth and LoRa | $45 | Micropython |
| Arduino Uno | 32 MHz | 1 kB/32 kB | 0.2 W | 14/6 | 6.8 × 5.4 cm/100 g | -/- | $25 | Processing-based |
| Arduino Nano | 16 MHz | 512 B/16 kB | 0.2 W | 14/6 | 4.2 × 1.8 cm/20 g | -/- | $10 | Processing-based |

Operating Systems for IoT Microcomputers

The choice of an adequate Operating System is a very important issue in the design of a sensor since it will shape the corresponding software architecture. Thus, we now proceed by analysing the different operating systems available for the microcomputers referred above.

**Raspbian** is the most common operating system designed for Raspberry Pi. It is supported by all Raspberry Pi versions, and it has two versions: (i) a complete version with a graphical interface and many development tools which facilitate the development of solutions, but that consumes a lot of resources; and (ii) a lite version, without graphical interface, and with just a basic set of preinstalled software, allowing to add only those packages that are actually required.

Since it is a Linux-based operating system, it supports several programming languages like C, C++, Java, Scratch, Python, or BASH.

**Ubuntu MATE** is an option for Raspberry Pi microcomputers that is supported by model 2 and model 3. This operating system attempts to be simple from the end user perspective, integrating several entertainment applications, although it is also possible to add development tools to it.

**Angstrom** is a modified Linux optimised for Beaglebone microcomputers. It has no graphical interface, and it supports several programming languages such as Python, C, Java, or BASH. In addition, it has its own programming language called BoneScript, which is based on the node.js language.

**Yocto Project** is a complete embedded Linux development environment with tools and methods to facilitate the creation of embedded systems. It can be configured to run Arduino-based or Linux-based programs, thereby offering a great flexibility.

**Micropython** [79] is a Python 3.5 implementation optimised for running in microcontrollers. It allows interacting via a prompt, executing commands or running scripts in an autonomous way. It is entirely compatible with python, and it includes modules for accessing low-level hardware.

**Table 4.** Comparison of different operating systems.

| Operating System | Booting Time (s) | Min. Memory (MB) | Graphical Interface | Type | Programming Languages |
|---|---|---|---|---|---|
| Raspbian | 25 | 150 | Yes | Multi-thread | C Java Python Scratch |
| Raspbian Lite | 15 | 50 | No | Multi-thread | C Java Python Scratch |
| Ubuntu MATE | 30 | 200 | Yes | Multi-thread | C Java Python |
| Yocto Project | 20 | 150 | No | Multi-thread | C Java Python Arduino-based |
| Angstrom Linux | 15 | 100 | No | Multi-thread | C Java Python Bonescript |
| Windows 10 IoT | 35 | 250 | No | Multi-thread | Visual Studio C |
| Micropython | <1 | <1 | No | Multi-thread | Python - |
| Processing-based | <1 | <1 | No | Single-thread | Wiring - |

**Windows 10 IoT Core** is a Windows 10 based operating system oriented to Internet of Things projects using small devices. It is supported by Raspberry Pi 2 and 3, Arrow DragonBoard 410c,

and MinnowBoard MAX. Windows 10 IoT Core relies on the rich, extensible Universal Windows Platform (UWP) API for building solutions. It can be used together with the Visual Studio environment for programming.

Table 4 shows a brief comparison of the different Operating Systems currently available.

### 5.2. Air Pollution Sensors

Nowadays, we can find a wide variety of sensing technologies (see Figure 18) for gas detection (Metal oxide semiconductor, polymer, carbon nanotubes, moisture absorbing materials, Optics, Acoustics, etc.) as shown in [80]. Each technology has different properties, calibration processes and costs, among other characteristics, and so a comparison between these different technologies is required.

**Figure 18.** Different types of air pollution sensors.

To evaluate the different sensing technologies, we must mainly consider some characteristics, especially when focusing on the design of small mobile sensors: (i) sensibility, which refers to the range of values that the sensor can measure; (ii) selectivity, which is the capability of reacting only to the target gas; (iii) linearity, which is the rate of change with respect to gas variations; (iv) response time, which is the time required to start measuring correctly; (v) power consumption; and (vi) price.

In the market, we can easily find various gas sensors for air pollution monitoring, being the three following types of sensors the most common: electrochemical, semiconductor, and infrared. These have a wide range of prices and characteristics. Below we provide more details about each of these sensor types:

**Electrochemical gas sensors** measure the concentration of some air pollutant by oxidizing or reducing its internal porous membrane, thereby producing current changes. Usually these sensors behave quite linearly, allowing to make accurate measurements. They typically operate at 5 V, having a power consumption of about 600 mW; most pollution sensors in this category cost between $100 and $400.

**Semiconductor gas sensors** are the most common gas sensors because of their low cost and high sensitivity. They have an internal conductive material that increases their conductivity level in the presence of a specific air pollutant. These sensors are nonlinear and have a low selectivity, adding difficulty to the monitoring process. Usually they operate at 5 V, having a power consumption in the 500–900 mW range, and they cost between $10 and $35.

**Moisture absorbing material sensors** are used for measuring temperature and humidity. Their dielectric constant varies according to the water content in the environment. They operate at 5 V, having a power consumption of about 0.5 W. In addition, they are very cheap, with a price of about $5.

**Infrared sensors** measure gas/element variations by detecting interferences in an infrared laser. They are specially adequate for monitoring pollutants such as fine particulate matter sized less than 10 micrometers ($PM_{10}$), or fine particulate matter sized less than 2.5 micrometers ($PM_{2.5}$). They usually operate at 5 V, having a power consumption of about 1W, and their cost is about $40.

Table 5 provides a brief summary of the most significant aspects of these different types of sensors for comparison purposes.

**Table 5.** Comparison of different sensing module components.

| Sensor Type | Sensitivity | Selectivity | Linearity | Response Time | Power Comp. | Size (cm) | Price |
|---|---|---|---|---|---|---|---|
| Electrochemical | Medium | Medium | High | Medium | 0.6 W | 2.0 × 4.0 | $200 |
| Semiconductor | High | Low | Low | Low | 0.5 W | 2.0 × 4.0 | $10–$35 |
| Moisture Absorbing | High | Medium | High | High | 0.5 W | 2.0 × 4.0 | $5 |
| Infrared | High | Low | Medium | High | 1.0 W | 15.0 × 10.0 | $40 |

*5.3. Communications Modules*

Although the RFID [81] standard was developed for IoT solutions, there are currently several options available for providing communications between the sensors and the mobile terminal (usually a smartphone), as shown in Section 3. Figure 19 shows some communication modules examples. We now proceed to analyse the technical characteristics associated to the different available options.

The main characteristics to consider are: (i) distance, that is, the wireless coverage range; (ii) communication type, which refers to the characteristics of the channel over which messages are transmitted—usually two types are considered, i.e. serial-based, when a communications channel is open to transmit a stream of data, or message-based, when the data are transmitted via a unique message; (iii) message size, which refers to the maximum size of the message when the communication is message-based; (iv) power consumption; and (v) price.

**Figure 19.** Example of some communication modules.

Table 6 shows a brief summary of the most significant aspects to consider in terms of communication modules.

**Table 6.** Comparison of different network module components.

| Module | Communication Type | Max. Message Size | Data Rate | Distance | Power Comp. | Price |
|--------|-------------------|-------------------|-----------|----------|-------------|-------|
| Wi-Fi | Serial-based | - | +54 Mbps | 100 m | 0.5 W | $10 |
| NFC | Message-based | 32 kB | 424 Kbps | 0.04 m | 0.1 W | $35 |
| ZigBee | Message-based | 128 bits | 250 Kbps | 100 m | 0.1 W | $40 |
| SIGFOX | Message-based | 96 bits | 140 msg/day | +1 km | 0.3 W | $60 |
| LoRa | Message-based | 0.1, 1 or 10% TimeOnAir | 250–5400 bps | 2–15 km | 0.1 W | $45 |
| Bluetooth | Serial-based | - | +2.1 Mbps | 10 m | 0.2 W | $10 |
| Bluetooth LE | Message-based | 160 bits | 1 Mbps | 10 m | 0.05 W | $15 |

We can observe that all options are relatively cheap, with prices in the range from $40–$50, but the more widely used in the market, i.e., Wi-Fi and both Bluetooth and Bluetooth Low Energy, are the cheapest ones, with prices in the range $10–$15. In terms of power consumption, the best option is Bluetooth Low Energy (0.05 W), although ZigBee, LoRA and NFC also exhibit low power consumption levels. In terms of bitrate, the best performing technology is Wi-Fi, being that typical wireless sensing technologies, such as LoRA, ZigBee, or SIGFOX, have a low bitrate. Regarding the communications type, there are two options: (i) Wi-Fi and Bluetooth, which open a serial communications channel for transmitting a stream of bytes; and (ii) Bluetooth Low Energy, ZigBee, Sigfox, LoRA, and NFC, which transmit a message per iteration.

## 6. Mobile Air Pollution Sensor Design

The design of a small and cheap mobile sensor is a basic requirement for air pollution monitoring. After analysing the main technical characteristics of hardware components available in the market, it quickly becomes evident that there are several options for creating a mobile air quality sensor for crowdsensing in the Internet of Things context.

If focusing on the mobile sensor/smartphone wireless connection, the best option is using Bluetooth Low Energy since it is able to fulfil all requirements (low power consumption, low price, and small-sized modules) whereas all other options have different drawbacks. For instance, the ZigBee technology, despite being the most extended technology for wireless sensor networks due to its low power consumption, is not supported by current smartphones. The SIGFOX technology has very strict restrictions regarding the number of messages that it is possible to generate per time slot, thus having little applicability to our aims. The Wi-Fi technology, although being widely used and having a large coverage range, consumes more power and is typically used for Internet connectivity. Finally, the major problem of the NFC technology is its coverage range (only about 4 cm).

In terms of sensor device prototyping, the best option for creating a small and cheap mobile sensor for air pollution monitoring is the Semiconductor gas option. Notice that it is cheaper than the electrochemical sensor, and even though the latter is more accurate, the error introduced can be mitigated by combining information from other nearby users. Concerning infrared sensors, they are only applicable to a specific type of pollutant (fine particulate), while moisture absorbing sensors are mostly used to measure temperature and humidity.

Next, we propose and evaluate some possible configurations for the different studied boards using a semiconductor gas sensor for pollution monitoring, and a Bluetooth Low Energy interface as the transmission technology. For all the proposed hardware configurations, it is possible to use an external USB charger as a power supply to offer more autonomy, while maintaining the support for mobility.

Figure 20 shows the four solutions we have developed to evaluate the different hardware options, and Table 7 shows a comparison between these solutions.

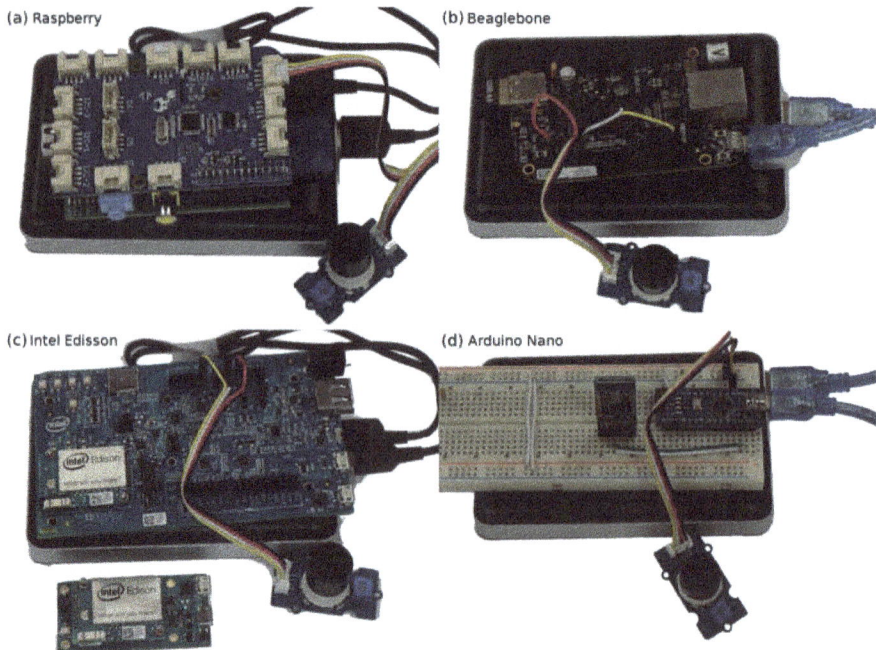

**Figure 20.** Proposed mobile sensing solutions for air quality monitoring.

**Table 7.** Comparison of the four different solutions proposed.

| Module | Extras | Network | Power Comp. | Weight | Price | Flexib. | Develop. | Comp. Power |
|--------|--------|---------|-------------|--------|-------|---------|----------|-------------|
| RPi 3 | +converter | – | 2000 mW | 200 g | €90 | ★ ★ ★ | ★ ★ ★ | ★ ★ ★ |
| Beaglebone | – | +BLE usb | 1500 mW | 150 g | €110 | ★ ★ ★ | ★ ☆ ☆ | ★ ★ ★ |
| Intel Edison | +breakout +expansion | – | 1000 mW | 100 g 200 g | €130 | ★ ★ ★ | ★ ★ ☆ | ★ ★ ★ |
| Arduino | +circuit | +BLE uart | 600 mW | 60 g | €55 | ★ ☆ ☆ | ★ ★ ★ | ★ ☆ ☆ |

The first hardware option we propose is based on the Raspberry Pi platform (see Figure 20a). Since it has no analog ports, it has to be provided with an analog/digital converter. For this purpose, we propose using GrovePi [82], which is an extension board that allows connecting several analog/digital grove ports to a Raspberry Pi in an easy way. Furthermore, since the Raspberry Pi has several USB ports, we propose using a standard USB Bluetooth module for Raspberry Pi 2, or the built-in Bluetooth module for Raspberry Pi 3. With this solution, it becomes possible to run several programming languages, as it is possible to install a Linux or a Windows 10 IoT operating system, and there are a lot of development efforts around it. This configuration has a power consumption of about 2000 mW, a total weight of 200 g, and it costs about $90. Overall, it is the most power-hungry solution among the four proposed, but it becomes the best option for quick prototyping due to its flexibility and large community of developers.

The second hardware approach relies on the BeagleBone board (see Figure 20b). It has several analog ports, allowing us to connect the sensor directly to this board without any intermediate device. In addition, since the BeagleBone has a USB port, it is possible to use a standard USB Bluetooth module. It runs a Linux-based operating system, allowing it to run different programming languages, but there are not too many developments or projects focusing on this solution. This configuration has a power consumption of about 1500 mW, a total weight of 150 g, and it costs about $110, thus being one of the most expensive options. It is not a very useful prototyping platform since it has characteristics that are similar to the Raspberry Pi, but it has a smaller developer community and less support.

The third hardware solution we propose is based on the Intel Edison platform (see Figure 20c). It has an embedded Bluetooth interface, but it does not provide an analog port to directly connect the sensor. For this purpose, it is possible to use: (i) the Arduino expansion board; or (ii) the Breakout board [83]. The first sensor connection option is very simple, however the sensor becomes excessively large. Regarding the second option, the overall size remains small, but it is necessary to make an ad hoc circuit to connect the sensor. The Edison board supports a Linux-based operating system (Yocto) including the possibility to run Arduino-based scripts. For this last configuration, the power consumption is of about 1000 mW, the total weight is 200 g, and it costs about $130, making it the most expensive option. It is useful mostly for end solutions due to its price.

The last embedded solution we propose is based on the Arduino platform (see Figure 20d). Since it was designed for these types of solutions, it becomes easy to connect a sensor via the existing analog ports; nevertheless, USB ports are not available, and so a Bluetooth module must be connected via an UART port for both Arduino Uno and Arduino Nano boards. This solution only runs Arduino-based scripts, reducing the programming flexibility, but we can find a lot of developments using this platform.

For the Arduino Uno solution, the power consumption is about 600 mW, the total weight is 150 g, and it costs about $55. This option is improved by the Arduino Nano solution, which has a power consumption of about 600 mW, a total weight of 60 g, and it costs about $50. The latter option is better than all others in terms of consumption, weight, and price, having as its only drawback the limited memory/CPU resources. It is also useful for restricted environments where the power consumption is very limited.

## 7. Conclusions and Future Work

Smart Cities is a trending topic, and many research efforts are being made worldwide to progress towards that new paradigm which encompasses several areas including smart government, smart transport, smart environment, and smart grid. Since air pollution is considered to be one of the most significant health risks worldwide, smart environment obviously becomes a very important area in the Smart City context.

Although air monitoring stations have been sparsely deployed in most large cities for controlling pollution levels, these are not enough to provide a detailed view of the pollutants' distribution in a city. In this sense, crowdsensing emerges as a good option to monitor the different pollutants by combining small mobile sensors and smartphones. Since current smartphones have much memory and high computing capabilities, mobile sensors can be kept minimal, focusing on data acquisition tasks alone.

In this paper, we have analysed the most relevant Internet of Things architectures and protocols, along with the requirements of an embedded mobile sensor platform from a crowdsensing perspective, identifying the basic tasks the sensor must be able to perform. Besides, an analysis of the hardware architecture requirements has been done, and candidate off-the-shelf hardware components have been analysed. Finally, several complete hardware configurations meeting all the design requirements have been developed and compared in terms of power consumption, weight, and cost. Overall, we have found that the Arduino Nano platform, despite having very limited resources, is able to fulfil the established requirements, thus being the most recommendable alternative in terms of price, weight, and power consumption features.

Regarding other hardware alternatives, microcomputers such as Raspberry Pi, BeagleBone, or Intel Edison are more powerful and flexible by supporting a standard Operating System, thereby allowing to quickly deploy any application. We believe that the Raspberry Pi solution can be the best option for quick prototyping. For more professional solutions, requiring higher processing capacity, the Intel Edison becomes a better option, although imposing a higher overhead in terms of development time. Finally, Arduino becomes an option for very restricted environments.

Overall, we find that a hardware solution applicable to all IoT contexts, and meeting low size and low power requirements, along with adequate communication interfaces and battery capacity, is still missing, although in years to come many more products are expected.

The next steps in this research line are to test the adequacy of the proposed solutions not only in the mobile user context, but also in other contexts including coupling these sensors to vehicles such as cars, buses, bikes, or even flying vehicles (multicopters).

**Acknowledgments:** This work was partially supported by the *Ministerio de Economía y Competitividad, Programa Estatal de Investigación, Desarrollo e Innovación Orientada a los Retos de la Sociedad, Proyectos I+D+I 2014*, Spain, under Grant TEC2014-52690-R, the *Generalitat Valenciana*, Spain, the Secretaría Nacional de Educación Superior, Ciencia, Tecnología e Innovación del Ecuador (SENESCYT), and the Universidad de Cuenca.

**Conflicts of Interest:** The authors declare no conflict of interest.

## References

1. Glasmeier, A.; Christopherson, S. Thinking about smart cities. *Camb. J. Reg. Econ. Soc.* **2015**, *8*, 3–12.
2. Borgia, E. The Internet of Things vision: Key features, applications and open issues. *Comput. Commun.* **2014**, *54*, 1–31.
3. Gubbi, J.; Buyya, R.; Marusic, S.; Palaniswami, M. Internet of Things (IoT): A vision, architectural elements, and future directions. *Future Gener. Comput. Syst.* **2013**, *29*, 1645–1660.
4. Lee, E.A. Cyber physical systems: Design challenges. In Proceedings of the 2008 11th IEEE International Symposium on Object Oriented Real-Time Distributed Computing (ISORC), Orlando, FL, USA, 5–7 May 2008; pp. 363–369.
5. Wolf, W. Cyber-physical systems. *Computer* **2009**, *42*, 88–89.
6. Baheti, R.; Gill, H. Cyber-physical systems. *Impact Control Technol.* **2011**, *12*, 161–166.
7. Rimal, B.P.; Choi, E.; Lumb, I. A Taxonomy and Survey of Cloud Computing Systems. In Proceedings of the Fifth International Joint Conference on INC, IMS and IDC, NCM '09, Seoul, Korea, 25–27 August 2009; Volume 9, pp. 44–51.
8. Tsai, W.T.; Sun, X.; Balasooriya, J. Service-oriented cloud computing architecture. In Proceedings of the 2010 Seventh International Conference on Information Technology: New Generations (ITNG), Las Vegas, NV, USA, 12–14 April 2010; pp. 684–689.
9. Dinh, H.T.; Lee, C.; Niyato, D.; Wang, P. A survey of mobile cloud computing: Architecture, applications, and approaches. *Wirel. Commun. Mob. Comput.* **2013**, *13*, 1587–1611.
10. Ahmed, A.; Ahmed, E. A survey on mobile edge computing. In Proceedings of the 2016 10th International Conference on Intelligent Systems and Control (ISCO), Coimbatore, India, 7–8 January 2016; pp. 1–8.
11. Zhu, J.; Chan, D.S.; Prabhu, M.S.; Natarajan, P.; Hu, H.; Bonomi, F. Improving web sites performance using edge servers in fog computing architecture. In Proceedings of the 2013 IEEE 7th International Symposium on Service Oriented System Engineering (SOSE), Redwood City, CA, USA, 25–28 March 2013; pp. 320–323.
12. Celino, I.; Kotoulas, S. Smart Cities [Guest editors' introduction]. *IEEE Internet Comput.* **2013**, *17*, 8–11.
13. Zanella, A.; Bui, N.; Castellani, A.; Vangelista, L.; Zorzi, M. Internet of things for smart cities. *IEEE Internet Things J.* **2014**, *1*, 22–32.
14. United States Environmental Protection Agency. Air Quality Index. Available online: http://cfpub.epa.gov/airnow/index.cfm?action=aqibasics.aqi (accessed on 29 November 2015).
15. Fishman, J.; Ramanathan, V.; Crutzen, P.; Liu, S. Tropospheric ozone and climate. *Nature* **1979**, *282*, 818–820.
16. Guo, B.; Wang, Z.; Yu, Z.; Wang, Y.; Yen, N.Y.; Huang, R.; Zhou, X. Mobile crowd sensing and computing: The review of an emerging human-powered sensing paradigm. *ACM Comput. Surv. (CSUR)* **2015**, *48*, 7.
17. Ganti, R.K.; Ye, F.; Lei, H. Mobile crowdsensing: Current state and future challenges. *IEEE Commun. Mag.* **2011**, *49*, 32–39.
18. Ma, H.; Zhao, D.; Yuan, P. Opportunities in mobile crowd sensing. *IEEE Commun. Mag.* **2014**, *52*, 29–35.
19. Cheng, Y.; Li, X.; Li, Z.; Jiang, S.; Li, Y.; Jia, J.; Jiang, X. AirCloud: A cloud-based air-quality monitoring system for everyone. In Proceedings of the 12th ACM Conference on Embedded Network Sensor Systems—SenSys '14, Memphis, Tennessee, 3–6 November 2014; ACM: New York, NY, USA, 2014; pp. 251–265.

20. Zheng, Y.; Liu, F.; Hsieh, H.P. U-Air: When urban air quality inference meets big data. In Proceedings of the 19th ACM SIGKDD International Conference on Knowledge Discovery and Data Mining—KDD '13, Chicago, IL, USA, 11–14 August 2013; pp. 1436–1444.

21. Hasenfratz, D.; Saukh, O.; Sturzenegger, S.; Thiele, L. Participatory air pollution monitoring using smartphones. In Proceedings of the 2nd International Workshop on Mobile Sensing, Beijing, China, 16–20 April 2012; pp. 1–5.

22. Brković, M.; Sretović, V. Urban Sensing–Smart Solutions for Monitoring Environmental Quality: Case Studies from Serbia. In Proceedings of the 48th ISOCARP—International Society of City and Regional Planners World Congress: Fast Forward: Planning in a (Hyper) Dynamic Urban Context, Perm, Russia, 10–13 September 2012.

23. Libelium. Available online: http://www.libelium.com/ (accessed on 27 November 2017).

24. Devarakonda, S.; Sevusu, P.; Liu, H.; Liu, R.; Iftode, L.; Nath, B. Real-time air quality monitoring through mobile sensing in metropolitan areas. In Proceedings of the 2nd ACM SIGKDD International Workshop on Urban Computing, Chicago, IL, USA, 11–14 August 2013; p. 15.

25. Mead, M.; Popoola, O.; Stewart, G.; Landshoff, P.; Calleja, M.; Hayes, M.; Baldovi, J.; McLeod, M.; Hodgson, T.; Dicks, J.; et al. The use of electrochemical sensors for monitoring urban air quality in low-cost, high-density networks. *Atmos. Environ.* **2013**, *70*, 186–203.

26. Manna, S.; Bhunia, S.S.; Mukherjee, N. Vehicular pollution monitoring using IoT. In Proceedings of the Recent Advances and Innovations in Engineering (ICRAIE), Jaipur, India, 9–11 May 2014; pp. 1–5.

27. Hjiri, M.; El Mir, L.; Leonardi, S.; Pistone, A.; Mavilia, L.; Neri, G. Al-doped ZnO for highly sensitive CO gas sensors. *Sens. Actuators B Chem.* **2014**, *196*, 413–420.

28. Borini, S.; White, R.; Wei, D.; Astley, M.; Haque, S.; Spigone, E.; Harris, N.; Kivioja, J.; Ryhanen, T. Ultrafast graphene oxide humidity sensors. *ACS Nano* **2013**, *7*, 11166–11173.

29. Chen, X.; Wong, C.K.; Yuan, C.A.; Zhang, G. Nanowire-based gas sensors. *Sens. Actuators B Chem.* **2013**, *177*, 178–195.

30. Zaatar, Y.; Zaouk, D.; Bechara, J.; Khoury, A.; Llinaress, C.; Charles, J.P. Fabrication and characterization of an evanescent wave fiber optic sensor for air pollution control. *Mater. Sci. Eng. B* **2000**, *74*, 296–298.

31. Khan, W.Z.; Xiang, Y.; Aalsalem, M.Y.; Arshad, Q. Mobile phone sensing systems: A survey. *IEEE Commun. Surv. Tutor.* **2013**, *15*, 402–427.

32. Ricquebourg, V.; Menga, D.; Durand, D.; Marhic, B.; Delahoche, L.; Loge, C. The smart home concept: Our immediate future. In Proceedings of the 2006 1st IEEE International Conference on E-Learning in Industrial Electronics, Hammamet, Tunisia, 18–20 December 2006; pp. 23–28.

33. Lenior, D.; Janssen, W.; Neerincx, M.; Schreibers, K. Human-factors engineering for smart transport: Decision support for car drivers and train traffic controllers. *Appl. Ergon.* **2006**, *37*, 479–490.

34. Demirkan, H. A smart healthcare systems framework. *IT Prof.* **2013**, *15*, 38–45.

35. Gao, W.; Tian, Y.; Huang, T.; Ma, S.; Zhang, X. The IEEE 1857 standard: Empowering smart video surveillance systems. *IEEE Intell. Syst.* **2014**, *29*, 30–39.

36. Hewitt, C. Security without IoT Mandatory Backdoors. Available online: https://ssrn.com/abstract=2795682 (accessed on 27 November 2017).

37. McFarlane, D.; Giannikas, V.; Wong, A.C.; Harrison, M. Product intelligence in industrial control: Theory and practice. *Annu. Rev. Control* **2013**, *37*, 69–88.

38. Atzori, L.; Iera, A.; Morabito, G. The internet of things: A survey. *Comput. Netw.* **2010**, *54*, 2787–2805.

39. Ejaz, W.; Anpalagan, A.; Imran, M.A.; Jo, M.; Naeem, M.; Qaisar, S.B.; Wang, W. Internet of Things (IoT) in 5G wireless communications. *IEEE Access* **2016**, *4*, 10310–10314.

40. Gupta, A.; Jha, R.K. A survey of 5G network: Architecture and emerging technologies. *IEEE Access* **2015**, *3*, 1206–1232.

41. Baronti, P.; Pillai, P.; Chook, V.W.; Chessa, S.; Gotta, A.; Hu, Y.F. Wireless sensor networks: A survey on the state of the art and the 802.15. 4 and ZigBee standards. *Comput. Commun.* **2007**, *30*, 1655–1695.

42. Asherson, S.; Kritzinger, P.; Pileggi, P. *Wireless Standards and Mesh Networks*; Computer Science Department, University of Cape Town: Cape Town, South Africa, 2007.

43. Sornin, N.; Luis, M. LoRa Specification. Available online: https://www.rs-online.com/designspark/rel-assets/ds-assets/uploads/knowledge-items/application-notes-for-the-internet-of-things/LoRaWAN%20Specification%201R0.pdf (accessed on 26 November 2017).

44. Sigfox. Available online: http://www.sigfox.com (accessed on 25 November 2017).

45. Haselsteiner, E.; Breitfuß, K. Security in near field communication (NFC). In Proceedings of the Workshop on RFID Security, Graz, Austria, 1 August 2006; pp. 12–14.
46. Bluetooth Core Specification Version 4.0. Available online: https://www.bluetooth.org/docman/handlers/downloaddoc.ashx?doc_id=229737 (accessed on 26 November 2017).
47. Gomez, C.; Oller, J.; Paradells, J. Overview and evaluation of bluetooth low energy: An emerging low-power wireless technology. *Sensors* **2012**, *12*, 11734–11753.
48. Mulligan, G. The 6LoWPAN architecture. In Proceedings of the 4th workshop on Embedded Networked Sensors, Cork, Ireland, 25–26 June 2007; pp. 78–82.
49. Fielding, R.T. REST: Architectural Styles and the Design of Network-Based Software Architectures. Ph.D. Thesis, University of California, Oakland, CA, USA, 2000.
50. Shelby, Z. Constrained RESTful Environments (CoRE) Link Format. Available online: https://tools.ietf.org/html/rfc6690.html (accessed on 26 November 2017).
51. Locke, D. *MQTT V3. 1 Protocol Specification*; Technical Report; International Business Machines Corporation (IBM) and Eurotech: Udine, Italy, 2010; p. 42.
52. Hunkeler, U.; Truong, H.L.; Stanford-Clark, A. MQTT-S—A publish/subscribe protocol for Wireless Sensor Networks. In Proceedings of the 3rd International conference on Communication Systems Software and Middleware and Workshops, COMSWARE 2008, Bangalore, India, 6–10 January 2008; pp. 791–798.
53. Shelby, Z.; Hartke, K.; Bormann, C. The Constrained Application Protocol (CoAP). Available online: https://tools.ietf.org/html/rfc7252 (accessed on 27 November 2017).
54. Saint-Andre, P. Extensible Messaging and Presence Protocol (XMPP): Core. Available online: https://tools.ietf.org/html/rfc6120.html (accessed on 27 November 2017).
55. Waher, P. XEP-0323: Internet of Things—Sensor Data. Available online: https://xmpp.org/extensions/xep-0323.pdf (accessed on November 2017).
56. Dawson-Haggerty, S.; Jiang, X.; Tolle, G.; Ortiz, J.; Culler, D. sMAP: A simple measurement and actuation profile for physical information. In Proceedings of the 8th ACM Conference on Embedded Networked Sensor Systems, Zürich, Switzerland, 3–5 November 2010; pp. 197–210.
57. Bray, T.; Paoli, J.; Sperberg-McQueen, C.M.; Maler, E.; Yergeau, F. Extensible markup language (XML). *World Wide Web J.* **1997**, *2*, 27–66.
58. Bray, T. The JavaScript Object Notation (JSON) Fata Interchange Format. Available online: https://buildbot.tools.ietf.org/html/rfc8259 (accessed on 27 November 2017).
59. Schneider, J.; Kamiya, T.; Peintner, D.; Kyusakov, R. Efficient XML Interchange (EXI) Format 1.0. W3C Proposed Recommendation. Available online: https://www.w3.org/TR/exi/ (accessed on 27 November 2017).
60. Furuhashi, S. MessagePack: It's Like JSON. But Fast and Small. Available online: http://msgpack.org (accessed on 26 November 2014).
61. Botts, M.; Robin, A.; Greenwood, J.; Wesloh, D. *OGC SensorML: Model and XML Encoding Standard*; OGC Geospatial Consortium: Wayland, MA, USA, 2014.
62. Compton, M.; Neuhaus, H.; Taylor, K.; Parashar, A. Semantic Sensor Network Ontology. In *Web Semantics: Science, Services and Agents on the World Wide Web*; CSIRO: Victoria, Australia, 2012; pp. 25–32.
63. Kajimoto, K.; Matsukura, R.; Hund, J.; Kovatsch, M.; Nimura, K. *Web of Things (WoT) Architecture (Unofficial Draft)*; WoT W3C Interest Group: Cambridge, MA, USA, 2016.
64. Yick, J.; Mukherjee, B.; Ghosal, D. Wireless sensor network survey. *Comput. Netw.* **2008**, *52*, 2292–2330.
65. Akyildiz, I.F.; Su, W.; Sankarasubramaniam, Y.; Cayirci, E. Wireless sensor networks: A survey. *Comput. Netw.* **2002**, *38*, 393–422.
66. Munir, S.A.; Ren, B.; Jiao, W.; Wang, B.; Xie, D.; Ma, J. Mobile wireless sensor network: Architecture and enabling technologies for ubiquitous computing. In Proceedings of the 21st International Conference on Advanced Information Networking and Applications Workshops, AINAW'07, Niagara Falls, ON, Canada, 21–23 May 2007; Volume 2, pp. 113–120.
67. Tilak, S.; Abu-Ghazaleh, N.B.; Heinzelman, W. A taxonomy of wireless micro-sensor network models. *ACM SIGMOBILE Mob. Comput. Commun. Rev.* **2002**, *6*, 28–36.
68. Alvear, O.; Zamora, W.; Calafate, C.; Cano, J.C.; Manzoni, P. An Architecture Offering Mobile Pollution Sensing with High Spatial Resolution. *J. Sens.* **2016**, *2016*, 1458147.

69. Tsujita, W.; Yoshino, A.; Ishida, H.; Moriizumi, T. Gas sensor network for air-pollution monitoring. *Sens. Actuators B Chem.* **2005**, *110*, 304–311.

70. Kamionka, M.; Breuil, P.; Pijolat, C. Calibration of a multivariate gas sensing device for atmospheric pollution measurement. *Sens. Actuators B Chem.* **2006**, *118*, 323–327.

71. Alvear, Ó.; Calafate, C.T.; Cano, J.C.; Manzoni, P. Calibrating low-end sensors for ozone monitoring. In *International Internet of Things Summit*; Springer: Cham, Switzerland, 2015; pp. 251–256.

72. Wong, D.W.; Yuan, L.; Perlin, S.A. Comparison of spatial interpolation methods for the estimation of air quality data. *J. Expo. Sci. Environ. Epidemiol.* **2004**, *14*, 404–415.

73. Lane, N.D.; Miluzzo, E.; Lu, H.; Peebles, D.; Choudhury, T.; Campbell, A.T. A survey of mobile phone sensing. *IEEE Commun. Mag.* **2010**, *48*, 140–150.

74. Seeedstudio. Available online: http://www.seeedstudio.com/ (accessed on 26 November 2017).

75. Adafruit. Available online: https://www.adafruit.com/ (accessed on 26 November 2017).

76. TSN Systems . Available online: http://www.tst-sistemas.es/en/ (accessed on 26 November 2017).

77. Pycom. Available online: https://pycom.io/ (accessed on 26 November 2017).

78. Ben Fry, C.R. Processing. Available online: https://www.processing.org/ (accessed on 26 November 2017).

79. Micropython. Available online: https://micropython.org/ (accessed on 26 November 2017).

80. Liu, X.; Cheng, S.; Liu, H.; Hu, S.; Zhang, D.; Ning, H. A survey on gas sensing technology. *Sensors* **2012**, *12*, 9635–9665.

81. Landt, J. The history of RFID. *IEEE Potentials* **2005**, *24*, 8–11.

82. Seeed-Studio. Grovepi Extension Board. Available online: http://www.seeedstudio.com/depot/GrovePi-Starter-Kit-for-Raspberry-Pi-p-2240.html (accessed on 26 November 2017).

83. Intel-Corp. Intel Edison Compute Module. Available online: http://www.intel.eu/content/www/eu/en/do-it-yourself/edison.html (accessed on 26 November 2017).

*sensors*

MDPI

*Article*

# A Strain-Based Method to Detect Tires' Loss of Grip and Estimate Lateral Friction Coefficient from Experimental Data by Fuzzy Logic for Intelligent Tire Development

Jorge Yunta [1],*, Daniel Garcia-Pozuelo [1], Vicente Diaz [1] and Oluremi Olatunbosun [2]

[1] Department of Mechanical Engineering, Research Institute of Vehicle Safety (ISVA), Universidad Carlos III de Madrid, Avd. De la Universidad, 28911 Leganés (Madrid), Spain; dgramos@ing.uc3m.es (D.G.-P.); vdiaz@ing.uc3m.es (V.D.)

[2] School of Mechanical Engineering, University of Birmingham, B15 2TT Edgbaston, UK; o.a.olatunbosun@bham.ac.uk

* Correspondence: jyunta@ing.uc3m.es; Tel.: +34-91-624-8840

Received: 20 December 2017; Accepted: 24 January 2018; Published: 6 February 2018

**Abstract:** Tires are a key sub-system of vehicles that have a big responsibility for comfort, fuel consumption and traffic safety. However, current tires are just passive rubber elements which do not contribute actively to improve the driving experience or vehicle safety. The lack of information from the tire during driving gives cause for developing an intelligent tire. Therefore, the aim of the intelligent tire is to monitor tire working conditions in real-time, providing useful information to other systems and becoming an active system. In this paper, tire tread deformation is measured to provide a strong experimental base with different experiments and test results by means of a tire fitted with sensors. Tests under different working conditions such as vertical load or slip angle have been carried out with an indoor tire test rig. The experimental data analysis shows the strong relation that exists between lateral force and the maximum tensile and compressive strain peaks when the tire is not working at the limit of grip. In the last section, an estimation system from experimental data has been developed and implemented in Simulink to show the potential of strain sensors for developing intelligent tire systems, obtaining as major results a signal to detect tire's loss of grip and estimations of the lateral friction coefficient.

**Keywords:** intelligent tires; strain gauges; fuzzy logic; tire slip; lateral friction coefficient

---

## 1. Introduction

As the only part that keeps the contact between the vehicle chassis and the road, tires are a key factor for electronic control systems as well as for comfort and fuel consumption, among others. Despite the fact that current tires perform well in a huge variety of situations, they are just passive elements that do not contribute actively to driver or vehicle control systems to improve the driving experience and traffic safety. All these reasons encourage researches and companies to invest time and efforts in the development of an intelligent tire that works as a source of information for drivers and active control systems such as the Traction Control System (TCS) or the Electronic Stability Program (ESP) to intervene before a crash occurs.

Up to now, the Tire Pressure Monitoring System (TPMS), which was introduced in the market in 2002 [1] and provides information about tire inflation pressure, is the only system installed in tires that gives the drivers some kind of information about tire working conditions while driving. Nevertheless, real-time monitoring of other parameters like forces at the tire-road interface, wear and tire-road friction, which also depend on uncontrollable environmental factors, is the main obstacle as well as the ultimate goal of the intelligent tire development.

During the past 20 years, many researches have been made on the field of the intelligent tire [2], concluding that, although the TPMS was a good advance, the scope of the intelligent tire is much more ambitious than TPMS, as shown Figure 1.

**Figure 1.** Intelligent tire's potential benefits [2,3].

As is well known, driver actions in emergency situations can subject tires to forces greater than they can handle. As a consequence, the tire saturates and is unable to transfer enough lateral or longitudinal tractive effort, so the vehicle becomes unstable and probably uncontrollable. Due to nonlinear tire-road contact features, this situation may result in a traffic accident, more so if unsuitable road conditions are encountered like wet or muddy surfaces where friction is changing [4]. Therefore, the possibility of tires playing an active role in in mitigating this kind of situation is subject to the potential ability of intelligent tires to monitor some essential parameters about their working conditions.

Regarding parameters that are restricted by the driver's action, there are two important angles to take especially into account: the wheel steer angle ($\gamma$), which is usually explained as the angle between the tire moving direction and the longitudinal center line of the vehicle, and the slip angle ($\alpha$), which is the angle between a rolling wheel's actual direction of travel and the direction towards which it is pointing [5]. These angles and the different forces generated in the contact patch area are represented in Figure 2. In this paper, the tire slip angle will be used in the last section as a function of the wheel steer angle as a tool to detect that the tire is about to lose adhesion.

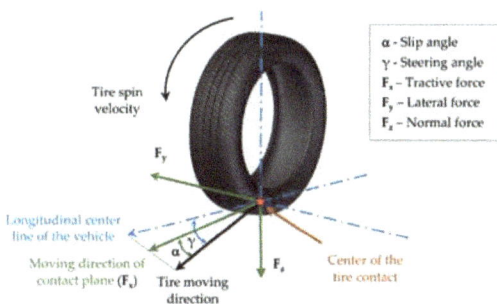

**Figure 2.** Tire slip angle and steering angle.

Several researches have concluded that the estimation of the slip angle is a vital point to improve the behavior of the tire and the vehicle [6,7]. It was found that the slip angle is directly related to the lateral force, which is usually defined as the force that the tire transmits to the road in the lateral direction, generating the required grip [8,9]. For this reason, a large and growing body of literature has also investigated their estimation [10–13]. Overall, these researches indicate that lateral and longitudinal tire-road forces depend on different parameters regarding the working conditions or the tire properties. For this reason, the relation between them and the lateral force is quite complicated and nonlinear. In cornering conditions, a car's tires deform laterally and longitudinally in the contact patch area [14]. Thus, tires must generate lateral forces to drive the car along a certain path. If the tire is capable of transferring these forces to the road, the tire does not slide and, as a result, the car follows the desired path.

On the other hand, forces generated in the tire-road interface are part of the parameters whose values cannot be controlled solely by the driver's action, because they depend on other uncontrollable factors, for instance, temperature or road conditions. The relation between those forces, which are represented in Figure 2, and slip angle, determines the vehicle's dynamics. Regarding lateral dynamics, the typical tire lateral characteristic curve as a function of the slip angle is usually divided into three regions: linear/elastic, transitional, and frictional [14]. In the linear section, the tire tread is not sliding on the road at any point in its contact. However, at higher slip angle some parts of the contact patch area are sliding, generating lower lateral force. For better understanding of this work, it is important to clarify that the term "sliding" does not imply that the driver realizes that the tire is not working at the limit of adhesion or the vehicle is not following the desired path, since during the driving experience the influence of the loss of adhesion is not easily observable when it has just started. For this reason, to detect that the lateral force is decreasing could be a key information for the driver and the active safety control system before the situation becomes more dangerous.

In addition, lateral force depends mainly on vertical load as well as slip angle [14]. Although these curves are specific and different for each tire, it is generally observed that lateral force increases as vertical load increases, and therefore, the required adhesion. Besides, lateral force stabilizes when it reaches its maximum at a certain slip angle. In this work, some lateral force vs. slip angle curves have been experimentally obtained in order to get a mathematical expression that allows the lateral force to be calculated as a function of the vertical load and the slip angle.

In the same manner as for vertical load, lateral force and slip angle, the influence of other mechanical tire features such as inflation pressure or rolling speed on tire strain behavior, which can make possible the estimation of tire behavior and provide information for vehicle active control systems and drivers, is also analyzed in this paper. Most of these tire features have been analyzed in several works related to the intelligent tire. Overall, they analyze the tire's deformation by measuring sidewall deflection, tire tread deformation, acceleration, etc. by means of different sensors such as Surface Acoustic Wave (SAW) sensors [15–17], optical sensors [18], piezoelectric sensors [13,19], accelerometers [20,21] or capacitive sensors [22–25]. Although all of them demonstrated the capability to provide information about tire working conditions from tire behavior, they usually have some limitations such as energy consumption, robustness or dependency of the tire-rim relative position.

There is a large volume of published studies indicating that strain sensors meet the requirements to achieve an advanced intelligent tire system from strain measurement [26–28]. For instance, Morinaga et al. [9] installed strain sensors on the inner liner of the tire tread, obtaining significant, stable and precise information about tire forces generated in the tire-road contact. In addition, strain sensors are less expensive than other sensors to measure tire deformation, which is associated with most of the tire working conditions. The price of strain gauges as well as the robustness thereof have demonstrated that they are suitable for developing the intelligent tire.

In addition to the discussion about what sensor technology is more suitable for developing the intelligent tire, it should not be forgotten that its future viability is subject to the real-time application. To do this, many obstacles must be surmounted, such as the compatibility of the sensors with tire rubber characteristics (for example, stiffness issues), data transmission [29] or economic issues relating to the cost of the sensors, but the main obstacle has always been to meet the power requirements of all the electronic components. This aspect has been explored recently in some researches [30]. Yilmazoglu et al. [31] proposed an optimized material system that could reduce the power for sensors implementation. Moreover, Jousimaa et al. [32] recently used a piezoelectric energy harvesting system, storing energy to supply energy for low-power electronic devices.

Finally, in addition to the multi-disciplinary knowledge about sensor technology, data analysis, tire characteristics, etc. that is required for the intelligent tire development, the involvement of computational methods is absolutely necessary to take advantage of the sensor's measurements. Overall, most researches focus on estimating the tire-road friction coefficient, first, by characterizing the road surface from tire behavior or using tire or vehicle models, and then implementing them in estimation tools such as Kalman filters [33–37].

In this paper, a series of experiments by means of strain sensor equipment and an indoor tire test rig have been carried out. These devices were selected to measure strain dynamic behavior based on tire working conditions, which could help towards developing intelligent tire systems in the near future. Thus, this paper is devoted to provide a strong experimental base with test results under different tire working conditions. This work is a continuation of the experiments carried out to measure the tire's deformation at the inner surface of the tire tread and estimate tire working conditions by means of fuzzy logic [26,28,38]. It concentrates on the influence of pressure, rolling speed, vertical load and slip angle on maximum strain values and lateral force measurements to elucidate how maximum strain data and lateral force are interconnected. In the last section, an experimental model of the lateral force behavior and previous estimations of slip angle and vertical load [38] are used to detect the tire's loss of grip and calculate the lateral friction coefficient, showing the proof of concept of a proposed Slip Detection System (SDS) as well as the potential of strain sensors for the development of intelligent tires.

## 2. Materials and Methods

### 2.1. Strain-Based Experiment Setup

An indoor tire test rig which allows the speed, vertical load and slip angle to be varied has been used for the experiments. The test system setup (including the indoor tire test rig equipped with the tire prototype) is shown in Figure 3a. The drum's curved surface has an insignificant effect on the results because of its large diameter (2.44 m) which ensures that there is only slight difference between a flat road and the large drum for tire/road contact. The error in contact length is less than 0.1% [39]. The test rig allows the lateral force to be monitored when the slip angle is changed, as shown in Figure 3b.

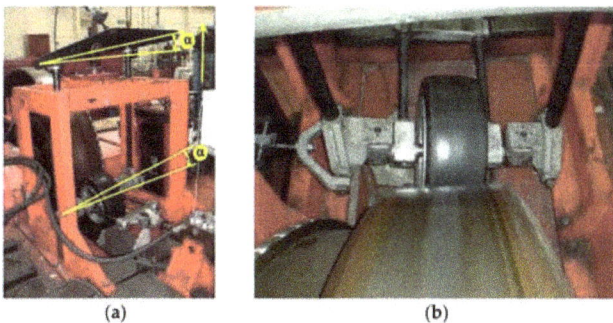

| (a) | (b) |

**Figure 3.** (**a**) Indoor tire test rig setup; (**b**) Tire working under cornering conditions.

Some strain-based intelligent tire systems that use strain sensors have been suggested in the past [9,26,28]. They were attached on the inner liner of the tire tread, demonstrating that a strain sensor installed near the contact patch provides more detailed and prompt information about local frictional condition between the tire and the road. In this study, some strain gauges are mounted on the inner liner surface of the tire tread in longitudinal and lateral directions, as shown in Figure 4a.

A data acquisition system with high resolution is especially necessary because of the possibility that the tread thickness attenuates the strain measurements. Figure 4b shows the installation of the rectangular rosette strain sensors. The manufacturer recommended adhesive for rubber materials was used, therefore, local stiffening effect caused by the strain gage can be neglected. The strain gage's length is 2 mm with gage resistance 120 $\Omega$. The resolution of the strain measurement is 0.001 $\mu\varepsilon$.

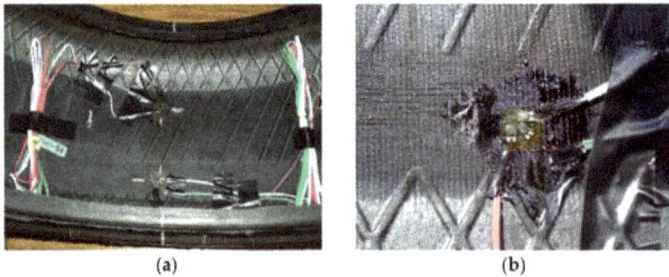

(a)  (b)

**Figure 4.** (a) Strain gauges' setup; (b) Multiaxial gage example.

These strain gauges, which are widely used for large strain measurements, were attached at different points of the inner liner of the tire tread in order to measure lateral ($\varepsilon_{y2}$—Channel 1 and $\varepsilon_{y1}$—Channel 3) and longitudinal ($\varepsilon_x$—Channel 2) strain, as shown in Figure 5. Regarding the position of the strain gauges, note that there are three multiaxial strain gauges located symmetrically to tire central line, two of them in the same cross section and the third one separated by 123.75° of angular rotation. The distances "d" and "l" are about 0.040 m and 0.515 m, respectively.

In order to pack and route the lead wires from the strain sensors to the outside of the rim, three additional valves were installed on the rim, as shown in Figure 6a. The lead wires were routed through the valves and all the holes and joints were sealed with adhesive to avoid any possible air leaks, leaving the system for 72 h to check the seal reliability. These valves allow the connection between strain sensors and data acquisition system. Although several channels are illustrated in Figure 5, not all of them were analysed in detail due to two principal reasons: some wires were damaged when passing them through the valves and others showed high similarity in the measurements and do not seem to provide interesting additional information.

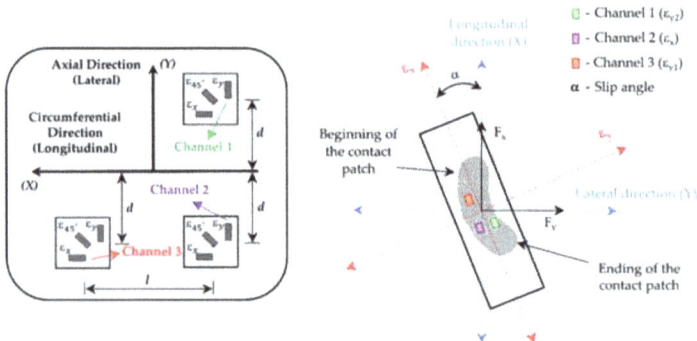

**Figure 5.** Strain gauges' location scheme.

The SoMat® 2000 Field Computer (Somat Corporation, Urbana, IL, USA), which is suitable for portable data collection, was used as a data acquisition system. The hardware of the system consists of a Processor module, which has the microprocessor data acquisition system, and a Power/Communication module, equipped with batteries, as shown in Figure 6b.

**(a)**      **(b)**

**Figure 6.** (a) Installed valves on the rim; (b) Installed SoMat® 2000 in the wheel.

SoMat® 2000 has one Wheatstone strain bridge for each strain sensor, which can be used in a quarter, half or full bridge configuration. The device was driven by TestPoint®software (Capital Equipment Corporation, Norton, MA, USA) for Windows (WinTCS). When a test starts, the data acquisition module (i.e., the strain gauges module) is connected to the strain sensors. Secondly, the user can download the test specification, designed using WinTCS, from a computer to the SoMat® 2000 to initialize and run it. By sampling the electrical analogue signals, the data acquisition system can store and manipulate digital data as bytes of binary digits. Although in this work the sampling frequency was set to be 1000 Hz, the SoMat® 2000 Field Computer can provide a sampling frequency from 0.0005 to 5000 Hz, which ensures that the test strain data resolution is adequate to monitor enough strain points per tire revolution.

However, it should not be forgotten that the data acquisition can be influenced by some factors which are usually divided into two groups: properties of the data acquisition system and the tire properties (such as tire radius or speed). Regarding properties of the data acquisition system, a large number of data channels would increase the amount of data acquired, however, more memory would be needed and the available test time would decrease. The maximum speed used in this work was 50 km/h, which is suitable for the sampling frequency used. The working range of the SoMat® 2000 strain gage module used covered from $-5000$ to $5000$ µε.

The test carried out consisted of studying the tread's dynamic behavior by changing test conditions in order to determine the influence of pressure, vertical load, speed and slip angle on tire strain. To do this, measurements from strain sensors were downloaded to a computer after the tests and analyzed. A DUNLOP SP SPORT 175/505 R13 (tubeless) slick radial tire was used in the tests. This type of tire, which usually works under low vertical loads and inflation pressures, is used for the Formula SAE vehicles.

*2.2. Test Conditions*

The operational range of parameters used for the cornering conditions were:

- Tire inflation pressure: 0.8–1.4 bar;
- Tire preload: 250–1000 N;
- Tire speed: 10–50 km/h;
- Tire slip angle: 0–14°.

It should be underlined that trials between 0° and 10° have been performed every 2 degrees, but it was decided not to include data at 12° as the potentiometer, which supplies the system, did not provide a stable signal for that slip angle. Instead, it was decided to conduct trials at 13° as exception.

Finally, the effect of temperature or inflation pressure changes in the tire between non-rotating and rotating phases are not considered, being the recommendations of the strain gauges' manufacturer 23 °C and 50% relative humidity.

## 3. Results

### 3.1. Experimental Data Analysis

The influence of lateral force and the tire working conditions considered during the experiments on strain data is studied in this work by the detailed analysis of experimental data.

First of all, it should be mentioned that although some of those working conditions, mainly tire parameters, could be measured using several sensors, the ultimate goal of the proposed tire prototype is to provide a platform for reliable and accurate measurement of more tire operating parameters/variables using the minimum number of sensors possible.

The variation in the obtained strain data may also be useful information to identify the variation of tire working conditions. Since the tire's deformation occurs due to the requirements of the working conditions, the variation and the characteristics of the obtained strain data related to the operating conditions play a significant role in the estimation of the tire's dynamic behavior, tire tread wear and other characteristics [40].

Before starting the experimental data analysis, representative data curves were calculated considering the average strain for each point of the tire perimeter by averaging the separate cycles, which were obtained under the same driving conditions. Then, lateral force, slip angle and tire surface deformation can be contrasted.

Strain curves have some characteristics that must be mentioned. As indicated above, data from three different channels were obtained, one in the longitudinal direction ($\varepsilon_x$) and two in the lateral direction ($\varepsilon_{y1}$ and $\varepsilon_{y2}$), as shown in Figure 5. Note that the channels 2 and 3 are located on the outer part of the contact patch while the channel 1 is on the inner part of the contact patch (see Figure 5). Nevertheless, past researches [38] showed that the most significant information is measured by channels 2 and 3. Figure 7a,b show examples of strain curves measured by these channels for different slip angles.

It was also revealed in [28] that in cornering conditions the representative points are the maximum tensile points in channel 2, which correspond to the center of the contact patch, as well as the maximum tensile values at the beginning and ending of the contact patch in channel 3. In addition, Figure 7a shows that the offset value of the signal is different depending on the conditions, which is also interesting as a source of information when the sensor is far from the contact patch.

Finally, although it seems that the maximum compressive points (see Figure 7b) could also be useful to extract information about tire working conditions, previous studies reveal that these maximum points are not linear for all test conditions and fluctuations exist that hinder the extraction of useful information, thus, it is not considered to be as useful as the others are.

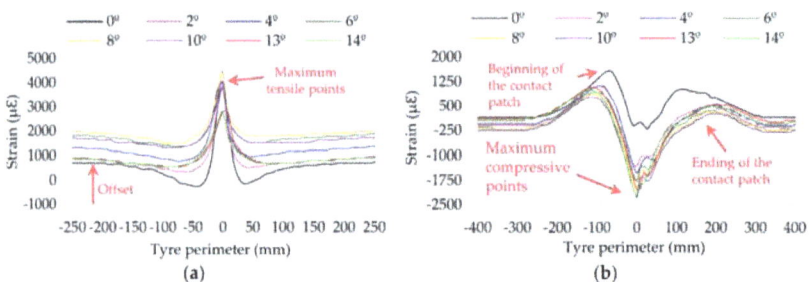

**Figure 7.** (a) Strain data curves in channel 2 ($\varepsilon_x$) at 1 bar, 750 N, 30 km/h; (b) Strain data curves in channel 3 ($\varepsilon_{y1}$) at 1 bar, 750 N, 30 km/h.

In the following sections, the influence of rolling speed, inflation pressure and vertical load related to slip angle and lateral force is shown, considering the maximum strain peaks and the offset values of the strain signal.

### 3.2. Influence of Speed

Figure 8 shows the variation of lateral force as slip angle increases at 10, 30 and 50 km/h. As expected, it can be clearly split into three parts. From 0° to 3°, which correspond to the elastic region (see Figure 8), the curves are practically overlapped. The second part covers from 3° to 9°. The maximum lateral force is reached at 8°, approximately.

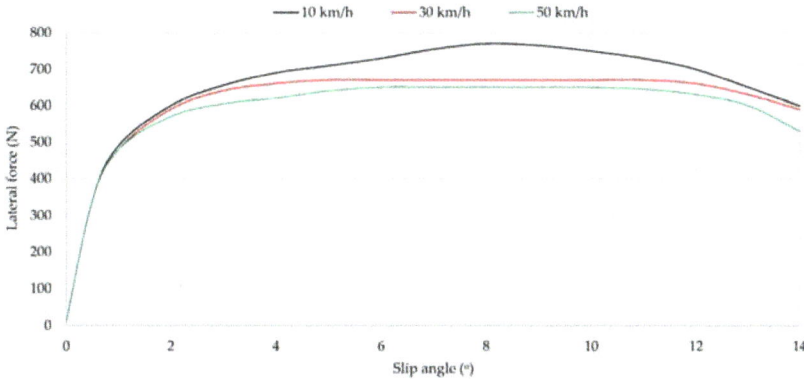

**Figure 8.** Influence of speed at 0.8 bar, 750 N. N: Newtons.

The three characteristic regions can be easily identified in Figure 9a,b, which represent the variation of maximum tensile strain data as slip angle increases. Strain data decrease drastically from 10° approximately for any speed. In the same way, data trend also changes in Figure 9b from 10° but in this case strain data increase. This fact reveals that the Poisson effect appears clearly in the tread tire, since when maximum strain values decrease in longitudinal direction from 10° (channel 2), they increase in lateral direction (channel 3). Note that this phenomenon occurs both at the beginning and ending of the contact patch.

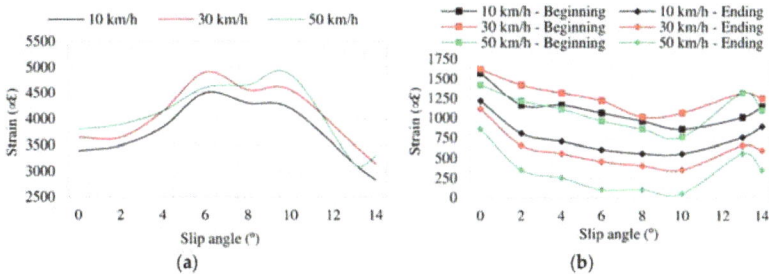

**Figure 9.** (**a**) Maximum strain peaks in channel 2 ($\varepsilon_x$) at 0.8 bar, 750 N; (**b**) Maximum strain peaks in channel 3 ($\varepsilon_{y1}$) at 0.8 bar, 750 N.

It is also interesting to underline from Figure 9a that between 6° and 10° maximum strain data values are reached. Since these points belong to the transitional range (i.e., the optimal working range), it can be assumed that when maximum strain values are higher, the maximum required grip is obtained. However, in contrast to Figure 9, the variation of maximum tensile strain is not monotonic, especially in the longitudinal direction where, for instance, at 10 and 30 km/h the maximum tensile peaks occur at 6° while at 50 km/h it is found at 10°. It may be due to the behaviour of the tire, a non-linear system, which is influenced by the rubber tribology behaviour, the study of which is not an objective of this paper. Moreover, there appears to be some relation between maximum strain values and maximum lateral force, since from 10° the lateral force reduces and maximum strain values change notably.

### 3.3. Influence of Inflation Pressure

Figure 10 shows the influence of the slip angle on the lateral force at 0.8, 1, 1.2 and 1.4 bar of inflation pressure. Note that, curves from 4° to 8° are practically overlapped, so inflation pressure does not have too much effect in this region. However, the influence of tire pressure could be important from 10°, when the tire is not working at the limit of adhesion.

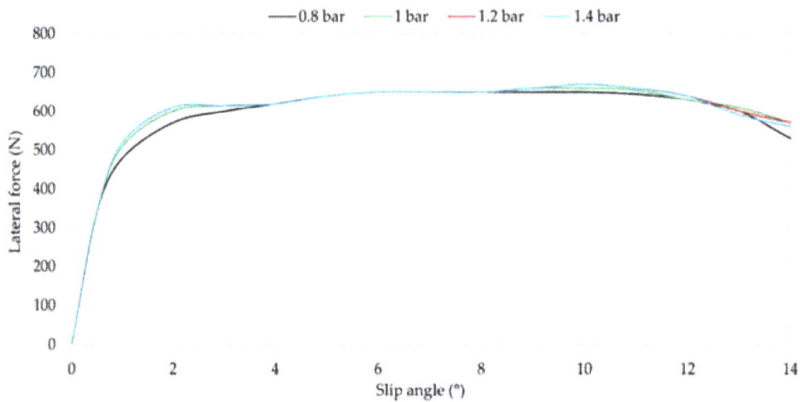

**Figure 10.** Influence of inflation pressure at 750 N, 50 km/h.

Maximum strain values are represented in Figure 11 for channels 2 and 3. Although there are fluctuations to 10°, from that point the data trend changes notably. These points correspond to the points at which the tire begins sliding.

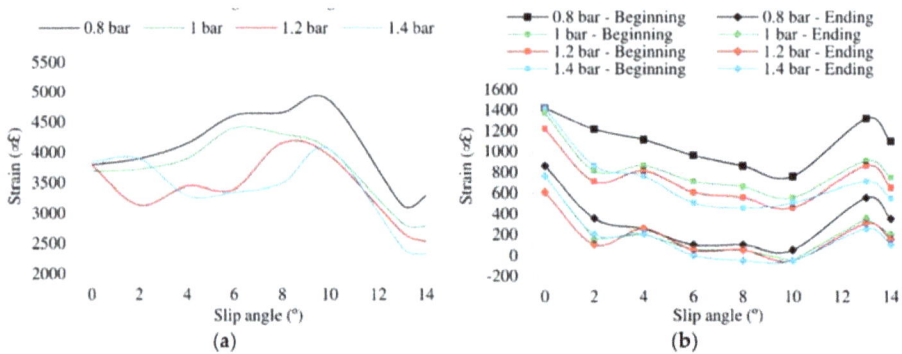

**Figure 11.** (a) Maximum strain peaks in channel 2 ($\varepsilon_x$) at 750 N, 50 km/h; (b) Maximum strain peaks in channel 3 ($\varepsilon_{y1}$) at 750 N, 50 km/h.

Figure 12 represents the offset variation for inflation pressure changes. In this case, there seems to be a relation between offset values in longitudinal direction (see Figure 12a) and offset values in lateral direction (see Figure 12b) and the maximum lateral force, that is reached around 9° in Figure 10, just before the tire starts sliding. In addition, the relationship between Figure 12a,b can be explained by the Poisson effect, especially when the tire is sliding.

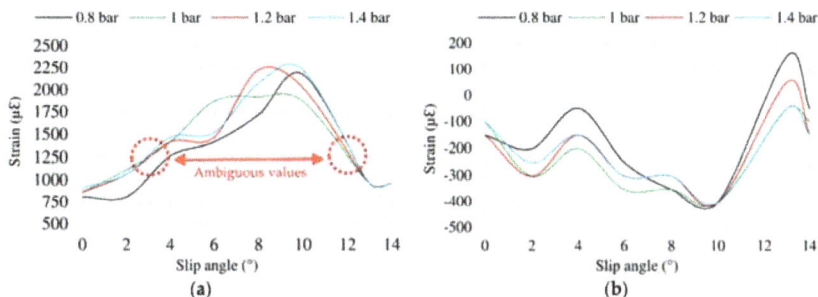

**Figure 12.** (a) Offset values in channel 2 ($\varepsilon_x$) at 750 N, 50 km/h; (b) Offset values in channel 3 ($\varepsilon_{y1}$) at 750 N, 50 km/h.

Finally, it is observed in Figure 12a that some strain data are the same for different slip angles, for instance, for 3° and 12° (i.e., ambiguous values). This fact will play a key role for the fuzzy logic implementation in the last section of this paper.

### 3.4. Influence of Vertical Load

Figure 13 illustrates the influence of the slip angle on lateral force for vertical loads of 750 and 1000 N. It should be mentioned that, although tests at 250, 500, 750 and 1000 N were carried out for obtaining results from strain sensors, the vertical load actuator that applied a specific value of vertical load was not capable of keeping a constant value when it was not very high, so lateral force values were not as accurate as in case of 750 and 1000 N. For this reason, lateral force results at 250 and 500 N have been neglected in this case.

Different from Figures 8 and 10, the curves are much more stable and are well separated in Figure 13. In addition, they reach the transitional region at 6°. Regarding the frictional region, when 1000 N is applied, the lateral force starts to decrease at 12°. However, when 750 N is applied, it starts to decrease at 11° approximately. The obtained results reveal that the vertical load directly affects the tire's slip.

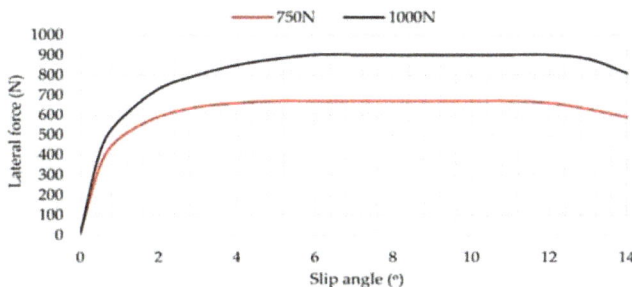

**Figure 13.** Lateral force vs. slip angle at 0.8 bar, 30 km/h.

Figure 14 shows that the maximum strain data changes notably from 10°. In the case of channel 2, the maximum strain values practically halve in relation to channel 3.

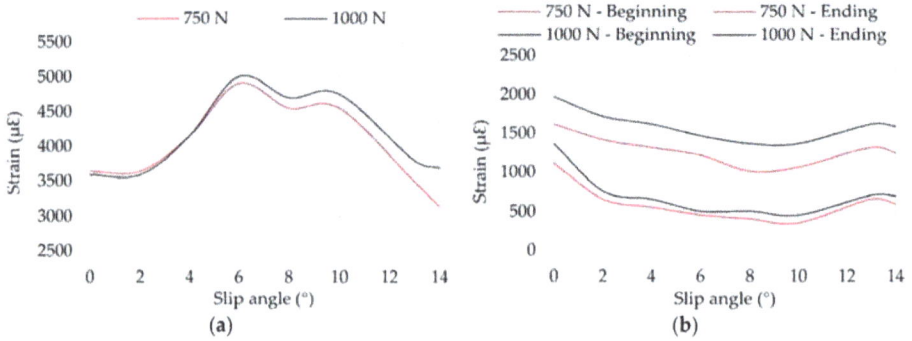

**Figure 14.** (**a**) Maximum strain peaks in channel 2 ($\varepsilon_x$) at 0.8 bar, 30 km/h; (**b**) Maximum strain peaks in channel 3 ($\varepsilon_{y1}$) at 0.8 bar, 30 km/h.

Figure 15 illustrates the variation of offset values related to vertical load changes. In case of Figure 15b, the trends of strain curves change drastically from 10° for 750 N and from 12° for 1000 N. However, in longitudinal direction the trend's change occurs from 10° in both cases.

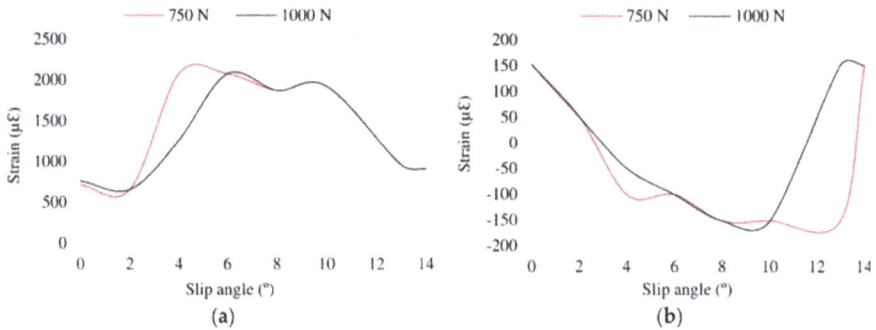

**Figure 15.** (**a**) Offset values in channel 2 ($\varepsilon_x$); (**b**) Offset values in channel 3 ($\varepsilon_{y1}$) for different vertical loads at 0.8 bar, 30 km/h.

## 4. Practical Approach to Estimate the Required Lateral Friction Coefficient.

In this section, a practical approach to show the potential of tire strain data to estimate the requested lateral friction coefficient (LFC) by the tire and detect the loss of grip is developed. Previous works have demonstrated that some key parameters of tire performance such as vertical load and slip angle can be accurately estimated by means of fuzzy logic.

In this work, these estimated parameters are used to develop an estimation system from some input parameters, demonstrating that the tire's loss of adhesion can be detected before reaching a critical situation for the driver and the required LFC by the tire can also be accurately estimated.

First of all, experimental $F_y$ data shown in Figure 13 is used to define an experimental model to calculate $F_y$ as a function of vertical load and slip angle by a single equation. Then, the fuzzy logic estimation system developed previously [38] and the experimental model are implemented in the environment MATLAB/Simulink® (The Math-Works, Natick, MA, USA,) providing a signal to inform the driver and active control systems of the tire's loss of grip. Finally, the requested LFC from fuzzy logic estimations and the estimated calculation of $F_y$ from the experimental model are used to calculate the required LFC from a theoretical equation.

### 4.1. Experimental Modelling of Lateral Force Behaviour

As is widely known, the lateral force depends mainly on vertical load and slip angle [14] , which have been accurately estimated from strain data in previous work [38], so the need now is to estimate the lateral force at the point where the tire loses grip. For this reason, the development of an experimental model to characterize and calculate the behavior of lateral force from vertical load and slip angle is especially interesting and it is essential to use the estimated values of those parameters in a real-time control system in the near future. In this section, the obtained $F_y$ curves (see Figure 13) for 750 N and 1000 N have been used to develop an experimental model of $F_y$ for the tire used in the tests. This equation, which defines the behavior of the lateral force through the linear and transitional regions (i.e., the range in which the tire is working at the limit of adhesion), is written as a function of $F_z$ and $\alpha$ as follows:

$$F_y = C_1 + (C_2 \cdot F_z) \cdot \left(1 - e^{(C_3 \cdot \alpha)}\right) \tag{1}$$

where $C_1 = 10$, $C_2 = 0.889$ and $C_3 = -1.06$.

This equation has been obtained by taking an exponential function as reference and then using regression techniques to adjust it to the $F_y$ experimental curves. This equation can be improved and adapted depending on the tested tire, but serves as a first approach to show the potential of the proposed methodology.

Looking at the $F_y$ curves shown in Figure 16, they have an exponential shape that makes it possible to obtain an experimental model that fits quite well with the experimental data without considering too many parameters (the combined coefficient of correlation of the experimental model is 0.98). However, the behavior of $F_y$ for commercial tires can show curves with a shape more complicated to model mathematically.

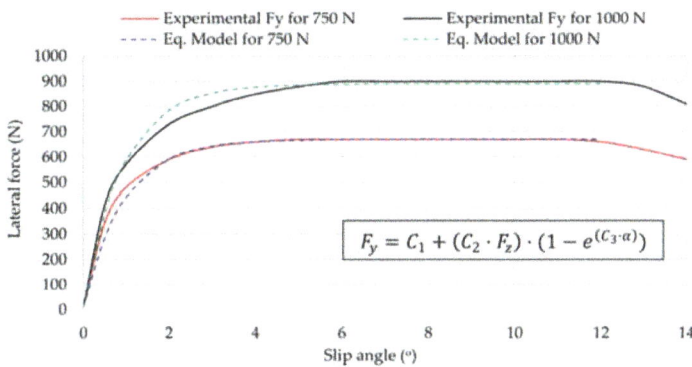

**Figure 16.** Experimental model adjusting to experimental $F_y$ data at 30 km/h, 0.8 bar.

Figure 16 also shows that the experimental equations do not fit with experimental data above 12°, where the tire starts sliding. Nevertheless, since the estimation system has been developed to detect the tire's loss of grip before using this equation to obtain an estimated calculation of the required LFC, when the system detects the tire is sliding, it does not use the equation and returns a value of 0 for the LFC. In this way, the lack of fitting between the experimental model and the experimental data above 12° is not crucial. It should be highlighted that the 0 value of the LFC does not mean that the tire does not have good friction conditions to keep the current working conditions, but the tire is not working at the maximum lateral force hence the loss of control might be near if the conditions become more demanding, so the SDS could work as a warning system for drivers and active safety control systems to be ready to take actions.

*4.2. Simulink Implementation of Slip Detection System and Lateral Friction Coefficient Estimation*

The fuzzy logic system developed is implemented in MATLAB/Simulink® to use the estimations of the vertical load and slip angle provided by the fuzzy logic system for detecting the tire's slip by the SDS and estimating the required LFC, as shown Figure 17.

First, the fuzzy logic blocks developed are used to estimate vertical load and slip angle from several inputs. As a reminder, since the inflation pressure is an information accurately provided today by the Tire Pressure Monitoring System (TPMS) and rolling speed is also a known information, they are used as inputs. In addition, the fuzzy logic system uses the maximum strain peaks as well as the offset values (see Figure 7) as the main inputs to estimate vertical load and slip angle.

Secondly, the wheel steer angle (gamma) is used as an input to calculate the tire steer angle, assuming that they could be monitored and they keep a relation of 1:15. Taking advantage of the fact that offset values for 13° and 14° give similar results as in case of 0°, 2° or 4° (see Figure 12a as an example), the fuzzy logic system estimates a value that does not correspond to the fixed wheel steer angle. In this way, as the estimations for the slip angle are quite accurate, it has been established that when the estimated slip angle differs by more than 10% of the tire steer angle, the SDS returns a value of 1 for the SDS output. As a first approach, it is assumed that the tire slip angle is directly related to the wheel steer angle, without considering the contribution of the other tires of the vehicle. This fact limits the application of the results taking into account the behavior of a whole vehicle, but this analysis serves as a proof of concept for the estimation of the vehicle slip angle from the corresponding equations of state. This simplification is not being considered in current developments. Finally, note that the relationship between lateral force and strain data is very advantageous, since strain data reduces drastically when a slight loss of grip occurs, therefore, the loss of grip can be detected before the situation becomes critical.

In the last stage, the estimated calculation of $F_y$ obtained by the experimental model defined in Equation (1) and the estimated value of $F_z$ obtained by fuzzy logic are used to calculate the required LFC by the following expression [41]:

$$\text{LFC} = \frac{F_y}{F_z} \tag{2}$$

Finally, it should not be forgotten that when the required LFC is calculated, the status of the tire's sliding is known. Thus, when the SDS detects that the tire is losing grip (signal = 1), the Equation (2) is not used, and the value of the required LFC is automatically 0.

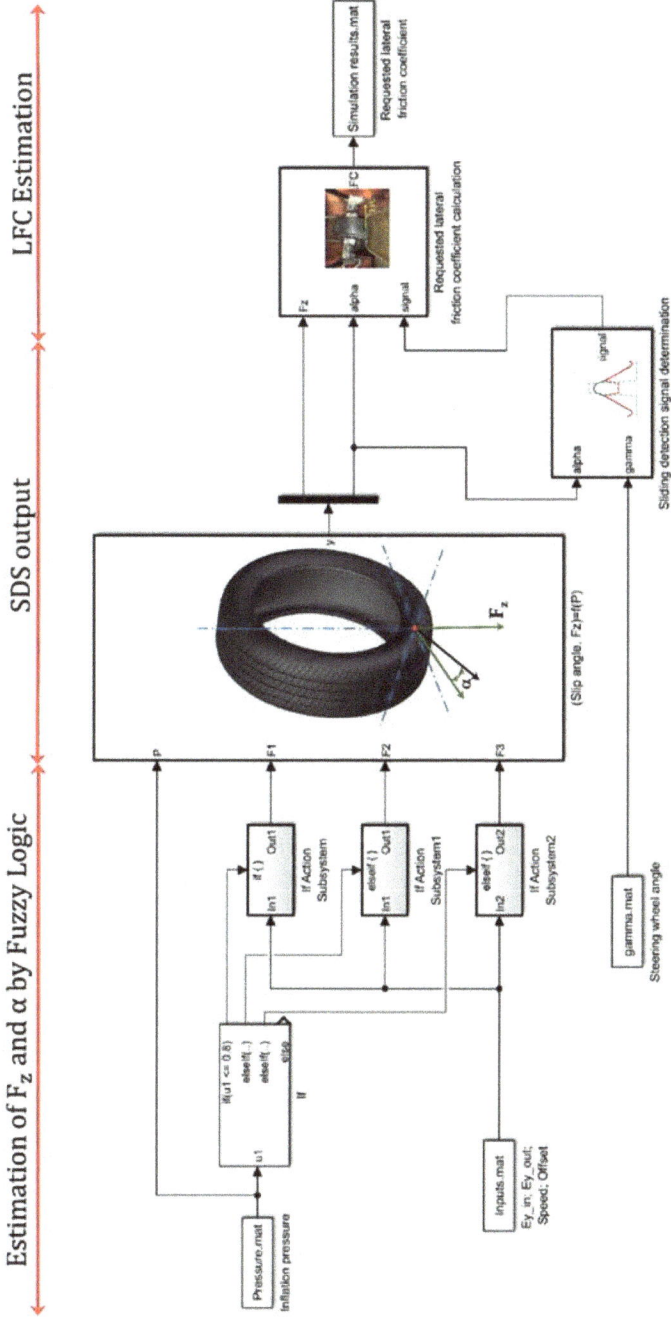

**Figure 17.** Slip Detection System implementation in MATLAB/Simulink® and required lateral friction coefficient calculation.

### 4.3. Simulation Results

In this section, the results obtained in each stage of the implemented system in MATLAB/Simulink® are illustrated. Figure 18 shows the slip angle estimated by fuzzy logic compared to the experimental slip angle, which is obtained using the relation that keeps with the wheel steer angle. As it is observed, when the tire is sliding, the estimated slip angle drops to 5° while the experimental slip angle has values of 13° and 14°. In these cases, the SDS output that indicates whether the tire is working or not is equal to 1.

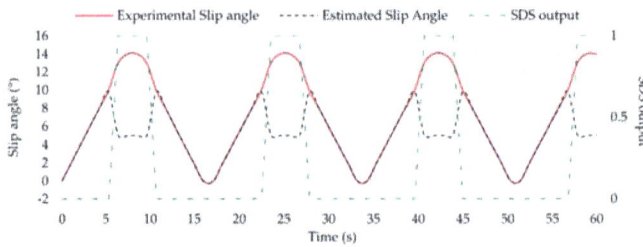

**Figure 18.** Tire slip detection from the slip angle estimation.

During situations under slip working conditions (i.e., SDS output equal to 1), $F_z$ estimations by fuzzy logic compared to experimental $F_z$ data are not as accurate as when the tire is not sliding, as shown in Figure 19.

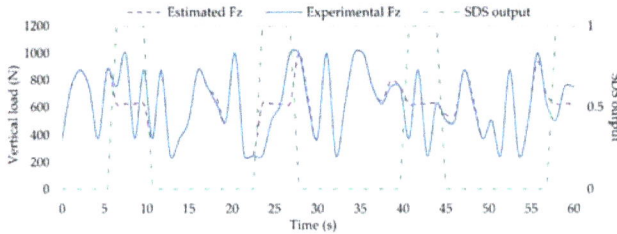

**Figure 19.** Simulation results of $F_z$ compared to experimental values.

In the same way, the estimated $F_y$ values using Equation (1) are worse in this situation, as shown in Figure 20. However, when the tire is working under optimum conditions (i.e., SDS output equal to 0), estimated $F_y$ and $F_z$ values are quite accurate compared to experimental data. Overall, the differences of estimated $F_z$ and $F_y$ compared with experimental data under optimum conditions are 1.64% and 3.11%, respectively.

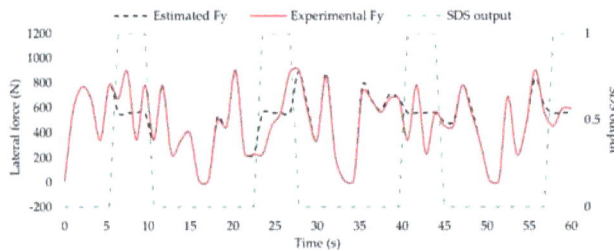

**Figure 20.** Simulation results of $F_y$ compared to experimental values.

Results obtained for the required LFC using the Equation (2) are shown in Figure 21. As they have been calculated from estimated $F_y$ and $F_z$ values, it is not considered as a direct calculation. However, results show that experimental LFC (obtained from experimental $F_z$ and $F_y$ data) and estimated LFC practically overlapped when the tire is working under optimum conditions (i.e., SDS output equal to 1), the percentage difference being around 1.5%. Finally, as mentioned before, the system implemented in Simulink returns a value of 0 for the required LFC when it detects the tire is losing grip, so the percentage differences between experimental and estimated values increase notably.

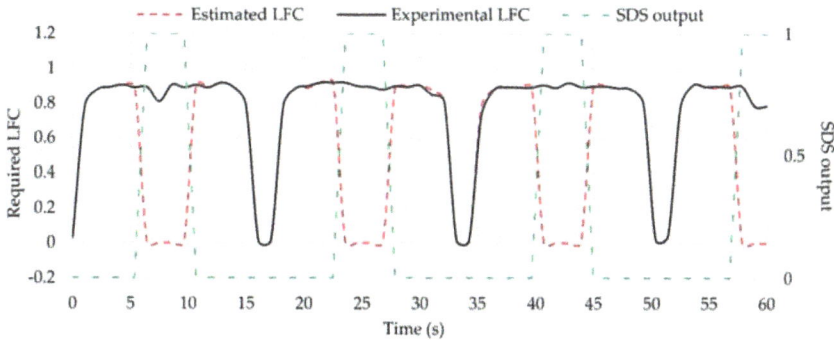

**Figure 21.** Estimated LFC compared to theoretical values.

Finally, although it might be thought that the good slip angle estimations obtained (see Figure 18) when the tire is working within the limit of adhesion are not very useful, the sideslip angle of a vehicle, which is the ratio of forward and lateral velocities in the form of an angle, depends to a large extent on the tires' slip angle [42]. The sideslip angle is essential for active safety control systems and its estimation from the tires' slip angle is a work that needs to be further developed.

## 5. Discussion

In this work, tests in cornering rolling conditions at steady state (without evaluating demanding situations under acceleration or braking processes) have been carried out. The main equipment used was: data acquisition system, indoor test rig and a tire with attached strain sensors. The monitored information has been related not only with tire parameters such as vertical load, speed, pressure and slip angle, but also with the generated lateral force.

The obtained results reveal that tire working conditions affect, notably, the tire's deformation. The variations of the considered test parameters imply not only changes in longitudinal and lateral maximum strain values, but also in the lateral force when slip angle increases. In addition, the tire's loss of grip is shown in maximum tensile strain values, to such an extent that when the tire begins sliding, maximum strain value decreases or increases radically depending on the measuring direction.

The intelligent tire used in this study has demonstrated that strain sensors are suitable for measuring tire dynamic behavior, particularly maximum tensile and compressive strain limits. Furthermore, the accuracy as well as the robustness of the strain gauges are demonstrated. Based on the results, strain gauges have demonstrated that they can measure with high accuracy the tire tread behavior under dynamic conditions.

The size of the strain gauges and the manufacturing processes of commercial tires gives a certain possibility to integrate the sensors into the tire layers, improving the robustness of the system and introducing the suggested strain-based methodology into mass production.

The possibility of detecting that the tire is working at the limit of adhesion before an accident occurs has been demonstrated by means of the implementation of the fuzzy logic systems in MATLAB/Simulink®. In addition, the experimental model to estimate the lateral friction coefficient

*Sensors* **2018**, *18*, 490

has shown a good reliability compared to the theoretical calculation when the tire is working at the limit of grip. However, regarding the computational time, the developed estimation method must be optimized in order to meet the requirements of the real-time data acquisition. In addition, one of the limitations of the proposed model is that it would only work properly at the first stage of the understeer, since the relationship between the wheel steer angle and the tire slip angle remains constant only during this stage. Therefore, the analysis of the variation between these angles from certain values is part of the future works, since this information would provide more accurate estimations when the tire is working at the limit of adhesion. Moreover, in the same manner that other researches based on tire test rig did, it has been assumed that the wheel slip angle and the tire slip angle (that appears in the tire-road contact) are equivalent and have the same behavior. However, in order to develop a reliable system using the proposed methodology, it will be necessary to assess if it is possible to consider this assumption as correct, especially when tests under real conditions are carried out.

On the other hand, it needs to be emphasized that the model defined in Equation (1) is only valid for the steel drum. Different models should be applied for different road/tire combinations, which means that this work serves just as a proof of concept, but a reliable way must be found to predict the LFC in real time (for instance, by means of road image recognition).

Future work will be focused on extending the range of some test conditions and testing other kind of tires, especially commercial tires, whose behavior is very different in comparison with slick tires. Secondly, although the proposed model serves as a proof of concept, tests considering different road conditions (i.e., friction coefficients) and considering other parameters (for instance, the tire and road temperatures) are necessary to corroborate some of the conclusions of this work and get a complete set of data that could allow us to estimate directly the lateral force instead of calculating it from the estimations of vertical load and slip angle. In addition, strain sensors attached to the sidewall could give us some useful information about tire working conditions. Finally, test under combined longitudinal and lateral slip conditions are necessary to decide if the proposed algorithm is suitable or other estimation methods such as Kalman filtering are more adequate before carrying out tests under real conditions.

**Acknowledgments:** We acknowledge the University of Birmingham for the facilities and the Universidad Carlos III de Madrid for the financial support that covers the costs to publish in open access.

**Author Contributions:** Daniel Garcia-Pozuelo defined the research topic and performed the experiments; Oluremi Olatunbosun provided direction for experimental methods; Jorge Yunta, Daniel Garcia-Pozuelo, Vicente Diaz and Oluremi Olatunbosun analysed the data and Jorge Yunta developed the MATLAB/Simulink® implementation. All authors have contributed to the production of the paper and have approved the manuscript.

**Conflicts of Interest:** The authors declare no conflict of interest.

## References

1. 106th Congress of USA, USA. Transportation Recall Enhancement, Accountability and Documentation (TREAD) Act, 2000. Available online: https://www.congress.gov/106/plaws/publ414/PLAW-106publ414.pdf (accessed on 25 January 2018).

2. Apollo Project Report. Final Report, Including Technical Implementation Plan (Annex), Deliverable 22/23, Finland. http://virtual.vtt.fi/virtual/proj3/apollo/deliverables/apollo_final%20report.pdf (accessed on 26 January 2018).

3. Modernracer, Volkswagen Golf GTI Cutway. 2005. Available online: http://www.modernracer.com (accessed on 25 January 2018).

4. Austin, L.; Morrey, D. Recent advances in antilock braking systems and traction control systems. *Proc. Inst. Mech. Eng. Part D* **2000**, *214*, 625–638. [CrossRef]

5. Pacejka, H.B. Basic tyre modelling considerations. In *Tire and Vehicle Dynamics, Butterworth-Heinemann*, 2nd ed.; Butterworth-Heinemann: Oxford, UK, 2005; pp. 1–60, ISBN 978-0-0805-4333-8.

6.  Phanomchoeng, G.; Rajamani, R.; Piyabongkarn, D. Real-Time Automotive Slip Angle Estimation with Nonlinear Observer. In Proceedings of the 2011 American Control Conference, San Francisco, CA, USA, 29 June–1 July 2011; pp. 3942–3947.

7.  Li, B.; Du, H.; Li, W.; Zhang, Y. Side-slip angle estimation based lateral dynamics control for omni-directional vehicles with optimal steering angle and traction/brake torque distribution. *Mechatronics* **2015**, *30*, 348–362. [CrossRef]

8.  Hong, S.; Erdogan, G.; Hedrick, K.; Borrelli, F. Tyre–road friction coefficient estimation based on tyre sensors and lateral tyre deflection: Modelling, simulations and experiments. *Veh. Syst. Dyn.* **2013**, *51*, 627–647. [CrossRef]

9.  Morinaga, H.; Wakao, Y.; Hanatsuka, Y.; Kobayakawa, A. The Possibility of Intelligent Tire (Technology of Contact Area Information Sensing). In Proceedings of the FISITA 2006 World Automotive Congress, Yokohama, Japan, 22–27 October 2006.

10. Baffet, G.; Charara, A.; Lechner, D. Estimation of vehicle sideslip, tire force and wheel cornering stiffness. *Control Eng. Pract.* **2009**, *17*, 1255–1264. [CrossRef]

11. Kim, J. Identification of lateral tyre force dynamics using an extended Kalman filter from experimental road test data. *Control Eng. Pract.* **2009**, *17*, 357–367. [CrossRef]

12. Jayachandran, R.; Ashok, S.D.; Narayanan, S. Fuzzy-logic based modelling and simulation approach for the estimation of tire forces. *Procedia Eng.* **2013**, *64*, 1109–1118. [CrossRef]

13. Erdogan, G.; Alexander, L.; Rajamani, R. A Novel wireless piezoelectric tire sensor for the estimation of slip angle. *Meas. Sci. Technol.* **2010**, *21*, 015201. [CrossRef]

14. Milliken, W.F.; Milliken, D.L. *Race Car Vehicle Dynamics*; SAE International: Troy, MI, USA, 1994.

15. Cyllik, A.; Strothjohann, T.; Scholl, G. *The Intelligent Tire-Applications of the Tread Sensor*; VDI Berichte: Hannover, Germany, 2001.

16. Zhang, X.; Wang, F.; Wang, Z.; Li, W.; He, D. Intelligent Tires Based on Wireless Passive Surface Acoustic Wave Sensors. In Proceedings of the 7th International IEEE Conference on Intelligent Transportation Systems (IEEE Cat. No. 04TH8749), Washington, WA, USA, 3–6 October 2004; pp. 960–964.

17. Eom, J.; Lee, H.; Choi, B. A study on the tire deformation sensor for intelligent tires. *Int. J. Precis. Eng. Manuf.* **2014**, *15*, 155–160. [CrossRef]

18. Tuononen, A.J. Optical position detection to measure tyre carcass deflections. *Veh. Syst. Dyn.* **2008**, *46*, 471–481. [CrossRef]

19. Yi, J. A Piezo-sensor-Based "Smart Tire" system for mobile robots and vehicles. *IEEE/ASME Trans. Mechatron.* **2008**, *13*, 95–103. [CrossRef]

20. Niskanen, A.; Tuononen, J.A. Three three-Axis IEPE accelerometers on the inner liner of a tire for finding the tire-road friction potential indicators. *Sensors* **2015**, *15*, 19251–19263. [CrossRef] [PubMed]

21. Anghelache, G.; Negrus, E.M.; Sorohan, S. *Radial Vibrations of Tyre Tread-Band at Different Rolling Speeds*; Paper No. F2000G354; Seoul 2000 FISITA World Automotive Congress: Seoul, Korea, 2000; pp. 1–5.

22. Sergio, M.; Manaresi, N.; Tartagni, M.; Guerrieri, R.; Canegallo, R. On Road Tire Deformation Measurement System using a Capacitive-Resistive Sensor. In Proceedings of the IEEE Sensors 2003 (IEEE Cat. No. 03CH37498), Toronto, ON, Canada, 22–24 October 2003; Volume 2, pp. 1059–1063.

23. Matsuzaki, R.; Todoroki, A. Intelligent tires based on measurement of tire deformation. *J. Solid Mech. Mater. Eng.* **2008**, *2*, 269–280. [CrossRef]

24. Matsuzaki, R.; Todoroki, A. Wireless strain monitoring of tires using electrical capacitance changes with an oscillating circuit. *Sens. Actuators A Phys.* **2005**, *119*, 323–331. [CrossRef]

25. Cao, S.; Pyatt, S.; Anthony, J.C.; Kubba, I.A.; Kubba, E.A.; Olatunbosun, O. Flexible bond wire capacitive strain sensor for vehicle tyres. *Sensors* **2016**, *16*, 929. [CrossRef] [PubMed]

26. Yang, X.; Olatunbosun, O.; Garcia-Pozuelo Ramos, D.; Bolarinwa, E. Experimental investigation of tire dynamic strain characteristics for developing strain-based intelligent tire system. *SAE Int. J. Passeng. Cars Mech. Syst.* **2013**, *6*, 97–108. [CrossRef]

27. Yang, X.; Olatunbosun, O.; Garcia-Pozuelo, D.; Bolarinwa, E. FE-Based Tire Loading Estimation for Developing Strain-Based Intelligent Tire System; SAE International: Detroit, MI, USA, 2015.

28. Garcia-Pozuelo, D.; Olatunbosun, O.; Yunta, J.; Yang, X.; Diaz, V. A Novel strain-based method to estimate tire conditions using Fuzzy Logic for intelligent tires. *Sensors* **2017**, *17*, 350. [CrossRef] [PubMed]

29. Matsuzaki, R.; Todoroki, A. Wireless monitoring of automobile tires for intelligent tires. *Sensors* **2008**, *8*, 8123–8138. [CrossRef] [PubMed]

30. Singh, K.B.; Bedekar, V.; Taheri, S.; Priya, S. Piezoelectric vibration energy harvesting system with an adaptive frequency tuning mechanism for intelligent tires. *Mechatronics* **2012**, *22*, 970–988. [CrossRef]

31. Yilmazoglu, O.; Brandt, M.; Sigmund, J.; Genc, E.; Hartnagel, H.L. Integrated InAs/GaSb 3D magnetic field sensors for "the intelligent tire". *Sens. Actuators A Phys.* **2001**, *94*, 59–63. [CrossRef]

32. Jousimaa, O.J.; Xiong, Y.; Niskanen, A.J.; Tuononen, A.J. Energy Harvesting System for Intelligent Tyre Sensors. In Proceedings of the 2016 IEEE Intelligent Vehicles Symposium (IV), Gotenburg, Sweden, 19–22 June 2016; pp. 578–583.

33. Boada, M.J.L.; Boada, B.L.; Garcia-Pozuelo, D.; Diaz, V. Neural-empirical tyre model based on recursive lazy learning under combined longitudinal and lateral slip conditions. *Int. J. Automot. Technol.* **2011**, *12*, 821–829. [CrossRef]

34. Li, B.; Du, H.; Li, W. Comparative study of vehicle tyre–road friction coefficient estimation with a novel cost-effective method. *Veh. Syst. Dyn.* **2014**, *52*, 1066–1098. [CrossRef]

35. Wang, B.; Guan, H.; Lu, P.; Zhang, A. Road surface condition identification approach based on road characteristic value. *J. Terramech.* **2014**, *56*, 103–117. [CrossRef]

36. Liu, Y.; Li, T.; Yang, Y.; Ji, X.; Wu, J. Estimation of tire-road friction coefficient based on combined APF-IEKF and iteration algorithm. *Mech. Syst. Signal Process.* **2017**, *88*, 25–35. [CrossRef]

37. Chen, L.; Luo, Y.; Bian, M.; Qin, Z.; Luo, J.; Li, K. Estimation of tire-road friction coefficient based on frequency domain data fusion. *Mech. Syst. Signal Process.* **2017**, *85*, 177–192. [CrossRef]

38. Garcia-Pozuelo, D.; Yunta, J.; Olatunbosun, O.; Yang, X.; Diaz, V. A strain-based method to estimate slip angle and tire working conditions for intelligent tires using Fuzzy Logic. *Sensors* **2017**, *17*, 874. [CrossRef] [PubMed]

39. Wei, C.; Olatunbosun, O.A. Transient dynamic behavior of finite element tire traversing obstacles with different heights. *J. Terramech.* **2014**, *56*, 1–16. [CrossRef]

40. Anghelache, G.; Moisescu, R.; Sorohan, Ş.; Bureţea, D. Measuring system for investigation of tri-axial stress distribution across the tyre–road contact patch. *Measurement* **2011**, *44*, 559–568. [CrossRef]

41. Rajamani, R. Tire-Road Friction Measurement on Highway Vehicles. In *Vehicle Dynamics and Control*, 2nd ed.; Springer US: MA, USA, 2012; pp. 397–425. ISBN 978-1-4614-1433-9.

42. Li, J.; Zhang, J. Vehicle sideslip angle estimation based on hybrid Kalman filter. *Math. Probl. Eng.* **2016**, *2016*, 3269142. [CrossRef]

![sensors logo] *sensors*

MDPI

*Article*

# Customizable Optical Force Sensor for Fast Prototyping and Cost-Effective Applications

Jorge A. Díez [1,*], José M. Catalán [1], Andrea Blanco [1], José V. García-Perez [1], Francisco J. Badesa [2] and Nicolás Gacía-Aracil [1]

[1] Departamento de Ingeniería de Sistemas y Automática, Universidad Miguel Hernández de Elche, 03202 Elche, Spain; jcatalan@umh.es (J.M.C.); ablanco@umh.es (A.B.); j.garciap@umh.es (J.V.G.-P.); nicolas.garcia@umh.es (N.G.-A.)

[2] Departamento de Ingeniería en Automática Electrónica, Arquitectura y Redes de Computadores, Universidad de Cádiz, 11510 Puerto Real, Spain; javier.badesa@uca.es

[*] Correspondence: jdiez@umh.es; Tel.: +34-965222505

Received: 15 January 2018; Accepted: 3 February 2018; Published: 7 February 2018

**Abstract:** This paper presents the development of an optical force sensor architecture directed to prototyping and cost-effective applications, where the actual force requirements are still not well defined or the most suitable commercial technologies would highly increase the cost of the device. The working principle of this sensor consists of determining the displacement of a lens by measuring the distortion of a refracted light beam. This lens is attached to an elastic interface whose elastic constant is known, allowing the estimation of the force that disturbs the optical system. In order to satisfy the requirements of the design process in an inexpensive way, this sensor can be built by fast prototyping technologies and using non-optical grade elements. To deal with the imperfections of this kind of manufacturing procedures and materials, four fitting models are proposed to calibrate the implemented sensor. In order to validate the system, two different sensor implementations with measurement ranges of $\pm 45$ N and $\pm 10$ N are tested with the proposed models, comparing the resulting force estimation with respect to an industrial-grade load cell. Results show that all models can estimate the loads with an error of about 6% of the measurement range.

**Keywords:** optical force sensor; printable sensor; cost-effective sensor; hysteresis correction; Prandtl–Ishlinskii model; perceptron

## 1. Introduction

### 1.1. Background

In the framework of the AIDE H2020 European Project [1], a hand exoskeleton has been developed as a part of the multimodal system that will assist disabled people during the realization of a wide range of activities of daily living [2]. This device must be able to pick up various everyday objects such as glasses, adapted cutlery or handles, bottles, dishes, or a toothbrush. Hence, the grip must be safe in order to provide real independence to the user. Additionally, a rehabilitation version of that device is being developed in parallel [3]. For both purposes, it is interesting to have a measurement of the interaction force between user and hand exoskeleton. This information could be useful from different points of view:

- Feedback in an impedance control system,
- Comparison with a threshold to limit the interaction force between user and device for safety reasons,
- Detection of movement intention,
- Measurements to develop objective clinical assessment systems.

*1.2. State of the Art*

In the current literature, we can find a wide diversity of sensor technologies that have been implemented in wearable and exoskeleton robots for the same purpose, such as torque sensors [4,5], strain gauges [6–8], miniature load cells [9–11], force-sensitive resistors [12–14], and Hall effect force sensors [15,16]. However, these technologies present some problems for its use in exoskeletons and wearable devices:

- In case of torque sensors, they measure the load in the motor shaft, which in over-constrained mechanisms implies not measuring all the interaction forces. Additionally, they are limited to be used with rotary actuators.
- The main shortcoming of strain gauges is the difficulty in their placement and fixation. Strain gauges must be glued firmly to a surface that deforms in a concrete way, resulting in relatively complex geometries or miniaturization hardly achievable without specialized material assets.
- There is a wide variety of miniature load cells; however, most of them can only measure compression forces, while compression–extension sensors may result unaffordable or not sufficiently miniaturized.
- Force-sensing resistors depend heavily on the surface contact and are not sufficiently reliable to be considered for this application.
- Other technologies like Hall effect force sensors may result in high-consumption and bulky solutions that can be integrated only in the biggest wearable devices.

Optical force sensors may suppose a promising alternative to overcome these shortcomings that make the force measurement in wearable devices difficult. The wide variety of optical and opto-electronical components commercially available allow multiple designs and optical arrangements that may be easily adapted to the requirements of these kinds of devices.

With optical technology, high precision and resolution can be achieved by using optical fiber sensors, which rely on determining the distortion produced in the fiber due to an external load. This strain may be measured with multiple techniques, such as Fabry–Perot interferometers [17,18] or Bragg gratings [19,20]. These solutions have potential applications in high performance applications and are not affected by the electromagnetic fields produced by actuators or medical instrumentation; however, they might result in being too expensive for most wearable applications.

More affordable solutions can be implemented by using light emitting diodes (LED) and photo-receivers that measure the incident light. The light can be guided with optical fibers to conserve the robustness to strong electromagnetic fields [21,22]. Other implementations work by direct interaction between the emitter and receiver, without any intermediate optical element [23–25] resulting in miniaturized devices. The addition of intermediate optical components can increment the design flexibility with a sacrifice of miniaturization capability [26–28].

*1.3. Objectives*

The goal of the presented research consists in developing the architecture of a force sensor with compression–extension measurement capability, which might be built completely with fast-prototyping techniques and materials, so that it can be easily shaped in a custom form to be integrated into a design still in the prototype stage. Additionally, to make its use more flexible, the proposed sensor architecture is also designed to allow the adjustment of its performance once the sensor is built and placed by only replacing some simple elements.

Despite the design being intended to be as general-purpose as possible, one of the short-term goals of the research is to obtain a device that can be integrated into a small wearable device, concretely the hand-exoskeleton mentioned in Section 1.1.

## 2. Materials and Methods

### 2.1. Hardware Description

The proposed architecture consisted of optical sensors based on the use of photo-detectors. These kinds of sensors can measure micrometric displacements by sensing fluctuations in the incidence of a light beam over a photo-detector. These variations in the light flux may be induced by occlusion of the light inciding on a single photo-detector (Figure 1, Left) or by modifying the position of the light focus over a segmented detector (Figure 1, Right).

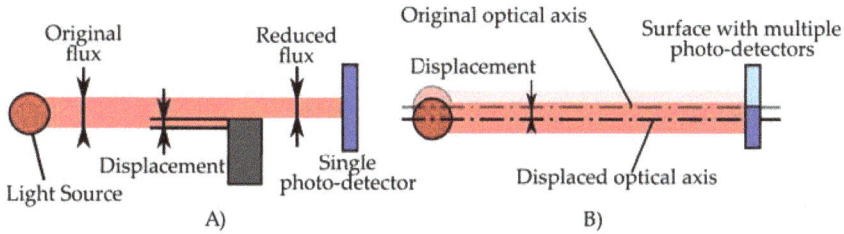

**Figure 1.** Optical micrometric sensors strategies. (**A**) light incidence is modified by occlusion of the photo-detector. (**B**) displacement is measured by modifying the position of the focus point in a matrix of photo-detectors.

This working principle allows the measurement of small displacements. Thus, if the measured displacement corresponds to a deformation of an elastic material, this geometrical distortion may be easily related to a stress and thus a force producing it.

Hence, a properly designed optical system in conjunction with a correctly sized deformable elastic structure may allow the measurement of forces with a concrete measurement range and sensitivity.

### 2.1.1. Optical Architecture

Starting from the two methods to measure light fluctuations exposed above, it can be concluded that, in order to measure small displacements, the occlusion method would require a photo-detector with a high sensitivity since variation of the light flux is directly proportional to the displacement. However, the strategy based on the modification of the focus point in the light sensitive surface allows the addition of optical elements that can refract the light beam, amplifying the induced deviation and resulting in an increase of the sensitivity of the system.

In particular, the proposed optical assembly, shown in Figure 2, consists of:

- A point light source, which emits a directionless light front.
- A pinhole to narrow the light front, reducing the stray light and obtaining the light beam.
- A lens that refracts the light beam and collimates it over the light sensitive surface. This element will be the element subdued to the displacement to be measured. It will have a focal length short enough to induce a high deviation of the light beam when a small misalignment is applied (Figure 2, Bottom).
- A light sensing surface, composed of a matrix of photo-detectors so the position in which the light is focused can be determined by the difference of the light measured in each element of the matrix.

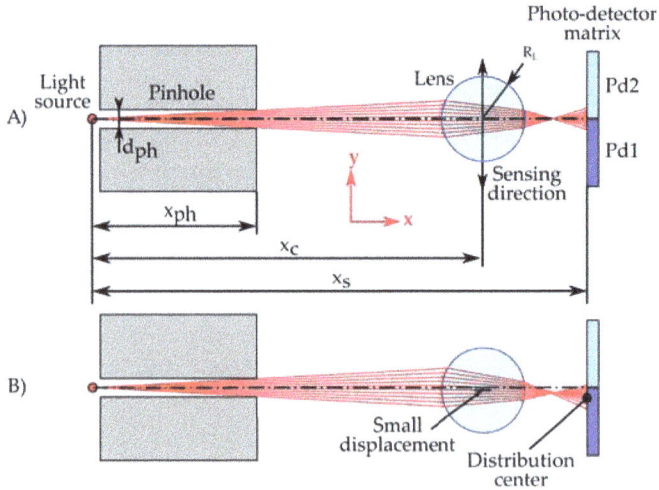

**Figure 2.** (**A**) proposed optical architecture in nominal position (perfect alignment) and geometrical parameters that determine the sensor response. (**B**) resulting light distribution in the photo-detecting surface after applying a small displacement in the lens in the *y*-direction (direction of the force to be measured).

This optical scheme has five geometrical design parameters, corresponding to the distance of the pinhole output ($x_{ph}$), its diameter ($d_{ph}$), the distance of the lens center ($x_c$), and the distance of the photo-sensitive surface ($x_s$), all of them measured from the center of the light source. The refractive index of the lens is an additional parameter that will determine the response of the system. However, since it depends on the material constituting the lens, only a very limited set of values can be chosen.

In order to achieve a system with the desired behavior, a theoretical model has been constructed, in order to be able to check which parameter configurations fit better with the intended performance. Since the focal length of the lens has the same magnitude order than the rest of the dimensions, the paraxial approximation cannot be applied to this system. Thus, a 2D ray-tracing model [29] has been stated to simulate the path followed by a certain number of rays from the light source to the photo-sensitive surface:

1.  First of all, a set of $q_{rays}$ rays are generated from the light-source-pinhole assembly. Each ray ($i$) is defined by its origin ($x_i^o$, $y_i^o$) and the direction of propagation expressed by the angle $\alpha_i$, measured clockwise from the optical axis. These parameters are computed using the Equations (1) to (3):

$$x_i^o = 0, \tag{1}$$

$$y_i^o = 0, \tag{2}$$

$$\alpha_i = \left( i - \frac{q_{rays}}{2} \right) atan \left( \frac{d_{ph}}{2x_{ph}} \right). \tag{3}$$

2.  Once generated, the first intersection ($x_i^r, y_i^r$) between each ray and the spherical lens is computed solving the Equations (4) and (5). In each intersection, the normal vector to the surface ($\vec{u}_i$) and

a vector ($\vec{n}_i^{in}$) with the same direction than the incident ray and modulus equal to the refractive index ($n^{in}$) of the propagation medium are computed using Equations (6) and (7):

$$R_L^2 = (x_i^r - x_c)^2 + (tan(\alpha_i)(x_i^r - x_i^o) + y_i^o - y_c)^2,\tag{4}$$

$$y_i^r = tan(\alpha_i)(x_i^r - x_i^o) + y_i^o,\tag{5}$$

$$\vec{u}_i = (x_i^r - x_c, y_i^r - y_c),\tag{6}$$

$$\vec{n}_i^{in} = n^{in}\frac{(x_i^r - x_i^o, y_i^r - y_i^o)}{\sqrt{(x_i^r - x_i^o)^2 + (y_i^r - y_i^o)^2}}.\tag{7}$$

3. The refraction of the rays is modeled by applying the Snell's law in its vectorial form (Equations (8) and (9)), to obtain the vector $\vec{n}_i^{out}$, which has the same direction than the refracted ray and a modulus equal to the refractive index of the new propagation medium ($n_i^{out}$):

$$(\vec{n}_i^{in} \times \vec{u}_i) \cdot (\vec{n}_i^{out} \times \vec{u}_i) = n_i^{in}n_i^{out}cos(\theta),\tag{8}$$

$$|\vec{n}^{in} \times \vec{u}_i| = |\vec{n}^{out} \times \vec{u}_i|.\tag{9}$$

4. Steps 2 and 3 solve the refraction of a light beam (composed by multiple rays) that interacts with an arbitrary surface that separates two mediums with different refractive indexes. This procedure generates a new set of rays, which can be described using the Equations (10) to (12), by their origin ($x_i^{\prime o}$, $y_i^{\prime o}$) and the angle that defines the direction of propagation ($\alpha_i'$), similarly to the original set:

$$x_i^{\prime o} = x_i^r,\tag{10}$$

$$y_i^{\prime o} = y_i^r,\tag{11}$$

$$\alpha_i' = acos\left(\frac{\vec{n}_i^{out} \cdot \vec{u}_i}{|\vec{n}_i^{out}||\vec{u}_i|}\right).\tag{12}$$

5. Steps 2 to 4 are performed again with the new refracted beam in order to compute the second refraction, which happens when it crosses the interface that separates the lens medium and the air. The resulting set of rays ($x_i^{\prime\prime o}$, $y_i^{\prime\prime o}$, $\alpha_i^{\prime\prime}$) corresponds to the beam focused by the lens.

6. Finally, the intersection of each resulting ray with the photo-sensitive surface ($y_i^{sensor}$) is determined by Equation (13). The individual photo-detector excited by each ray can be determined by the computed coordinate $y_i^{sensor}$, so the histogram of the number of rays per each photo-cell can be easily obtained:

$$y_i^{sensor} = tan(\alpha_i^{\prime\prime})(x_s - x_i^{\prime\prime o}) + y_i^{\prime\prime o}.\tag{13}$$

One point that must be underlined is that, during the execution of the described algorithm, some equation systems can lead to multiple solutions. The relevant ones can be easily isolated by applying conditional sentences. Additionally, the Snell's law in vectorial form also returns solutions corresponding to reflections, which have been neglected since this effect is expected to have low influence in the results.

This mathematical model allows the execution of simulations, by just evaluating it with the same input parameters ($x_{ph}, x_c$, $x_s$, $d_{ph}$, $R_L$, $q_{rays}$, $n^{air}$ and $n^{lens}$) varying the position of the lens ($y_c$). The resulting histograms may help to estimate the performance of the system for different input configurations.

Since the response of many photo-detectors is usually given in terms of incident light power, the number of rays that fall upon each photo-cell can be translated into these terms by just dividing the total power that passes throughout the pinhole ($P^{ph}$) by the number of computed rays. Thus, the power ($P_j^{pd}$) that receives each photo-cell ($j$) of the photo-sensitive surface can be computed by knowing the number of rays that falls on it ($q_{rays}^i$) as shown in Equation (14). Knowing the responsivity ($r$) of the photo-detector, this power can be easily translated to a current. Furthermore, with a proper transimpedance and amplifying circuitry (with conversion factor $A$), a voltage value for each photo-detector can be obtained (Equation (15)):

$$P_j^{pd} = \frac{P^{ph}}{q_{rays}} q_{rays}^i,$$ (14)

$$V_j^{pd} = Ar P_j^{pd}.$$ (15)

Summing up, with the optical arrangement shown in Figure 2, it is possible to estimate the displacement ($y_c$) of the lens as a function of the difference of electric response of a photo-sensitive electronic device (Equation (16)). This expression may be determined by applying a polynomial fitting to the results of this simulation method for several $y_c$ values:

$$y_c = f(V_2^{pd} - V_1^{pd}).$$ (16)

To increase the robustness of the optical system to misalignment, a cylindrical geometry has been chosen for the lens so that it will project the point light source as a diffused line in the photo-detector matrix. This fact together with the two-dimensional distribution of photo-cells will allow for stacking the information of each row to achieve a better signal-noise ratio (SNR) and a more uniform light distribution that can balance optical deviations, as shown in Figure 3.

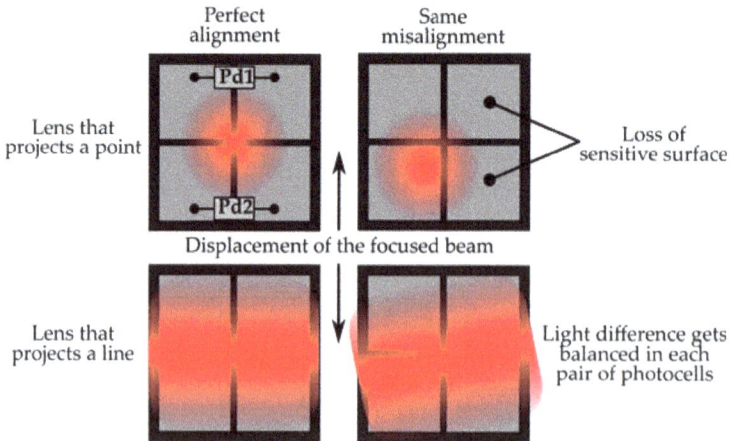

**Figure 3.** Scheme that shows the light focused by a lens in a 2 × 2 photo-detector matrix and the effect that the misalignment of the optical train produces in the light distribution.

### 2.1.2. Elastic Frame

Once the relation between the displacement of the lens ($y_c$) and the response of the photo-detector is known, it is necessary to relate this shift with the force that is going to be measured.

For this purpose, the proposed solutions consist of attaching the lens to an elastic element with a very high stiffness in all directions of movement and rotations except in that one in which the force wants to be measured. In this last direction, the type of elastic element should allow a fine tuning of the stiffness, so the behavior of the sensor can be adapted to the working conditions.

The most simple architecture would be the use of helical springs or elastic washers. The former solution has the inconvenience that helical springs should be guided in order to avoid displacements in other directions than the desired one, resulting in complex and not-compact structures, additionally, the commercially available range of stiffness and sizes might not be suitable for the desired application. Elastic washers might be a smaller solution; however, the range of possible stiffness is limited.

The proposed arrangement achieves a balance between flexibility and ease of design and size. It consists of attaching the lens between two beam-mounted parallel elastic plates as shown in Figure 4A. Plates have a dimension much smaller than the other two, so they have a preferential direction of deformation, especially when subdued to bending actions. Thus, they do not require any kind of guidance to assure that the displacement of the lens is performed in the desired direction. However, a plate does not have enough stiffness when a torsion moment is applied, so a not-perfectly centered force could cause a non-desired rotation of the lens. This issue is solved by adding a second plate parallel to it, so the stiffness is highly increased, especially when the distance between plates has the same (or higher) order of magnitude than the big dimensions of the plate.

The deformation of a plate is governed by complex equations, which, in most cases, cannot be analytically solved, so numerical methods must be used. Since it is expected that countless imperfections stem from the fabrication process, the authors do not consider that a complex solving method would necessarily produce accurate results. Instead of this, the classical beam theory would be used to model the elastic behavior of the plates, which can be solved by analytic means and leads to handy expressions useful for sizing process. Hence, each plate will be solved as a beam with the cross section shown in Figure 4B.

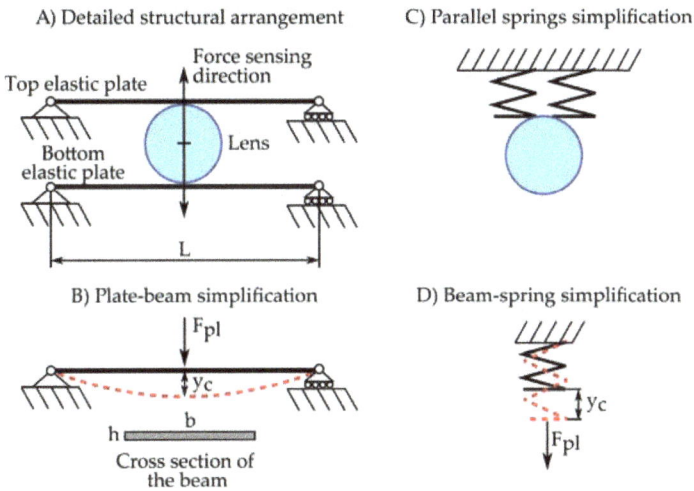

**Figure 4.** (**A**) proposed architecture to associate the lens displacement with the force to be measured; (**B**) beam to be solved to estimate the behavior of the elastic plate and the beam's cross section; (**C**) both plates can be simplified to a parallel spring association; (**D**) spring equivalent to the beam shown in B.

In particular, the deformation of the center of the beam, which can be considered equal to the studied displacement of the lens ($y_c$), can be computed as shown in the Equation (17), where $L$, $b$ and $h$ are the dimensions specified in the Figure 4, $E$ is the Young's modulus of plate material and $F_{pl}$ the part of the total force that deforms that plate. Grouping the geometrical and material parameters in a constant, the beam can be simplified to a spring (Figure 4D) with the stiffness ($k_{pl}$) described in Equation (18). Since both plates are attached to the sensor's frame and deform in the same direction, the equivalent springs can be associated in parallel as shown in Figure 4C. Thus, the relation between the force to be measured and the displacement of the lens can be computed by applying the one-dimensional form of Hooke's law (Equation (19)):

$$y_c = \frac{L^3}{48Ebh^3}F_{pl},$$ (17)

$$k_{pl} = \frac{48Ebh^3}{L^3},$$ (18)

$$F = 2k_{pl}y_c.$$ (19)

As can be concluded by the Equations (17) to (19), for a determinate optical arrangement, multiple sensing ranges and sensitivities can be achieved by just changing the geometrical parameters of the plate. Special attention must be paid in the effect of the thickness ($h$). Since stiffness presents a cubic dependence on it, small changes lead to great variations in the stiffness, allowing a fast tuning of the range of measurement. Conversely, the width of the plate ($b$) has a proportional relation with the spring constant, so a fine adjustment of the sensor's performance can be achieved with it. The length of the plate ($L$) does also have an important effect on the response of the system; however, its modification is discouraged since it normally implies a modification in the dimensions of the optical system, with hardly-predictable results.

### 2.1.3. Manufacturing Process

One of the main objectives of this research consists of achieving a sensor, which can be mainly built with prototyping technologies, so its manufacture can be completely carried out in a research laboratory.

Regarding the optical system, all the required photo-electronic devices can be easily handled without special machinery or requirements. The light source can be implemented with a simple board holding a LED sensor, a simple voltage divider and two supply terminals. The photo-sensitive detectors may present different manufacturing difficulties according to the chosen technology. Hence, the authors propose the use of a photo-diode matrix with common-cathode arrangement so that only individual anodes and the common-cathode need to be outputted. These electronic boards can be easily machined in a small numerical control milling machine and soldered manually.

As for the lens that focuses the light beam, optical grade elements normally require a careful handling and special tools in order not to damage its optical properties. Thus, instead of using these kinds of components, a general purpose acrylic (PMMA) glass is cut to obtain custom-length cylindrical lenses. PMMA rods are an inexpensive industrial supply that can be easily found in a wide variety of diameters ranging from about 1 mm to several centimeters, providing a broad range of possible focal lengths. Since the used material is not optical-grade, properties such as refractive index or surface quality may present flaws. However, the desired feature of this architecture is the capability of working in such coarse conditions.

In respect of the frame of the sensor and the elastic elements that define the displacement of the lens, they are designed to be built by fused-filament 3D-printing process with the most extended materials used in this technology, i.e., Polylactic acid (PLA) and Acrylonitrile butadiene styrene (ABS). The most affordable 3D printers do not guarantee uniform and repeatable mechanical properties of the print since these depend on the printing parameters as well as the different imperfections that might appear due to the chemical composition of the material and dimmensional tolerances [30]. However, with a suitable oversizing, printed parts can work in strain–stress conditions far from the limit points. This is especially suitable when designing the rigid frame that places each element such as light source, photo-detector and plate supports.

As for the elastic plates, they also can be 3D printed under certain conditions. Concretely, they must be printed with layers parallel to the plate plane and complete infill. A conservative piece of criteria to choose an adequate size for plates of PLA (considering a Young's modulus $E = 2000 \text{ N/mm}^2$ [30]) composition consists in seeking that the required displacement ($y_c$) do not exceed half the width of the individual plates. When this criteria cannot be achieved and the required plate width must be lower than the displacement, other materials can be used for building the plates. During this research, the possibility of using leaf springs made of AISI 301 Steel ($E$ = 20,000 N/mm$^2$) has been studied to produce a very sensitive sensor for a reduced range of measurement.

### 2.1.4. Particular Implementation

As exposed in Section 1.3, this architecture was devised in order to obtain a sensor that could be easily integrated inside the frame of a hand exoskeleton. Thus, the prototype implemented to check its feasibility and performance is aimed at satisfying the requirements and loads of this application, concretely a range of measurement of $\pm40$ N should be suitable for measuring the interaction between the exoskeleton and the hand. A more detailed description of the problem can be found in [31].

According to the theoretical model developed above, a sensor with PLA plates can be built for the proposed application with the characteristics described below. Additionally, the same architecture will be tested replacing the PLA plates with thin AISI 301 Steel leaf springs, which will allow a wider lens displacement under lower loads in order to obtain a more sensitive sensor to measure smaller loads. The expected response for both studied configurations is shown in Figure 5.

- Optical assembly

    - Light source: LED Kingbright APTD1608LSECK/J3-PF (Taipei, China), Red Color.
    - Pinhole data: $x_{ph} = 4$ mm, $d_{ph} = 1.5$ mm, $P^{ph} \approx 7$ μW
    - Lens data: PMMA Cylindrical lens, $x_c = 9.5$ mm, $R_L = 1$ mm, $n_{lens} = 1.49$
    - Phodo-detector: OPR5911 Quad Photodiode, $2 \times 2$ Matrix. Phodo-diode size: 1.27 mm, responsivity $r = 0.45$ mA/mW, $x_s = 11.5$ mm.

- Elastic frame

    - Effective plate length: $L = 10$ mm
    - Plate width: $b = 10$ mm
    - Plate thickness for PLA case: $h = 1$ mm
    - Plate thickness for AISI 301 case: $h = 0.1$ mm
    - Expected range of measurement for PLA case: $\pm45$ N
    - Expected range of measurement for AISI 301 case: $\pm10$ N

The geometry of the sensor has been designed to be as similar as possible to the definitive implementation that will be integrated into the exoskeleton; however, several modifications have been performed to adapt it to the test bench that will be used for the purpose of this study. As can be seen in Figure 6, there are two different rigid frames that host the photo-electronic devices (Blue parts) and lens (Red part). These frames are attached by the two elastic plates that will allow a relative displacement in the direction of the applied loads, disturbing the light beam.

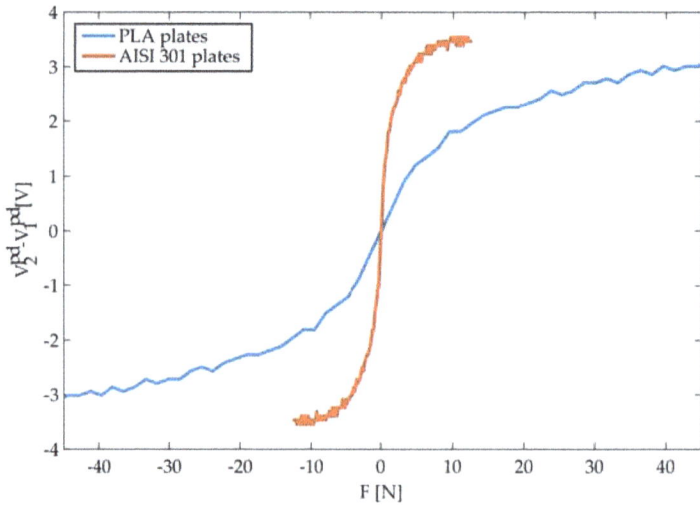

**Figure 5.** Computed response for the implemented sensor according to the kind of elastic plates used. Signals present a slight ripple due to the discrete nature of the ray tracing simulation.

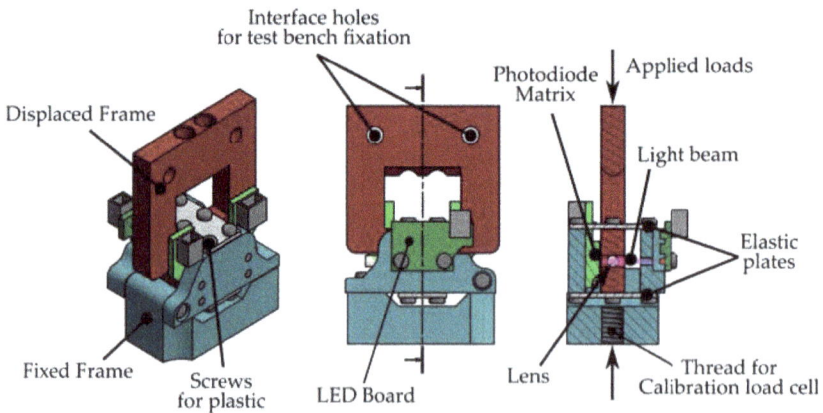

**Figure 6.** Mechanical drawings of the implemented force sensor. Blue parts are rigidly attached among them and hold the LED and Photo-Detector. The red part holds the lens and presents a slight relative displacement with respect to the blue parts when a load is applied.

## 2.2. Signal Conditioning, Acquisition and Processing

The chosen device to measure the incident light is a quadruple (2 × 2) photo-diode matrix. This sensor features four outputs that correspond to the current that passes throughout each anode, which is proportional to the light power that receives that diode. Following the cost-effective philosophy, the signal conditioning and acquisition circuitry have been designed to use common electronic elements.

In particular, for each photo-diode, the anode current is converted to a voltage value by a non-inverting transimpedance circuit using general purpose operational amplifiers. Afterwards, this voltage value is amplified by a noninverting amplifier stage, as shown in Figure 7. Since the circuit will be supplied with the 5 V ($V_{cc}$) output of a micro-controller, all operational amplifiers are asymmetrically powered. Additionally, this fact will improve the range of measurement since it is not expected (nor desired) that conditioned signals ($V_j^{pd}$) reach negative values.

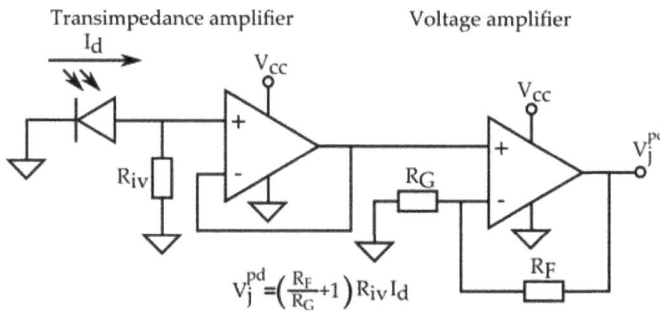

**Figure 7.** Signal conditioning circuit implemented for each individual photo-diode. The expression of the relation between output voltage ($V_j^{pd}$) and current in the photodiode ($I_d$) is shown.

Despite the outputs ($V_j^{pd}$) being combined using analogical components in order to get only one analog output, for this study, the sensor will output four voltages corresponding to each photo-diode. Output signals are read by the micro-controller throughout a multiplexed analog-to-digital converter (ADC) so that they can be added and subtracted digitally reducing the risk signal saturation. The micro-controller implements a digital filter to reduce the ripple introduced by the ADC.

For the studied implementation, the technical details can be summarized as follows:

- Operational amplifiers: 2× Quadruple general-purpose TI LM324 (One device per pair of photo-diodes).
- Resistance values: $R_{iv} = 30$ kΩ, $R_F = 120$ kΩ, $R_G = 1.5$ kΩ.
- Resulting conversion factor: $A = 2.43$ V/mA
- Micro-controller: ATmega1280 that features $V_{cc} = 5$ V and 10 bit ADC.
- Digital filter: Mean filter with a time window of eight values.

*2.3. Fitting Models*

All the previous mathematical expressions are useful for design and sizing stages; however, the performed approximations cannot model many aspects such as color sensitivity of photo-diodes, optical misalignment and imperfection during the manufacture process or hysteresis of constructive materials. Thus, this equation cannot be used to accurately predict the performance of the proposed sensor. Hence, to deal with these effects, four fitting models will be used in order to approximate the real behavior of the sensor.

2.3.1. Polynomial Fit

The most simple model of all that will be studied is a third order polynomial fit (Equation (20)) of a set of pairs voltage–force values. This approximation is expected to be able to deal with the nonlinearity of the optical model as well as reproduce the deviation from the ideal model due to optical misalignment. However, polynomials does not offer any mathematical way to model hysteresis cycles.

Despite being the less complex of the studied models, it has an advantage that can make it the most suitable solution for certain applications. Concretely, this approximation only needs a minimum of samples equal to the order of the polynomial plus one. Therefore, it can be fitted by recording the sensor response with a simple set of weights. Several samples for a certain weight can be acquired in order to use a least-square fitting method to obtain an averaged sensor curve:

$$F = \sum_{i=0}^{3} a_i (V_2^{pd} - V_1^{pd})^i. \tag{20}$$

### 2.3.2. Generalized Prandtl–Ishlinksii Model

The Generalized Prandtl–Ishlinskii model [32] (GPI) is one of the most extended approximations to model systems that exhibit nonlinear, asymmetrical, and saturated hysteresis cycles [33–36]. This model represents the response of the system as a weighted sum of different play operators ($S_{r_i}$) as shown in Equation (21), where $V$ corresponds to the output of the sensor ($V = V_2^{pd} - V_1^{pd}$), $F(t)$ is the applied force and $p(r_i)$ are the different weights computed as shown in Equation (23). Regarding the play operators ($S_{r_i}$), they are defined by two envelope functions ($\gamma_r, \gamma_l$ that correspond to the rising and decreasing curves of an hysteresis cycle (Equations (25) and (26)). As it can be seen in Equation (22), all the play operators have the same envelope functions but differ in the parameter $r_i$ that determines the width of the hysteresis loop. These width parameters are established in Equation (24). The stated GPI model presents a series of free parameters ($\alpha, \rho, \tau, a_j, b_j$) that can be determined by using optimization techniques from a set of samples ($V(t), F(t)$). The number of finite sums ($n$) can be manually stated since weight ($p(r_i)$) rapidly will become null for higher $i$ values:

$$V(F(t)) = \sum_{i=0}^{n} p(r_i) S_{r_i}[F(t)], \tag{21}$$

$$S_{r_i} = \begin{cases} \max(\gamma_r - r_i, S_{r_{i-1}}) & \text{for} \quad \dot{F}(t) \geq 0, \\ \min(\gamma_l + r_i, S_{r_{i-1}}) & \text{for} \quad \dot{F}(t) < 0, \end{cases} \tag{22}$$

$$p(r_i) = \rho e^{-\tau r_i}, \tag{23}$$

$$r_i = \alpha i, \tag{24}$$

$$\gamma_r = a_1 \tanh(a_2 F(t) + a_3) + a_4, \tag{25}$$

$$\gamma_l = b_1 \tanh(b_2 F(t) + b_3) + b_4. \tag{26}$$

The computed model allows for computing the response of the sensor under certain load conditions. However, the desired result consists of the determination of the force that produces the measured response of the sensor ($F(V(t))$). Thus, the inverse model must be obtained, which can be analytically inferred from the direct model as shown in Equations (27) to (32):

$$F(V(t)) = V^{-1}(F(t)) = \gamma^{-1} \left( \sum_{i=0}^{n} g_i R_{r_i}[V(t)] \right), \tag{27}$$

$$R = \begin{cases} \max(V(t) - q_i, R_{r_{i-1}}) & \text{for} \quad \dot{V}(t) \geq 0, \\ \min(V(t) + q_i, S_{r_{i-1}}) & \text{for} \quad \dot{V}(t) < 0, \end{cases} \tag{28}$$

$$\gamma^{-1} = \begin{cases} \gamma_r^{-1} & \text{for} \quad \dot{V}(t) \geq 0, \\ \gamma_l^{-1} & \text{for} \quad \dot{V}(t) < 0, \end{cases} \tag{29}$$

$$g_j = -\frac{p_j}{(p(r_0) + \sum_{i=1}^{j} p(r_i))(p_{r_0} + \sum_{i=1}^{j-i} p(r_i))}, \tag{30}$$

$$g_0 = \frac{1}{p_{r_0}}, \tag{31}$$

$$q_j = \sum_{i=0}^{j} p(r_i)(r_j - r_i). \tag{32}$$

### 2.3.3. Artificial Neural Networks

Artificial Neural Networks (ANN) are a powerful tool for fitting complex models [37] and have already been successfully used in hysteresis modelling and compensation applications [38–41]. Rather than stating a series of equations with physical meaning, this solution relies on optimizing an established network of interconnected simple operators, or neurons, with a relatively big set of free parameters (at least more numerous than the alternatives exposed above).

For the presented application, the chosen network architecture is the Multilayer Perceptron [42]. According to Irie et al. [43], a three-layer perceptron, with a sufficient number of neurons, should be able to fit any arbitrary mapping function. Due to the foreseeable hysteretic behaviour of the sensor, the relation between the applied force ($F(t)$) and the measured signal of the sensor ($V(t)$) is not a proper mathematical function. However, if it is assumed that the estimated force may be univocally computed as a function of the current sensor measure and those taken in previous time steps ($F(t_i) = f(V(t_i), V(t_{i-1}), V(t_{i-2}))$), then Irie's conditions are accomplished. In particular, the three-layer perceptron (3LP) that will be trained has the next characteristics (illustrated at the top of Figure 8):

- Input layer: three inputs corresponding to $V(t_i)$, $V(t_{i-1})$, $V(t_{i-2})$ to be able to model the temporal dependency of the hysteresis.
- Hidden layer: seven neurons with symmetric sigmoid as transfer functions.
- Output layer: one output neuron with linear transfer function that returns the estimated $F(t)$.
- Training algorithm: Levenberg–Marquardt backpropagation algorithm.

Additionally, since it is expected to obtain a reasonable amount of training data the five-layer perceptron (5LP), shown at the bottom of Figure 8, will be also trained in order to check whether it can increase the performance of the sensor or it will overfit the model.

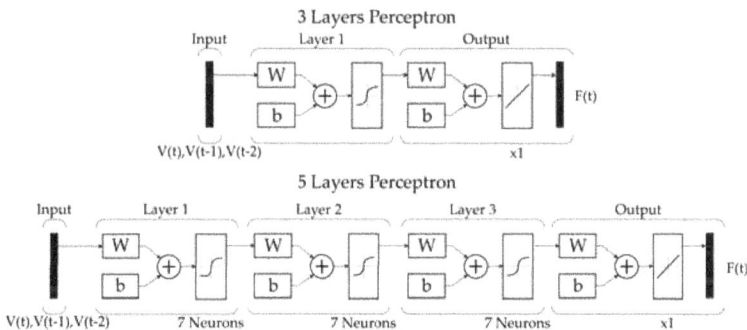

**Figure 8.** Schematic view of both 3 and 5-layer perceptrons used to model the hysteretic behaviour of the sensor. Both neural networks feature three inputs corresponding to the actual sensor voltage and the two previous values; hidden layers of seven neurons and an output layer that returns a single value, corresponding to the estimated force measurement.

### 2.4. Experimental Setup

The aim of the proposed experimentation is the evaluation of the performance of the new optical force sensor presented in this paper. In order to perform this assessment, a comparison of the sensor response using two different materials for the elastic plates, PLA and spring-steel, is carried out. Moreover, the different fitting algorithms, exposed in the previous section, are evaluated for both versions of the optical force sensor.

For this purpose, the test bench shown in Figure 9 has been developed. This setup consists of the following elements:

- Optical force sensor. The developed sensor that is going to be validated with the experiment.
- Commercial load cell. A calibrated and industrial grade force sensor, which is used as a reference for the fitting model procedure and validation. The selected load cell model is LCM201–100N manufactured by Omega (Stamford, CT, USA).
- Active element to apply force. A double-acting pneumatic cylinder (Festo DFK-16-40-P (Esslingen, Germany)) is used to apply the desired forces to the force sensors during the experiment. Two proportional pressure control valves (Festo MPPE-3-1/8-10-010) have been selected to control the pressure applied on the cylinder.
- Structural frame. Structure where all elements are mounted.

**Figure 9.** Test bench for validation experiment.

The experiment was conducted in two different steps. The first one is a calibration phase, where the different sensor models are computed by training the four fitting approaches exposed in the previous section. For this aim, the pneumatic cylinder is driven with the input pressure signal ($p_{input}$) exposed in Equation (33), with frequencies in the range of [0, 2] Hz and an amplitude ($A$) in the range of $\pm 3.5$ Bar for the sensor based on PLA elastic plates and $\pm 1$ Bar for the sensor based on spring-steel elastic plates. The response of the calibrated load cell is used as input for training the fitting algorithms in order to obtain the model of the optical force sensor:

$$p_{input}(t) = A \cdot sin\left(\frac{2\pi t}{127}\right) \cdot sin\left(\frac{2\pi}{0.6} \cdot sin\left(\frac{2\pi t}{9}\right)\right). \tag{33}$$

Once the different sensor models are computed in the calibration phase, the second step is the validation of the these models. For this purpose, four different signals were selected to excite the optical force sensor with as much heterogeneous inputs as possible:

- 1 Hz sine wave input. Sine signal with a frequency of 1 Hz with the same maximum amplitude ($A$) previously used in the calibration phase.
- Multi-Frequency wave input. Signal composed of different sine waves in order to obtain a signal with frequency and amplitude variations.
- Random wave input. Signal with random values with a maximum frequency of 1.5 Hz and the maximum amplitude used in the calibration phase ($A$).
- Human interaction. Signal obtained by direct interaction with the test bench without using the pneumatic cylinder.

Each trained sensor model is used to obtain an estimation of the force applied to the system. This response is compared with the one obtained with the commercial load cell, which is considered the objective behavior. In the following sections, all of these results are presented and explained in detail.

## 3. Results and Discussion

### 3.1. Real Sensor Performance

In Figure 10, the response of the optical force sensor prototype for both PLA and spring-steel elastic plates is presented. These responses are measured with the input signal $p_{input}$, expressed in Equation (33). In these plots, the existence of hysteresis can be appreciated for both versions of elastic plates. It must be pointed out that the hysteresis cycle of the steel based sensor is significantly asymmetrical, presumably due to the presence of bendings and residual burrs product of the machining process. In contrast, the PLA based sensor has a smoother behavior, due to a more accurate fabrication process, but with a wider hysteresis.

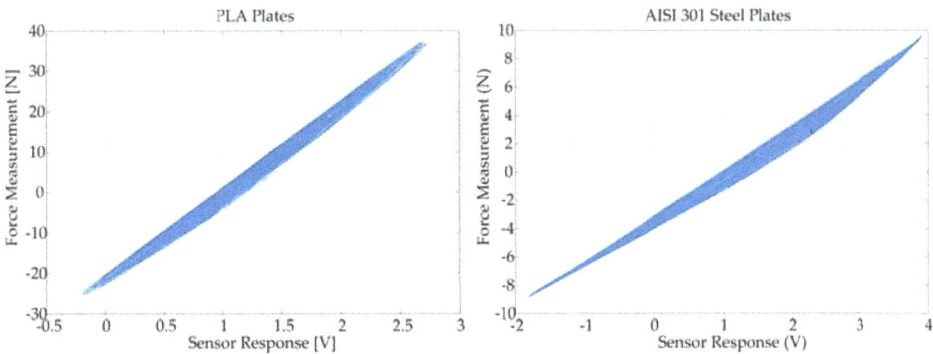

**Figure 10.** Sensor voltage output compared to the force measured in the commercial load cell, for both tested configurations. The shown graphs are the result of overlapping all the load–unload curves applied during the calibration process, revealing the shape of the hysteresis cycle for each sensor.

The comparison between the real and the theoretical performance (Figure 4) shows that the real behaviour differs from the expected sigmoid response. This fact results from the assumptions made in the optical simulations, such as homogeneous light distribution outputting from the LED source or neglected secondary reflections and refractions that can lead to a different light power distribution. In spite of this, the theoretical model gives a correct order of magnitude of the measured voltage and force.

Regarding the PLA-based sensor, the real performance shows that the optical system is heavily misaligned, resulting in a sensor response of about 1 V when no force is applied. Additionally, the sensor response covers a range of about 3 V in contrast to the 6 V that spans the theoretical model. This fact implies that the real PLA plates are almost twice as stiff as expected. This is consistent with the range of values for the elastic modulus of (2020 to 3550 MPa) established by Lanzotti et al. [30]. Another source of error in this aspect can be that the real behaviour of the elastic interfaces in somewhat between the simply supported and fixed beam approximations due to the significant thickness of the plate.

As for the steel-based implementation, the initial optical misalignment is similar to the measured one for the PLA plates. Thus, the main source of misalignment lies in the fixation of the optical elements in the common frame for both sensors rather than in the plates. In this case, the span of the range of measurement of the sensor is quite consistent with the beam-theory approximation, since the mechanical properties of this normalized material are more accurately determined. Additionally, the low thickness of the plate allows the rotation of the beam axis around the frame edge that acts like the support, resulting in a better approximation to the ideal conditions.

## 3.2. Model Fitting Performance

These responses are used as training data for the four proposed fitting algorithms: polynomial fit, generalized Prandtl–Ishlinskii model (GPI), 3-layer perceptron (3LP) and 5-layer perceptron (5LP). The statistical analysis of the performance for each model and signal is summarized in Table 1 and Figure 11. Additionally, Figures 12 and 13 show a visual example of the performance of each model applied on a small sample of the random signal input. A series of more detailed figures that help to understand the particularities of each model can be found in Appendix A.

**Table 1.** This table contains statistical information about the results obtained in the validation experiment. MAE refers to the Mean Absolute Error, MEDIAN to the Median of the absolute error, SD to the Standard Deviation, Corr Coeff to the Cross correlation coefficient, Outliers to the percentage of samples that lie out of the MEDIAN ± 2SD interval, and RCT to the relative computation time with respect to the time required to compute the Polynomial Fit.

| Input | Variable | Polynomial Fit | | GPI Model | | 3LP | | 5LP | |
|---|---|---|---|---|---|---|---|---|---|
| | | PLA | STEEL | PLA | STEEL | PLA | STEEL | PLA | STEEL |
| Sine Wave | MAE (N) | 2.3147 | 0.9379 | 2.5134 | 0.7312 | 1.9853 | 0.4288 | 1.9974 | 0.4091 |
| | MEDIAN (N) | 1.8432 | 0.9299 | 2.4267 | 0.7378 | 1.9999 | 0.4639 | 1.9372 | 0.37341 |
| | SD (N) | 1.2147 | 0.4754 | 0.8571 | 0.2227 | 0.4484 | 0.2059 | 0.4209 | 0.2232 |
| | Corr Coef | 0.9979 | 0.9969 | 0.9991 | 0.9994 | 0.9998 | 0.9997 | 0.9998 | 0.9997 |
| | Outliers (%) | 0.1050 | 0.0253 | 0.1050 | 0.1519 | 0.4202 | 0.0506 | 0.7353 | 0.0253 |
| Multi-Frequency Wave | MAE (N) | 2.2222 | 0.2916 | 2.1924 | 0.5386 | 2.2024 | 0.1076 | 2.2551 | 0.1033 |
| | MEDIAN (N) | 2.1415 | 0.2383 | 2.1531 | 0.5879 | 2.1800 | 0.0790 | 2.2220 | 0.0753 |
| | SD (N) | 0.9141 | 0.2490 | 0.4228 | 0.3327 | 0.6876 | 0.0985 | 0.7718 | 0.1130 |
| | Corr Coef | 0.9972 | 0.9948 | 0.9994 | 0.9927 | 0.9985 | 0.9993 | 0.9970 | 0.9992 |
| | Outliers (%) | 1.1416 | 9.3394 | 2.4099 | 0.0506 | 4.3379 | 5.5176 | 5.7585 | 4.6064 |
| Random Wave | MAE (N) | 1.2371 | 0.5313 | 1.1349 | 0.4083 | 1.1916 | 0.6022 | 1.2634 | 0.6155 |
| | MEDIAN (N) | 1.0303 | 0.6371 | 1.1378 | 0.3845 | 1.0863 | 0.6653 | 1.1042 | 0.6788 |
| | SD (N) | 0.9150 | 0.2740 | 0.4989 | 0.2593 | 0.6863 | 0.2144 | 0.8307 | 0.2179 |
| | Corr Coef | 0.9976 | 0.9992 | 0.9995 | 0.9988 | 0.9990 | 0.9994 | 0.9980 | 0.9995 |
| | Outliers (%) | 0.0761 | 0.0253 | 0.1015 | 0.0253 | 2.8412 | 0.0253 | 6.7986 | 0.0253 |
| Human Interaction | MAE (N) | 2.2618 | 0.6221 | 2.2643 | 0.4325 | 2.4103 | 0.6754 | 2.4798 | 0.7165 |
| | MEDIAN (N) | 2.2304 | 0.5953 | 2.3151 | 0.3345 | 2.3873 | 0.6594 | 2.4370 | 0.7032 |
| | SD (N) | 1.0011 | 0.2475 | 0.7065 | 0.3425 | 0.9520 | 0.2239 | 1.0353 | 0.2043 |
| | Corr Coef | 0.9966 | 0.9991 | 0.9988 | 0.9979 | 0.9973 | 0.9989 | 0.9960 | 0.9995 |
| | Outliers (%) | 0.4225 | 0.0253 | 4.1127 | 0.0253 | 0.1690 | 1.8983 | 0.4225 | 0.2278 |
| Overall Statistics | RCT | 1.0 | | 345.87 | | 109.56 | | 137.56 | |
| | MAE (N) | 1.9271 | 0.5958 | 1.9011 | 0.5276 | 1.9236 | 0.4535 | 1.9841 | 0.4611 |
| | MEDIAN (N) | 1.8601 | 0.5753 | 2.0568 | 0.5359 | 2.1305 | 0.4762 | 2.2193 | 0.4608 |
| | SD (N) | 1.0751 | 0.3995 | 0.7849 | 0.32001 | 0.9160 | 0.2915 | 0.9945 | 0.3048 |
| | Corr Coef | 0.9964 | 0.9956 | 0.9984 | 0.9945 | 0.9977 | 0.9985 | 0.9967 | 0.9985 |
| | Outliers (%) | 0.2341 | 2.4487 | 0.6378 | 0.03164 | 0.7831 | 0.0063 | 1.6712 | 0.0190 |

### 3.2.1. Polynomial Fit Analysis

Polynomial fit cannot model the hysteretic behaviour of the sensor, but it can fit a mean curve in order to uniformly distribute the error for all force values. As it can be seen in Figure A1, for the PLA sensor, it is able to compute a mean curve along the cycle, while in the steel sensor it does fit the lower envelope of the hysteresis.

This fact produces a high error dispersion for any kind of input signal, which is displayed in Figure A2. However, for the hysteresis cycle that features the PLA sensor, this error dispersion becomes closer as the force values increase and the cycle becomes narrower (Figure A3).

Since this approximation does not model the dynamic behaviour of the sensor, the error is rate-dependent. Figure A4 shows a opposite behaviour depending on the shape of the hysteresis cycle: the PLA sensor features an increase in the dispersion of the error for fast variations in the force measurement, while the steel sensor presents the opposite reaction.

Figures A5 and A6 illustrate the performance of the polynomial model for an example signal. It can be seen that it outputs an estimation that has a time response quite similar to the load cell's one. This is numerically reflected in the cross correlation coefficient (Corr Coef) of Table 1, which is almost 1.

### 3.2.2. Generalized Prandtl–Ishlinskii Model Analysis

The generalized Prandtl–Ishlinskii model (GPI) has been able to partially model both hysteresis cycles of each kind of sensor. Figure A1 shows a quite approximate loop for the PLA sensor, but an excessively narrow response for the steel one.

In contrast to the polynomial model, this approximation is able to obtain a centered model in both cases, resulting in the tight error distributions displayed in Figure A2. For the PLA case, this model results in a quite uniform error distribution independently to the force values, while in the steel implementation it behaves similarly to the polynomial case (Figure A3) since the obtained hysteresis loop is almost negligible. For both sensor implementations, the error distribution is rate independent, as illustrated in Figure A4.

Examples shown in Figures A5 and A6 reveal that this model has a quite approximate response for the PLA case, while in the steel case it tends to return higher absolute values than the load cell.

### 3.2.3. Multilayer Perceptron Analysis

Figure A1 shows that both kinds of perceptron can model the hysteresis cycle of both sensors, resulting in an apparently more suitable response than the previous models.

Analyzing the corresponding error distributions plotted in Figure A2, it can be seen that in most cases they are comparable to the polynomial approximation, only outstanding for the "Frequency Wave Input" case, which is the most similar one to the training case. A similar trend to the polynomial approximation can be seen when evaluating the error for different force ranges (Figure A3). As for the rate dependency, they only present marginal differences for the steel case (Figure A4).

The time response represented in Figures A5 and A6 reveals also a behaviour similar to the polynomial fit. Moreover, 5LP can lead to wrong dynamic evolution as it can be seen between seconds 60 and 65 in the 5LP representation of Figure A5.

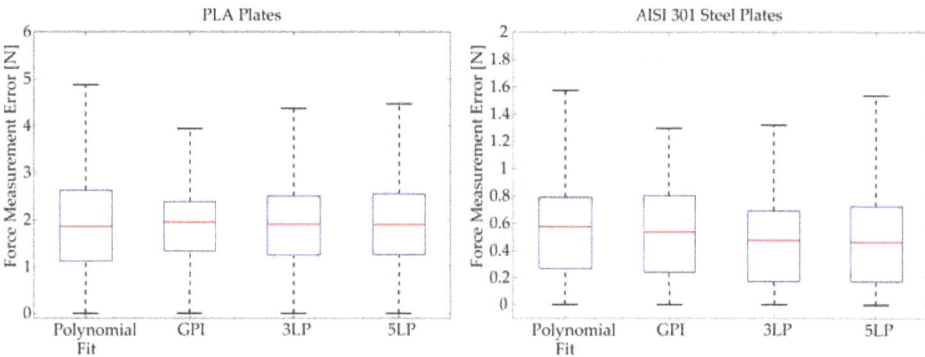

**Figure 11.** Error distribution of each fitting model evaluated over all the test signals. The red line indicates the mean value, blue boxes enclose the samples between percentiles 25 and 75, and black whiskers covers an standard deviation of $\pm 2.7SD$.

### 3.2.4. Comparative Analysis

When comparing the overall performance of each model taking into account all the analyzed signals (Figure 11), it is possible to conclude that there is not a model that stands out from the rest of the studied approximations. As shown in the last row of Table 1, all of the models have a similar mean

absolute error (MAE) and median of the absolute error (MEDIAN), which are about the 5% and 6% of
the maximum absolute force measurable for the PLA (45 N) and steel (10 N) sensors, respectively.

The only significant differences can be found in the dispersion of the error, quantified by the
standard deviation (SD). In this regard, the GPI model produces narrower deviations than neural
networks, which in turn are tighter than the polynomial fit.

As for the dynamical response of the models, they all produce signals that are quite similar to
the expected one. This fact can be visually checked in Figures 12 and 13 and numerically evaluated
with the cross correlation coefficients (Corr Coef) of the table, which are extremely close to the unit
(identical signal).

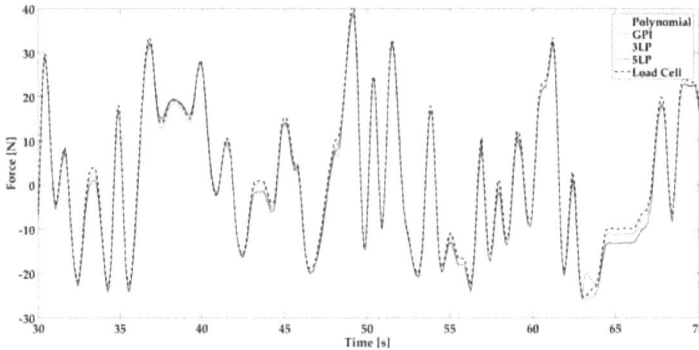

**Figure 12.** Time response of all fitting models and load cell signal for a 40 s sample of the Random
Signal test for PLA based sensor.

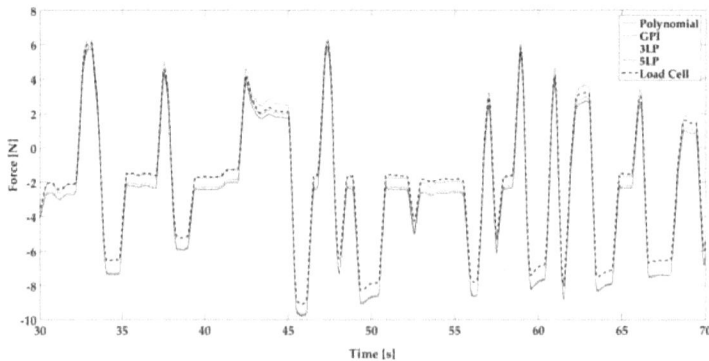

**Figure 13.** Time response of all fitting models and load cell signal for a 40 s sample of the Random
Signal test for steel based sensors.

In short, all models have certain features and shortcomings that can make them more or less
suitable for certain working conditions, depending on the selection of each one of the concrete
applications or available material resources. The characteristics and suitability of each model can be
summarized as follows:

- Polynomial fit: It is the most simple model, which does not need a complex infrastructure
  to be calibrated. It normally achieves a centered solution that, despite not correcting the
  hysteresis, can distribute the error across all measurement ranges. Additionally, due to the
  simplicity of the required mathematical operations, this is the fastest model to evaluate, which

allows its implementation in real-time applications without requiring much computing power. This approximation may be the suitable for applications where only the dynamic evolution of the force is interesting, such as movement intention detection, endstops or coarse force controllers.

- Generalized Prandtl–Ishlinskii model: This is the most balanced model, which has enough free parameters to be able to model an hysteresis cycle while avoiding overfitting artifacts. It must be pointed out that, due to limitations of the optimization process, this alternative was trained with only the 2.5% of the acquired calibration data. Therefore, one can conclude that this model is outstanding concerning the generalization capability, being able to successfully extrapolate non-trained inputs. Its main disadvantage is the computational cost that requires its optimization. Additionally, when evaluating individual samples, it is almost 350 times slower (RCT) than the polynomial counterpart, which does not imply that it cannot be implemented for real-time applications, but it will require more computing power for the same sampling frequency. With well-chosen envelope functions, this might be the most suitable model for applications where the applied force pattern is unforeseeable, such as accurate measurements of human–machine interaction.
- Multilayer Perceptrons: Perceptrons are halfway solutions between polynomials and GPI. Like most artificial neural networks, they suffer from lack of generalization capability, so they can lead to wrong extrapolation or overfitting and attention must be paid when choosing their architecture. However, they are easy to train, require moderate computing power, and provide outstanding results when the working conditions are similar to the training ones.

## 4. Conclusions

In conclusion, the proposed sensor architecture has proven to have a coherent measurable response when an external compression–extension force is applied. By using cheap and common fabrication techniques together with the application of one of the proposed models, it is possible to obtain a sensor with a reasonable performance (in comparison with an industrial-grade load cell) that is completely customizable in force measurement and sensitivity and whose geometry can be adapted to be integrated into a device that is in the design process. Once built and integrated, this architecture allows a fast and handy modification of the desired range of measurement.

Once known the discrepancy between the theoretical model and the real response, future research lines will try to improve the approximations taking into account further effects to obtain more accurate expressions that could help to predict the behavior of the designed sensor. This refined model together with the flexibility that provides the 3D printing technologies would allow for automatically generating and building a sensor with the desired performance, thereby incorporating the benefits of the fast prototyping techniques to the force sensing technology. It is important to state that, ultimately, a fitting model would be still mandatory in order to ensure the robustness of the sensor to manufacture imperfections.

As for the measurement error that these models try to minimize, there are several strategies to explore. On one side, the use of better manufactured steel plates might lead to narrower hysteresis cycles, reducing the dependency on the fitting model and their training issues; however, this fact may increase the fabrication cost and reduce the flexibility of design. Alternatively, the use of the Generalized Prantl-Ishlinskii model is encouraging and seems to have a good generalization capability. The implemented GPI model, however, was limited to symmetrical sigmoid envelopes, which do not fit the real shape of the hysteresis well. Additionally, the optimization method used to compute this model is not the fastest and does not allow using all the available data to train the system, resulting in a loss of its potential to compensate for the error. Therefore, the implementation of a Prandtl–Ishlinskii model with asymmetrical envelopes and more specialized optimization methods, such as particle swarm optimization [44], will allow a more accurate model identification that can significantly reduce the estimation error due to insufficient hysteresis compensation.

**Acknowledgments:** This work has been founded by the European Commission through the project AIDE: Adaptive Multimodal Interfaces to Assist Disabled People in Daily Activities (Grant Agreement No. 645322); by the Spanish Ministerio de Economía y Competitividad through the project DPI2015-70415-C2-2-R; and by Conselleria d'Educació, Cultura i Esport of Generalitat Valenciana through the grants ACIF2016/216 and APOTIP 2016/021.

**Author Contributions:** J.A.D. conceived and developed the system and experiments, J.M.C. and F.J.B. coded the data analysis software, A.B. and J.V.G.-P. implemented the experimental setup, and N.G.-A. and F.J.B. performed guiding and coordination tasks.

**Conflicts of Interest:** The authors declare no conflict of interest. The founding sponsors had no role in the design of the study; in the collection, analyses, or interpretation of data; in the writing of the manuscript, and in the decision to publish the results.

## Appendix A. Experimental Validation: Detailed Figures

*Appendix A.1. Comparison of the Response of the Computed Fitting Models*

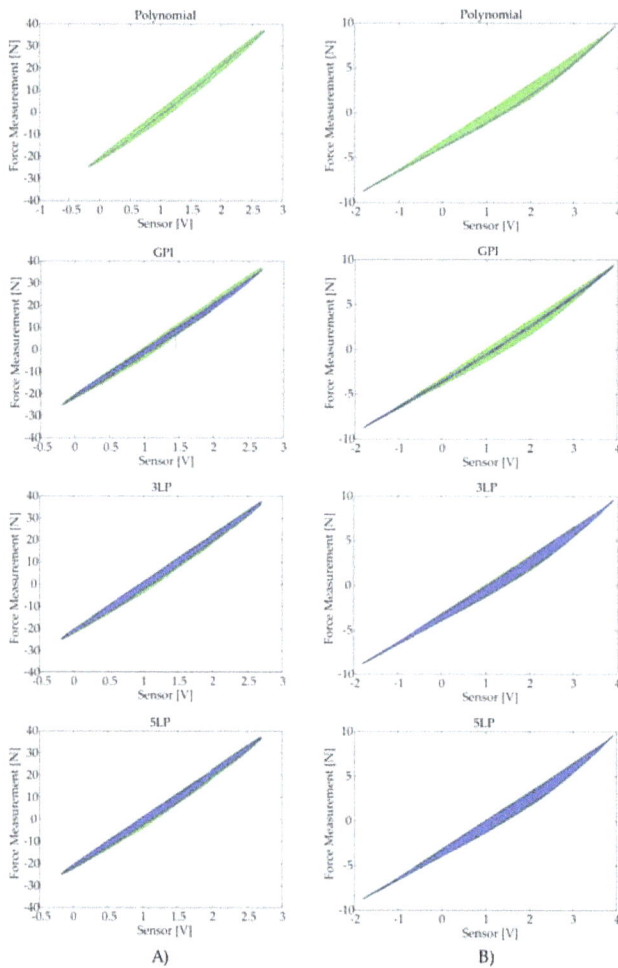

**Figure A1.** Response of each computed model (blue) superposed to the calibration data (green), where (**A**) corresponds to the data of the PLA based sensor and (**B**) to the steel based sensor.

*Appendix A.2. Error Distributions of the Fitted Models*

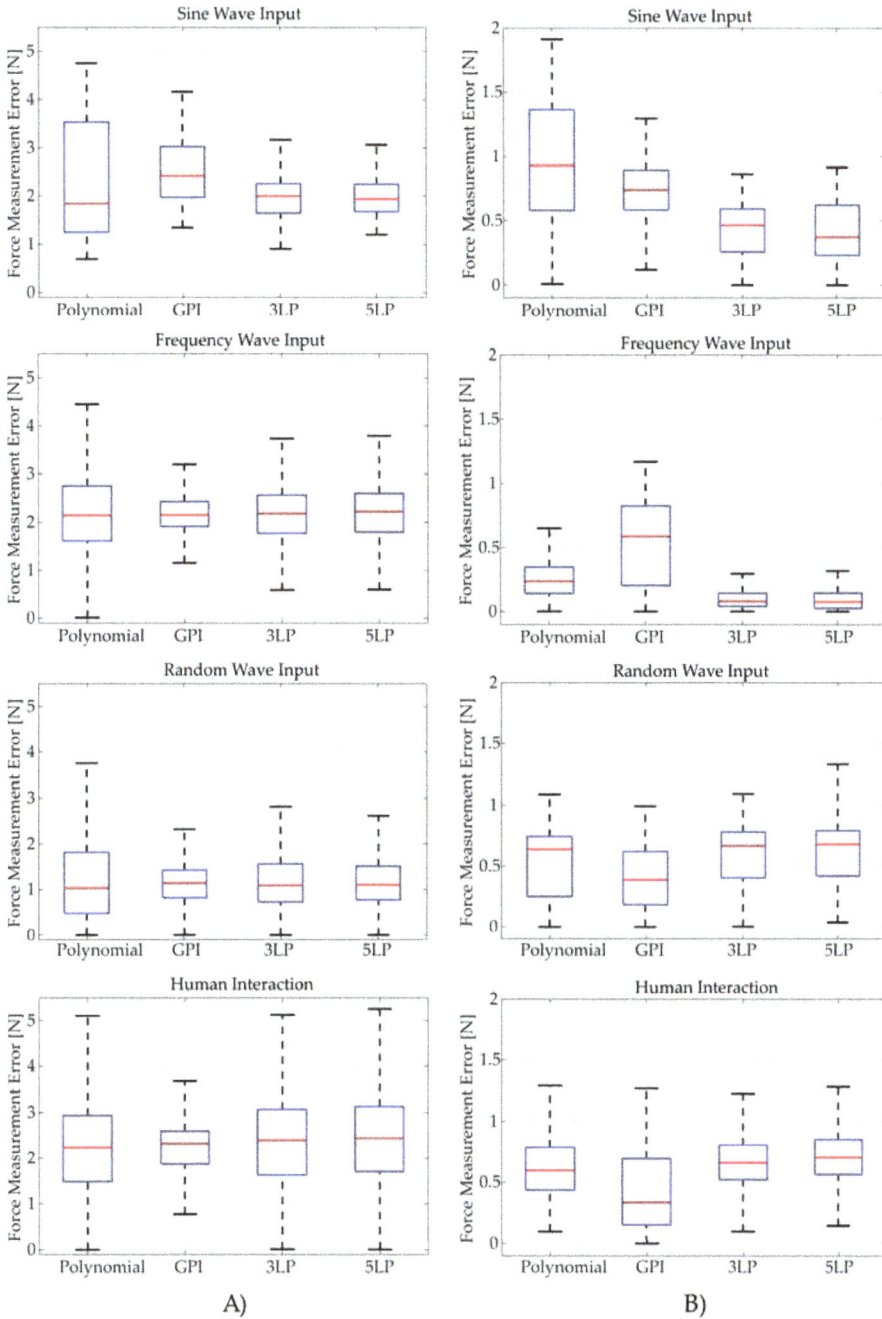

**Figure A2.** Error distribution computed for each tested kind of signal and model, where (**A**) corresponds to the data of the PLA based sensor and (**B**) to the steel based sensor.

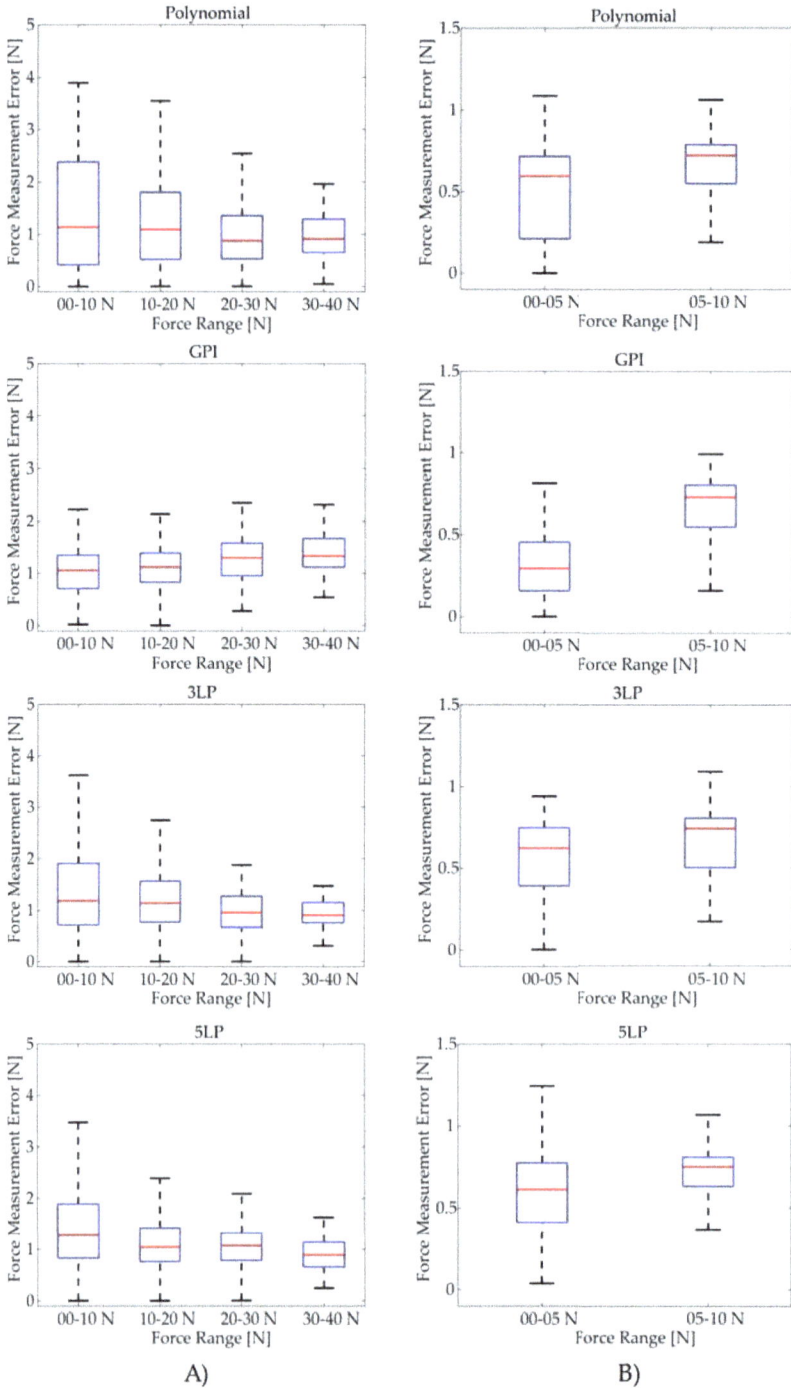

**Figure A3.** Error distribution computed for the Random Signal test and range of force. Where (**A**) corresponds to the data of the PLA based sensor and (**B**) to the steel based sensor.

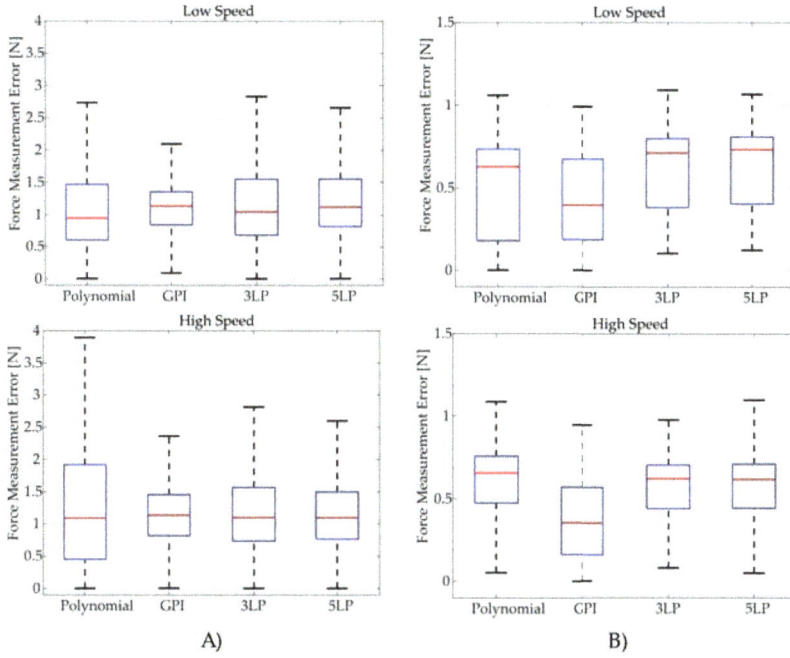

**Figure A4.** Error distribution computed for the Random Signal test and range of variation of the force. (Low Speed ≤ 5 N/s; High Speed> 5 N/s). Where (**A**) corresponds to the data of the PLA based sensor and (**B**) to the steel based sensor.

*Appendix A.3. Error of Computed Sensor Models with a Human Interaction Input Signal*

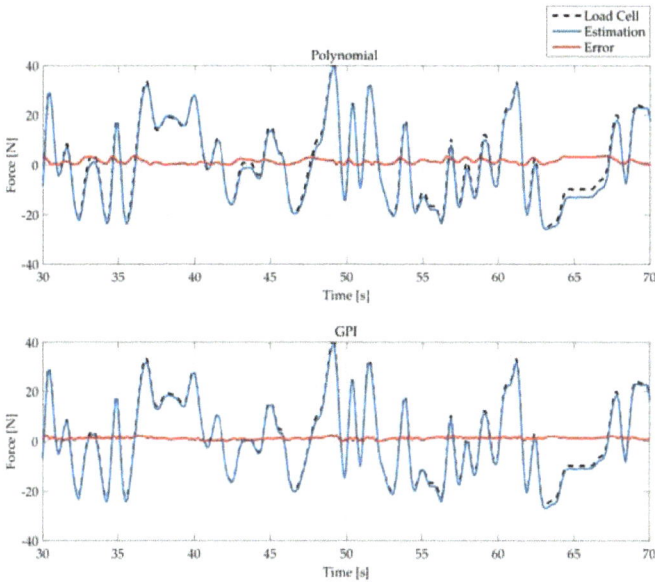

**Figure A5.** *Cont.*

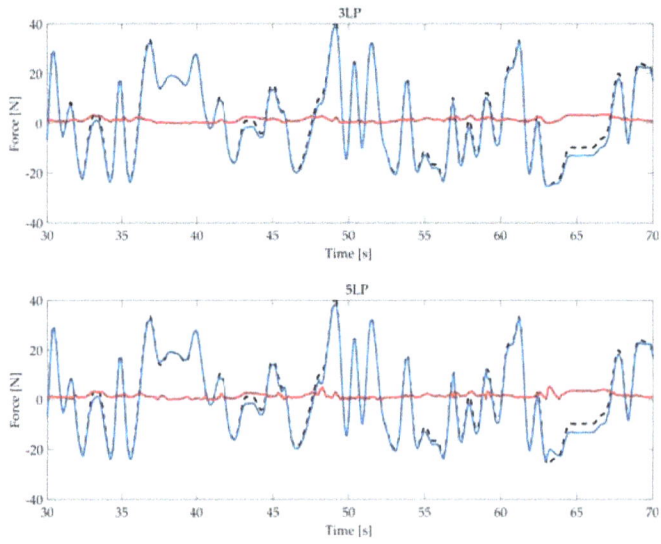

**Figure A5.** Detailed view of the time response shown in Figure 12 for each model.

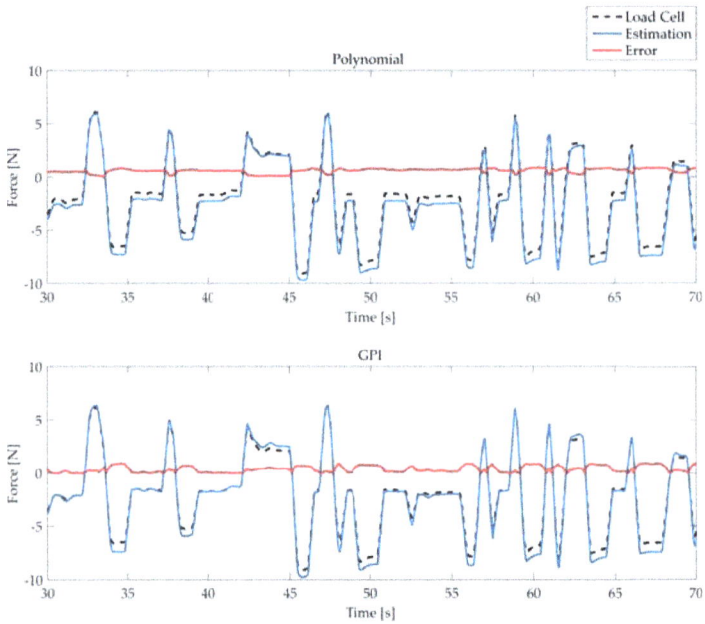

**Figure A6.** *Cont.*

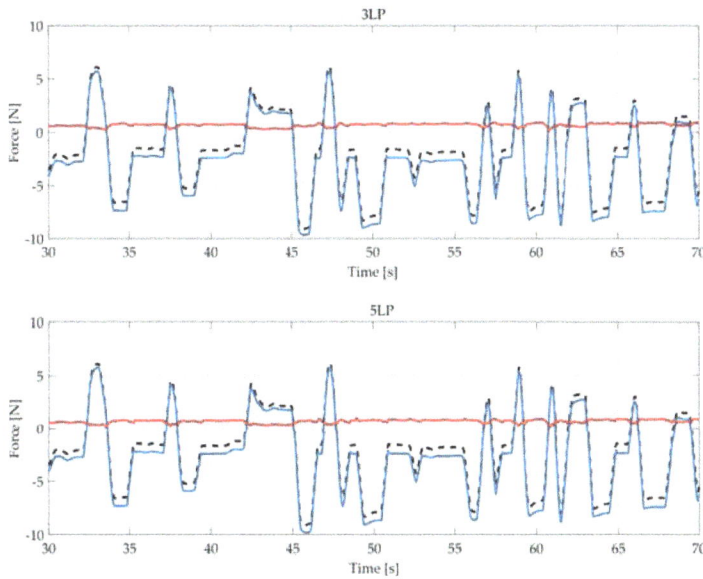

**Figure A6.** Detailed view of the time response shown in Figure 13 for each model.

## References

1. AIDE: Adaptive Multimodal Interfaces to Assist Disabled People in Daily Activities (GA 645322). Available online: http://cordis.europa.eu/project/rcn/194307_en.html (accessed on 10 January 2018).
2. Diez, J.A.; Blanco, A.; Catalan, J.M.; Bertomeu-Motos, A.; Badesa F.J.; Garcia-Aracil, N. Mechanical Design of a Novel Hand Exoskeleton Driven by Linear Actuators. In *Advances in Intelligent Systems and Computing, Proceedings of the ROBOT 2017: Third Iberian Robotics Conference, Seville, Spain, 22–24 November 2017*; Ollero, A., Sanfeliu, A., Montano, L., Lau, N., Cardeira, C., Eds.; Springer: Cham, Switzerland, 2018; Volume 694.
3. Díez, J.A.; Catalán, J.M.; Lledó, L.D.; Badesa, F.J.; Garcia-Aracil, N. Multimodal robotic system for upper-limb rehabilitation in physical environment. *Adv. Mech. Eng.* **2016**, *8*, 1–8, doi:10.1177/1687814016670282.
4. Schabowsky, C.N.; Godfrey, S.B.; Holley, R.J.; Lum, P.S. Development and pilot testing of HEXORR: Hand EXOskeleton rehabilitation robot. *J. NeuroEng. Rehabil.* **2010**, *7*, 36.
5. Hyun, D.J. On the Dynamics and Control of a MedicalExoskeleton. Ph.D. Thesis, UC Berkeley, Berkeley, CA, USA, 2012.
6. Shields, B.L.; Main, J.A.; Peterson, S.W.; Strauss, A.M. An anthropomorphic hand exoskeleton to prevent astronaut hand fatigue during extravehicular activities. *IEEE Trans. Syst. Man Cybern. Part A Syst. Hum.* **1997**, *27*, 668–673.
7. Yahud, S.; Dokos, S.; Morley, J.W.; Lovell, N.H. Experimental validation of a tactile sensor model for a robotic hand. In Proceedings of the Annual International Conference of the IEEE on Engineering in Medicine and Biology Society (EMBC 2009), Minneapolis, MN, USA, 3–6 September 2009; pp. 2300–2303.
8. Carrozza, M.C.; Massa, B.; Micera, S.; Lazzarini, R.; Zecca, M.; Dario, P. The development of a novel prosthetic hand-ongoing research and preliminary results. *IEEE/ASME Trans. Mechatron.* **2002**, *7*, 108–114.
9. Fontana, M.; Marcheschi, S.; Salsedo, F.; Bergamasco, M. A Three-Axis Force Sensor for Dual Finger Haptic Interfaces. *Sensors* **2012**, *12*, 13598–13616.
10. Fontana, M.; Fabio, S.; Marcheschi, S.; Bergamasco, M. Haptic hand exoskeleton for precision grasp simulation. *J. Mech. Robot.* **2013**, *5*, 1–9.
11. Chiri, A.; Vitiello, N.; Giovacchini, F.; Roccella, S.; Vecchi, F.; Carrozza, M.C. Mechatronic design and characterization of the index finger module of a hand exoskeleton for post-stroke rehabilitation. *IEEE/ASME Trans. Mechatron.* **2012**, *17*, 884–894.

12. Rodriguez-Cheu, L.E.; Gonzalez, D.; Rodriguez, M. Result of a perceptual feedback of the grasping forces to prosthetic hand users. In Proceedings of the 2nd IEEE RAS & EMBS International Conference on Biomedical Robotics and Biomechatronics (BioRob 2008), Scottsdale, AZ, USA, 19–22 October 2008; pp. 901–906.

13. Baker, M.D.; McDonough, M.K.; McMullin, E.M.; Swift, M.; BuSha, B.F. Orthotic hand-assistive exoskeleton. In Proceedings of the 2011 IEEE 37th Annual Northeast Bioengineering Conference (NEBEC), Troy, NY, USA, 1–3 April 2011; pp. 1–2.

14. Wege, A.; Hommel, G. Development and control of a hand exoskeleton for rehabilitation of hand injuries. In Proceedings of the 2005 IEEE/RSJ International Conference on Intelligent Robots and Systems (IROS 2005), Edmonton, AB, Canada, 2–6 August 2005; pp. 3046–3051.

15. Johnston, D.; Zhang, P.; Hollerbach, J.; Jacobsen, S. A full tactile sensing suite for dextrous robot hands and use in contact force control. In Proceedings of the 1996 IEEE International Conference on Robotics and Automation, Minneapolis, MN, USA, 22–28 April 1996; Volume 4, pp. 3222–3227.

16. Pons, J.L.; Rocon, E.; Ceres, R.; Reynaerts, D.; Saro, B.; Levin, S.; Van Moorleghem, W. The MANUS-HAND dextrous robotics upper limb prosthesis: Mechanical and manipulation aspects. *Auton. Robot.* **2004**, *16*, 143–163.

17. Su, H.; Zervas, M.; Furlong, C.; Fischer, G.S. A miniature MRI-compatible fiber-optic force sensor utilizing fabry-perot interferometer. In *MEMS and Nanotechnology*; Springer: New York, NY, USA, 2011; Volume 4, pp. 131–136.

18. Xiao, L.; Yang, T.; Huo, B.; Zhao, X.; Han, J.; Xu, W. Impedance control of a robot needle with a fiber optic force sensor. In Proceedings of the IEEE 13th International Conference on Signal Processing (ICSP), Chengdu, China, 6–10 November 2016; pp. 1379–1383.

19. Fernandez, A.F.; Berghmans, F.; Brichard, B.; Mégret, P.; Decréton, M.; Blondel, M.; Delchambre, A. Multi-component force sensor based on multiplexed fibre Bragg grating strain sensors. *Meas. Sci. Technol.* **2001**, *12*, 810.

20. Park, Y.L.; Ryu, S.C.; Black, R.J.; Chau, K.K.; Moslehi, B.; Cutkosky, M.R. Exoskeletal force-sensing end-effectors with embedded optical fiber-Bragg-grating sensors. *IEEE Trans. Robot.* **2009**, *25*, 1319–1331.

21. Su, H.; Fischer, G.S. A 3-axis optical force/torque sensor for prostate needle placement in magnetic resonance imaging environments. In Proceedings of the IEEE International Conference on Technologies for Practical Robot Applications (TePRA 2009), Woburn, MA, USA, 9–10 November 2009; pp. 5–9.

22. Hara, M.; Matthey, G.; Yamamoto, A.; Chapuis, D.; Gassert, R.; Bleuler, H.; Higuchi, T. Development of a 2-DOF electrostatic haptic joystick for MRI/fMRI applications. In Proceedings of the IEEE International Conference on Robotics and Automation (ICRA'09), Kobe, Japan, 12–17 May 2009; pp. 1479–1484.

23. Hirose, S.; Yoneda, K. Development of optical six-axial force sensor and its signal calibration considering nonlinear interference. In Proceedings of the 1990 IEEE International Conference on Robotics and Automation, Cincinnati, OH, USA, 13–18 May 1990; pp. 46–53.

24. Jeong, S.H.; Lee, H.J.; Kim, K.R.; Kim, K.S. Design of a miniature force sensor based on photointerrupter for robotic hand. *Sens. Actuators A Phys.* **2018**, *269*, 444–453.

25. Palli, G.; Hosseini, M.; Melchiorri, C. A simple and easy-to-build optoelectronics force sensor based on light fork: Design comparison and experimental evaluation. *Sens. Actuators A Phys.* **2018**, *269*, 369–381.

26. Takahashi, N.; Tada, M.; Ueda, J.; Matsumoto, Y.; Ogasawara, T. An optical 6-axis force sensor for brain function analysis using fMRI. In Proceedings of the IEEE on Sensors, Toronto, ON, Canada, 22–24 October 2003; Volume 1, pp. 253–258.

27. Tada, M.; Kanade, T. An MR-compatible optical force sensor for human function modeling. In *Medical Image Computing and Computer-Assisted Intervention–MICCAI 2004*; Springer: Berlin/Heidelberg, Germany, 2004; pp. 129–136.

28. Tada, M.; Kanade, T. Design of an MR-compatible three-axis force sensor. In Proceedings of the 2005 IEEE/RSJ International Conference on Intelligent Robots and Systems (IROS 2005), Edmonton, AB, Canada, 2–6 August 2005; pp. 3505–3510.

29. Glassner, A.S. (Ed.) *An Introduction to Ray Tracing*; Elsevier: Amsterdam, The Netherlands, 1989.

30. Lanzotti, A.; Grasso, M.; Staiano, G.; Martorelli, M. The impact of process parameters on mechanical properties of parts fabricated in PLA with an open-source 3-D printer. *Rapid Prototyp. J.* **2015**, *21*, 604–617.

31. Díez, J.A.; Blanco, A.; Catalán, J.M.; Badesa, F.J.; Lledó, L.D.; Garcia-Aracil, N. Hand exoskeleton for rehabilitation therapies with integrated optical force sensor. *Adv. Mech. Eng.* **2018**, *10*, 1–11, doi:10.1177/1687814017753881..

32. Al Janaideh, M.; Rakheja, S.; Su, C.Y. An analytical generalized Prandtl–Ishlinskii model inversion for hysteresis compensation in micropositioning control. *IEEE/ASME Trans. Mechatron.* **2011**, *16*, 734–744.

33. Sayyaadi, H.; Zakerzadeh, M.R.; Zanjani, M.A.V. Accuracy evaluation of generalized Prandtl-Ishlinskii model in characterizing asymmetric saturated hysteresis nonlinearity behavior of shape memory alloy actuators. *Int. J. Res. Rev. Mechatron. Des. Simul.* **2011**, *1*, 59–68.

34. Zhang, J.; Merced, E.; Sepúlveda, N.; Tan, X. Modeling and inverse compensation of hysteresis in vanadium dioxide using an extended generalized Prandtl–Ishlinskii model. *Smart Mater. Struct.* **2014**, *23*, 125017.

35. Sánchez-Durán, J.A.; Oballe-Peinado, Ó.; Castellanos-Ramos, J.; Vidal-Verdú, F. Hysteresis correction of tactile sensor response with a generalized Prandtl–Ishlinskii model. *Microsyst. Technol.* **2012**, *18*, 1127–1138.

36. Liu, S.; Su, C.Y.; Li, Z. Robust adaptive inverse control of a class of nonlinear systems with Prandtl-Ishlinskii hysteresis model. *IEEE Trans. Autom. Control* **2014**, *59*, 2170–2175.

37. Maren, A.J.; Harston, C.T.; Pap, R.M. *Handbook of Neural Computing Applications*; Academic Press: Cambridge, MA, USA, 2014.

38. Islam, T.; Saha, H. Hysteresis compensation of a porous silicon relative humidity sensor using ANN technique. *Sens. Actuators B Chem.* **2006**, *114*, 334–343.

39. Lin, F.J.; Shieh, H.J.; Huang, P.K. Adaptive wavelet neural network control with hysteresis estimation for piezo-positioning mechanism. *IEEE Trans. Neural Netw.* **2006**, *17*, 432–444.

40. Wang, H.; Song, G. Innovative NARX recurrent neural network model for ultra-thin shape memory alloy wire. *Neurocomputing* **2014**, *134*, 289–295.

41. Zhou, M.; Wang, Y.; Xu, R.; Zhang, Q.; Zhu, D. Feed-forward control for magnetic shape memory alloy actuators based on the radial basis function neural network model. *J. Appl. Biomater. Funct. Mater.* **2017**, *15*, 25–30.

42. Murtagh, F. Multilayer perceptrons for classification and regression. *Neurocomputing* **1991**, *2*, 183–197.

43. Irie, B.; Miyake, S. Capabilities of three-layered perceptrons. In Proceedings of the IEEE International Conference on Neural Networks, San Diego, CA, USA, 24–27 July 1988; Volume 1, p. 218.

44. Yang, M.J.; Gu, G.Y.; Zhu, L.M. Parameter identification of the generalized Prandtl–Ishlinskii model for piezoelectric actuators using modified particle swarm optimization. *Sens. Actuators A Phys.* **2013**, *189*, 254–265.

*sensors*

MDPI

*Article*

# A COTS-Based Portable System to Conduct Accurate Substance Concentration Measurements

Juan Aznar-Poveda [1], Jose Antonio Lopez-Pastor [1], Antonio-Javier Garcia-Sanchez [1,*], Joan Garcia-Haro [1] and Toribio Fernández Otero [2]

[1]  Department of Information and Communication Technologies (TIC), Technical University of Cartagena, ETSIT, Campus Muralla del Mar, E-30202 Cartagena, Spain; juan.aznarp@gmail.com (J.A.-P.); joseantonio.lopez@upct.es (J.A.L.-P.); joang.haro@upct.es (J.G.-H.)
[2]  Center for Electrochemistry and Intelligent Materials (CEMI), Technical University of Cartagena, ETSII, Campus Alfonso XIII, E-30203 Cartagena, Spain; toribio.fotero@upct.es
*  Correspondence: antoniojavier.garcia@upct.es; Tel.: +34-968-326-538

Received: 22 December 2017; Accepted: 7 February 2018; Published: 10 February 2018

**Abstract:** Traditionally, electrochemical procedures aimed at determining substance concentrations have required a costly and cumbersome laboratory environment. Specialized equipment and personnel obtain precise results under complex and time-consuming settings. Innovative electrochemical-based sensors are emerging to alleviate this difficulty. However, they are generally scarce, proprietary hardware and/or software, and focused only on measuring a restricted range of substances. In this paper, we propose a portable, flexible, low-cost system, built from commercial off-the-shelf components and easily controlled, using open-source software. The system is completed with a wireless module, which enables the transmission of measurements to a remote database for their later processing. A well-known PGSTAT100 Autolab device is employed to validate the effectiveness of our proposal. To this end, we select ascorbic acid as the substance under consideration, evaluating the reliability figure and obtaining the calibration curves for both platforms. The final outcomes are shown to be feasible, accurate, and repeatable.

**Keywords:** screen-printed electrode; potentiostat; cyclic voltammetry

## 1. Introduction

Advances in accurate substance concentration measurements depend on the progress of new devices/systems, which apply emerging technologies. Traditional procedures and methods involve the use of complex laboratory equipment and exhaustive control of multiple external variables, as with high performance liquid chromatography (HPLC) [1] or spectrophotometry [2], among others. They take longer and require more effort to develop specific treatments and processes aimed at obtaining appropriate samples for further analysis. Despite being the most popular methods in areas as diverse as water quality [3], food safety [4], medical diagnosis [5], or drug control [6], their high cost and the need for specialized staff and proprietary licenses restrict their use to big companies, public institutions, or other similar organizations. Furthermore, the size of the equipment limits portability, making it, therefore, advisable to install these systems in specialized laboratories.

Recent developments in sensors based on electrochemical techniques [7–10] and their associated electronics have favorably impacted on this research field, since they offer appreciable advantages over the laboratory procedures. Many innovative biosensor types, such as nanowire arrays, FET devices or three-electrode systems are emerging. The latter are the most widely used for their simplicity. Three-electrode sensors become key components of the so-called points-of-care (henceforth, PoC). A PoC is a small, portable device, developed principally by the pharmaceutical industry, capable of determining very low levels of given substances (e.g., proteins) such as glucose [11]. Its purpose is to avoid inconvenient displacements to perform blood tests or other analyses. In addition, the latest advances also allow the delivering of test results through a mobile network [12]. Other areas, such as food quality control, are also interested in similar devices. So, in [13], a miniature PoC-like instrument that includes hand-made electrodes of graphite, accurately measures, "in situ" (e.g., in a crop) concentrations of substances such as vitamins, acids or carbohydrates present in fruit, vegetables, cereals and beverages, and can indicate whether the product fulfills the expected quality. In other cases, the aforementioned electrodes are printed on ceramics, and are denoted screen-printed electrodes (hereinafter referred to as SPE). These electrodes are a well-suited solution due to their low-cost, easy production, use, maintenance, disposability and handling. Diverse companies such as Metrohm-Autolab B.V. [14], DropSens [15] or PalmSens [16], provide both SPE and the measuring instrument, called potentiostat, within their respective product catalogs.

Nowadays, these systems are proprietary. They are only able to conduct tests for a small number of specific substances, and are not affordable for the general public due to their cost. A more flexible solution is therefore necessary, to fully satisfy a higher number of target customers, while keeping costs low and avoiding proprietary designs. In this sense, there are recent projects intended to design open-source and highly accurate potentiostats, though they are conceived for non-portable laboratory environments [17,18].

To overcome this, diverse portable and open-source potentiostat prototypes are being developed. So, in the education field, we find the proposal in [19]—a straightforward device mainly focused on teaching students. It lacks wireless communication interfaces and a formal validation studio is not provided. In [20,21], multimodal electrochemical methods are implemented for different applications using complex and custom-made sensors; such as improving the specificity of ethanol detection in multiple analyte and low concentration solutions. Furthermore, in these works, users must acquire specific skills, in order to operate the equipment. From the design of the silicon board to the acquisition and welding of the required components the entire electronics must be implemented, which clearly reduces their reproducibility. To face these drawbacks, commercial-off-the-shelf (COTS) components should be used, offering to developer's abilities in their integration and mounting. This strategy is followed for the electrochemical sensing of cortisol [22,23]. However, the sensor requires an insightful understanding of electrochemistry and materials to be manufactured. Finally, they miss the portability feature because a PC connection is required.

In this paper, we contribute with the design and development of a portable, low-cost, customizable sensor platform, based on a potentiostat, and entirely made up of COTS components, including the sensor, in order to facilitate the reproducibility to end-users. Additionally, our platform includes a communication module to allow remote data acquisition and to facilitate the information reading for users and specialists located outside of the measurement site. The solution proposed here has been exhaustively validated with a well-known professional potentiostat [24]. We employ ascorbic acid as the substance under consideration, due to its extensive, well-documented record [25], and it is typically used for checking and verifying traditional methods and procedures. Our results reveal the appropriate operation of our system in terms of applicability, usability, and reliability.

The rest of this paper is organized as follows. Section 2 introduces the methodology employed in this research, and describes the main parts of the proposed device regarding system architecture and software structure. Then, the results obtained are presented in Section 3, as well as the validation of the device designed, by means of its calibration line and a repeatability study. Section 3 also explores the reusability of the screen-printed electrodes. Section 4 elaborates on the results discussion. Finally, Section 5 summarizes major conclusions.

## 2. Materials and Methods

### 2.1. Reagents and Validation Instruments

All chemicals employed were of analytical grade and without any further processing. Sample solutions were prepared with 99% pure, solid state L-ascorbic acid ($C_6H_8O_6$; 250 g), 95%–98% liquid sulfuric acid ($H_2SO_4$; 1000 mL), and MilliQ water (25 °C, $\Omega$ = 18.2 M$\Omega$/cm). A set of PGSTAT100 Potentiostat/Galvanostat, from Metrohm-Autolab B.V. controlled by NOVA 1.11 electrochemical software [14] was used to validate the system proposed, due to its recognized robustness and reliability, compared to other commercial potentiostats.

### 2.2. Sample Preparation

Environmental conditions such as light, temperature, and humidity quickly degrade the ascorbic acid (AA), causing unwanted, uncontrolled, and irreversible processes. The resulting substance is an oxided form of AA, denoted as dehydroascorbic acid (DHA). Considering this chemical reaction, there are two basic approaches to quantify the AA into a substance, namely: (i) ignoring the possible presence of DHA, which negatively impacts on the measurement accuracy; and (ii) calculating the AA concentration as the sum of both AA and DHA, thus employing the HPLC technique. Concerning the latter, a specific method was proposed and developed in [26], where AA is deliberately oxidized to a more stable DHA substance using a sulfuric acid solution. Thereby, the DHA measurement obtained will correspond to the AA concentration. To this end, a set of AA aqueous solutions was prepared, using 0.5 M sulfuric acid as a solvent and a fresh solution for every experiment. In particular, sixteen different AA solutions, ranging $2 \times 10^{-3}$–$1 \times 10^{-6}$ M were used as samples to study the functionality of our device.

### 2.3. System Architecture

Figure 1 shows the system architecture of the device, designed, and developed to accurately determine the AA concentration in a solution through COTS components. The core consists of an LMP91000EVM potentiostat [27] regulated by a Raspberry Pi 2 Model B microcontroller (Raspberry Pi Foundation, Cambridge, UK) [28]. A commercial SPE from DropSens [29] (ref. DRP-110) may be inserted through an SPE adapter from IORodeo, which is connected to a handmade 2-mm jack interface. The system is powered by a lithium battery through a power pack expansion board (3800 mAh 5 V/1.8 A). This battery is directly connected to the microcontroller and a 10.1-inch HDMI LCD touchscreen, which facilitates the device setting and reading by means of a specific graphic user interface (GUI). Finally, a Wi-Fi module (Wireless N300 Nano USB Adapter) from D-Link is responsible for remotely dispatching the results obtained, which will also be stored in a database installed in a Tomcat server.

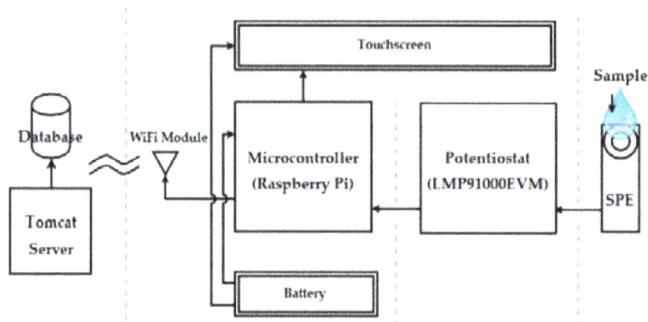

**Figure 1.** Hardware connections and data flowchart.

From a technological point of view, the process to determine the AA concentration through electrochemical methods based on an SPE is as follows. Firstly, a 50 µL drop of AA is poured over the carbon electrode surface to generate a potential difference between electrodes. It is important to cover the 3 carbon electrodes completely, and advisable to employ a suitable micropipette (in our case, a micropipette ranging 1–100 µL). Two electrodes are involved in this process: (i) the working electrode (WE), in which the redox reaction of the sample containing AA takes place, resulting in the electric current of interest calculating the concentration value; (ii) The counter electrode (CE) allows this current flow owing to a potential sweep applied to the WE related to the reference electrode. The applied potential originates the AA oxidation on the WE and the concomitant current flow between the WE and the CE as the experiment progresses. In the used SPE, from DropSens, both WE and CE are printed with carbon. The mentioned reference electrode (RE) is a non-polarized electrode keeping a constant potential ($E_h$) and having a high entry impedance hindering any significant current flow through. In our case, RE is printed using silver in order to obtain these properties. It is required to control the potential applied to the WE. Secondly, the microcontroller forwards specific commands to the potentiostat to stimulate and control the analyte reaction that occurs in the WE of the SPE surface. The microcontroller also amplifies the generated current across the WE, to be processed later for the test. In our particular case, the stimulation process comprises a potential cycle, that is, a scan/sweep in electric potential (both direct and inverse) between two voltage limits. The described process is known as cyclic voltammetry (CV, henceforth), which will ease the relevant validation due to its popularity and implementation in the most of commercial potentiostats. Furthermore, the CV traces provides valuable information about reduction and oxidation phases, and thus the reversibility and irreversibility of the process. As a final step, a precise AA concentration is eventually obtained through a calibration line, as will be explained in more detail below.

Concerning the core of the device, and considering that open-source, low-cost, and high computing performance features should be satisfied, a Raspberry Pi 2 model B has been selected because of its built-in microcontroller and peripherals. The same procedure has been applied to choose the potentiostat. The resulting unit was the LMP91000 chip evaluation module (LMP91000EVM), which includes a chip with the same name. The LMP91000 is a programmable analog front-end (AFE) potentiostat, configured to perform CV measurements with 3-lead biosensors, making use of SPEs and the solution under test. An analog-to-digital converter (ADC161S626, Texas Instruments, Dallas, TX, USA) has also been integrated into the evaluation module, enabling the digitalization of the output voltage. This module is pre-designed to be attached through a GPSI16 connector to an SPIO4 board, which, operating with the required software, allows practitioners and users to obtain the CV results. However, this option was discarded due to the cost of the SPIO4 and the inability to check the state of the connections. In this way, it was decided to study how the Raspberry Pi and LMP91000EVM could be interfaced, using some female–female jumper wires. For this purpose, we configured the LMP91000 chip through the $I^2C$ protocol, and received the digitalized ADC output by means of the SPI protocol. The end connection scheme of the proposed interface is specified in Table 1.

**Table 1.** Physical pin numbers used to interface Raspberry Pi and LMP91000EVM.

| Raspberry Pi | Connection Name | Protocol | LMP91000EVM |
|---|---|---|---|
| 19 | MOSI | SPI | 7 |
| 21 | MISO | SPI | 5 |
| 23 | SCLK | SPI | 3 |
| 24 | CE0 | SPI | 1 |
| 3 | SDA | $I^2C$ | 11 |
| 5 | SCL | $I^2C$ | 12 |
| 9, 14, 6, 39 * | GND * | - | 2, 4, 8, 10 * |
| 2 | 5V | - | 14 |
| 1 | 3.3V | - | 13 |

* Order is not relevant in ground pins. Pins not shown might be open.

## 2.4. Software Description

Figure 2 illustrates a complete diagram of the software structure. Commands and instructions are handled by the microcontroller through scripts that were written in Python programming language. The main script is denoted as cvgit.py, and is run each time its desktop shortcut is clicked, activating an intuitive and understandable GUI, programmed through Python's TkInter widgets package. Simultaneously, GPIO channels, handled by the RPi. GPIO module in Python, turn on a green LED, indicating an idle status. This *cvgit.py* file loads, on the one hand, the *var.py* script, which contains all the variables, register addresses, and configuration parameters needed to enable SPI and I²C communications, and, on the other hand, the *settings.py* file, which encapsulates functionalities of these protocols, such as the acquisition and processing of data from the ADC.

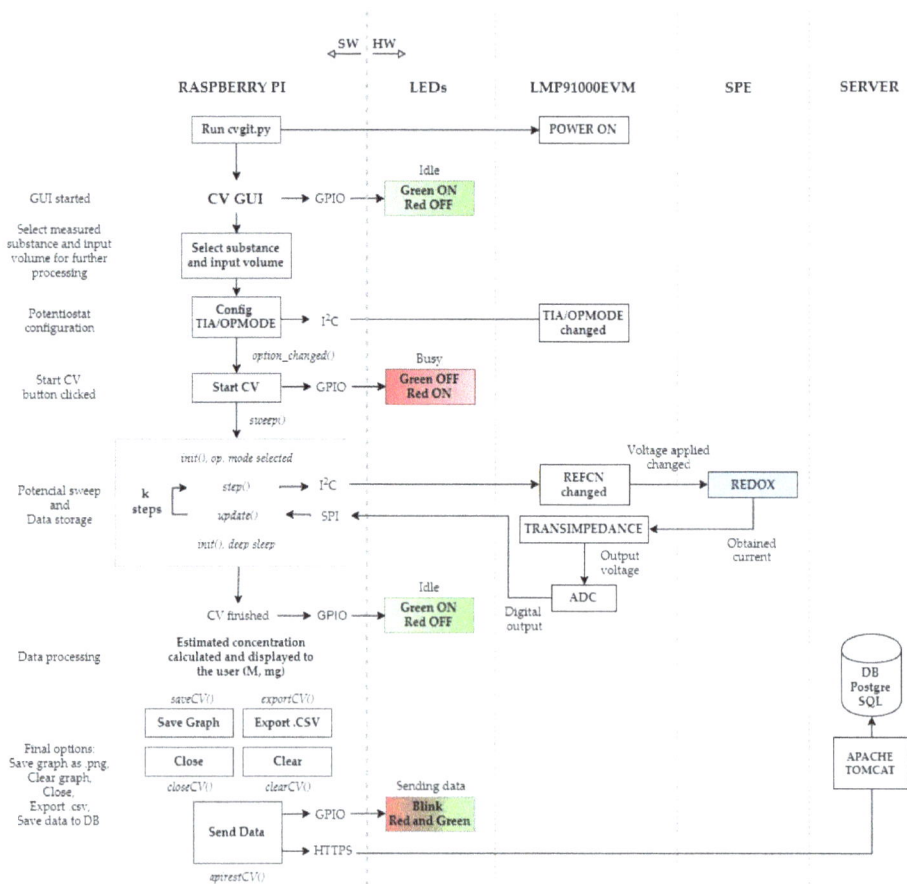

**Figure 2.** Functionality of hardware components and their associated software.

Following the logical flow of the CV procedure, the configuration of the LMP91000EVM by means of the I²C *smbus* library is a key aspect to consider. In Figure 3, we observe that the LMP91000 chip consists mainly of two operational amplifiers, configurable voltage dividers/resistances, and a temperature sensor.

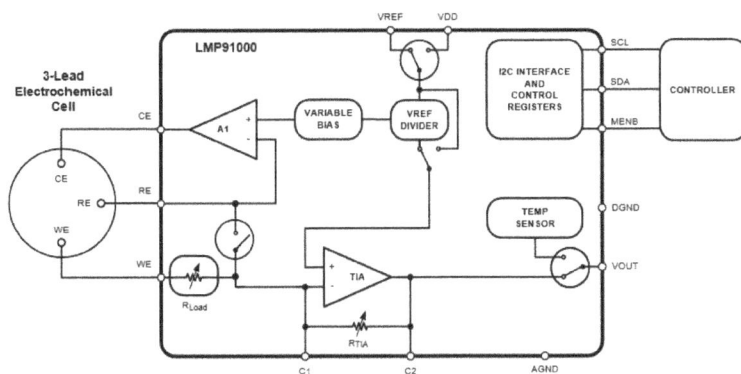

**Figure 3.** LMP91000 Simplified application schematic [30] (© Texas Instruments Incorporated).

In detail, the first amplifier, called A1, connects to the RE and CE electrodes. This amplifier stabilizes the redox reaction occurring in these electrodes, while the potential sweep is applied. The sweep is performed through the *variable bias*, which allows the configuration of the voltage, divided into steps ranging from 0% to 24% of the $V_{REF}$ (reference voltage). The LMP91000 datasheet [30] specifies that these steps are not regular, using steps of 1% until 2% of the $V_{REF}$ is achieved, and then steps of 2% until 24% of the $V_{REF}$, resulting in a total of 14 steps. Another issue to consider is the REFCN register, which is used to modify the *variable bias*, to tune the value of the voltage applied to the electrodes. This register also allows the setting up of the bias sign (positive or negative), the internal zero, that is, the voltage at the output of the LMP91000 chip ($V_{OUT}$) when no reaction occurs in the electrodes, and to select the reference source used, external (from another device), or internal. $R_{LOAD}$ is a fixed 10 $\Omega$ resistance connected to the working electrode. The second operational amplifier, called TIA, is responsible for converting the current obtained by the WE into voltage through a transimpedance $R_{TIA}$, which may be configured by the user. Both $R_{TIA}$ and $R_{LOAD}$ resistances are controlled by the TIACN register exclusively dedicated to this purpose. Different values of transimpedance allow for the adjustment of the output current range. For instance, if the electric current level is very small, due to a much-diluted solution or a very low responsive sensor, a high $R_{TIA}$ will be required to maintain the output voltage in the ADC range. On the contrary, a very high current level will result in a low $R_{TIA}$ value.

According to the mentioned datasheet, different operation modes are possible. However, we only employ the 3-lead amperometric cell, temperature sensor, standby, and deep-sleep modes. The first mode allows to provide a complete CV procedure for 3-electrode sensors. The temperature sensor is used to check the suitable operation of the ADC software libraries; if the correct temperature is registered, the correct digitalization process work is assumed. The standby mode is the state set among scans to reduce power consumption. On the other hand, the deep-sleep state occurs when the chip is not performing any electrochemical activity during a long-time period. Note that both $R_{TIA}$ and the operational modes may be changed by the end-user, via the developed GUI. In the event of successful modification, an acknowledgement message is displayed in the GUI console. Users can also configure other optional parameters such as (i) the desired title of the measurement, with which the program will save the .csv and .png files; (ii) the desired substance under test, in order to compute the molarity and weight through its corresponding calibration line (currently only AA calibration is available); and (iii) the volume required to perform the conversion from molarity to weight (mg). Once the GUI has been configured, to start a CV procedure, the end-user only needs to activate the start button of the available menu, switching the LED status of the device to busy (red). Then, the *sweep* function is executed. Besides the start and end configurations for the standby and deep-sleep states, this method incorporates two additional functionalities. On the one hand, the *sweep* function calls the *step* method, which, as described, modifies the REFCN register. On the other hand, it invokes the *update* function to acquire the digital output data from the ADC by means of the SPI protocol, which is managed by the *spidev* library. This data acquisition fulfills the ADC and LMP91000 specifications, in accordance with Equations (1) and (2) below,

$$V = ((V_{RAW} \cdot SPAN)/BR) + V_{REF} \text{ (V)}, \tag{1}$$

$$I = ((V - (V_{REF}/2))/R_{TIA}) \times 10^6 \text{ (µA)}, \tag{2}$$

where $V_{RAW}$ is the ADC raw read voltage, SPAN is the full-scale voltage range calculated from $(V_A - (V_{REF}/2^{16}))$, with $V_A$ as the single analog supply (5 V). BR is the maximum decimal value for an *n*-bit binary code, i.e., restricted to 16-bit resolution (as in our case), BR is $2^{16} - 1 = 65,635$. Finally, the $V_{REF}$ is set to 2.5 V to later perform the sweep of interest. To calculate the current, the Ohms law equation is applied, considering the resistance and voltage involved in the process. Both expressions, as well as their specific nomenclature, are taken as input parameters of the *update* function, which presents the results as a current/voltage plot on the GUI console, updating them as the CV sweep advances. To this end, the *update* function is supported by *matplotlib*, a plotting library for Python. The CV process conducts a complete sweep, ranging 0–0.6 V (24% of $V_{REF}$) in steps of 50 mV per second, that is, steps of 2% of the $V_{REF}$, as shown in Table 2.

For this purpose, the REFCN register points to an external source as voltage reference, and sets 50% of the $V_{REF}$ (1.25 V, since the $V_{REF}$ is fixed at 2.5 V) as the internal zero. Under this configuration, for example, a potential difference of 0.3 V between WE and RE is regulated by 12% of the $V_{REF}$ REFCN register value. Regarding the internal zero, when no electrochemical activity takes place, the output voltage is a fixed value of 1.25 V. In short, the step and update functions should be sequentially called on 25 occasions, 13 times from the direct sweep (0 V → 0.6 V), and 12 more times from the inverse sweep (0 V ← 0.6 V), thus avoiding measuring the 0.6 V point twice. A better resolution can be achieved by adding an external DAC able to provide multiple voltage values that acts as reference sources. This advanced configuration will be developed in the next versions of the device.

The CV procedure finishes when the sweep returns to zero. Then, functions such as save graph into a *.png* file (*saveCV*), save the data array into a *.csv* file (*exportCV*), clear graph (*clearCV*), close program (*closeCV*), or send data to a database will pop up in the GUI menu. It should also be noted that our device is able to remotely send information thanks to a Wi-Fi interface. In this sense, current and voltage data may be transmitted to a PC, which maintains a PostgreSQL database and an Apache Tomcat server to store and manage data results. Substance concentration under consideration may be calculated by interpolating the received current on a calibration line. This function was obtained by means of a linear regression of dataset $(x, y)$ comprised by the maximum registered current ($y$-axis) in each resulting CV of a set of solutions with a well-known molarity ($x$-axis), both in logarithmic scales. Thereby, when an unknown molarity solution of AA is measured, only the regression expression is necessary to compute it. This feature is included in the software developed for the proposed device and is displayed to the end-user in the GUI. In other words, once the CV is performed, the molarity 'M' (mol/L) measured in the sample being tested and the corresponding weight 'm' (mg) related to an input volume are directly displayed on the GUI console. Molarity is obtained from calibration line, as explained above. Then, the weight (mg) is calculated by means of the relation among the molecular weight W of the substance, the input volume V (1 L by default), and the obtained molarity M, as is showed in Equation (3).

$$m \text{ (mg)} = [M \text{ (mol/L)} \cdot V \text{ (L)} \cdot W \text{ (g/mol)}] / 1000 \tag{3}$$

Figure 4 shows the end prototype and its main hardware components. Both software and hardware documentation is available online [31] to facilitate the best project understanding and reproducibility to the audience.

*Sensors* **2018**, *18*, 539

**Table 2.** Procedure of the potential sweep performed. Percentage of the reference source applied, the equivalent voltage, and the binary code necessary to configure the REFCN register.

| % of $V_{REF}$ | 0 | 2 | 4 | 6 | 8 | 10 | 12 | 14 | 16 | 18 | 20 | 22 | 24 |
|---|---|---|---|---|---|---|---|---|---|---|---|---|---|
| Potential Step (V) | 0.00 | 0.05 | 0.10 | 0.15 | 0.20 | 0.25 | 0.30 | 0.35 | 0.40 | 0.45 | 0.50 | 0.55 | 0.60 |
| REFCN Register (Binary) | 10110000 | 10110010 | 10110011 | 10110100 | 10110101 | 10110110 | 10110111 | 10111000 | 10111001 | 10111010 | 10111011 | 10111100 | 10111101 |

**Figure 4.** Final potentiostat prototype: (**a**) front view (with the application running on the screen); (**b**) detail of an inserted commercial Dropsens DS110 SPE; (**c**) zoom of an achieved concentration value a $1 \times 10^{-3}$ M molarity; (**d**) rear view, focusing on the SPE plug and three electrode connections from the LMP91000EVM unit; and (**e**) general rear view of the entire device in an idle state (green LED turned on) with all components labeled.

## 3. Results

### 3.1. Calibration Line, Range and Reliability

A comparative study with a PGSTAT100 potentiostat was conducted to validate the proper operation of our device. The same adapter as that of the proposed device [32] has been used to connect the commercial screen-printed electrodes (CE, WE, and RE) with the corresponding PGSTAT100 jacks, by means of three two-millimeter jack clamps. The distinctive shape of the resulting CV curve should be similar in each sweep for both pieces of equipment. To this end, several CV were obtained for different AA concentrations, resulting in the plots depicted in Figure 5. As can be observed, the higher the molarity, the greater the registered current. From these results, we selected the maximum current value (peak) of each AA solution (commonly five to six values are enough, as indicated in chemical analysis basics [33]) to obtain the calibration curve, as shown in Figure 6. In particular, sixteen AA solutions were prepared. The most highly diluted among them presented a high signal to noise ratio. This is shown by a very unstable output current, due to many undesired peaks. In contrast, the most concentrated solutions led to ADC saturation. Therefore, only nine solutions were considered to have the appropriate behavior, that is, noise (too low molarities) or saturation (too high molarities) phenomena were not detected. These correctly responding solutions were used to further extract the peak current and draw the calibration curve. To increase the visibility of lower concentrations and currents, we decided to represent the results in a logarithmic scale. Following these premises, our device obeys a linear log–log relationship, according to the following expression:

$$\log_{10}(I(\mu A)) = 0.90 \times \log_{10}(M(mol/L)) + 4.16, \tag{4}$$

with an $R^2 = 0.988$ regression coefficient. The same procedure was replicated in order to calibrate the PGSTAT100 unit; deriving Equation (5):

$$\log_{10}(I(\mu A)) = 0.93 \times \log_{10}(M(mol/L)) + 4.11, \tag{5}$$

with, in this case, the regression coefficient as $R^2 = 0.989$. This means that our device operates on a wide concentration range, $10^{-6}$–$10^{-2}$ M. Despite the clear parallelism between both calibration lines, the reliability of our device was also calculated. For this purpose, the sensitivity figure was compared, which related the slope of both lines in terms of the coefficient ($(0.90/0.92) \times 100$). The result shows a reliability greater than 97%.

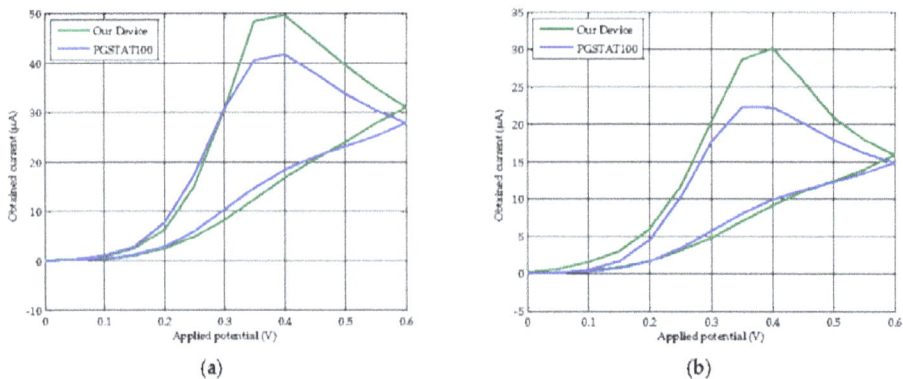

(a)

(b)

**Figure 5.** *Cont.*

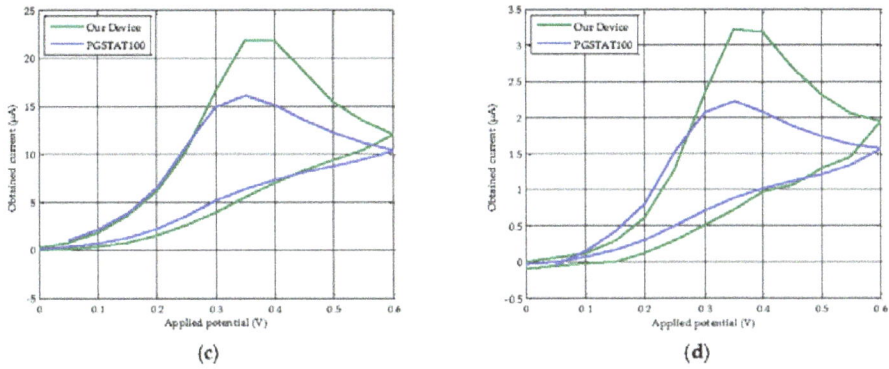

(c)  (d)

**Figure 5.** CV results comparative using commercial screen-printed electrodes (CE, WE and RE) between our device and PGSTAT100 for AA solutions of different molarity (a) $2 \times 10^{-3}$ M; (b) $1 \times 10^{-3}$ M; (c) $6 \times 10^{-4}$ M; (d) $1 \times 10^{-4}$ M.

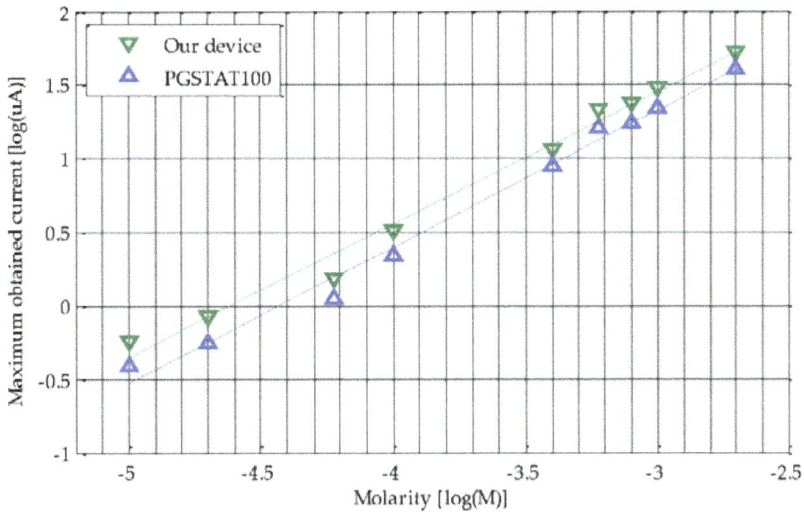

**Figure 6.** Calibration comparison between our device and the PGSTAT100 potentiostat.

## 3.2. Repeatability

Another important positive feature which characterizes and validates our device is repeatability. This term refers to how much a measurement changes in relation to a new one, repeating the test using the same type of SPE sensors (a new one for each measurement), solution samples, and, to the extent possible, under the same environmental conditions (temperature, humidity, pressure, etc.). It is well known that the set PGSTAT100 potentiostat and DS110 sensors ensure high repeatability (provided that appropriate storage, maintenance, and use are guaranteed). Therefore, our efforts were addressed to obtaining results as close as possible to this set. In practice, $2 \times 10^{-3}$ M and $6 \times 10^{-5}$ M solutions of AA were prepared and further measured with ten different DS110 sensors, each. These concentrations were selected to evaluate the repeatability at the ends of the calibration line, since intermediate values should also be reasonably compliant. Figure 7 shows ten superimposed CV curves measured for both devices. As expected, the current average value of our device is higher than the one measured with the

PGSTAT100 potentiostat, with these data in accordance with the CV curves shown in Figure 5, above. Regarding our device, dispersion in respect to the average denoted as the relative standard deviation (RSD) is 4.60% and 4.27% for the electric current peaks corresponding to the CV curves for $2 \times 10^{-3}$ M and $6 \times 10^{-5}$ M respectively. On the other hand, RSD values of 1.81% and 1.20% were obtained with the PGSTAT100 equipment for $2 \times 10^{-3}$ M and $6 \times 10^{-5}$ M respectively. More details on the average and RSD values of the maximum electric currents are indicated in Table 3.

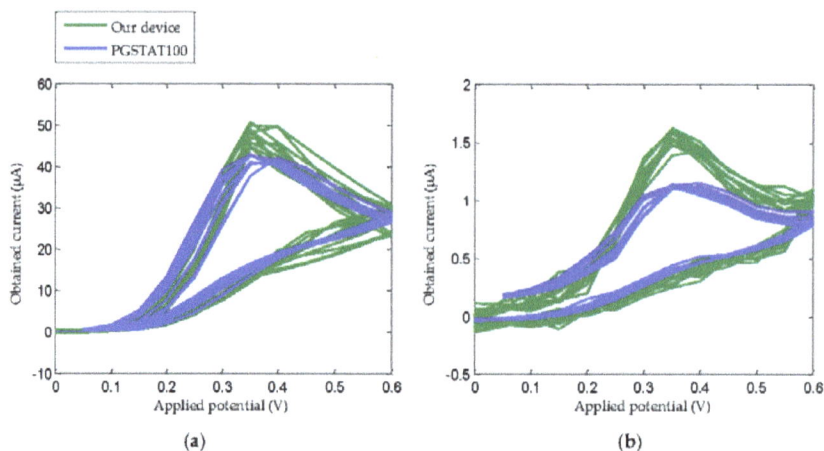

**Figure 7.** Device repeatability curves for ten samples of solutions of AA (**a**) $2 \times 10^{-3}$ M and (**b**) $6 \times 10^{-5}$ M.

**Table 3.** Repeatability values between PGSTAT100 and our device.

| Molarity (M) | PGSTAT100 x̄ (μA) ± RSD (%) | Our Device x̄ (μA) ± RSD (%) |
|---|---|---|
| $2 \times 10^{-3}$ | 41.99 μA ± 1.81% | 47.51 μA ± 4.60% |
| $6 \times 10^{-5}$ | 1.12 μA ± 1.20% | 1.55 μA ± 4.26% |

*3.3. Sensor Reusability*

Reusability is becoming an increasingly important metric in this specific scientific area, in which SPE sensors are consumables that are quickly used and discarded in each test, thus increasing the cost of the experiment. So, it is highly advisable to perform a reusability study, especially for applications where, on the one hand, a reduction of the cost factor is mandatory, and, on the other hand, commercial screen-printed carbon electrodes are subject to biochemical degradation when using them more than once. To this end, a battery of electric current measurements was carried out, operating a PGSTAT100 and implementing a straightforward procedure: the same SPE (DS110) sensor is immediately washed after the CV with MilliQ water, and its surface dried with a clean disposable paper towel to straightaway take a new measurement, extracting a sample of the same solution of AA (in this case, a $2 \times 10^{-3}$ M solution). This process was repeated five times to estimate the sensor behavior after each additional re-use. To guarantee the reliability of the experiment, three different SPEs were employed, obtaining, for each use, the average of the three measurements, together with the deviation of this average with respect to the value obtained for the first use. These current outcomes are shown both in Table 4 and in Figure 8, and discussed in the next section.

**Table 4.** Reusability study of DS110 SPE.

| Uses | Max. Current (µA) | Deviation in Respect to the 1st Use (µA) |
|------|-------------------|------------------------------------------|
| 1st  | 40.02             | 0                                        |
| 2nd  | 39.29             | 0.73                                     |
| 3rd  | 38.93             | 1.09                                     |
| 4th  | 38.14             | 1.88                                     |
| 5th  | 40.62             | 0.60                                     |

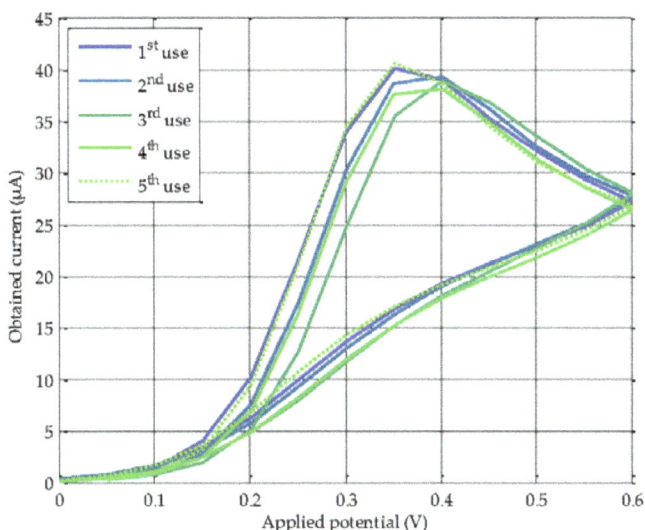

**Figure 8.** Repeatability study performed by means of a PGSTAT100. CV average for five consecutive uses and three different DS110 screen printed carbon electrodes.

## 4. Discussion

As can be observed in Figure 5, most of the electric current values measured with Autolab equipment match the ones obtained by our solution, except in the 0.4–0.6 V range of the direct sweep, where a significant difference is found. This deviation is basically produced by the differences in the electronic components employed in their respective developments. In particular, the transimpedance amplifier has a limited number of internal gain resistors. Under these circumstances, in comparison with the PGSTAT100, a lower adjustment to the entry voltage of the analog-to-digital converter (ADC) is reached. These particular points around 0.4–0.6 V coincide with those steps of the potential difference sweep where a higher current is generated. For instance, for a $1 \times 10^{-3}$ M solution CV, the PGSTAT100 registers a maximum current of 22.24 µA, while our device measures 30.20 µA. A priori, this deviation is a drawback in terms of accuracy. However, analyzing the calibration curves (see Figure 6), the results reveal that the maximum difference of current between the PGSTAT100 outcomes and our potentiostat development for each CV remains constant. An average difference of 28.17% between the maximum obtained currents in PGSTAT100 and in our device, is experienced for the entire studied range. Another fact worth noting when operating with electrochemical-based sensors, is the slightly growing variability in low concentration solutions, in comparison with the higher ones for the CV conducted in the same way (Figures 5 and 7). In any case, an optimal fitting with a regression coefficient higher than the 0.95 standard is achieved. This value, together with a reliability of around 97% between calibration plots, guarantees the excellent proportionality between both platforms, which in turn, becomes a "quasi" constant offset. Therefore, from these parallel calibration lines, the following two facts can be inferred:

1) Our calibration curve is thoroughly valid for interpolating values different from the proposed ones in this work. That is, if the maximum current measured is 0.5 µA, a molarity of $1.13 \times 10^{-4}$ M is reached, as shown in Figure 6.

2) The current peaks obtained for the PGSTAT100 and our device can be introduced in Equations (3) and (4) to derive the same molarity.

To sum up, the results achieved in this work show that our potentiostat implementation is comparable to the costly Autolab machine. This affirmation is further verified when analyzing the repeatability results. As could be expected, our device has a greater RSD than the PGSTAT100 potentiostat. However, these values differ by less than 5% (µA), which translates into good performance by our device.

Some interesting considerations can be also drawn from the reusability study (see Figure 8). Taking into consideration the calibration plot for our solution, modifying the value of the generated current in 1 µA involves a change of $2.45 \times 10^{-4}$ M in the molarity of the measured substance. Therefore, depending on the application under study and target sensitivity, a single DS110 SPE can be reused (four times for an AA solution measurement), which results in an additional reduction in consumable cost. For instance, three uses of the same sensor lead, approximately to an average deviation (error) with respect to the first use of 1.88 µA, which in molarity means: $1.88 \text{ µA} \times 2.45 \times 10^{-4}$ M/ µA $= 4.61 \times 10^{-4}$ M. This value can be negligible or significant, depending on the specific application and substance to be measured, since it will have an impact on the SPE modification, environmental conditions, required accuracy, etc. From the fourth and subsequent re-uses, the registered current increases, giving rise to undesirable effects, such as carbon surface distortion in the SPE, therefore, jeopardizing reliability.

## 5. Conclusions

The main contribution of this paper is the design and development of a portable, electrochemical-based solution to accurately measure the AA concentration outside of the usual laboratory environment. Therefore, our proposal is suitable for outdoor measurement campaigns, enabling new perspectives and innovative applications in yet unexplored fields (e.g., to know the level of AA concentration in fruit and vegetables, directly in the field). The system designed satisfies the following three aspects: (i) it is disposable and cheap, and unmodified general-purpose screen carbon printed electrodes may be employed as sensors; (ii) it is made from commercial off-the-shelf (COTS) components, which can be easily acquired and integrated, providing portability and low-cost; and (iii) it incorporates communications to transmit acquired data to remote ends. In addition to the hardware system integration, robust software to control the entire device and the sensor measurements has also been developed. Raw acquired data are further processed to extract useful information that is presented to the end-user. For this purpose, specific GUI-controlled software is programmed as a means of providing seamless interaction between the device and the end-user (e.g., molarity in milligrams and number of moles are directly calculated from the CV). Finally, the device is able to connect to a remote database to store all the acquired information, ready for future queries and for automated big data analysis.

A well-known, standard, but costly, laboratory potentiostat, such as the PGSTAT100, has been employed to validate the functionality of our device. To this end, a thorough performance comparison between both platforms has been accomplished. Results obtained reveal the positive performance of our proposal in terms of CV curves and repeatability. As an additional contribution, a reusability study was carried out, indicating a potential cost reduction in consumable supplies, if several consecutive measurements are allowed for the same screen carbon printed electrode. However, this will finally depend on the specific application and required accuracy. At this moment, only AA was considered as a test substance, our purpose being to broaden the scope in the future, increasing the number of target substances. The CV approach presented here is observed as a solid base to analyze other substances. This will only require an adequate SPE sensor type for the new substance under consideration, and minor changes to the device (to improve CV dynamic modification of the settings and parameters and extend the system to additional amperometric techniques), highlighting the versatility of the system proposed.

**Acknowledgments:** This research has been supported by the project AIM, ref. TEC2016-76465-C2-1-R (AEI/FEDER, UE). The authors would also like to express their gratitude to A. Caparrós for her assistance and wise advice.

**Author Contributions:** J.A.L.-P., A.-J.G.-S. and J.G.-H. conceived the device concept. J.A.-P. carried out the hardware and software implementation. J.A.L. and J.A.-P. designed and performed the experiments. T.F.O. contributed in the rigorousness of the electrochemical tests conducted. Finally, all authors analyzed the results and contributed to write the paper.

**Conflicts of Interest:** The authors declare no conflict of interest.

## References

1. Lykkesfeldt, J. Determination of Ascorbic Acid and Dehydroascorbic Acid in Biological Samples by High-Performance Liquid Chromatography Using Subtraction Methods: Reliable Reduction with Tris[2-carboxyethyl]phosphine Hydrochloride. *Anal. Biochem.* **2000**, *282*, 89–93. [CrossRef] [PubMed]
2. Measuring Protein Concentration through Absorption Spectrophotometry. Available online: http://w3.marietta.edu/~spilatrs/biol309/labexercises/Spectrophotometry.png (accessed on 25 October 2017).
3. Albert, D.B.; Martens, C.S. Determination of low-molecular-weight organic acid concentrations in seawater and pore-water samples via HPLC. *Mar. Chem.* **1997**, *56*, 27–37. [CrossRef]
4. Conte, L.S.; Moret, S.; Purcaro, G. HPLC in food analysis. *Chromatogr. Sci. Ser.* **2011**, *101*, 561–660. [CrossRef]
5. Yang, J.; Xu, G.; Zheng, Y.; Kong, H.; Pang, T.; Lv, S.; Yang, Q. Diagnosis of liver cancer using HPLC-based metabonomics avoiding false-positive result from hepatitis and hepatocirrhosis diseases. *J. Chromatogr. B Anal. Technol. Biomed. Life Sci.* **2004**, *813*, 59–65. [CrossRef] [PubMed]
6. Floriani, G.; Gasparetto, J.C.; Pontarolo, R.; Gonçalves, A.G. Development and validation of an HPLC-DAD method for simultaneous determination of cocaine, benzoic acid, benzoylecgonine and the main adulterants found in products based on cocaine. *Forensic Sci. Int.* **2014**, *235*, 32–39. [CrossRef] [PubMed]
7. Kang, X.; Wang, J.; Wu, H.; Liu, J.; Aksay, I.A.; Lin, Y. A graphene-based electrochemical sensor for sensitive detection of paracetamol. *Talanta* **2010**, *81*, 754–759. [CrossRef] [PubMed]
8. Lazerges, M.; Bedioui, F. Analysis of the evolution of the detection limits of electrochemical DNA biosensors. *Anal. Bioanal. Chem.* **2013**, *405*, 3705–3714. [CrossRef] [PubMed]
9. Lautner, G.; Gyurcsányi, R.E. Electrochemical Detection of miRNAs. *Electroanalysis* **2014**, *26*, 1224–1235. [CrossRef]
10. Hsu, P.F.; Ciou, W.L.; Chen, P.Y. Voltammetric study of polyviologen and the application of polyviologen-modified glassy carbon electrode in amperometric detection of vitamin C. *J. Appl. Electrochem.* **2008**, *38*, 1285–1292. [CrossRef]
11. Hasslacher, C.; Kulozik, F.; Platten, I. Accuracy of self monitoring blood glucose systems in a clinical setting: Application of new planned ISO-Standards. *Clin. Lab.* **2013**, *59*, 727–733. [CrossRef] [PubMed]
12. Nemiroski, A.; Christodouleas, D.C.; Hennek, J.W.; Kumar, A.A.; Maxwell, E.J.; Fernandez-Abedul, M.T.; Whitesides, G.M. Universal mobile electrochemical detector designed for use in resource-limited applications. *Proc. Natl. Acad. Sci. USA* **2014**, *111*, 11984–11989. [CrossRef] [PubMed]
13. Barberis, A.; Fadda, A.; Schirra, M.; Bazzu, G.; Serra, P.A. Detection of postharvest changes of ascorbic acid in fresh-cut melon, kiwi, and pineapple, by using a low cost telemetric system. *Food Chem.* **2012**, *135*, 1555–1562. [CrossRef] [PubMed]
14. Metrohm-Autolab. Available online: www.metrohm-autolab.com/ (accessed on 25 October 2017).
15. DropSens. Available online: http://www.dropsens.com/en/home.html (accessed on 8 February 2018).
16. PalmSens. Available online: https://www.palmsens.com/ (accessed on 22 December 2017).
17. Dryden, M.D.M.; Wheeler, A.R. DStat: A versatile, open-source potentiostat for electroanalysis and integration. *PLoS ONE* **2015**, *10*, e0140349. [CrossRef] [PubMed]

18. Meloni, G.N. Building a microcontroller based potentiostat: A inexpensive and versatile platform for teaching electrochemistry and instrumentation. *J. Chem. Educ.* **2016**, *93*, 1320–1322. [CrossRef]

19. Rowe, A.A.; Bonham, A.J.; White, R.J.; Zimmer, M.P.; Yadgar, R.J.; Hobza, T.M.; Honea, J.W.; Ben-Yaacov, I.; Plaxco, K.W. CheapStat: An Open-Source, "Do-It-Yourself" Potentiostat for Analytical and Educational Applications. *PLoS ONE* **2011**, *6*, e23783. [CrossRef] [PubMed]

20. Jalal, A.H.; Umasankar, Y.; Gonzalez, P.J.; Alfonso, A.; Bhansali, S. Multimodal technique to eliminate humidity interference for specific detection of ethanol. *Biosens. Bioelectron.* **2017**, *87*, 522–530. [CrossRef] [PubMed]

21. Umasankar, Y.; Jalal, A.H.; Gonzalez, P.J.; Chowdhury, M.; Alfonso, A.; Bhansali, S. Wearable alcohol monitoring device with auto-calibration ability for high chemical specificity. In Proceedings of the BSN 2016—13th Annual Body Sensor Networks Conference, San Francisco, CA, USA, 14–17 June 2016; pp. 353–358.

22. Cruz, A.F.D.; Norena, N.; Kaushik, A.; Bhansali, S. A low-cost miniaturized potentiostat for point-of-care diagnosis. *Biosens. Bioelectron.* **2014**, *62*, 249–254. [CrossRef] [PubMed]

23. Kaushik, A.; Yndart, A.; Jayant, R.D.; Sagar, V.; Atluri, V.; Bhansali, S.; Nair, M. Electrochemical sensing method for point-of-care cortisol detection in human immunodeficiency virus-infected patients. *Int. J. Nanomed.* **2015**, *10*, 677–685. [CrossRef]

24. Makiewicz, P.; Matias, D.; Jaskuła, M.; Biegun, M.; Penkala, K.; Mijowska, E.; El Fray, M.; Podolski, J. Accuracy analysis of measurements in electrochemical biosensing. In *IFMBE Proceedings*; Springer: Singapore, 2016; Volume 55, pp. 370–373.

25. Pisoschi, A.M.; Pop, A.; Serban, A.I.; Fafaneata, C. Electrochemical methods for ascorbic acid determination. *Electrochim. Acta* **2014**, *121*, 443–460. [CrossRef]

26. Deutsch, M.J.; Weeks, C.E. Micro-fluorometric assay for vitamin C. *J. Assoc. Off. Agric. Chem.* **1965**, *48*, 1248–1256.

27. Texas Instruments. Available online: http://www.ti.com/tool/LMP91000EVM (accessed on 5 December 2017).

28. Raspberry Pi 2 Model B. Available online: https://www.raspberrypi.org/products/raspberry-pi-2-model-b/ (accessed on 9 November 2017).

29. Dropsens DS110 SPE. Available online: http://www.dropsens.com/en/pdfs_productos/new_brochures/110-c110.png (accessed on 8 February 2018).

30. LMP91000 Sensor AFE System: Configurable AFE Potentiostat for Low-Power Chemical Sensing Applications. Available online: http://www.ti.com/lit/ds/symlink/lmp91000.png (accessed on 8 February 2018).

31. J Aznar Poveda Software Repository—OPEN SOURCE CVGIT. Available online: https://github.com/GITUPCT/CVGIT.git (accessed on 16 December 2017).

32. IO Rodeo SPE Adapter. Available online: https://iorodeo.com/collections/cheapstat-open-source-potentiostat/products/cheapstat-screen-printed-electrode-adapters?variant=12478343814 (accessed on 16 December 2017).

33. Harris, D.C. *Quantitative Chemical Analysis*; WH Freeman: New York, NY, USA, 2007; Volume 42, ISBN 0716770415.

*sensors*

MDPI

Article

# Analysis of the High-Frequency Content in Human QRS Complexes by the Continuous Wavelet Transform: An Automatized Analysis for the Prediction of Sudden Cardiac Death

Daniel García Iglesias [1,2], Nieves Roqueñi Gutiérrez [3], Francisco Javier De Cos [3] and David Calvo [1,2,3,*]

[1]    Arrhythmia Unit, Hospital Universitario Central de Asturias, 33011 Oviedo, Spain;
      danigarciaiglesias@gmail.com
[2]    Instituto de Investigación Sanitaria del Principado de Asturias, 33011 Oviedo, Spain
[3]    Grupo para la Modelización Matemática Avanzada (MOMA), Universidad de Oviedo, 33004 Oviedo, Spain;
      nievesr@uniovi.es (N.R.G.); fjcos@uniovi.es (F.J.D.C.)
*    Correspondence: dcalvo307@secardiologia.es; Tel.: +34-985108000; Fax: +34-985274688

Received: 11 January 2018; Accepted: 7 February 2018; Published: 12 February 2018

**Abstract:** Background: Fragmentation and delayed potentials in the QRS signal of patients have been postulated as risk markers for Sudden Cardiac Death (SCD). The analysis of the high-frequency spectral content may be useful for quantification. Methods: Forty-two consecutive patients with prior history of SCD or malignant arrhythmias (patients) where compared with 120 healthy individuals (controls). The QRS complexes were extracted with a modified Pan-Tompkins algorithm and processed with the Continuous Wavelet Transform to analyze the high-frequency content (85–130 Hz). Results: Overall, the power of the high-frequency content was higher in patients compared with controls (170.9 vs. 47.3 $10^3 nV^2 Hz^{-1}$; $p = 0.007$), with a prolonged time to reach the maximal power (68.9 vs. 64.8 ms; $p = 0.002$). An analysis of the signal intensity (instantaneous average of cumulative power), revealed a distinct function between patients and controls. The total intensity was higher in patients compared with controls (137.1 vs. 39 $10^3 nV^2 Hz^{-1} s^{-1}$; $p = 0.001$) and the time to reach the maximal intensity was also prolonged (88.7 vs. 82.1 ms; $p < 0.001$). Discussion: The high-frequency content of the QRS complexes was distinct between patients at risk of SCD and healthy controls. The wavelet transform is an efficient tool for spectral analysis of the QRS complexes that may contribute to stratification of risk.

**Keywords:** electrocardiographic analysis; wavelet transform; high-frequency content; sudden cardiac death

---

## 1. Introduction

### 1.1. Physiological Basis

Cardiac malignant arrhythmias (MA) are the major cause of sudden cardiac death (SCD) in the general population, leading to great concerns in patients, physicians and health care systems everywhere [1]. Despite extensive research on arrhythmia mechanisms and treatment, which leads to highly effective therapies like implantable defibrillators, risk stratification continues to be a matter under debate. Thousands of patients are implanted a defibrillator every year worldwide, jet a minority of them will develop MA and benefits from the therapy. In addition, a significant part of them will develop adverse reactions like infections or lead fractures. In an attempt to maximize the benefit from the therapy and minimize adverse reactions, physicians would benefit from knowing a reliable source of data from the patients that allows for predicting future occurrence of MA, thus applying preventive measurements (i.e., implantable defibrillators) with the maximum of the sensitivity and specificity. Among a variety of technological tools for quantifying the risk, the stratification methods based on the surface electrocardiogram (ECG) have always been preferred by clinicians due to its widespread use, ease of access, simplicity in application and low associated cost. Under this perspective, the spectral analysis of the ECG emerges as potentially useful, with the opportunity to contribute to the risk stratification.

On the time domain analysis of the surface ECG, the fragmentation of the QRS complexes is defined as notches that disrupt the linear progression of the time-voltage series recorded from the body surface of the patients with an ECG sensor. Importantly, QRS fragmentation is gaining increasing attention as a risk marker for the development of SCD or Malignant Arrhythmias (MAs) that might help to stratification of risk in the clinical scenario [2,3]. In previous studies, the physiological basis of fragmentation has been related to delays in conduction of the electrical wavefronts through the myocardium, mostly secondary to the interposition of scar and fibrotic tissue [4,5]. The latter mechanism also explain the late potentials usually recorded from the myocardium of patients with ischemic cardiomyopathy [6,7] and other clinical entities with a high risk of SCD such as Hypertrophic Cardiomyopathy [8] or Dilated Cardiomyopathy [9]. The Brugada Syndrome (BrS) [10,11] is a particular case in which functional mechanisms promoting delayed conduction explain the occurrence of delay/late potentials in the absence of fibrotic tissue [12]. Both fragmentation of the QRS and delayed potentials have also been postulated as risk markers for SCD and MA in BrS patients [3,13].

Some studies have demonstrated a robust link between QRS fragmentation and late potentials [14]. Overall, fragmentation would represent a more prominent phenomenon that will allow for their identification by visual inspection of the time domain record of the QRS complex, while less prominent forms will be masked within the total energy/voltage of the normal myocardium, thus requiring signal processing for appropriate identification. The signal average is a classical method applied to time domain records [15,16]. Briefly, in this method a signal average is computed to increase the signal to noise ratio. Thereafter, a high pass filter allows the detection of low voltage/high-frequency potentials contained at the ending of the QRS complexes. However, the signal averaged ECG is highly dependent on noise, requires long time records, and displays low sensitivity for the detection of high-frequency content within the QRS (not at the end). In contrast, a variety of spectral methods may provide efficient analysis of the QRS signal, enabling identification of late potentials by their subrogates in the frequency domain: the high-frequency content [17]. As will be stated below, the Wavelet continuous transform might also have the advantage to efficiently locate the high-frequency content along the QRS complexes. The latter will be particularly interesting in characterizing dynamic process in which the electrical conduction delay in the myocardium varies according to physiological or interventional procedures [18].

*1.2. The Wavelet Continuous Transform*

As described by Roesch, the use of the Wavelet transform is a reasonable option to study periodic phenomena in time series, particularly in the presence of frequency changes over time [19]. The medical application of this technique for signal processing has been widespread since the early 1980s. The Morlet wavelet was first described in the early 1980s [20–22]. It is based on the Gabor transform [23], a Gaussian window sinusoid that allows decomposition of a signal in its frequency and phase content over time [19]. Unlike the Gabor transform, the Morlet wave maintains its shape through frequency changes, providing a separation of the contributions of different frequency bands without loss in temporal resolution [22]. In summary, the Morlet wavelet is defined by the following equation, where $w$ represents the angular velocity or rotation, in radians per unit time:

$$\psi(t) = \pi^{-1/4} e^{i\omega t} e^{\frac{-t^2}{2}}, \tag{1}$$

The Wavelet transform using the Morlet wavelet of a time series ($x_t$) is defined as the convolution of the series with a set of daughter wavelets generated by the mother wavelet by time translation and scaled by $s$:

$$Wave(\tau, s) = \sum_t x_t \frac{1}{\sqrt{s}} \psi \times \left( \frac{t - \tau}{s} \right), \tag{2}$$

The position of the daughter wavelet in the time domain is determined by the displacement of the time parameter in increments of $dt$. The local amplitude of any periodic component of the time series and how it evolves over time can be obtained from the module of its wavelet transform [19]. The square of the amplitude constitutes the density of the wavelet energy in the time-frequency domain and is called the energy spectrum of the wave [24]:

$$Power(\tau, s) = \frac{1}{s} \times \left| Wave(\tau, s)^2 \right|, \tag{3}$$

### 1.3. The Use of the Wavelet Transform for the Study of the Surface Electrocardiogram

The Wavelet transform applied to the surface ECG has been analyzed in variety of clinical scenarios. In the works by Gramatikov et al., the continuous Wavelet transform was used to detect high-frequency content along the QRS in patients with acute ischemia, either provoked during a coronary angioplasty procedure (mechanical occlusion of the artery lumen), or spontaneously occurring in patients with acute ischemic heart disease (angina pectoris) [25,26]. An increase in the high-frequency content during angioplasty (frequency range: 16–200 Hz) and during the acute ischemic episodes (frequency range: 20–100 Hz) was observed in these studies, and subsequently normalized once the ischemia disappeared. Similarly, Magrans et al., studied patients during a coronary angioplasty procedure [27]. They found that, at the time of the angioplasty, the high-frequency content increased. Remarkably, those changes were more pronounced at the terminal part of the QRS signal, thus suggesting displacements of the high-frequency content along the QRS that must be consider in terms of the signal processing and the interpretation of data. Another interesting field for study, in which the wavelet transform of the QRS complexes has been successfully used, is to analyze patients with specific electrical conduction delays requiring resynchronization therapy. With this regard, it has been postulated that the use of the Wavelet transform might help to discern patients that will be unresponsive to the therapy, thus avoiding costs and potential adverse events [28,29].

However, the number of patients included for analysis in those studies is low and the possibility of implementing automatic systems for the analysis of QRS complexes has not been entirely evaluated. The utility of the Wavelet transform to analyze the high-frequency content and to predict the risk of MA or SCD have been poorly studied. Only the work of Murata et al., has shown an increase in the high-frequency content of patients with MA or SCD compared to those without [17,30,31].

### 1.4. Pan-Tompkins Algorithm for Automatic Detection of QRS Complexes

The detection of QRS complexes is the first step for any automatic ECG analysis. There are many methods for this purpose, with generally good performance, although each method has situations where it fails. The algorithm published in 1985 by J. Pan and W. Tompkins is one of the most used and allows the correct detection of the dominant peaks (R wave) of the QRS complexes [32]. In later works the effectiveness of this algorithm has been evaluated and compared with other algorithms [33], just confirming an adequate performance [34,35]. Briefly, the Pan-Tompkins algorithm is based on 4 sequential steps: (i) differentiation, (ii) squared elevation, (iii) integration with a moving window and finally, (iv) selection of the detection threshold to select the R waves of the QRS complexes.

### 1.5. Hypothesis and Objectives

Hypothesis: The high frequency content of the surface ECG display distinctive characteristics in patients under a risk of MA and SCD.

Objective 1: To evaluate the ability of the Wavelet transform for the analysis, quantification and temporal localization of high frequency content in QRS complexes, both in healthy controls and in patients with a history of SCD or MA.

Objective 2: To analyze the contributing power and time distribution of the high frequency content within the QRS complexes in patients and healthy controls in order to study differential patterns between them.

## 2. Materials and Methods

### 2.1. Study Population

The group of patients was obtained from a consecutive registry of 42 patients with prior history of Sudden Cardiac Arrest (SCA) or MA (patients), who came to our cardiac electrophysiology laboratory for diagnostic or therapeutic purposes. The term SCA was used to describe sudden death cases in which specific resuscitation records were available or the individual had survived the cardiac arrest event [35]. SCD was defined as an unexpected death without obvious extracardiac cause, occurring with a rapid witnessed collapse, or if un-witnessed, occurring within 1 h after the onset of symptoms. SCD was considered probable if occurred unexpectedly without obvious extracardiac cause within the previous 24 h. In any situation, the death should not occur in the setting of a prior terminal condition, such as a malignancy that was not in remission or end-stage chronic obstructive lung disease [35]. MA were defined as sustained life-threatening ventricular arrhythmia (Ventricular Tachycardia or Ventricular Fibrillation), which finished spontaneously or with treatment (pharmacological or electrical). A control group of 120 healthy individuals (controls) were analyzed. In all of them cardiac examination was performed to rule out any cardiac disease.

For the collection of electrocardiographic data the ethical committee approval was first obtained (Asturias regional ethical committee for clinical research, project number 35/2013). In addition, the investigations were carried out following the rules of the Declaration of Helsinki of 1975, revised in 2008.

### 2.2. Collection of Electrocardiographic Data: The ECG Sensor

The standard ECG (12 leads) of the patients included in the study were digitally collected with the commercially available ECG sensor EPTracer® from CardioTek© Electrophysiological Measurement System, Maastricht (here in after the EPTracer sensor). The EPTracer sensor is a computerized system designed for both clinical and experimental electrophysiological studies, allowing real-time visualization of the ECG signal and continuous storage (12 bit resolution, up to 1 kHz sampling rate, and a maximal differential input-voltage of +/−5 mV recording). As displayed in Figure 1, panels A & B, the potential at different sites of the body surface was explored by 10 exploring electrodes distributed according to clinical standards. The opposite extremes of the electrodes were connected to a multi-channel amplifier (panel C), where potential differences (ECG signals or leads) were computed between pairs of electrodes or calculated against an indifferent electrode (central terminal of Wilson) [36,37]. ECG signals underwent filtering at the EPTracer sensor (linear-phase digital band-pass filter: 0.05–150 Hz to suppress low-frequency noise that results from baseline wander, movement, and respiration, and higher-frequency noise that results from muscle artifact, and power-line or radiated electromagnetic interference) [37] and amplification. A notch filter was also applied at the 50 Hz frequency, in order to eliminate the power-line interference.

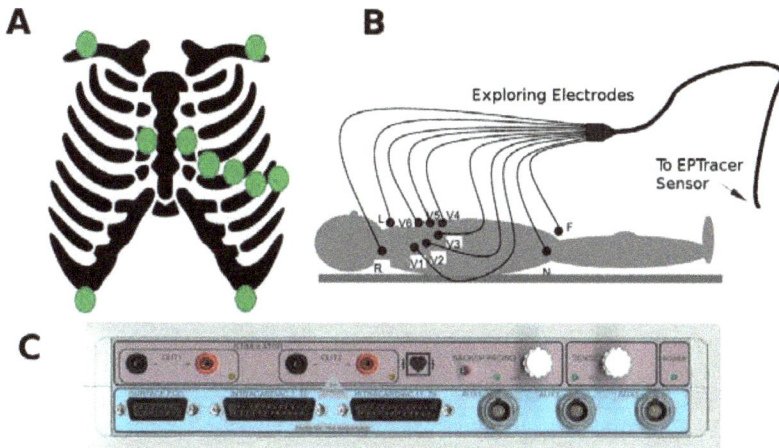

**Figure 1.** Electrocardiogram (ECG) recording and characteristics of the EPTracer sensor. (**A**) Schematic diagram representing the standard position of the exploring electrodes (green dots) on the body surface. (**B**) According to the position of the exploring electrodes on the body surface, and the configuration of pairs for computing differences in potential, the ECG signals receive standardized names. L, R, F, V1, V2, V3, V4, V5, and V6 constitute the differences in potential between the corresponding electrode and an indifferent electrode (central terminal of Wilson). In addition, three ECG signals are computed as the algebraic sum of potentials at different pairs of electrodes (lead I = L + R; lead II = R + F; Lead III = L + F). (**C**) EPTracer sensor front panel.

### 2.3. Extraction of QRS Complexes Using a Modified Pan-Tompkins Algorithm

For the purpose of the study, the ECG records were performed continuously for at least 12 consecutive seconds (Figure 2A). To extract the QRS complexes, we propose a modified algorithm of the originally described by Pan-Tompkins [32]. As previously described, the latter consisted of four consecutive steps: differentiation, squared elevation, detection threshold calculation, and correction by local maxima. However, in order to make spectral comparisons between subjects that affected the amplitude and the timing of the high frequency content, there is a need for: (i) normalizing the amplitude of the QRS signal; and (ii) select a stable reference for timing. After normalization, we would need also to select appropriate new thresholds for QRS detection. The latter is the rationale for the modifications that we introduced in the originally described Pan-Tompkins algorithm, summarized as follows (see also Figure 2):

First, we normalized the ECG signal to allow appropriate comparisons between leads and patients.

In order to determine the threshold for detection of the R-wave in the QRS complexes (after differentiation and squared elevation of the signal), the 99.5 percentile of the signal (*P99.5*) was calculated. All the extreme values (defined as those greater than *P99.5*) were removed and the signal was typified over the value of *P99.5* (Figure 2B). A threshold of 0.6 defined the time points where there were QRS complexes (*tQRS*; red line in Figure 2B).

A temporal correction was performed for the morphology of the QRS complexes. To do this, the point of the QRS complex with a higher positive voltage value in the V6 derivation was selected (*tMaxQRS*; red line in Figure 2C), because clinically it is the lead in which the peak of the R wave is better defined.

Finally, to subtract the QRS complexes for analysis a window of 145 ms around the *tMaxQRS* point was selected (from 60 ms before *tMaxQRS* to 85 ms after *tMaxQRS*; Figure 2C).

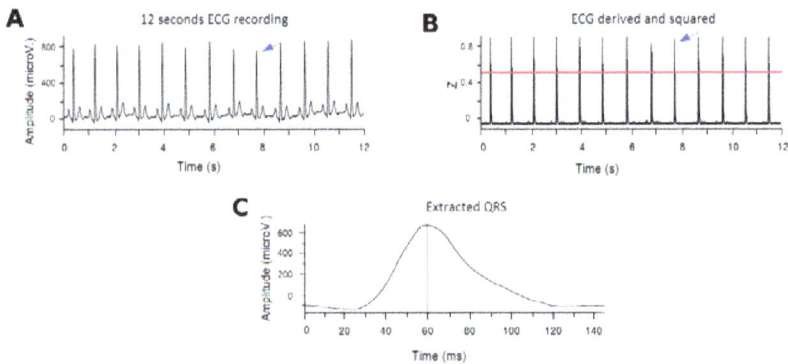

**Figure 2.** Example of the ECG processing and the QRS extraction. (**A**): Surface ECG (12 consecutive seconds). The blue arrow indicates a QRS complex. (**B**) Signal after derivation, square elevation and typification. The 0.6 threshold to detect the QRS complexes is denoted (red line). The blue arrow indicates a QRS complex after processing. (**C**) Example of an extracted QRS complex. The *tMaxQRS* is indicated by the red line. ECG: electrocardiogram.

In order to evaluate the algorithm, we compared our modified algorithm's performance against other published algorithms. For this purpose we used the MIT-BIH Arrhythmias Database [38], and calculated the Sensitivity (Se), Positive Predictive Value (PPV), Error (Er), Total Beats (TB), False Positives (FP) and False Negatives (FN), as described in previous methods [32,39–42].

### 2.4. Wavelet Continuous Transformation for the Analysis of the High-Frequency Content

The time-frequency data of each QRS complex were collected using the Wavelet transform (Morlet wavelet). According to previous definitions of high frequency content in other works [14–17], data were analyzed in the defined range of high frequencies (85–130 Hz), with an upper period of 11.5 ms and lower period of 7.7 ms (Figure 3A). A temporal definition of 1 kHz and a frequency resolution of 1/125 suboctaves were used. Calculations for the Wavelet Continuous Transform were performed with the WaveletComp library [19] for R [43].

### 2.5. Quantification of High-Frequency Content in the QRS Complexes

To analyze the distribution of the high-frequency content along the QRS signal, we computed the cumulative power contained at each time epoch of the QRS complex (Figure 3B). From the obtained distribution we defined (i) the *Peak Power* as the highest cumulative power of the high-frequency content (red dotted line in Figure 3B); (ii) the *Time to Peak Power* as the time epoch where Peak Power was reached (green dotted line in Figure 3B); and (iii) the *Total Power* as the area under the curve of the whole power function.

In addition, we calculated the relative contribution of the high-frequency content to the total power of the QRS complexes as a function of time: (i) at the initial part of the QRS complex (prior to *tMaxQRS*) named the *Initial High Frequency Contribution*; and (ii) at the final part of the QRS (after *tMaxQRS*) named the *Final High Frequency Contribution*. The ratio between the initial and final contribution was defined as the *High Frequency Contribution Ratio*.

### 2.6. Intensity Analysis

For a better description of the distribution of the high-frequency content along the QRS complex, the spectral intensity of the high-frequency content throughout the entire QRS complex is calculated. Intensity was defined as the instantaneous average of cumulative power and computed according to the following equation:

$$Intensity(t) = \frac{1}{dt} \times \sum_{i=1}^{t}(Power_i),\qquad(4)$$

where $Power_i$ is the average high-frequency power in $t = i$ and $dt$ is the time elapsed until each time epoch. On the spectral intensity function we defined: (i) the *Peak Intensity* as the highest intensity in the intensity function (red dotted line in Figure 3C); (ii) the *Time to Peak Intensity* as the time epoch where *Peak Intensity* was reached (green dotted line in Figure 3C), (iii) the *Final Intensity* as the intensity value at the ending point of the intensity function (blue dotted line in Figure 3C); and (iv) the *Total Intensity* as the area under the curve of the whole function.

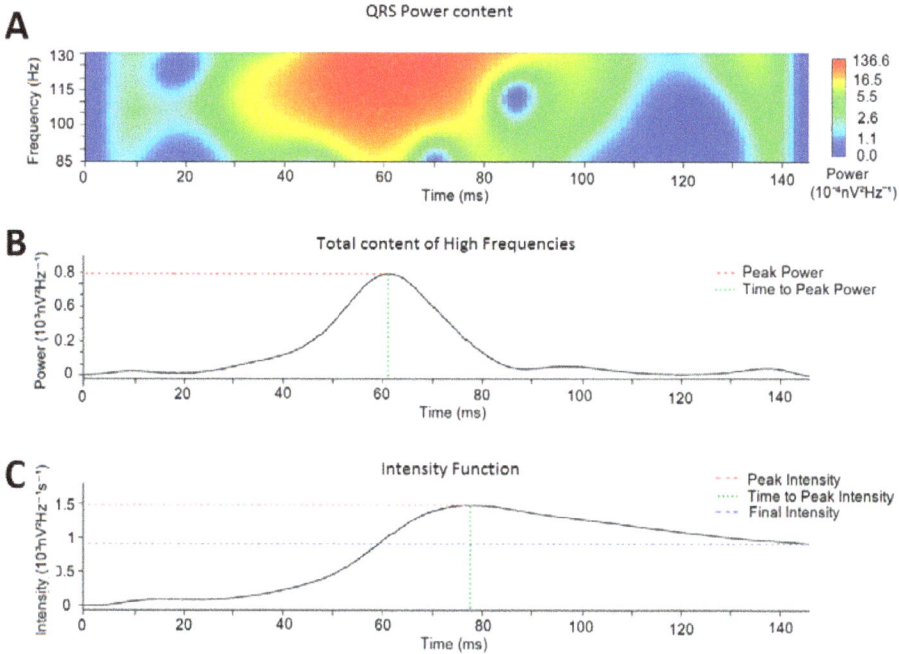

**Figure 3.** Example of a Wavelet continuous transform on a QRS complex (frequency range: 85–130 Hz). (**A**) Power spectrum of the QRS complex. (**B**) Cumulative power of the high-frequency content at each time epoch. The red dotted line marks the *Peak Power* and the green dotted line marks the *Time to Peak Power*. (**C**) Intensity Function. *Final Intensity* is marked by blue dotted line; *Time to Peak Intensity* is marked by the green dotted line; and *Peak Intensity* is marked by the red dotted line.

## 2.7. Statistical Analysis

Initially, a descriptive statistical analysis of the baseline variables of the patients was performed to characterize the study population. Categorical variables are expressed as percentage and the continuous quantitative as Mean ± Standard Error. For the contrast between patients and controls, a Chi square test was performed for the categorical variables and the Student's T test for the continuous variables, in all cases with a significance level of 0.05. The quantification of the parameters used to analyze the high-frequency content along the entire QRS complex, as well as the spectrogram intensity in the high-frequency range is expressed as Mean ± Standard Deviation. For the contrast between the patients and the controls we used a Student's T test. Statistical calculations were performed with the statistical software R [43].

## 3. Results

### 3.1. Population Characteristics

We analyzed data from 162 individuals (120 healthy controls and 42 patients). Table 1 summarizes clinical characteristics. No statistical differences between both groups were found. Ischemic Heart Disease (28.57%) and BrS (26.19%) were the most frequently observed pathology in patients. All of them had suffered at least an episode of MA, and 11.9% had experienced an episode of SCA.

**Table 1.** Population characteristics.

| Variables | Controls (*n* = 120) | Patients (*n* = 42) |
|---|---|---|
| Age (years) | 53.3 ± 14.37 | 59.6 ± 18.26 |
| Male (N/%) | 67 (55.8) | 34 (80.95) |
| Cardiomyopathy (N/%) | 0 (0) | 42 (100) |
| Ischemic | NA | 12 (28.57) |
| Idiopathic | NA | 4 (9.52) |
| BrS * | NA | 11 (26.19) |
| Others | NA | 15 (35.71) |
| Arrhythmic Events (N/%) | 0 (0) | 42 (100) |
| SCA * | NA | 5 (11.9) |
| MA * | NA | 42 (100) |

* BrS: Brugada Syndrome; SCA: Sudden Cardiac Death; MA: Malignant Arrhythmias.

### 3.2. Comparative Analysis between Different QRS Detection Algorithms

As previously stated, the proposed modifications in the Pan-Tompkins algorithm were designed to allow appropriate comparisons of the high frequency content between different subjects. Although improving detection features was out of our goal, we consider appropriate to display the performance of the modified algorithm in comparison with others. Table 2 display that Sensitivity and Positive Predictive Values of our modified algorithm remain in the range of those previously published for other methods. However, the Sensitivity and the Positive Predictive Value are slightly lower for our modified Pan-Tompkins in comparison with the originally described.

**Table 2.** Comparative analysis between different QRS detection algorithms.

| Algorithm | Database | Se (%) | PPV (%) | Er | TB | FP | FN |
|---|---|---|---|---|---|---|---|
| J. Pan et al. (1985) [32] | MITDB | 99.75 | 99.53 | 0.675 | 116137 | 507 | 277 |
| J.P. Martinez et al. (2004) [42] | MITDB | 99.8 | 99.86 | 0.34 | 107567 | 153 | 220 |
| Z. Zidelmal et al. (2012) [39] | MITDB | 99.64 | 99.82 | 0.54 | 109494 | 193 | 393 |
| R. Tafresi et al. (2014) [40] | PTBDB | 99.06 | 98.9 | N/A | N/A | N/A | N/A |
| M. Yochum et al. (2016) [41] | CinCC11 | 99.87 | 91.71 | N/A | N/A | N/A | N/A |
| Present Work | MITDB | 98.45 | 96.67 | 3.53 | 114654 | 2567 | 1125 |

* Se: Sensitivity; PPV: Positive Predictive Value; Er: Error; TB: Total Beats; FP: False Positives; FN: False Negatives.

### 3.3. Analysis of the High-Frequency Content

Overall, patients had significantly higher high-frequency content along the QRS complex than controls (data summarized in Table 3 & Figure 4). The *Total Power* and the *Peak Power* were significantly higher in patients compared with controls. The *Time to Peak Power* was also prolonged in patients with regard to controls, thus demonstrating differences not only in the absolute values, but also in the distribution of the high-frequency content along the QRS. However, when analyzing the relative contribution of the high-frequency content there were no differences in the *High Frequency Contribution Ratio* between patients and controls. That probably means a global displacement of the total frequency content to more distal positions in patients compared with controls. The latter is well exemplified in a patient with BrS in which, at different states of the pathology, the time domain record of the QRS complex displayed progressively delayed components (moving waves; Figure 5, black arrow) at the terminal part of the QRS. This case illustrates that as the ECG record becomes more pathological, the position of the moving wave is located at more delayed positions of the QRS. With regression to normal ECG tracings, the moving wave returns to more proximal positions.

**Table 3.** Summary of high-frequency analysis.

| Variables | Controls (*n* = 120) | Patients (*n* = 42) | *p* |
|---|---|---|---|
| Peak Power ($10^3 nV^2 Hz^{-1}$) | 1.709 ($\pm$ 1.13) | 7.033 ($\pm$ 15.09) | 0.028 |
| Time to Peak Power (ms) | 64.768 ($\pm$ 5.868) | 68.952 ($\pm$ 7.609) | 0.002 |
| Total Power ($10^3 nV^2 Hz^{-1}$) | 47.298 ($\pm$ 26.129) | 170.782 ($\pm$ 282.714) | 0.007 |
| Initial High Frequency Contribution ($10^3 nV^2 Hz^{-1}$) | 2.012 ($\pm$ 1.21) | 5.409 ($\pm$ 4.97) | <0.001 |
| Final High Frequency Contribution ($10^3 nV^2 Hz^{-1}$) | 45.287 ($\pm$ 25.601) | 165.463 ($\pm$ 280.608) | 0.008 |
| High Frequency Contribution Ratio | 0.053 ($\pm$ 0.034) | 0.069 ($\pm$ 0.049) | 0.059 |

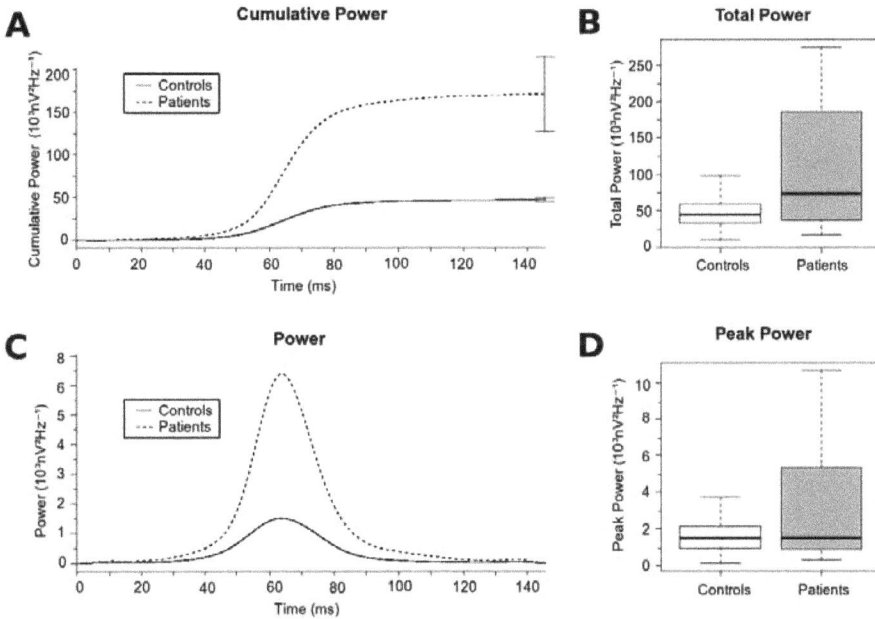

**Figure 4.** High-frequency content along the QRS Complex. (**A**) Cumulative high-frequency content for each group at each time epoch. The final error bars represents the Total Power with its standard error for each group. (**B**) Boxplot comparing the Total Power. (**C**) Time distribution of the high-frequency content for each group at each time epoch. (**D**) Boxplot comparing the *Peak Power*.

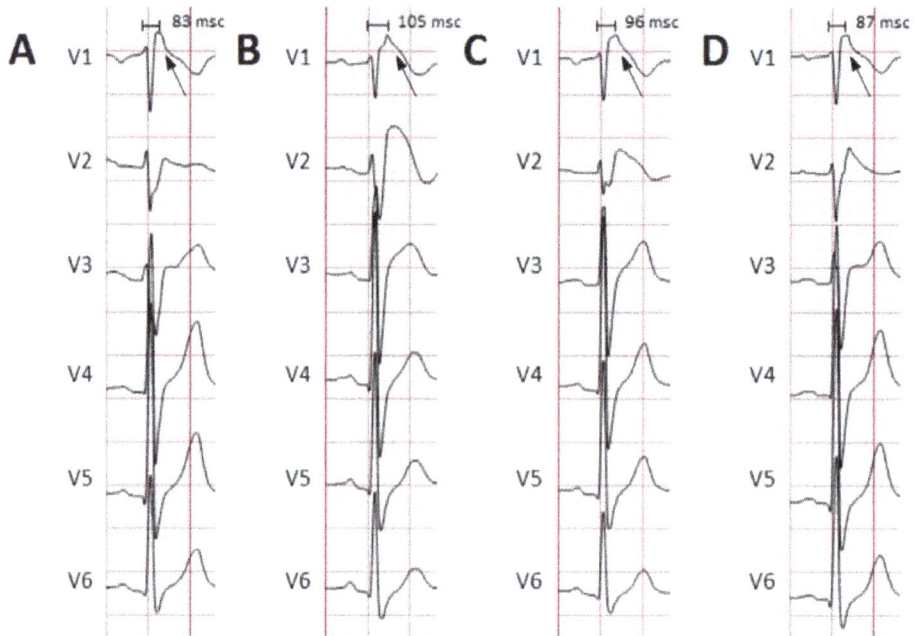

**Figure 5.** Example of different QRS complexes at different times in a patient with BrS. The precordial leads (V1 through V6) are shown. (**A**) Basal ECG, before Flecainide administration. Flecainide is an antiarrhythmic drug with major effects on cardiac cell's depolarization. Due to its effects on the cardiac sodium channel (main alteration in BrS), it is used in patients with suspected BrS to unmask the characteristic electrocardiographic pattern. No Brugada pattern is observed. (**B,C**) ECG showing a Brugada pattern immediately and a few minutes after Flecainide administration. A moving wave becomes apparent with increasing delay (black arrows and time intervals). (**D**) Resumption to basal conditions demonstrating the returning of the moving wave to more proximal parts of the QRS. BrS: Brugada syndrome. ECG: electrocardiogram.

*3.4. Intensity Analysis*

Overall, patients displayed higher intensity of the high-frequency content along the QRS complex than controls (data summarized in Table 4 & Figure 6). The *Final Intensity*, the *Peak Intensity* and the *Total Intensity* were significantly higher in patients compared with controls. In addition, the *Time to Peak Intensity* was reached later in patients when compared with controls.

**Table 4.** Summary of the intensity analysis.

| Variables | Controls ($n = 120$) | Patients ($n = 42$) | $p$ |
|---|---|---|---|
| Peak Intensity ($10^3 nV^2 Hz^{-1} s^{-1}$) | 0.506 ($\pm$ 0.296) | 1.854 ($\pm$ 3.389) | 0.014 |
| Time to Peak Intensity (ms) | 82.116 ($\pm$ 5.103) | 88.738 ($\pm$ 9.461) | <0.001 |
| Total Intensity ($10^3 nV^2 Hz^{-1} s^{-1}$) | 39.024 ($\pm$ 21.574) | 137.128 ($\pm$ 233.521) | 0.001 |
| Final Intensity ($10^3 nV^2 Hz^{-1} s^{-1}$) | 0.32 ($\pm$ 0.177) | 1.155 ($\pm$ 1.91) | 0.007 |

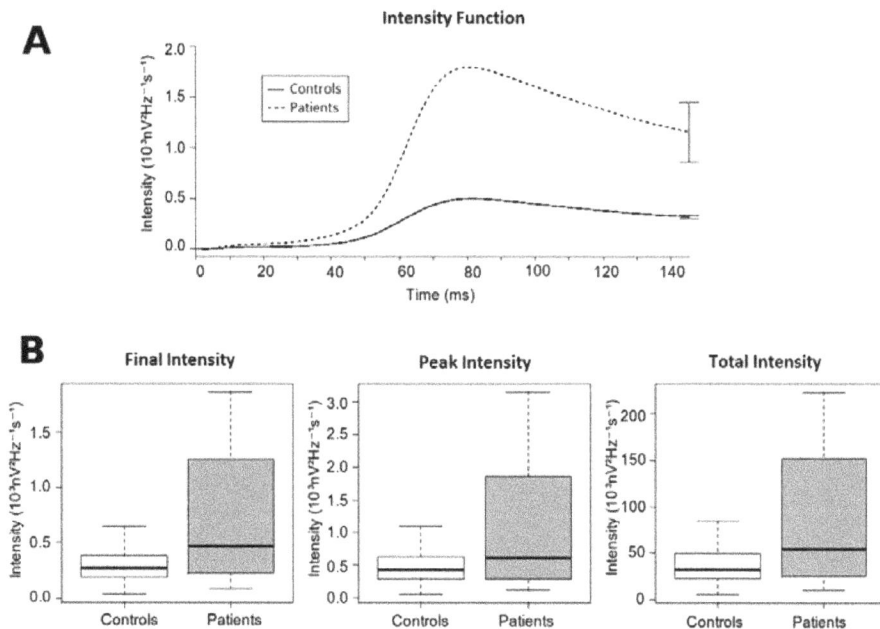

**Figure 6.** Intensity analysis. (**A**) Intensity function. It shows the value of the intensity function for each group in each time epoch. The final error bars represent the Final Intensity with its standard error for each group. (**B**) Boxplot comparing different measures of the intensity function between both groups. From left to right: *Final intensity*, *Peak intensity* and *Total Intensity*.

## 4. Discussion

The results of our study show that the wavelet transform of the QRS complexes allows for appropriate characterization of the high-frequency content that exert a differential behavior between healthy individuals and patients affected by severe cardiac arrhythmias leading to SCD. In our cohort of patients, we demonstrate that the relative contribution of the high frequencies to the spectral content is higher than in healthy controls, and behave with a slightly delay in their appearance at the QRS complexes. Overall, the absolute values and the distribution of the of high-frequency content may be of prognosis significance in patients with different pathologies, which in conjunction with the analysis of other clinical variables, like cardiac syncope and structural abnormalities, might contribute to better stratification of risk and more appropriate measurements to take in order to reduce the risk of SCD. In addition, the analysis proposed in our work is feasible for an automatized, on-line analysis of ECG records that would allow quick evaluation of patients.

The Wavelet continuous transform has been proposed as a useful technique for studying the frequency power spectrum, and especially to study the temporal distribution of the high-frequency content along QRS complexes. In this sense, it is important to remark how this constitutive part along the QRS may present a dynamic behavior with time and also change under different clinical situations even within the same patient. At different states of the pathology the morphology of the ECG may change, which might make diagnosis of those patients difficult [18]. In addition, those changes on the ECG morphology may also change the distribution of the frequency content along the QRS, depending on the severity of the disease or the administration of various drugs. For example, in Figure 5 we display the case of a patient diagnosed with BrS. The ECG records show how there is a displacement of late potentials towards a later position within the QRS complex as the ECG demonstrates more pathological conditions (from Brugada type III to Brugada Type I). This dynamic behavior makes it

*Sensors* **2018**, *18*, 560

important to consider analytical techniques that are able to track the displacements in the frequency content in order to increase precision and sensitivity. We also previously demonstrate that ventricular fibrillation, the most lethal arrhythmia causing SCD in patients, is characterized by a highly dynamical behavior in the frequency content, but with maintenance of a hierarchical organization at the phase spectrum [13]. The latter allows for spatial location of the sources that eventually maintain the arrhythmia, and pave the way for automatized analytical methods able to efficiently track those dynamical changes and displacements of the frequency content that would help for developing more efficient clinical procedures.

Compared with other techniques focused on the time or frequency domain, we consider that the wavelet transform may provide significant advantages. Although firstly reported, the analysis of QRS fragmentation on the time domain may be less useful than the analysis of the high-frequency content to detect patients at risk for SCD [2,8,9]. The main limitation of the analysis of fragmentation is that it is based on a subjective classification, with no absolute measurements, and thus it is more difficult to standardize. Also they are less sensitive than frequency domain analysis [44]. In addition, unlike other techniques for the analysis of the frequency power spectrum (i.e., Fast Fourier Transform), the time-frequency analysis (i.e., Wavelet transform) also allows for the location of high-frequency content along time [45], which can help to identify significant displacements as discussed before. There are other techniques for time-frequency analysis, mainly the signal-averaged electrocardiogram [45], which is one of the first described techniques for high-frequency analysis in QRS complexes [15,16]. As stated at the introduction section, this technique may fail in situations with high noise to signal ratio, it requires long ECG records, and it is only useful in the terminal part of the QRS, not in the whole QRS complex [46]. Spectral methods provide a more efficient analysis of the QRS signal [17] and we have found the Wavelet transform particularly useful in our patients.

## 5. Conclusions

The high-frequency content of the QRS complexes distinctively behaves between patients at high risk of SCD, and healthy controls. The wavelet transform is an efficient tool for spectral analysis of the QRS complexes that allows for quantifying the differential behavior between both studied groups. Future research is needed to clarify how the proposed technique may contribute to stratification of patient risk.

## 6. Limitations

Differences between patients with and without MA/SCA were not evaluated, thus we cannot ascertain the predictive capabilities in patients in primary prevention. Patient groups had a higher percentage of men than the control group. Because there may be some differences in QRS characteristics associated with gender, it could explain, to some extent, the differences observed between groups.

Overall, to analyze the global utility of the Wavelet transform, comparisons with other techniques are required. In order to develop an automatic method for quantification and characterization of high-frequency content there is need for combining the relevant metrics (high frequency content timing and magnitude) using a machine learning technique, and assessing its performance. Both will be desirable but were not tested/performed in our work.

**Acknowledgments:** We thank the people working at the Arrhythmia Unit of the University Hospital of Asturias. We also thank Marta Torres and Esther Villa for their assistance in the study. We have not received funds for covering the costs to publish in open access.

**Author Contributions:** D.C., D.G.I., N.R.G. and F.J.C. conceived and designed the experiments; D.G.I. performed the experiments; D.G.I. analyzed the data; D.G.I., D.C., N.R.G. and F.J.C. wrote the paper.

**Conflicts of Interest:** The authors declare no conflict of interest

*Sensors* **2018**, *18*, 560

## References

1. From, R.; Fishman, G.I.; Chugh, S.S.; Dimarco, J.P.; Albert, C.M.; Anderson, M.E.; Bonow, R.O.; Buxton, A.E.; Chen, P.-S.; Estes, M.; et al. Sudden Cardiac Death Prediction and Prevention. *Circulation* **2010**, *122*, 2335–2348.
2. Adler, A.; Viskin, S. Risk stratification in brugada syndrome: Clinical characteristics, electrocardiographic parameters and auxiliary testing. *Heart Rhythm* **2016**, *13*, 299–310. [CrossRef] [PubMed]
3. Morita, H.; Kusano, K.F.; Miura, D.; Nagase, S.; Nakamura, K.; Morita, S.T.; Ohe, T.; Zipes, D.P.; Wu, J. Fragmented QRS as a marker of conduction abnormality and a predictor of prognosis of Brugada syndrome. *Circulation* **2008**, *118*, 1697–1704. [CrossRef] [PubMed]
4. Das, M.K.; Suradi, H.; Maskoun, W.; Michael, M.A.; Shen, C.; Peng, J.; Dandamudi, G.; Mahenthiran, J. Fragmented Wide QRS on a 12-Lead ECG. Clinical perspective. *Circ. Arrhythm. Electrophysiol.* **2008**, *1*, 258–268. [CrossRef] [PubMed]
5. Das, M.K.; Zipes, D.P. Fragmented QRS: A predictor of mortality and sudden cardiac death. *Heart Rhythm* **2009**, *6*, S8–S14. [CrossRef] [PubMed]
6. Das, M.K.; Khan, B.; Jacob, S.; Kumar, A.; Mahenthiran, J. Significance of a fragmented QRS complex versus a Q wave in patients with coronary artery disease. *Circulation* **2006**, *113*, 2495–2501. [CrossRef] [PubMed]
7. Das, M.K.; Saha, C.; El Masry, H.; Peng, J.; Dandamudi, G.; Mahenthiran, J.; McHenry, P.; Zipes, D.P. Fragmented QRS on a 12-lead ECG: A predictor of mortality and cardiac events in patients with coronary artery disease. *Heart Rhythm* **2007**, *4*, 1385–1392. [CrossRef] [PubMed]
8. Kang, K.W.; Janardhan, A.H.; Jung, K.T.; Lee, H.S.; Lee, M.H.; Hwang, H.J. Fragmented QRS as a candidate marker for high-risk assessment in hypertrophic cardiomyopathy. *Heart Rhythm* **2014**, *11*, 1433–1440. [CrossRef] [PubMed]
9. Igarashi, M.; Tada, H.; Yamasaki, H.; Kuroki, K.; Ishizu, T.; Seo, Y.; Machino, T.; Murakoshi, N.; Sekiguchi, Y.; Noguchi, Y.; et al. Fragmented QRS Is a Novel Risk Factor for Ventricular Arrhythmic Events After Receiving Cardiac Resynchronization Therapy in Nonischemic Cardiomyopathy. *J. Cardiovasc. Electrophysiol.* **2017**, *28*, 327–335. [CrossRef] [PubMed]
10. Calvo, D.; Flórez, J.P.; Valverde, I.; Rubín, J.; Pérez, D.; Vasserot, M.G.; Rodríguez-Reguero, J.; Avanzas, P.; De La Hera, J.M.; Gómez, J.; et al. Surveillance after cardiac arrest in patients with Brugada syndrome without an implantable defibrillator: An alarm effect of the previous syncope. *Int. J. Cardiol.* **2016**, *218*, 69–74. [CrossRef] [PubMed]
11. Florez, J.P.; García, D.; Valverde, I.; Rubín, J.; Pérez, D.; González-Vasserot, M.; Reguero, J.; de la Hera, J.M.; Avanzas, P.; Gómez, J.; et al. Role of syncope in predicting adverse outcomes in patients with suspected Brugada syndrome undergoing standardized flecainide testing. *Europace* **2017**. [CrossRef]
12. Nademanee, K.; Raju, H.; De Noronha, S.V.; Papadakis, M.; Robinson, L.; Rothery, S.; Makita, N.; Kowase, S.; Boonmee, N.; Vitayakritsirikul, V.; et al. Fibrosis, connexin-43, and conduction abnormalities in the Brugada syndrome. *J. Am. Coll. Cardiol.* **2015**, *66*, 1976–1986. [CrossRef] [PubMed]
13. Calvo, D.; Atienza, F.; Saiz, J.; Martínez, L.; Ávila, P.; Rubín, J.; Herreros, B.; Arenal, Á.; García-Fernández, J.; Ferrer, A.; et al. Ventricular Tachycardia and Early Fibrillation in Patients with Brugada Syndrome and Ischemic Cardiomyopathy Show Predictable Frequency-Phase Properties on the Precordial ECG Consistent with the Respective Arrhythmogenic Substrate. *Circ. Arrhythm. Electrophysiol.* **2015**, *8*, 1133–1143. [CrossRef] [PubMed]
14. Couderc, J.-P.; Chevalier, P.; Fayn, J.; Rubel, P.; Touboul, P. Identification of post-myocardial infarction patients prone to ventricular tachycardia using time–frequency analysis of QRS and ST segments. *EP Eur.* **2000**, *2*, 141–153. [CrossRef] [PubMed]
15. Simson, M.B.; Untereker, W.J.; Spielman, S.R.; Horowitz, L.N.; Marcus, N.H.; Falcone, R.A.; Harken, A.H.; Josephson, M.E. Relation between late potentials on the body surface and directly recorded fragmented electrograms in patients with ventricular tachycardia. *Am. J. Cardiol.* **1983**, *51*, 105–112. [CrossRef]
16. Coto, H.; Maldonado, C.; Palakurthy, P.; Flowers, N.C. Late potentials in normal subjects and in patients with ventricular tachycardia unrelated to myocardial infarction. *Am. J. Cardiol.* **1985**, *55*, 384–390. [CrossRef]
17. Tereshchenko, L.G.; Josephson, M.E. Frequency content and characteristics of ventricular conduction. *J. Electrocardiol.* **2015**, *48*, 933–937. [CrossRef] [PubMed]

18. Calvo, D.; Rubín, J.M.; Pérez, D.; Gómez, J.; Flórez, J.P.; Avanzas, P.; García-Ruíz, J.M.; De La Hera, J.M.; Reguero, J.; Coto, E.; et al. Time-dependent responses to provocative testing with flecainide in the diagnosis of Brugada syndrome. *Heart Rhythm* **2015**, *12*, 350–357. [CrossRef] [PubMed]

19. Rosch, A.; Schmidbauer, H. WaveletComp: A Guided Tour through the R-Package. Available online: http://www.hs-stat.com/projects/WaveletComp/WaveletComp_guided_tour.png (accessed on 15 June 2017).

20. Morlet, J.; Arens, G.; Fourgeau, E.; Giard, D. Wave propagation and sampling theory—Part II: Sampling theory and complex waves. *Geophysics* **1982**, *47*, 222–236. [CrossRef]

21. Morlet, J.; Arens, G.; Fourgeau, E.; Glard, D. Wave propagation and sampling theory—Part I: Complex signal and scattering in multilayered media. *Geophysics* **1982**, *47*, 203–221. [CrossRef]

22. Goupillaud, P.; Grossmann, A.; Morlet, J. Cycle-octave and related transforms in seismic signal analysis. *Geoexploration* **1984**, *23*, 85–102. [CrossRef]

23. Gabor, D. Theory of communication. Part 3: Frequency compression and expansion. *J. Inst. Electr. Eng. Part III Radio Commun. Eng.* **1946**, *93*, 445–457. [CrossRef]

24. Carmona, R.; Hwang, W.L.; Torrésani, B. *Practical Time-Frequency Analysis: Continuous Wavelet and Gabor Transforms, with an Implementation in S*; Academic Press: San Diego, CA, USA, 1998; Volume 9, ISBN 0-12-160170-6.

25. Gramatikov, B.; Brinker, J.; Yi-Chun, S.; Thakor, N.V. Wavelet analysis and time-frequency distributions of the body surface ECG before and after angioplasty. *Comput. Methods Programs Biomed.* **2000**, *62*, 87–98. [CrossRef]

26. Gramatikov, B.; Iyer, V. Intra-QRS spectral changes accompany ST segment changes during episodes of myocardial ischemia. *J. Electrocardiol.* **2015**, *48*, 115–122. [CrossRef] [PubMed]

27. Magrans, R.; Gomis, P.; Voss, A.; Caminal, P. Effect of acute myocardial ischemia on different high-frequency bandwidths and temporal regions of the QRS. In Proceedings of the 2011 Annual International Conference of the IEEE Engineering in Medicine and Biology Society, EMBC, Boston, MA, USA, 30 August–3 September 2011; pp. 7083–7086.

28. Vassilikos, V.P.; Mantziari, L.; Dakos, G.; Kamperidis, V.; Chouvarda, I.; Chatzizisis, Y.S.; Kalpidis, P.; Theofilogiannakos, E.; Paraskevaidis, S.; Karvounis, H.; et al. QRS analysis using wavelet transformation for the prediction of response to cardiac resynchronization therapy: A prospective pilot study. *J. Electrocardiol.* **2014**, *47*, 59–65. [CrossRef] [PubMed]

29. Xia, X.; Couderc, J.-P.; Mcnitt, S.; Zareba, W. Predicting Effectiveness of Cardiac Resynchronization Therapy based on QRS Decomposition using the Meyer Orthogonal Wavelet Transformation. *Comput. Cardiol.* **2010**, *37*, 983–986.

30. Murata, H.; Ohara, T.; Kobayashi, Y.; Miyauchi, Y.; Katoh, T.; Mizuno, K. Detection of arrhythmogenic substrates in prior myocardial infarction patients with complete right bundle branch block QRS using wavelet-transformed ECG. *J. Nippon Med. Sch.* **2009**, *76*, 291–299. [CrossRef] [PubMed]

31. Tsutsumi, T.; Takano, N.; Matsuyama, N.; Higashi, Y.; Iwasawa, K.; Nakajima, T. High-frequency powers hidden within QRS complex as an additional predictor of lethal ventricular arrhythmias to ventricular late potential in postmyocardial infarction patients. *Heart Rhythm* **2011**, *8*, 1509–1515. [CrossRef] [PubMed]

32. Pan, J.; Tompkins, W.J. A Real-Time QRS Detection Algorithm. *IEEE Trans. Biomed. Eng.* **1985**, *BME-32*, 230–236. [CrossRef] [PubMed]

33. Okada, M. A Digital Filter for the ORS Complex Detection. *IEEE Trans. Biomed. Eng.* **1979**, *BME-26*, 700–703. [CrossRef]

34. Portet, F.; Hernández, A.I.; Carrault, G. Evaluation of real-time QRS detection algorithms in variable contexts. *Med. Biol. Eng. Comput.* **2005**, *43*, 379–385. [CrossRef] [PubMed]

35. Meyer, C.; Fernandez, J.; Harris, M. Combining algorithms in automatic detection of R-peaks in ECG signals. In Proceedings of the 18th IEEE Symposium on Computer-Based Medical Systems, Dublin, Ireland, 23–24 June 2005; Volume 10. [CrossRef]

36. Wilson, F.N.; Johnston, F.D.; Macleod, A.G.; Barker, P.S. Electrocardiograms that represent the potential variations of a single electrode. *Am. Heart J.* **1934**, *9*, 447–458. [CrossRef]

37. Kligfield, P.; Gettes, L.S.; Bailey, J.J.; Childers, R.; Deal, B.J.; Hancock, E.W.; van Herpen, G.; Kors, J.A.; Macfarlane, P.; Mirvis, D.M.; et al. Recommendations for the Standardization and Interpretation of the Electrocardiogram. *J. Am. Coll. Cardiol.* **2007**, *49*, 1109–1127. [CrossRef] [PubMed]

38. Moody, G.; Mark, R. The impact of the MIT-BIH Arrhythmia Database. *IEEE Eng. Med. Biol. Mag.* **2001**, *20*, 45–50. [CrossRef] [PubMed]

39. Tafreshi, R.; Jaleel, A.; Lim, J.; Tafreshi, L. Automated analysis of ECG waveforms with atypical QRS complex morphologies. *Biomed. Signal Process. Control* **2014**, *10*, 41–49. [CrossRef]

40. Zidelmal, Z.; Amirou, A.; Adnane, M.; Belouchrani, A. QRS detection based on wavelet coefficients. *Comput. Methods Programs Biomed.* **2012**, *107*, 490–496. [CrossRef] [PubMed]

41. Yochum, M.; Renaud, C.; Jacquir, S. Automatic detection of P, QRS and T patterns in 12 leads ECG signalbased on CWT. *Biomed. Signal Process. Control* **2016**, *25*, 46–52. [CrossRef]

42. Martinez, J.; Almeida, R.; Olmos, S.; Rocha, A.; Laguna, P. A wavelet-based ECGdelineator: Evaluation on standard databases. *IEEE Trans. Biomed. Eng.* **2004**, *51*, 570–581. [CrossRef] [PubMed]

43. R Development Core Team. *R: A Language and Environment for Statistical Computing*; The R Foundation for Statistical Computing: Vienna, Austria, 2008.

44. Amit, G.; Granot, Y.; Abboud, S. Quantifying QRS changes during myocardial ischemia: Insights from high frequency electrocardiography. *J. Electrocardiol.* **2014**, *47*, 505–511. [CrossRef] [PubMed]

45. Tsutsumi, T.; Okamoto, Y.; Takano, N.; Wakatsuki, D.; Tomaru, T.; Nakajima, T. High-frequency power within the QRS complex in ischemic cardiomyopathy patients with ventricular arrhythmias: Insights from a clinical study and computer simulation of cardiac fibrous tissue. *Comput. Biol. Med.* **2017**, *87*, 132–140. [CrossRef] [PubMed]

46. Kelen, G.J.; Henkin, R.; Starr, A.M.; Caref, E.B.; Bloomfield, D.; El-Sherif, N. Spectral turbulence analysis of the signal-averaged electrocardiogram and its predictive accuracy for inducible sustained monomorphic ventricular tachycardia. *Am. J. Cardiol.* **1991**, *67*, 965–975. [CrossRef]

*Article*

# A Neural Network Approach for Building An Obstacle Detection Model by Fusion of Proximity Sensors Data

Gonzalo Farias [1,*], Ernesto Fabregas [2], Emmanuel Peralta [1], Héctor Vargas [1], Gabriel Hermosilla [1], Gonzalo Garcia [3] and Sebastián Dormido [2]

[1]  Pontificia Universidad Católica de Valparaíso, Avenida Brasil 2147, Valparaíso 2362804, Chile; emmanuel.peraltah@gmail.com (E.P.); hector.vargas@pucv.cl (H.V.); gabriel.hermosilla@pucv.cl (G.H.)
[2]  Departamento de Informática y Automática, Universidad Nacional de Educación a Distancia, Juan del Rosal 16, 28040 Madrid, Spain; efabregas@bec.uned.es (E.F.); sdormido@dia.uned.es (S.D.)
[3]  Radar Research and Innovations, 11702 W 132nd Terrace, Overland Park, KS 66213, USA; garciagarreton@hotmail.com
*   Correspondence: gonzalo.farias@pucv.cl; Tel.: +56-(32)-2273673

Received: 20 December 2017; Accepted: 21 February 2018; Published: 25 February 2018

**Abstract:** Proximity sensors are broadly used in mobile robots for obstacle detection. The traditional calibration process of this kind of sensor could be a time-consuming task because it is usually done by identification in a manual and repetitive way. The resulting obstacles detection models are usually nonlinear functions that can be different for each proximity sensor attached to the robot. In addition, the model is highly dependent on the type of sensor (e.g., ultrasonic or infrared), on changes in light intensity, and on the properties of the obstacle such as shape, colour, and surface texture, among others. That is why in some situations it could be useful to gather all the measurements provided by different kinds of sensor in order to build a unique model that estimates the distances to the obstacles around the robot. This paper presents a novel approach to get an obstacles detection model based on the fusion of sensors data and automatic calibration by using artificial neural networks.

**Keywords:** proximity sensors; automatic calibration; neural networks

## 1. Introduction

Obstacle avoidance is one of the main challenges for any practical design in robotics. Many approaches to face this problem can be found in the literature. Some of them use optical sensors and computer vision for general purpose object detection tasks [1–3]. In particular, proximity sensors are a widely-implemented solution for obstacle detection and collision avoidance in mobile robotics. These sensors are normally located around the robot in order to detect and avoid objects when it navigates in a dynamic environment [4–6].

The proximity sensors are able to detect the presence of nearby objects without physical contact, providing an estimation of the distance to the obstacles. Commonly, this kind of sensor emits electromagnetic fields (e.g., infrared light) or sound pulses (e.g., ultrasonic), in order to detect changes in the field or returned signal [7]. For this reason, the surface properties of the objects have an important influence on the estimation of the distance. The other aspect that has to be taken into account is the time response of the sensor because the distance is estimated based on this value.

The infrared (IR) and ultrasonic (US) sensors are the most extensively used in robotics due to their low cost and fast time response, which make them an attractive option for real-time object detection [8]. These sensors use different technologies to estimate the distance to an object. On the one hand, IR sensors use light of wavelength in the range of 760 nm (IR spectrum) for sensing the reflected light intensity. On the other hand, US sensors use a high-frequency sound wave (40 kHz) in order to detect the returning pulses (echo) [9]. IR sensors are used to measure short distances (around 0.1 to 1.5 m) with a narrow (line) beam shape, while US sensors are used to measure long distances (around 0.15 to 6.5 m) with a conical beam shape. The width of the beam is a function of the surface area, frequency, and type of transducers [9,10].

In both cases, the distance estimation depends greatly on what the object is made of. For example, since the IR sensors are a vision-based measurement, the colour of the object and the environment light condition could affect their outcome [11,12]. On the other hand, because of the US sensors are sound-based, they are useful for transparent objects and able to work under poor lighting conditions, but this kind of sensors is much more sensitive to mirror-like surfaces, and the measurement can be affected by environmental parameters such as temperature, humidity, ambient noise, among others [8]. A more detailed description of this issue can be found in [9], where the authors make an analysis of the influence on the behaviour of both sensors for different types of materials such as: cardboard, paper, sponge, wood, plastic, rubber and tile.

Despite the popularity of proximity sensors, they need to be carefully calibrated before using since the measured distance between the robot and the obstacle is relative. A deficient calibration can produce significant errors in the distance estimation, and therefore, the performance of the obstacle avoidance algorithm is poor. If the sensors are included with the mobile robot, the manufacturer usually gives a model to convert raw data into distance, however such model provides an average performance of the distance estimation since this built-in model is not tuned for each sensor. Thus, in order to increase the accuracy of the distance estimation, the traditional method of calibration for a suitable distance model, turns frequently into a manual and time-consuming process [13].

Fusion of different kind of sensors is a fast developing area of research that shows many examples in the literature through the integration of multiple sensors [14,15]. Thus, the integration of the information supplied by US and IR sensors can be a natural extension of previous works in order to provide reliable distance measurements for robot navigation by using these proximity sensors, because the advantages of one sensor compensate for the disadvantages of the other. The fusion of sensors can be implemented by using fuzzy logic [16], neural networks [17–19], Kalman filtering [20], or support vector machines among other solutions [14]. Although we can find examples of sensor fusion for robot navigation in the literature, many of the them show only theoretical results [20,21] or quite ad hoc experiments [16–18], or they mix the distance model with the collision avoidance algorithm [17,19,22], implying that work on scientific experimental validation of sensor fusion methods for robot navigation still needs further research.

This paper presents a novel approach for modelling the distance to obstacles using the data provided by two types of proximity sensors. The proposed approach is applied on the fusion of IR and US sensor of the very popular Khepera IV robot [23]. Our approach also provides an automatic way to obtain and calibrate the model for distance estimation to obstacles by using artificial neural networks. After obtaining an accurate distance to the objects it is possible to use any kind of collision avoidance algorithm to perform a successful navigation of the robot in a dynamic environment. In order to validate the proposed approach, the results section of the paper provides real experiments of obstacle avoidance with a comparison between the traditional calibration method (provided by the manufacturer) versus the developed approach by using the same collision avoidance algorithm.

The remainder of the paper is organized as follows: Section 2 describes the traditional sensors calibration method and its drawbacks; Section 3 presents the novel method of automatic smart calibration of the sensors; Section 4 presents some results and experiments obtained with the developed method in comparison with the traditional calibration method; and finally, Section 5 shows the main conclusions and future work.

*Sensors* **2018**, *18*, 683

## 2. Traditional Calibration of Proximity Sensors

The traditional calibration method of proximity sensors is a manual repetitive process. It consists in placing some objects around the robot and measuring the raw values given by the sensors (usually voltages). With this values, a model of the sensors is built using a conventional identification method. The result of this procedure are the coefficients of the model (linear or nonlinear) that calculates the distances to the objects using the voltage of the sensors as inputs. Usually the same model is used for all the sensors of the same type, but in practice, not all the sensors work identically and this can introduce erroneous values.

To carry out the traditional calibration process in the laboratory, two Khepera IV robots [23] are used. Khepera IV is a mobile robot that has been designed for indoor pedagogical purposes and brings numerous features, for example, a colour camera, WI-FI and Bluetooth communications, an array of 8 infrared sensors (IR) for obstacle detection (2 mm to 20 cm) and 5 ultrasonic (US) sensors for long range object detection (20 cm to 40 cm), etc. Figure 1 represents the distribution and the ranges of the proximity sensors in the Khepera IV robot.

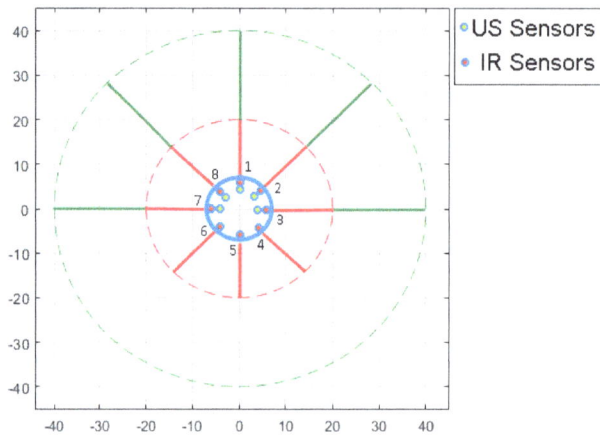

**Figure 1.** Khepera IV robot sensors distribution.

The IR sensors (red colour) are distributed every 45° around the robot (form 1 to 8). While the US sensors (green colour) are distributed every 45° in the front half of the robot (1, 2, 3, 7 and 8).

Figure 2 shows the set up configuration in the laboratory to carry out the calibration. To start the process, the robot is placed at the center of a "calibration carpet". Plastic obstacles of different colours conforming a wall concentric to the robot sensors are used to measure their values.

**Figure 2.** Setup for the traditional calibration method.

The robot stores 20 values per sensor and then the obstacles are displaced 1 cm over its radius. The process is repeated from 1 cm to 40 cm for all the sensors. If the distance is less than 40 cm the process is repeated. Otherwise, the data stored in the robot is sent to the PC to calculate the mean value for each sensor/distance. Then, the values are linearly interpolated. Figure 3 shows the flow diagram with the steps to carry out this process. The blocks represented in green colour are related to the robot. The blocks represented in orange colour are related to the PC. And, the pink colour blocks are related to the manually action taken by the user.

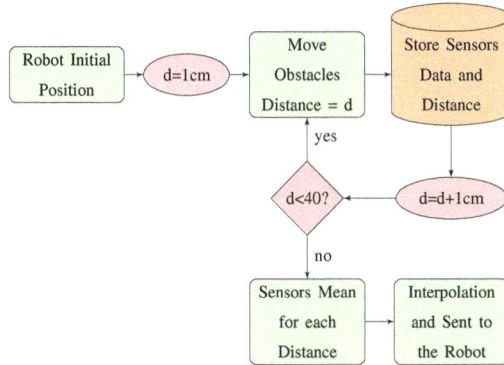

**Figure 3.** Flow diagram of the traditional calibration procedure.

Figure 4 shows the results of the previous process for the IR sensors: on the left side, the raw values from the sensors for the Robot 1; and on the right side, the raw values of the Robot 2. As can be seen, the sensors have a nonlinear behaviour for distance values less than 20 cm and the values are different for same sensor/distance (note that all these sensors are of the same type). For distance values greater than 20 cm the behaviour is almost the same. That is why it is important to uses the IR sensors only to measure distances less than 20 cm.

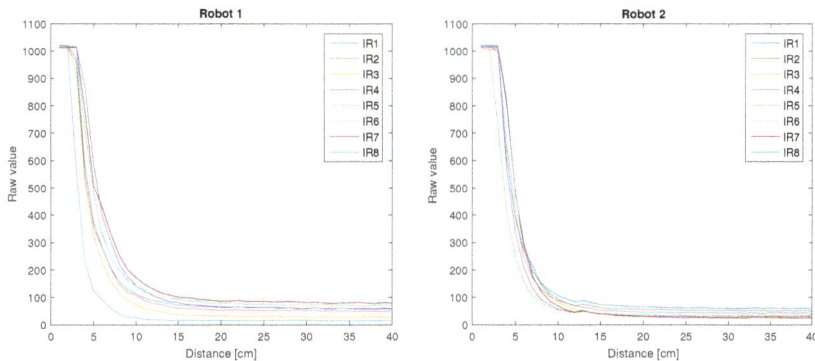

**Figure 4.** Robots 1 and 2 IR sensors values vs. distance.

Figure 5 shows the results of the previous process for the 5 US sensors of Robots 1 and 2. As can be seen, in both cases for distance values less than 20 cm the results are not good. But, for values greater than 20 cm, the results are almost linear. Note that like in the previous case, the values from the sensors in both robots are different, despite the fact that they are the same type of sensor.

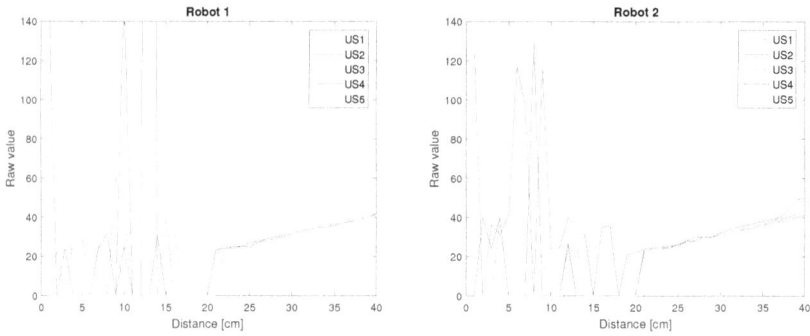

**Figure 5.** Robots 1 and 2 US sensors.

Following the traditional method the next step is to calculate the model of both type of sensors with the obtained values. Note that the model for each kind of sensor is different. You can obtain a model that combine these two models into one, like in [24].

In the resulting model, for values less than 20 cm the model of the IR sensors is used and for values grater than 20 cm the US model is used. Note that if you need more accuracy, you can obtain a different model for each combination of US and IR on each robot. This is a heavy work because you need to do this process manually in a repetitive way for each robot/sensor if you want to obtain accurate models for all sensors.

As was mentioned before, the results of the manual calibration shows that the information of the sensor must be fused to detect obstacles in the entire range (from 0 cm to 40 cm) in front half of the robot. While in the half back of the robot the obstacles can only be detected from 0 cm to 20 cm.

Figure 6 shows the mean value and the standard deviation of the sensors measurements for the Robot 1. On the left side are represented these values for the IR sensors. As can be seen, for short distances (between 2 cm and 15 cm) there are appreciable differences between the raw values of the sensors for the same distance. Which means that the use of the same model for each sensor may cause erroneous measurements in this range. For distances greater than 20 cm the differences are smaller.

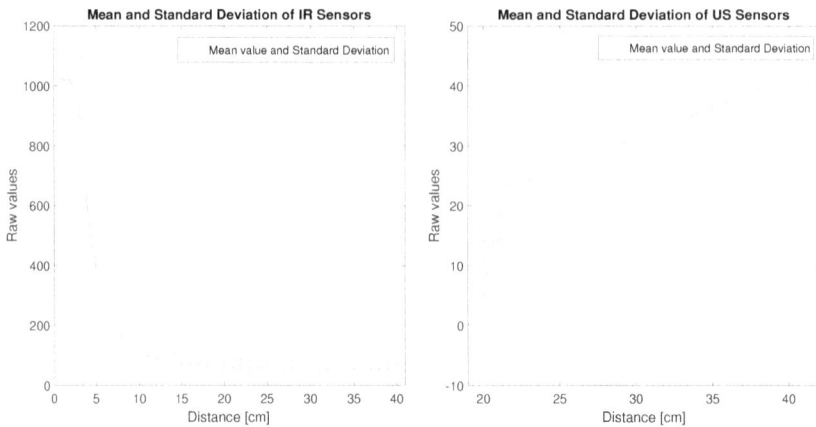

**Figure 6.** Mean and Standard Deviation of the IR and US sensors of Robot 1.

On the right side of Figure 6, the values of the US sensors are shown. As can be seen, for distances between 20 cm and 26 cm there are appreciable differences between the raw values of the sensors for the same distance. For distances greater than 27 cm the differences are smaller.

Next section presents the main contribution of this work: An automatic smart method to carry out the calibration of the sensors based on data and using an artificial neural network.

## 3. Automatic Smart Calibration Method

As was explained before, the traditional calibration is a manual and repetitive process that can be heavy depending of the accuracy that you need to obtain from the model of your sensors. In this section the novel automatic method based on data and using artificial neural networks is proposed.

### 3.1. Platform Used in the Laboratory

The method needs to obtain the absolute position of the robot each step execution. This task is carried out by the platform presented in [25], which is a previous work of the authors. This platform implements an IPS (Indoor Positioning System) to locate the robot. Figure 7 shows the architecture of the platform whose components are the following: (a) A personal computer (PC) with the Ubuntu Linux operating system that executes the software tools (Swistrack [26,27] and Monitor Module); (b) a PlayStation 3 (PS3) USB camera (fixed to the ceiling) connected via USB to the PC (The images of this camera are processed by *Swistrack* to obtain the position of the robot); (c) a Wi-Fi Router that communicates the Khepera IV robot with the PC; (d) an IP camera to show the performance of the experiments.

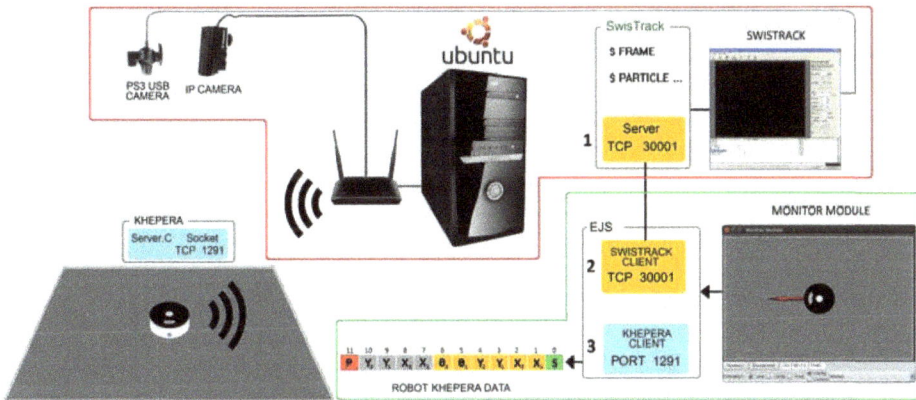

**Figure 7.** Hardware and software components of the platform.

### 3.2. Automatic Smart Calibration Method

The method consists of constructing the model of the proximity sensors automatically. For this, some obstacles are placed in known locations in the workspace. Then the robot navigates through the workspace storing the values from the proximity sensors. With the known positions of the obstacles and the information provided by the sensors of the robot, a neural network is built offline based on this data. The resulting model gets the raw values from the sensors as input and provides as output the distance to the obstacles around the robot. Note that this method needs to know the absolute position of the robot during the experiment.

The method consist of the following steps:

1. Set up of the workspace: A square shaped wall (80 cm) is placed in the Arena. The robot is placed in the center of the square with a known orientation. Note the model will work until 40 cm. The robot starts to move implementing the Braitenberg algorithm [28] to avoid the obstacles. Figure 8 shows the aerial view of the set up.

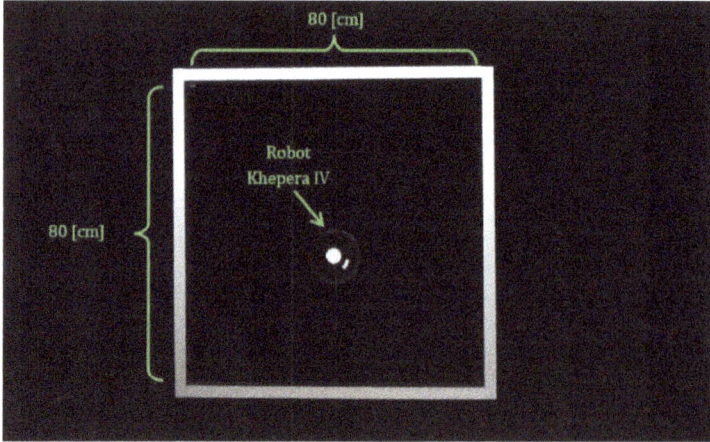

**Figure 8.** Set up of the workspace.

2. Data acquisition: The robot stores the data of the sensors ($S_1$, $S_2$, ..., $S_n$) and the time during its movement. On the other hand, the PC stores the position of the robot (coordinates and orientation) and also the time during the experiment. At the beginning of the process the time in the PC and the robot starts in 0. But the sample time is different for both because they run independently. Which means that the acquired data must be synchronized in time by interpolation.

   After the experiment finishes, the data of the robot is copied to the PC. Since the time stored in the robot and in the PC are not coincident this data needs to be adjusted and synchronized. This process is carried out of a code developed in MATLAB to do this task. Figure 9 shows an example of the data stored in the robot and in the PC.

| Robot Data | | | | | PC Data | | | |
|---|---|---|---|---|---|---|---|---|
| S1 | S2 | ... | Sn | Time | X | Y | θ | Time |
| 20 | 30 | ... | 40 | 0.1 | 10 | 5 | 90 | 0.15 |
| 25 | 87 | ... | 20 | 0.2 | 8 | 7 | 95 | 0.30 |
| . | . | ... | . | . | . | . | . | . |
| . | . | ... | . | . | . | . | . | . |
| . | . | ... | . | . | . | . | . | . |
| 50 | 50 | ... | 80 | 308.80 | -10 | 5 | 1 | 308.75 |
| 40 | 75 | ... | 20 | 308.95 | -10 | 7 | 2 | 309.15 |
| 30 | 80 | ... | 25 | 309.10 | -10 | 7 | 2 | 309.40 |

**Figure 9.** Data stored in the robot and in the PC.

Sensors **2018**, 18, 683

3. Data conditioning and synchronization: The data is adjusted by liner interpolation of the time, using Equation (1):

$$
\begin{cases}
X\left(t\right) = \left(\frac{X(T_1)-X(T_0)}{T_1-T_0}\right) * \left(t - T_0\right) + X\left(T_0\right) \\
Y\left(t\right) = \left(\frac{Y(T_1)-Y(T_0)}{T_1-T_0}\right) * \left(t - T_0\right) + Y\left(T_0\right), \quad T_0 < t \le T_1 \\
\theta\left(t\right) = \left(\frac{\theta(T_1)-\theta(T_0)}{T_1-T_0}\right) * \left(t - T_0\right) + \theta\left(T_0\right)
\end{cases}
\tag{1}
$$

where $T_0$ and $T_1$ are the initial and final time of the position data where the interpolation is made. $t$ is the time that is evaluated which corresponds with the data of the sensors. $X\left(t\right)$, $Y\left(t\right)$ and $\theta\left(t\right)$ are the pose of the robot in time $t$. Figure 10 shows the results of this process, where the data of the robot sensors and the robot positions are merged.

| Robot & PC Data | | | | | | | |
|---|---|---|---|---|---|---|---|
| S1 | S2 | ... | Sn | X | Y | θ | Time |
| 25 | 87 | ... | 20 | 9.3 | 5.6 | 91.7 | 0.2 |
| . | . | ... | . | . | . | . | . |
| . | . | ... | . | . | . | . | . |
| . | . | ... | . | . | . | . | . |
| 50 | 50 | ... | 80 | -10 | 5.3 | 1.1 | 308.80 |
| 40 | 75 | ... | 20 | -10 | 6.0 | 1.5 | 308.95 |
| 30 | 80 | ... | 25 | -10 | 6.8 | 1.9 | 309.10 |

**Figure 10.** Robot and PC data merged.

Once the position of the robot is obtained, it is necessary to obtain the positions of all sensors for each robot position. The positions of the sensors only depend on the pose of the robot. Using Equation (2) this position can be calculated.

$$
\begin{cases}
X_{sn} = radius_{robot} * \cos\left(angle_{sn} + \theta_{robot}\right) + X_{robot} \\
Y_{sn} = radius_{robot} * \sin\left(angle_{sn} + \theta_{robot}\right) + Y_{robot}
\end{cases}
\tag{2}
$$

Figure 11 represents the distances from the sensors to the walls (yellow colour). The black solid lines are the walls. The blue circle is the robot, the small rectangles are the sensors and the blue arrow represents the orientation of the robot. The dashed lines represent the detection directions of the senors. The dashed circle (40 cm of radius) represents the maximum detection range of the sensors.

The distance from each sensor to the walls is calculated by the intersection points (red) between the detection direction lines and the walls. Equation (3) shows how the coordinates of the intersection point $P(X_m;Y_m)$ are calculated for sensors 1, 2 and 3.

$$
\begin{cases}
X_m = 40 \\
Y_m = \left(\frac{Y_{robot}-Y_{sn}}{X_{robot}-X_{sn}}\right) * \left(X_m - X_{sn}\right) + Y_{sn}
\end{cases}
\tag{3}
$$

For sensors 7 and 8 is the same process only changing the axis because in this case the intersection is with the wall x = 40. Having these intersection points the distances and the angles are easily calculated by Equation (4). Finally the obtained data is filtered.

$$
\begin{cases}
\theta_m = atan\left(\frac{Y_{robot}-Y_m}{X_{robot}-X_m}\right) \\
d_m = \sqrt{\left(X_{sn} - X_m\right)^2 + \left(Y_{sn} - Y_m\right)^2}
\end{cases}
\tag{4}
$$

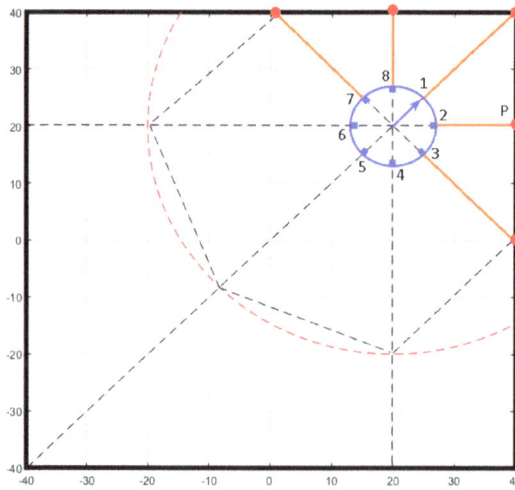

**Figure 11.** Distance from the sensors to known obstacles (walls).

4. Artificial Neural Network (ANN): With the calculated values and the raw data of the sensors, an ANN is built using the Neural Network Toolbox 9.0 of MATLAB [29], to obtain the model. Figure 12 shows a representation of the ANN implemented based on Levenberg-Marquardt method [30].

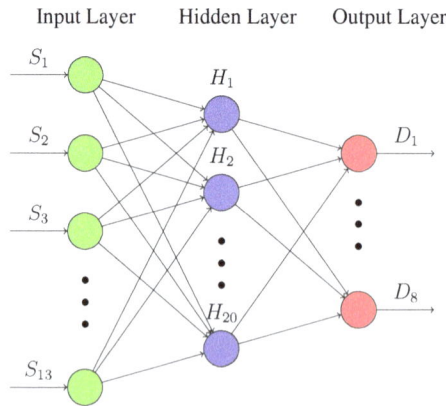

**Figure 12.** Artificial neural network implemented.

The input layer receives the 13 raw values ($S_1$–$S_{13}$) of the sensors (8 IR + 5 US). While the hidden layer has 20 neurons ($H_1$–$H_{20}$) with the Hyperbolic Tangent as activation function. The output layer has 8 neurons with the Identity as activation function. The outputs ($D_1$–$D_8$) are distances to the obstacles in the positions of the sensors (every 45° around the robot). The training was carried out with the 70% of the acquired data, the validation with another 15% and the testing with the remaining 15%. Note that the parameters of the ANN (number of neurons in the hidden layer and activation functions) have been determined empirically by trial and error after multiple tests.

5.  Testing the resulting model: Figure 13 shows a test developed in the workspace where the model was obtained. The sequence of images shows the displacement of the robot through the scenario. The black lines are the walls of the square (80 cm by 80 cm). The red small line represents the position and orientation of the robot. The performance of the models is drawn around the robot: (a) Blue lines: are the real distances to the obstacles (less than 40 cm and calculated with the intersection process described before); (b) Green lines: are the results of the the ANN model, taking and fusing the raw inputs from the sensors and providing the 8 distances; and (c) Red lines: are the distances to the obstacles with the traditional calibration method.

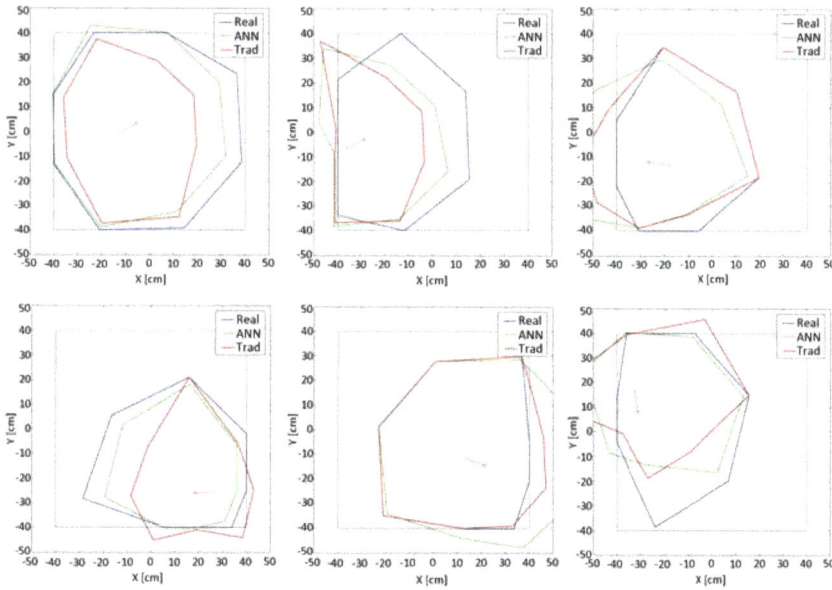

**Figure 13.** Images sequence of a test in the workspace.

As can be seen, all the models have different behaviors for different situations. The better results are obtained in general for distances less than 30 cm. But there are appreciable differences between the results of the half front part of the robot and the back half. This is due to the fact that the back part of the robots does not have US sensors. Even so, these results are more than acceptable compared with the majority of existing obstacle avoidance methods, and based on the maximum velocity of this robot and its steering characteristics.

Table 1 shows the Mean Absolute Percentage Error (MAPE) of the previous tests with respect to the real distances. The rows represent the distances ($D_1$–$D_8$) to the obstacles from the position of the sensors. The columns represent the results for both models (Artificial Neural Network (ANN) and Traditional (Trad)) for Robots 1 and 2.

**Table 1.** Mean Absolute Percentage Error for both models and robots.

| | MAPE (%) | | | |
| | Robot 1 | | Robot 2 | |
| | ANN | Trad | ANN | Trad |
|---|---|---|---|---|
| $D_1$ | 55.77 | **54.33** | **30.45** | 38.49 |
| $D_2$ | **67.06** | 79.01 | **25.56** | 40.09 |
| $D_3$ | **66.64** | 76.20 | **21.05** | 29.24 |
| $D_4$ | **50.71** | 66.46 | **12.68** | 23.44 |
| $D_5$ | **44.35** | 51.74 | **9.50** | 17.17 |
| $D_6$ | **47.61** | 51.12 | **11.76** | 21.54 |
| $D_7$ | **46.29** | 56.80 | **21.68** | 39.24 |
| $D_8$ | 58.09 | **55.37** | **32.03** | 42.45 |

In all the cases, the smallest error values are represented in bold. As can be seen, the results for both robots are different. In the case of Robot 2 the traditional method shows worse results than the ANN method. In the case of Robot 1, the behavior is similar to the previous one. Only distances 1 and 8 present worse results. In general, for both cases the ANN model presents better results than the traditional model.

Figure 14 shows the flow diagram of the developed method. The block related to the robot are represented in green colour. The orange colour represents the blocks related to the PC. Note that the ANN model is obtained offline with the acquired data by the robot and by the PC. After that the model is implemented in the robot. The model gets the raw values from the sensors and gives the distances to the objects in the 8 directions where the sensors are located in the robot.

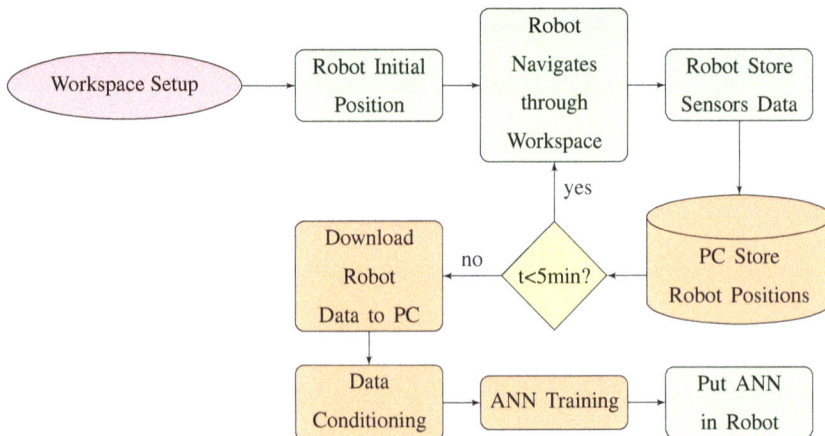

**Figure 14.** Flow diagram of the developed calibration procedure.

Note that this method has been developed for the Khepera IV robot. The properties of the robot and platform have a direct influence in the performance of the method. But it does not mean that the method cannot be implemented in other robots/platforms that use proximity sensors to estimate the distance to obstacles. To do that, one important element to be taken into account is the use of an indoor positioning system which gives the absolute position of the robot in running time. Other important aspect is the location of the sensors around the robot. This configuration has to be considered when the distances from each sensor to the walls (see Figure 11) are calculated by using Equations (2)–(4).

## 4. Experiment of Position Control with Obstacles Avoidance

This experiment is titled position control or point stabilization of a differential wheeled mobile robot. This problem has been widely studied mainly from an automatic control perspective [31]. The objective of the experiment is to drive the robot from the current position $C(x_c, y_c)$ and orientation $(\theta)$ to the target point $T_p(x_p, y_p)$. Figure 15 shows a representation of the variables involved in this experiment.

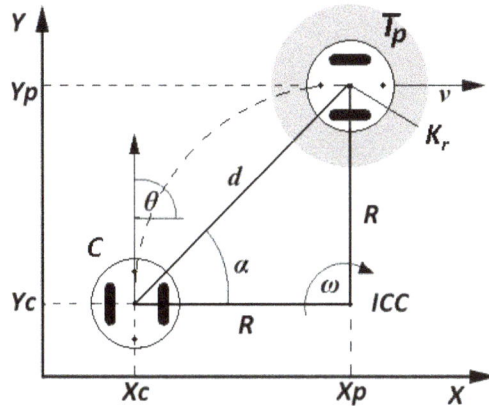

**Figure 15.** Position control problem.

Figure 16 shows the feedback control loop block diagram of this experiment but in this case, including the obstacle avoidance algorithm and the model obtained with the developed method.

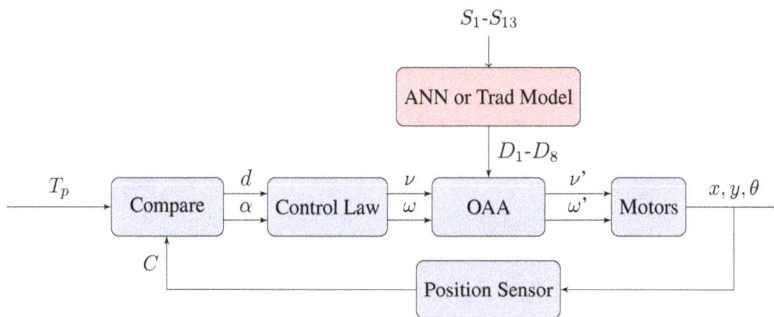

**Figure 16.** Block diagram of the position control problem.

The block Compare calculates the distance ($d$) and the angle $\alpha$ to the target point ($T_p$) from the current position of the robot ($C$). The block Control Law tries to minimize the orientation error, $\theta_e = \alpha - \theta$, and at the same time, to reduce the distance to the target point ($d = 0$) by manipulating the control signals (linear velocity ($v$) and angular velocity ($w$) of the robot). Equations (5) and (6) show the implementation of this control law based on [32].

$$v = \begin{cases} v_{max} & \text{if } |d| > K_r \\ d\left(\frac{v_{max}}{K_r}\right) & \text{if } |d| \leq K_r \end{cases} \tag{5}$$

$$\omega = \omega_{max} sin\left(\theta_e\right) \tag{6}$$

The block OAA represents the Obstacle Avoidance Algorithm, in this case: Braitenberg algorithm [28], in which the robot's sensors are tied directly to the motor controls and motor speeds respond to the sensor input directly. The Model block represents the developed algorithm which obtains the distances to the obstacles ($D_1$–$D_8$) from the raw data of proximity sensors ($S_1$–$S_{13}$). If no obstacle is detected, the output velocities of the block OAA are the same as its inputs ($v$, $\omega$).

The Braitenberg algorithm [28] creates a weighted matrix that converts the sensor inputs into motor speeds. This matrix is a two-dimensional array with the number of columns corresponding to the number of obstacle sensors (8) and the number of rows corresponding to the number of motors (2). The weights of the matrix are determined empirically depending on the location of the sensors in the robot. The 8 sensors of the Khepera IV robot are numbered clockwise beginning with the front sensor. Equation (7) represents the mentioned matrix where, for example, the term $W_{LS_1}$ represents the weight of the sensor $S_1$ in the speed of the left motor. Equation (8) represents the raw data of the proximity sensors at each time.

$$W = \begin{pmatrix} W_{LS_1} & W_{LS_2} & \cdot & \cdot & \cdot & W_{LS_8} \\ W_{RS_1} & W_{RS_2} & \cdot & \cdot & \cdot & W_{RS_8} \end{pmatrix} \tag{7}$$

$$S = \begin{pmatrix} S_1 & S_2 & \cdot & \cdot & \cdot & S_8 \end{pmatrix}^T \tag{8}$$

With these matrices, the velocities for each motor are calculated as is shown in Equation (9). Where ($S_{max}$) represents the maximum value of the sensor output.

$$v_{L,R} = W * \left(1 - (S/S_{max})\right) \tag{9}$$

Figure 17 shows the top view of the set-up of the scenario to carry out two experiments where the elements are the following: (a) red arrows represent the starting points and orientation of the robot; (b) green circle represents the target point for both cases; and (c) obstacles are walls (white colour) and cylinders (blue/red colours).

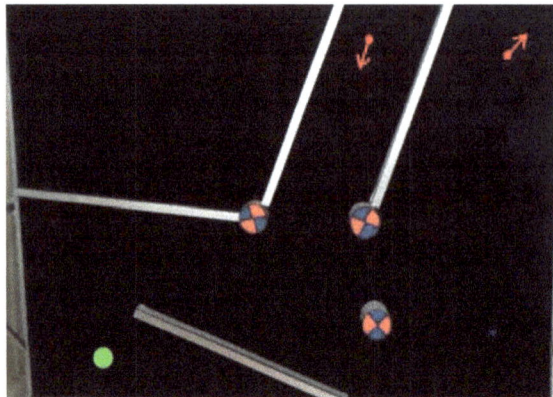

**Figure 17.** Scenario for obstacle avoidance experiments.

Figure 18 shows the results of both experiments with the ANN and Traditional models. The red lines and circles are the obstacles. In both cases, the robot begins the experiment starting at point 1 (cyan circle) and it must reach the target point 2 (green circle). The small blue points represent the position of the robot during the experiment for the traditional calibration model (Trad), while the small red points represent the position of the robot for the developed method (ANN).

In both cases the robot reaches the target point avoiding the obstacles that it finds on its way. As can be seen, the behavior of the robot with the ANN model has better performance than with the traditional model. These differences show that you can use the functions provided by the manufacturer or calibrate the sensors manually using the traditional method. In both cases you need to take into account that your obstacle avoidance algorithm can be working correctly for the values of distances that it is receiving. However, the real problem is that the sensors can be providing erroneous values due to bad calibration or the use of a poorly tuned model for all the sensors.

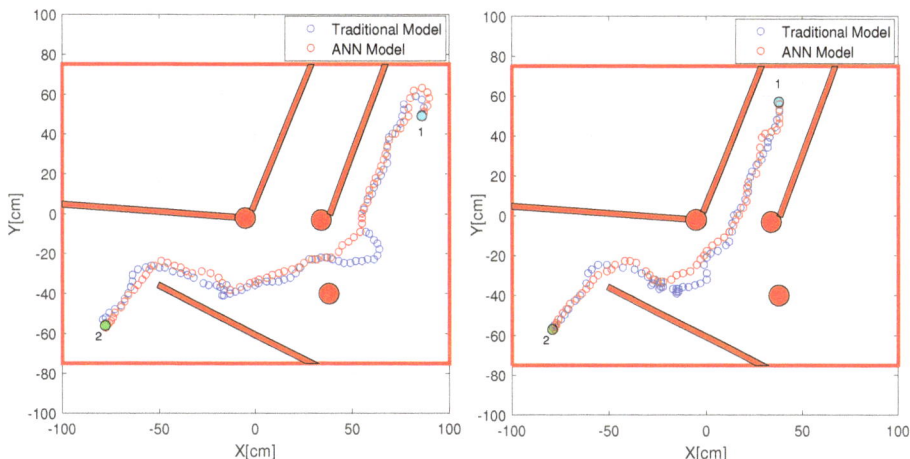

**Figure 18.** Obstacles avoidance experiment results for Robot 2.

Table 2 shows the Mean Absolute Percentage Error (MAPE %) of these experiments presented in Figure 18. The columns represent the following: (Exp) the number of the experiment, (Method) the model, (MAPE) the mean values of the distance errors (from $D_1$ to $D_8$) and (Time) the time to reach the target point.

**Table 2.** Results of three experiments for both models.

| Exp | Method | MAPE (%) | Time (s) |
|-----|--------|----------|----------|
| 1 | Trad | 83.9 | 59.4 |
| 1 | ANN | **65.5** | **50.8** |
| 2 | Trad | 88.8 | 60.4 |
| 2 | ANN | **58.4** | **51.3** |

The results represented in bold indicate the smallest values of the errors and time. As can be seen, for all experiments the ANN model presents less mean error than the traditional model. In addition, in both experiments the robot reaches the target point in less time, which means that the robot describes a smoother and more direct path to the target point due to better detection and avoidance of the obstacles.

Additionally, Figure 19 shows the angular velocity of the robot in the second experiment (right side of Figure 17) for both models: blue line represents the traditional model; and red line represents the ANN model.

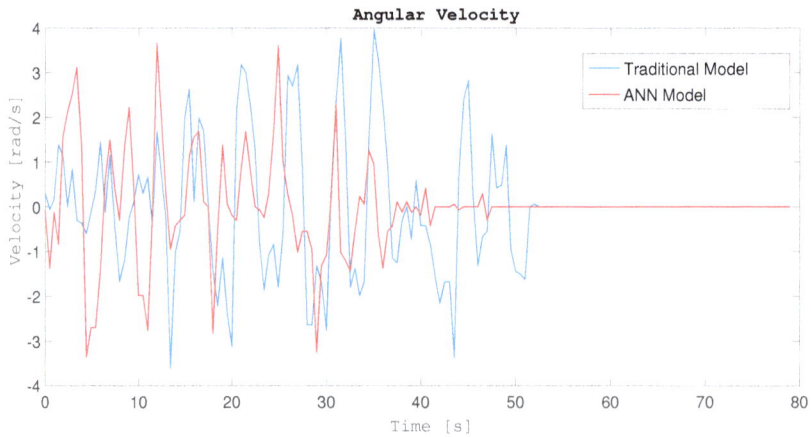

**Figure 19.** Angular velocities.

As can be seen, for the same experiment, with the ANN model the angular velocity has less oscillations than with the traditional model. Due to better detection and avoidance of the obstacles, the behavior of the robot is smoother because it can avoid the obstacles in a better way with less direction changes. Because it reaches the target point first, that means that the trajectory is more efficient. Figure 20 shows the linear velocities for the same experiments and models.

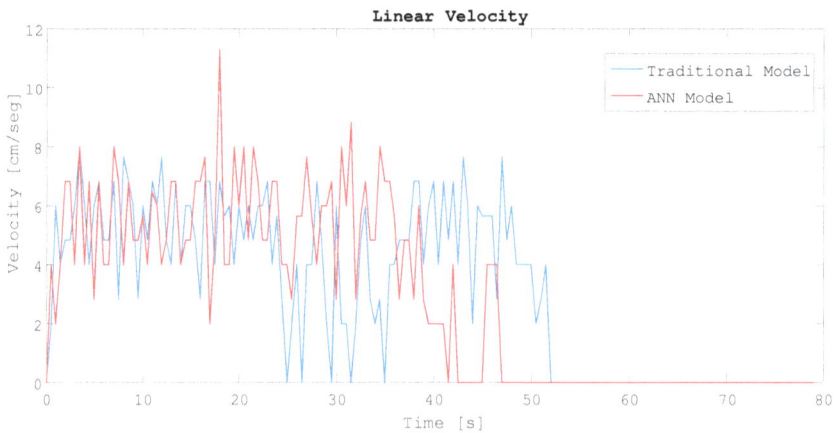

**Figure 20.** Linear velocities.

In the case of the linear velocities, also the ANN model has less oscillations and it becomes zero first. This means that the robot reaches the target point faster than with the traditional model and with a more constant velocity.

## 5. Conclusions

The most common proximity sensors in mobile robotics are the US and IR sensors. They are widely used for obstacle detection without physical contact. But they introduce uncertainties in the measurements due to their mode of operation: they need a feedback signal that depends on the surface of the object and the ambient conditions. Another important issue for these sensors is the calibration of their outcomes. Usually one has to obtain a model that converts the raw measurements provided by the sensors in distances. The method for building this model is a repetitive and lengthy process that usually is done manually. Moreover, the obtained model cannot provide the same distance even for two sensors of the same type (due to the uncertainties introduced by the surfaces of the objects). That is why in some situations it could be useful to gather all measurements provided by different sensors and combine them in order to get an accurate model to estimate the distances to the obstacles around the robot.

This paper presents a novel approach to get an obstacle detection model based on the combination of sensor data. The method is automatic and based on data using machine learning. The method implements an artificial neural network to obtain the model of the sensors. The model receives the raw values from the 13 sensors (8 IR + 5 US) and provides the distances to the obstacles around the robot (8 values). This means that the method combines the information from both kinds of sensors and builds a unique model for obstacle distances.

The developed method has been designed for the Khepera IV robot in an indoor environment with an IPS to obtain its absolute position in running time. Although it has been designed for these characteristics, this method can be implemented to different platforms. The fundamental points that have to be taken into account are the indoor positioning system and the location of the sensors around the robot.

To show the performance of the developed model, some results are provided and discussed. The experiment of position control with obstacles avoidance based on the Braitenberg algorithm was selected. To test it, two different scenarios were used for this experiment. In addition, the velocity of the robots was analyzed in both cases. The comparison between the traditional and the developed method shows the improvement in the results.

This method is undoubtedly an advance in the calibration of proximity sensors because it eliminates the traditional laborious method and provides better results. Future works are related to the implementation of this method in other platforms with other kinds of robots and sensors.

**Acknowledgments:** This work has been funded by the Chilean Ministry of Education under the Project FONDECYT 1161584, and the National Plan Project DPI2014-55932-C2-2-R of the Spanish Ministry of Economy and Competitiveness and funds from the Universidad Nacional de Educación a Distancia of Spain (UNED).

**Author Contributions:** Emmanuel Peralta designed and developed the platform and the work with the sensors of the robot, performed the experiments and analyzed the results. Ernesto Fabregas collaborated in the development of the platform and wrote the manuscript. Gonzalo Farias, Héctor Vargas and Gabriel Hermosilla performed the theoretical conception and planning of the system. Gonzalo Garcia revised the manuscript with Sebastián Dormido.

**Conflicts of Interest:** The authors declare no conflict of interest.

## References

1. Zhang, W.; Zelinsky, G.; Samaras, D. Real-time Accurate Object Detection using Multiple Resolutions. In Proceedings of the 11th IEEE International Conference on Computer Vision (ICCV), Rio de Janeiro, Brazil, 14–21 October 2007; pp. 1–8.
2. Ren, C.L.; Chun, C.L. Multisensor Fusion-Based Concurrent Environment Mapping and Moving Object Detection for Intelligent Service Robotics. *IEEE Trans. Ind. Electr.* **2014**, *61*, 4043–4051.
3. Kaijen, H.; Nangeroni, P.; Huber, M.; Saxena, A.; Ng, A.Y. Reactive grasping using optical proximity sensors. In Proceedings of the IEEE International Conference on Robotics and Automation (ICRA), Kobe, Japan, 12–17 May 2009; pp. 2098–2105.
4. Almasri, M.; Alajlan, A.; Elleithy, K. Trajectory Planning and Collision Avoidance Algorithm for Mobile Robotics System. *IEEE Sens. J.* **2016**, *12*, 5021–5028.
5. Alajlan, M.; Almasri, M.; Elleithy, K. Multi-sensor based collision avoidance algorithm for mobile robot. In Proceedings of the IEEE Long Island Systems, Applications and Technology Conference (LISAT), New York, NY, USA, 1 May 2015; pp. 1–6.

6. Almasri, M.; Elleithy, K.; Alajlan, A. Sensor Fusion Based Model for Collision Free Mobile Robot Navigation. *Sensors* **2016**, *16*, 24.

7. Webster, J.G.; Halit, E. *Measurement, Instrumentation, and Sensors Handbook: Spatial, Mechanical, Thermal, and Radiation Measurement*, 2nd ed.; CRC Press: Boca Raton, FL, USA, 2017.

8. Mustapha, B.; Zayegh, A.; Begg, R. Ultrasonic and infrared sensors performance in a wireless obstacle detection system. In Proceedings of the 1st International Conference on Artificial Intelligence, Modelling and Simulation (AIMS), Kota Kinabalu, Malasya, 3–5 December 2013; pp. 487–492.

9. Adarsh, S.; Kaleemuddin, S.M.; Bose, D.; Ramachandran, K.I. Performance comparison of Infrared and Ultrasonic sensors for obstacles of different materials in vehicle/robot navigation applications. In Proceedings of the IOP Conference Series: Materials Science and Engineering, Bangalore, India, 14–16 July 2016; pp. 1–7.

10. Shrivastava, A.K.; Verma, A.; Singh, S.P. Distance Measurement of an Object or Obstacle by Ultrasound Sensors using P89C51RD2. *Int. J. Comput. Theory Eng.* **2010**, *64*, 64–68.

11. Mohd Rusdy, Y.; Anuar, M.K.; Syahrim, N.; Anwar, N. Effect of glittering and reflective objects of different colours to the output voltage-distance characteristics of sharp GP2D120 IR. *Int. J. Electr. Power Eng.* **2012**, *2*, 6–10.

12. Benet, G.; Blanes, F.; Simó, J.E.; Pérez, P. Using infrared sensors for distance measurement in mobile robots. *Robot. Auton. Syst.* **2002**, *40*, 255–266.

13. Paunović, I.; Todorović, D.; Božić, M.; Đorđević, G.S. Calibration of ultrasonic sensors of a mobile robot. *Serbian J. Electr. Eng.* **2009**, *3*, 427–437.

14. Fung, M.L.; Chen, M.Z.; Chen, Y.H. Sensor fusion: A review of methods and applications. In Proceedings of the 29th Chinese Control And Decision Conference (CCDC), Chongqing, China, 28–30 May 2017; pp. 3853–3860.

15. Khaleghi, B.; Khamis, A.; Karray, F.O.; Razavi, S.N. Multisensor data fusion: A review of the state-of-the-art. *Inf. Fusion* **2013**, *14*, 28–44.

16. Martens, S.; Gaudiano, P.; Carpenter, G.A. Mobile robot sensor integration with fuzzy ARTMAP. In Proceedings of the Intelligent Control (ISIC) joinly with IEEE International Symposium on Computational Intelligence in Robotics and Automation (CIRA), Gaithersburg, MD, USA, 17 September 1998; pp. 307–312.

17. Guanshan, H. Neural Network Applications in Sensor Fusion for a Mobile Robot Motion. In Proceedings of the WASE International Conference on Information Engineering (ICIE), Qinhuangdao, China, 14–15 August 2010; pp. 46–49.

18. Akkaya, R.; Aydogdu, O.; Canan, S. An ANN based NARX GPS/DR System for Mobile Robot Positioning and Obstacle Avoidance. *J. Autom. Control* **2013**, *1*, 6–13.

19. Barton, A.; Volna, E. Control of autonomous robot using neural networks. In Proceedings of the International Conference of Numerical Analysis and Applied Mathematics (ICNAAM), Rhodes, Greece, 19–25 September 2016; p. 070002.

20. Safari, S.; Shabani, F.; Simon, D. Multirate multisensor data fusion for linear systems using Kalman filters and a neural network. *Aerosp. Sci. Technol.* **2014**, *39*, 465–471.

21. Myat, S.N.; Hla, M.T. Implementation of Multisensor Data Fusion Algorithm. *Int. J. Sens. Sens. Netw.* **2017**, *5*, 48–53.

22. Kam, M.; Zhu, X.; Kalata, P. Sensor fusion for mobile robot navigation. *Proc. IEEE* **1997**, *85*, 108–119.

23. KTeam. Khepera IV User Manual. 2015. Available online: http://ftp.k-team.com/KheperaIV/UserManual/ (accessed on 2 October 2017).

24. De Silva, O.; George, K.M.; Raymond, G.G. An ultrasonic and vision-based relative positioning sensor for multirobot localization. *IEEE Sens. J.* **2015**, *15*, 1716–1726.

25. Fabregas, E.; Farias, G.; Peralta, E.; Sánchez, J.; Dormido, S. Two Mobile Robots Platforms for Experimentation: Comparison and Synthesis. In Proceedings of the 14th International Conference on Informatics in Control, Automation and Robotics (ICINCO), Madrid, Spain, 26–28 July 2017; pp. 439–445.

26. Correll, N.; Sempo, G.; De Meneses, Y.L.; Halloy, J.; Deneubourg, J.L.; Martinoli, A. SwisTrack: A tracking tool for multi-unit robotic and biological systems. In Proceedings of the IEEE/RSJ International Conference on Intelligent Robots and Systems (IROS), Beijing, China, 9–15 October 2006; pp. 2185–2191.

27. Lochmatter, T.; Roduit, P.; Cianci, C.; Correll, N.; Jacot, J.; Martinoli, A. Swistrack—A flexible open source tracking software for multi-agent systems. In Proceedings of the Intelligent Robots and Systems (IROS), Nice, France, 22–26 September 2008; pp. 4004–4010.

28. Yang, X.; Patel, R.V.; Moallem, M. A fuzzy—Braitenberg navigation strategy for differential drive mobile robots. *J. Intell. Robot. Syst.* **2006**, *47*, 101–124.

29. Mathworks. MATLAB Neural Network Toolbox 9.0. 2017. Available online: https://es.mathworks.com/help/nnet/index.html (accessed on 2 October 2017).

30. Çelik, O.; Teke, A.; Yıldırım, H.B. The optimized artificial neural network model with Levenberg-Marquardt algorithm for global solar radiation estimation in Eastern Mediterranean Region of Turkey. *J. Clean. Prod.* **2016**, *116*, 1–12.

31. Kuhne, F.; Lages, W.F.; Da Silva, J.G. Point stabilization of mobile robots with nonlinear model predictive control. In Proceedings of the IEEE International Conference Mechatronics and Automation (ICMA), Niagara Falls, ON, Canada, 29 July–1 August 2005; pp. 1163–1168.

32. González Villela, V.J.; Parkin, R.; López Parra, M.; Dorador González, J.M.; Guadarrama Liho, M.J. A wheeled mobile robot with obstacle avoidance capability. *Ing. Mec. Tecnol. Desarro.* **2004**, *5*, 159–166.

*Article*

# Smart Sound Processing for Defect Sizing in Pipelines Using EMAT Actuator Based Multi-Frequency Lamb Waves

Joaquín García-Gómez [1], Roberto Gil-Pita [1,*], Manuel Rosa-Zurera [1], Antonio Romero-Camacho [2], Jesús Antonio Jiménez-Garrido [2] and Víctor García-Benavides [2]

[1] Department of Signal Theory and Communications, University of Alcalá, Ctra. Madrid-Barcelona, km. 33,600, 28805 Alcalá de Henares, Spain; joaquin.garciagomez@uah.es (J.G.-G.); manuel.rosa@uah.es (M.R.-Z.)

[2] Innerspec Technologies Europe S.L., Av. de Madrid 2, 28802 Alcalá de Henares, Spain; aromero@innerspec.com (A.R.-C.); jjimenez@innerspec.com (J.A.J.-G.); vgarcia@innerspec.com (V.G.-B.)

\* Correspondence: roberto.gil@uah.es; Tel.: +34-91-885-6751

Received: 15 January 2018; Accepted: 5 March 2018; Published: 7 March 2018

**Abstract:** Pipeline inspection is a topic of particular interest to the companies. Especially important is the defect sizing, which allows them to avoid subsequent costly repairs in their equipment. A solution for this issue is using ultrasonic waves sensed through Electro-Magnetic Acoustic Transducer (EMAT) actuators. The main advantage of this technology is the absence of the need to have direct contact with the surface of the material under investigation, which must be a conductive one. Specifically interesting is the meander-line-coil based Lamb wave generation, since the directivity of the waves allows a study based in the circumferential wrap-around received signal. However, the variety of defect sizes changes the behavior of the signal when it passes through the pipeline. Because of that, it is necessary to apply advanced techniques based on Smart Sound Processing (SSP). These methods involve extracting useful information from the signals sensed with EMAT at different frequencies to obtain nonlinear estimations of the depth of the defect, and to select the features that better estimate the profile of the pipeline. The proposed technique has been tested using both simulated and real signals in steel pipelines, obtaining good results in terms of Root Mean Square Error (RMSE).

**Keywords:** EMAT actuators; Lamb waves; pipeline inspection; defect sizing; smart sound processing

## 1. Introduction

Ultrasonic techniques have demonstrated over the years to be really useful for Non-Destructive Testing (NDT) examinations [1–3]. Conventional ultrasounds are primarily generated taking advantage of the piezoelectric effect. Although it is an efficient way of generating ultrasounds, a proper coupling between the transducer and test specimens is needed, which is a disadvantage. Therefore, materials inspected by conventional ultrasounds are covered with a thin layer of fluid. EMAT (Electro-Magnetic Acoustic Transducer) actuators are able to generate and receive ultrasonic waves without the need to have thorough contact with the surface of the material under investigation [4]. This technology is capable of generating multiple types of waves: Lamb, shear, longitudinal and Rayleigh. Besides, when EMAT technique is implemented with a meander-line-coil, the waves are generated in a directional way [5,6]. This fact is interesting since it allows differentiating between circumferential and axial scans.

A highlighted application of this technology is the pipeline inspection [7]. On the one hand, some manuscripts have focused on the defect detection and location in its circumferential path, mainly using shear waves [8–10]. However, it is important to obtain not only the position of the defect, but also its residual thickness. For the companies it is interesting to know this parameter, since it is a vital factor to make the decision to replace a section of the pipeline [11]. On the other hand, there have been some proposals of sizing techniques applied to pipeline inspection, mainly based on the analysis of the physical mode [12], but the distortion caused by the defects over the different modes strongly varies with the shape of the defect [12–14]. Smooth defects usually reflect less energy than abrupt ones, independently of the residual thickness of the pipe caused by the defect. Thus, the amplitude of the received echo is strongly related not only to the depth of the defect but also to its hardness, and both the amplitude and the time of arrival of the wrap-around signal vary with the length of the defect.

In general, the studies have followed the same line with regard to the information extracted from the wave modes. The most relevant and used parameters are the amplitude and the phase from the received signal [14,15]. Once this information is obtained, the use of Smart Sound Processing (SSP) techniques is suitable for solving sizing problems using EMAT guided waves. These methods involve extracting useful information from the sensed acoustic signals and applying nonlinear techniques to obtain estimations of useful parameters. Other proposals have applied this type of techniques in their studies, including: Artificial Neural Networks [16], Neural Networks with large number of neurons [17] or Adaptive Neuro-Fuzzy Inference Systems [18]. Using this type of methods is interesting to excite the coil at multiple frequencies, as the behavior of the Lamb modes is different depending on this parameter, and SSP allows to combine all this information and get better defect estimation results.

In this sense, this paper studies pipeline inspection mixing both the EMAT and SSP techniques. Specifically, EMAT-based Lamb waves will be generated at multiple frequencies. Axial scans will be developed and the circumferential path followed by waves will allow the analysis of the wrap-around signals received. These ultrasound signals will be related to the behavior of the pipeline depending on its profile, conditions and damage. Once these signals are measured, it will be possible to apply SSP techniques in order to get useful information from the amplitude and phase of the multi-frequency signals. In particular, feature selection techniques and Neural Networks-based estimators will be applied. Following this process, it will be feasible to obtain an approximated characterization of the residual thickness along the pipeline.

## 2. Materials and Methods

In this section the sensorization of Lamb waves through EMAT actuators will be described. Initially, EMAT technology and the fundamentals of Lamb waves will be introduced. A brief hardware description will be made at the end of the section as well.

### 2.1. Lamb Wave Generation Using EMAT Actuators

EMAT transducers consist of a coil wire and a magnet. The alternating electrical current flowing through the coil wire placed in a uniform magnetic field ($B$) near the surface of a ferromagnetic material, induces surfaces currents (Eddy Currents, $J$) in the material. The field generated by electrical coils interacts with the field generated by the magnet producing a Lorentz force ($F$) according to Equation (1).

$$F = J \times B \tag{1}$$

The disturbance is applied to the lattice of the material, producing an elastic wave. In a reciprocal process (reception of an ultrasonic wave), the interaction of elastic waves in the presence of a magnetic field induces currents in the EMAT receiver coil circuit. In Figure 1 a comparative between the generation of ultrasonic waves using conventional ultrasound methods and using EMAT technology is represented.

The advantages of using EMAT over piezoelectric transducers are: as the transduction process occurs within an electromagnetic depth skin, it is a couplant free technique; it is insensitive to surface conditions, being capable of inspecting rough, dirty (oily/wet), oxidized or uneven surfaces; inspection can be carried out on flat, curved or complex surfaces; it allows high speed inspections (up to 60 m/s), high temperature inspections and low temperature inspections, and it can generate Lamb, Shear

Horizontal (SH), Shear Vertical (SV), Longitudinal and Rayleigh waves due to its good selectivity in frequency. On the other hand, the challenges of EMAT are the high level of power required, the bigger size of the transducers and the lower Signal to Noise Ratio (SNR). Besides, the material under inspection needs to be conductive.

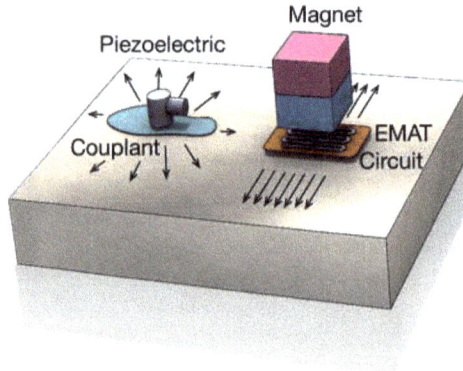

**Figure 1.** Conventional Ultrasound vs EMAT.

Guided Wave Testing is a NDT technique that employs ultrasonic stress waves that propagate along a structure while guided by its boundaries. Guided waves permit covering long distances from a single point with a limited number of sensors, being very effective for rapid scanning of pipelines and tanks. On relatively thin structures, it is possible to generate volumetric guided waves that fill up the material and permit a complete, volumetric inspection. The most common types of volumetric waves are SH and Lamb.

Lamb waves travel throughout the material with both vertical and forward motion in an elliptical pattern. These waves are dispersive by nature, and very sensitive to thickness variations. They can be classified in symmetric (also known as longitudinal) and asymmetric (also known as flexural) modes. The introduction of boundary conditions makes Lamb wave problems inherently more difficult than the more conventional bulk waves. Unlike the finite number of modes present in a bulk wave problem, there are an infinite number of modes associated with a given Lamb wave application. That is, a finite body can support an infinite number of different Lamb wave modes. Now the generation of the Lamb wave modes will be described. With this purpose, Lamé parameters will be defined. Lamé parameters are two material-dependent quantities denoted by $\lambda$ (Lamé's first parameter) and $\mu$ (Lamé's second parameter). They are defined by Equations (2) and (3).

$$\lambda = \frac{E\nu}{(1+\nu)(1-2\nu)} \tag{2}$$

$$\mu = \frac{E}{2(1+\nu)} \tag{3}$$

where $E$ is the Young's modulus, which measures the stiffness of a material, and $\nu$ is the Poisson's ratio, which is an elastic constant that measures how an elastic, linear and isotropic material is narrowed when it is longitudinally stretched.

Then the elastic wave equation needs to be taken into account.

$$\mu\nabla^2\mathbf{u} + (\lambda+\mu)\nabla\nabla\cdot\mathbf{u} = \rho\left(\frac{\partial^2\mathbf{u}}{\partial t^2}\right), \tag{4}$$

where $\rho$ represents the density of the material under inspection. Applying the Helmholtz decomposition, the displacement field **u** can be split into a rotational component $\nabla \times \mathbf{H}$ and an irrotational component $\nabla \phi$:

$$\mathbf{u} = \nabla \phi + \nabla \times \mathbf{H} \tag{5}$$

Then, the system of partial differential equations can be rewritten as:

$$c_L \nabla^2 \phi = \frac{\partial^2 \phi}{\partial t^2} \tag{6}$$

$$c_T \nabla^2 \mathbf{H} = \frac{\partial^2 \mathbf{H}}{\partial t^2}, \tag{7}$$

where $c_L = \sqrt{(\lambda + 2\mu)/\rho}$ and $c_T = \sqrt{\mu/\rho}$ represent the sound velocity for the longitudinal and transversal modes, respectively.

To continue with the analysis, an infinitely plate extended in the $x$ and $y$ directions will be assumed. Furthermore, it is considered that the wave propagates in the $x$ direction, the fields are uniform in the $y$ direction and boundary conditions at $z = -h/2$ and $z = +h/2$, where $h$ is the thickness of the plate, are considered traction free.

Assuming that the particle displacement is zero in the $y$ direction ($u_y = 0$) and the only rotation is about the $y$ axis ($H_x = H_z = 0$) in Equation (4), the Lamb wave equations are obtained [19,20].

$$\mu \left( \frac{\partial^2}{\partial x^2} + \frac{\partial^2}{\partial z^2} \right) u_x + (\lambda + \mu) \frac{\partial}{\partial x} \left( \frac{\partial u_x}{\partial x} + \frac{\partial u_z}{\partial z} \right) = \rho \left( \frac{\partial^2 u_x}{\partial t^2} \right) \tag{8}$$

$$\mu \left( \frac{\partial^2}{\partial x^2} + \frac{\partial^2}{\partial z^2} \right) u_z + (\lambda + \mu) \frac{\partial}{\partial z} \left( \frac{\partial u_x}{\partial x} + \frac{\partial u_z}{\partial z} \right) = \rho \left( \frac{\partial^2 u_z}{\partial t^2} \right) \tag{9}$$

Applying the restriction in the frontiers:

$$\mu \left( \frac{\partial u_x}{\partial z} + \frac{\partial u_z}{\partial x} \right) \Bigg|_{z = \pm \frac{h}{2}} = 0 \tag{10}$$

$$\lambda \frac{\partial u_x}{\partial x} + (\lambda + 2\mu) \frac{\partial u_z}{\partial z} \Bigg|_{z = \pm \frac{h}{2}} = 0 \tag{11}$$

Focusing on Lamb waves, they are composed of two waves (one longitudinal and one transversal) traveling at different angles $\theta_L$ and $\theta_T$, where the first one represents the longitudinal angle and the second one the transversal one. Therefore, the wave number $k$ is related to the component of the waves that propagates in the $x$ direction at velocity $c_p$:

$$k_L \cos \theta_L = k_T \cos \theta_T = k = \frac{2\pi f}{c_p} = \frac{\omega}{c_p}, \tag{12}$$

where $k_L = 2\pi f/c_L$, $k_T = 2\pi f/c_T$, $\omega$ is the angular velocity, $c_L$ is the longitudinal component of the velocity and $c_T$ is the transversal component of the velocity.

On the other hand, the displacement of each independent wave in the $z$ axis can be obtained using Equations (13) and (14).

$$\alpha_L = k_L \sin \theta_L = k_L \sqrt{1 - \cos^2 \theta_L} = \sqrt{k_L^2 - k^2} = \sqrt{\frac{\omega^2}{c_L^2} - k^2} = \omega \sqrt{\frac{1}{c_L^2} - \frac{1}{c_p^2}} \tag{13}$$

$$\alpha_T = k_T \sin \theta_T = k_T \sqrt{1 - \cos^2 \theta_T} = \sqrt{k_T^2 - k^2} = \sqrt{\frac{\omega^2}{c_T^2} - k^2} = \omega \sqrt{\frac{1}{c_T^2} - \frac{1}{c_p^2}} \tag{14}$$

where $\alpha_L$ and $\alpha_T$ represent the longitudinal and transversal displacements of the wave, and $\theta_L$ and $\theta_T$ are the angles related to these displacements. Taking into account that the wave is reflected in the surfaces, applying the boundary conditions and simplifying the equations, the dispersion equation of the Lamb modes is obtained. Equation (15) refers to the symmetric modes and Equation (16) refers to the asymmetric ones.

$$4k^2 \alpha_L \alpha_T \sin\left(\frac{\alpha_L h}{2}\right) \cos\left(\frac{\alpha_L h}{2}\right) + \sin\left(\frac{\alpha_L h}{2}\right) \cos\left(\frac{\alpha_L h}{2}\right) \left(\alpha_T^2 - k^2\right)^2 = 0 \qquad (15)$$

$$4k^2 \alpha_L \alpha_T \cos\left(\frac{\alpha_L h}{2}\right) \sin\left(\frac{\alpha_L h}{2}\right) + \cos\left(\frac{\alpha_L h}{2}\right) \sin\left(\frac{\alpha_L h}{2}\right) \left(\alpha_T^2 - k^2\right)^2 = 0 \qquad (16)$$

From the previous equations, it can be figured out that there exists a relation between the excited frequency $f$, the thickness of the pipe $z$ and the phase velocity $c_p$. More specifically, each mode will move at different $c_p$ depending on the other above-mentioned parameters. A similar relation can be obtained using the group velocity $c_g$, which is defined in Equation (17).

$$c_g = \frac{\partial \omega}{\partial k} \qquad (17)$$

Graphs showed in Figure 2 were obtained by means of the previous equations for different values of the product frequency by thickness, where phase velocity is represented in Figure 2a and group velocity in Figure 2b. As it can be observed, the relation between the velocities and the frequency is non-linear, so there exists dispersion in the Lamb wave propagation.

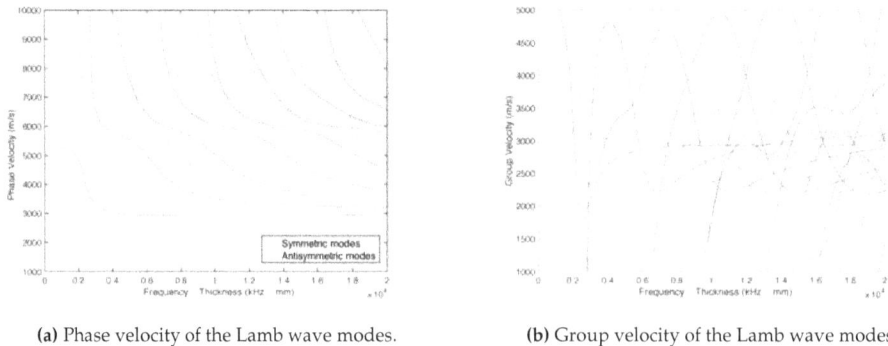

(a) Phase velocity of the Lamb wave modes.    (b) Group velocity of the Lamb wave modes.

**Figure 2.** Phase velocity and group velocity depending on the product frequency by thickness.

In Figure 2, black lines increasingly represent symmetric modes from left to right (S0, S1, S2...), while red lines represent in the same way antisymmetric modes (A0, A1, A2...). These graphs correspond to a steel pipe with the following parameters: Young's modulus $E = 200 \cdot 10^9$ N/m$^2$, Poisson's ratio $\nu = 0.3$ and density $\rho = 7700$ kg/m$^3$.

Now the methodology followed to generate and receive signals in the pipeline will be described. The transducer consists on a meander-line-coil which generates two signals per loop in the test piece (one per meander). These waves will be characterized by the wavelength which depends on the separation of the meanders.

The following equations are valid for one mode and then the same procedure will be applied iteratively for all the modes which appear at a set frequency. Thus, wave equation is set depending on the group and phase velocities. Considering $f$ as the excited frequency, the transmitted signal $s(x, t)$ will be generated according to Equation (18).

$$s(x,t) = \sin\left(2\pi f\left(\frac{x}{c_p} \pm t\right) + \phi\right),$$ (18)

where $\phi$ is a phase term that controls the phase of the transmitted signal.

In a real case, the transmitted signal includes an envelope $w(t)$ that generates the transmitted wave packet $p(x,t)$. This envelope limits the transmission time, and allows controlling the length of the transmitted pulse. Typically, the length of this envelope is described in function of $C$, the number of cycles included in the wave packet. That is, the length of $w(t)$ will be $C/f$, where $1/f$ is the time period corresponding to the excited frequency. This envelope will travel at an average velocity of $c_g$, and in general its shape will change with the distance due to dispersion effects.

The inclusion of the time envelope $w(t)$ in the transmitted signal causes the signal to be wider in the frequency domain. Thus, the number of cycles $C$ is related to the transmitted bandwidth, so that the lower the number of cycles, the wider the transmitted bandwidth. For instance, if a signal with $f = 300$ kHz and $C = 4$ cycles is transmitted, the 3 dB transmission bandwidth ranges from 252 kHz to 345 kHz. This must be taken into consideration, since the phase velocity at these frequencies might not vary linearly, causing dispersion in the wave packet.

Therefore, once the envelope is considered, the transmitted wave packet $p(x,t)$ will be expressed using Equation (19).

$$p(x,t) = s(x,t) \cdot w\left(\frac{x}{c_g} \pm t\right) = \sin\left(2\pi f\left(\frac{x}{c_p} \pm t\right) + \phi\right) \cdot \hat{w}\left(\frac{x}{c_g} \pm t\right)$$ (19)

Please note there that instead of using the transmitted envelope $w(t)$ we are using $\hat{w}(t)$, which changes its shape in function of the distance due to dispersion effects.

It is necessary to consider that under EMAT technology the excitation signal is generated in a set of $N$ loops of a coil, separated by a distance $L$, which will generate the propagation wave $y(x,t)$ using Equation (20).

$$y(x,t) = \sum_{m=1}^{2N} (-1)^m p(x - m \cdot L/2, t)$$ (20)

Please note here that each loop generates two signals (one per meander), and that the sign of their contribution to the propagation wave $y(x,t)$ is included in the term $(-1)^m$. Besides, the measure is sensed at a distance $D$, in another set of $N$ loops separated by a distance $L$. Therefore, the received signal $z(t)$ will be expressed using Equation (21).

$$z(t) = \sum_{n=1}^{2N} (-1)^n y(D - n \cdot L/2, t)$$ (21)

Again, the sign of each meander is represented by the term $(-1)^n$. Going back to the previous equations, the total signal sensed from each mode $z(t)$ is obtained.

$$z(t) = \sum_{m,n=1}^{2N} (-1)^{m+n} p(D - (m+n) \cdot L/2, t)$$ (22)

$$z(t) = \sum_{m,n=1}^{2N} (-1)^{m+n} \sin\left(2\pi f\left(\frac{D - (m+n) \cdot L/2}{c_p} \pm t\right) + \phi\right) \cdot \hat{w}\left(\frac{D - (m+n) \cdot L/2}{c_g} \pm t\right)$$ (23)

The signal received from each mode $z(t)$ has different values of $c_p$ and $c_g$, as it was concluded from Figure 2. Thus, each mode arrives at the receiver with different amplitude and envelope, depending on the attenuation of each mode and the difference of phase when the signal is received in the coil. Therefore, the amount of energy of the received signal can vary at different frequencies.

In order to find out more about the behavior of the modes in a set frequency range, a frequency sweep was made between 0 and 600 kHz with one coil and $C = 4$ cycles per wave packet. Figure 3 shows the phase velocity (Figure 3a) and group velocity (Figure 3b), where black color means the energy is maximum at that frequency. It is important to indicate that dispersion has been taken into account to carry out the experiments, since the signal has been decomposed with the envelope window $\hat{w}(t)$ through the Fourier Transform. Thus, the velocities and delays of the different frequencies which are part of the same pulse have been considered. As an example of the effects of dispersion over the wave packet, Figure 4 shows the dispersion suffered by the wave packet when traveling 0.8 m in the pipe (S0 mode, $f = 300$ kHz, $C = 4$ cycles).

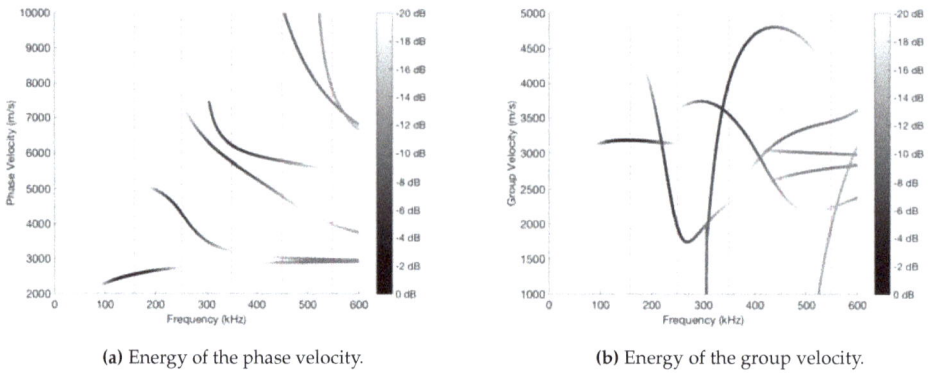

(a) Energy of the phase velocity.

(b) Energy of the group velocity.

**Figure 3.** Phase velocity and group velocity of the different modes represented according to the energy in a range of frequencies.

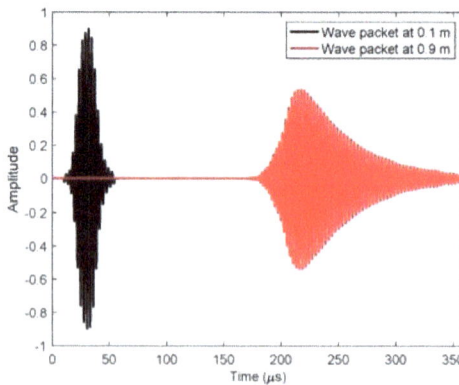

**Figure 4.** Dispersion suffered by the wave packet when traveling 0.8 m in the pipe (S0 mode, $f = 300$ kHz, $C = 4$ cycles).

The coil used in the experiments has the following parameters: distance between loops $L = 16.26$ mm and $N = 3$ loops. It can be observed in the graphs that the maximum energy appears in $f = 158$ kHz and mode A0. However, there exists a certain periodicity in the energy of the received signals. It implies that the same coil could be used to excite other frequencies, even if it has been designed to get the maximum energy in a set frequency. In fact, different frequencies will be excited in the experiments. Specifically, the frequencies indicated with red dashed lines in Figure 3 will be used, because the energy and the excited modes are different in each of them.

## 2.2. Hardware Description

The technology (Innerspec PowerBox H and the MRUT PMX scanner) and the pipe mock-ups needed to perform the empirical validation of the modeling results were provided by Innerspec Technologies S.L [21], a company which provides NDT solutions using EMAT technology.

The inspection instrument used is the Innerspec PowerBox H, a hand-held battery operated instrument. It is designed for ultrasonic applications that require very high voltages and/or long bursts of energy such as non-contact techniques (EMAT, Air-Coupled) and inspection of highly-attenuating materials. The instrument is capable of generating up to 1200V or 8kW of peak power at speeds of up to 300 Hz.

Guided waves can be used to cover distances ranging from a few millimeters to tens of meters. The two most common techniques for in-service inspections with guided waves are Long Range UT (LRUT) and Medium Range UT (MRUT). All the results showcased within this manuscript were obtained with the MRUT PMX scanner, which is used in both attenuation and reflection mode to cover shorter distances (0.1–5 m). The sensors are mounted on scanners to inspect long stretches of pipes or tanks. It typically works with frequencies from 100 kHz to 1 MHz, and can detect small pits ($\times 10$ more sensitivity than using LRUT).

The MRUT PMX scanner allows to scan axially with a single or double sensor on the pipe to measure attenuation and/or velocity changes in the signal due to corrosion, cracks or other defects around the circumference of the pipe. It is ideal for quick inspections of exposed pipe at speeds up to 150 mm/s (6 in/s).

Figure 5 shows the hardware equipment used in the manuscript.

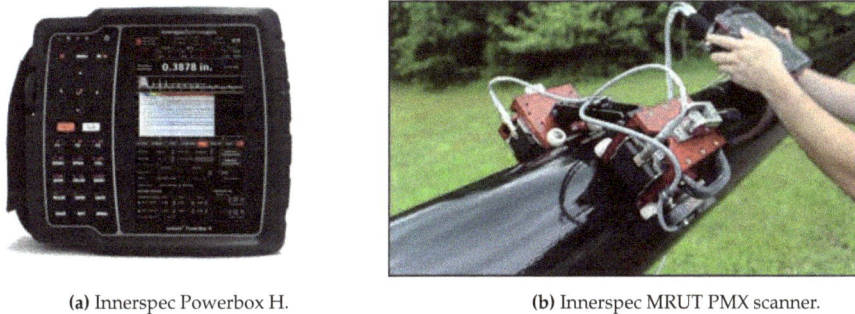

(a) Innerspec Powerbox H.    (b) Innerspec MRUT PMX scanner.

**Figure 5.** Measuring equipment used in the experiments.

## 3. Effects of the Defect Over the Lamb Waves

The modeling of the pipe by means of the ultrasonic waves is a non-trivial problem. The changing shape of the defects makes difficult to draw general conclusions about the relation between the defect and the received signals. The distortion caused by the defects over the different modes strongly varies with the shape of the defect [12,13]. For instance, the amplitude of the signal, the time of arrival (group velocity $c_g$) and the phase velocity $c_p$ of the wrap-around signal vary with the dimensions of the defect.

To study the relation between these parameters and the shape of the defects, the Finite Element Method (FEM) included in the Partial Differential Equations Toolbox of *Matlab* has been used. A database of 418 defects has been generated using this simulation tool. The defects have been characterized with three parameters: length ($l$), depth ($d$) and slope ($s$). Figure 6 depicts the defect dimensions using the three aforementioned parameters. If $s > l$ the defect is discarded. In the case studied, the thickness of the pipe is $z = 9.27$ mm. Table 1 shows the range of values that the parameters can take.

Likewise, the simulated coil has the following parameters: distance between loops $L = 16.26$ mm, $N = 3$ loops and distance to the receiver $D = 0.7$ m. Besides, each pulse of the signal contains 4 cycles in all the experiments.

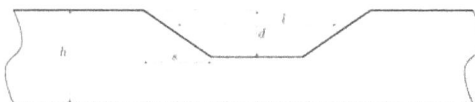

**Figure 6.** Model of the simulated defects.

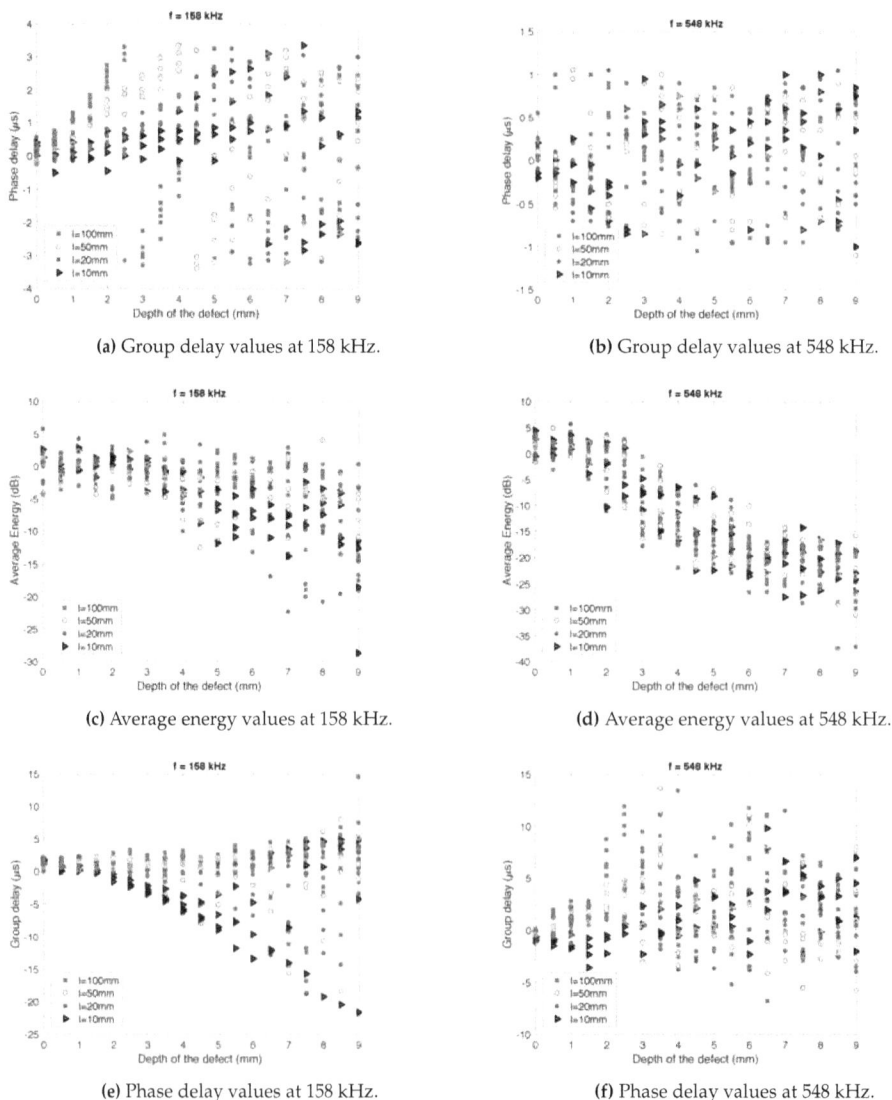

(a) Group delay values at 158 kHz.

(b) Group delay values at 548 kHz.

(c) Average energy values at 158 kHz.

(d) Average energy values at 548 kHz.

(e) Phase delay values at 158 kHz.

(f) Phase delay values at 548 kHz.

**Figure 7.** Feature values depending on the depth of the defect.

Figure 7 shows how the depth of the defect affects to the values of group delay (Figure 7a,b), average energy (Figure 7c,d) and phase delay (Figure 7e,f), at two frequencies: $f = \{158, 548\}$ kHz. For simplicity, the width of the defect was not modeled, that is to say, that dimension of the pipe was not considered. Each case has a few points since defects with different slope have been considered.

**Table 1.** Range of values for the different parameters of the defects.

| Parameter | Number of Possible Values | Values |
|---|---|---|
| Frequency (kHz) | 5 | 158, 250, 350, 450, 548 |
| Length (mm) | 4 | 10, 20, 50, 100 |
| Depth (mm) | 19 | 0, 0.5, 1,..., 9 |
| Slope (mm) | 7 | 1, 3, 5, 10, 50, 100 |

As it can be shown in the graphs, there exists a tendency in some of the considered features. Focusing on the average energy, calculated in $f = 548$ kHz, it is clear to see that, as the length of the defect increases, the average energy decreases drastically, especially between 2 and 6 mm of defect depth. This is exactly what would be expected when there exists a leak in the pipeline and the energy of the signal is scattered through it.

In the case of the group delay, considering the measurement taken at $f = 158$ kHz, it can be observed that the signal tends to be delayed (positive delay) when the defect increases. This does not happen when the length of the defect is very small ($l = 10$–20 mm), since the signal arrives earlier than in the non-defect case. In any event, the aim of this modeling work was to evaluate whether these features contain useful information to tackle the problem addressed.

There exists a difficulty of reaching a conclusion about the relation between the calculated features and the profile of the pipe. Because of that, it is necessary to apply advanced techniques which bring more information about what are the best features or how they should be mixed.

## 4. Smart Sound Processing (SSP) for Defect Sizing

It is necessary to apply SSP methods to solve the defect sizing problem in pipes that is being coped in the current manuscript. This type of methods usually follows the process described in Figure 8. First of all, it is important to extract useful information from the signals on the form of features. Once this is done, the next step is to select the ones that best work to solve the problem at hand. Finally, a predictor will construct a model capable of predicting the solution in an unknown case, such as a new pipeline.

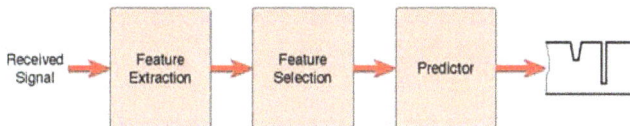

**Figure 8.** Scheme of an SSP system.

It is important to extract useful information from the received signal in order to be capable of detecting and sizing defects present in the pipes under inspection. With this purpose, different features were elicited:

- Maximum Amplitude (dB). This measure indicates the value of the maximum peak received from the signal. It is determined by looking for the value of the maximum peak around the expected point, which is the position of the maximum of the reference signal form in case of absence of any defects, $t_0$).
- Phase Delay (µs). This measure represents the time taken between the pulse shipment and its reception at the same pipe location. It is determined measuring the time difference between

the position of the maximum of the reference signal (signal without defect) $t_0$ and its closest maximum in the sensed signal. There may be considerable uncertainty in this feature if the delay is higher than half the period, since the nearest peak becomes the maximum of the signal.

- Average Energy (dB). It represents the average energy of the pulse. The interval considered for its calculation started 30 μs before $t_0$ and ends up 30 μs after it.
- Group Delay (μs). In order to calculate this measure, the centroid of the average energy of the pulse has been considered. It has been estimated using the centroid of the pulse around the expected maximum ($\hat{t_g}$), with Equation (24).

$$\hat{t_g} = \frac{\sum_{t=t_0-3\cdot10^{-5}}^{t_0+3\cdot10^{-5}} tz(t)^2}{\sum_{t=t_0-3\cdot10^{-5}}^{t_0+3\cdot10^{-5}} z(t)^2} \tag{24}$$

All these features have been extracted directly from the received signal at $D = 0.7$ m. Two additional features, extracted from the reflected signal, where included to study the importance of the analysis of the echos in the sizing problem.

- Maximum Amplitude of the echo (dB).
- Average Energy of the echoes (dB).

It is remarkable the fact that the reflected echoes are not always depicted in the gathered signals, since the position of their contribution depends on the relative position of the defect in the pipeline with respect to the EMAT actuator. Therefore, in some cases the echo is overlapped with the transmitted signal, and cannot be clearly identified. In the simulations two scenarios would be modeled: on the one hand, the case where the reflected echoes are present and on the other hand, the case where they are not, to study the importance of these echo dependent features in the performance of the sizing estimator.

In the problem at hand, 5 excitation frequencies were used. In total, 30 features were extracted taking into account that 6 features were obtained for each excitation frequency. Some of them will work better than others, so it is important to select the best ones and reduce the total amount of them, in order to properly estimate the size of the defect.

To select the features which better work in this experiment, a feature selection process was applied through evolutionary techniques. Evolutionary Algorithms (EAs) are inspired in natural evolution laws and allow to find the optimum solution from the solutions (denoted individuals) obtained in previous iterations [22]. In this paper, a tailored EA has been applied, searching for the best subset of features and trying to minimize the Root Mean Square Error (RMSE) of a Least Squares Linear Discriminant (LSLD). The use of more complex prediction methods has been avoided in the feature selection process, since the EA requires training and testing the predictor a large number of times. The considered constrains are to limit the number of frequencies used as well as the number of features selected.

The EA, which is described schematically in Figure 9, is composed of several steps:

1. A population of $N_p$ individuals is generated. Each solution consists of a binary vector with a length equal to the total number of features. Thus, ones indicate the features which are selected in the individual, while zeros indicate the features which remain outside.
2. The candidates of the population are restricted to the considered constrains. All of them are modified in order to randomly change the value of some bits until one or the two constrains are fulfilled, depending on the case.
3. A LSLD is designed with the subset of features of each candidate solution. The RMSE of the defect depth is calculated, which is the fitness function in this experiment. With this value, the population is ranked, keeping the best individuals in the top of the ranking.
4. After that, a selection process is applied, which consists in keeping the best 10% of the solutions, removing the remaining ones.

5. The removed solutions (90% of the population) are regenerated by crossover between the best candidates.

6. Mutations are applied to the new population. With this step, 1% of the bits is changed. Furthermore, this step is not applied to the best solution in order to ensure the convergence of the algorithm.

7. This process is repeated from step 2 during $N_g$ generations. The final best solution will be the best solution obtained in the last iteration.

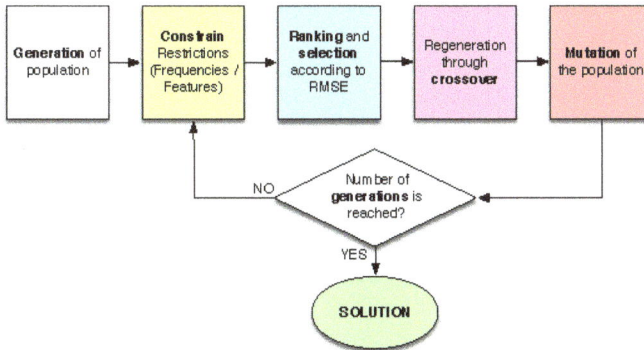

**Figure 9.** Scheme of the evolutionary algorithm applied in the experiments.

Sometimes the EAs do not reach a high convergency. In order to improve it, an elimination tournament of small EAs has been implemented. It consists in joining the winners of the EAs in pairs during several rounds ($N_r = 6$ in our case), until the best individual reaches the end and, consequently, it becomes the best solution to the problem. 32 small EAs were considered with a population of 100 individuals and 8 generations each, except for the last EA, where 16 generations are configured for convergency [22].

Once the best features have been selected for this specific problem, a non-linear predictor needs to be applied to get the final profile of the pipeline and to know the performance of the developed model. Neural Networks were applied, specifically the Multi Layer Perceptron (MLP) [23]. A perceptron is a neuron with a set of adjustable weights and an activation function by steps [24]. Levenberg-Marquardt algorithm, a method where the minimization function is a sum of quadratic terms [25], was applied for training purposes. The number of hidden neurons was a parameter in the experiments.

A $k$-fold cross validation has been applied in the generated database, being $k = 12$ [26]. This method allows to divide the database in $k$ groups so that the full process is repeated $k$ times, using one group as test subset and the remaining $k - 1$ groups as training subset. The advantage of this method is that the obtained results are more generalizable. It has been applied in both the feature selection process and the training of the neural network-based predictor.

## 5. Results

The research work showcased in this manuscript was carried out using both real and simulated measurements. This section of the paper will be opened discussing the results obtained during the simulation experiments.

### 5.1. Defect Sizing in Simulated Pipelines

An experiment was developed using the synthetic database described in Table 1. In this case, the RMSE was computed to evaluate the performance of the predictors. The objective is to know how well the received signal is estimated at different frequencies. The $k$-fold cross validation described in

*Sensors* **2018**, *18*, 802

Section 4 has been applied in this experiment, so all the results presented here have been obtained with this method.

It is important to indicate that noise from the real measurements was introduced to the signal, the amplitude and the temporal resolution in order to make the experiments as real as possible. A real pipe free from defects was taken in order to study the measuring system tolerance to noise. It was obtained an SNR of 32.93 dB, a variation in time with an standard deviation of 2.1434 samples at 10 MHz (0.2 μs) and a variation in scale of 0.80 dB.

As stated above, the relevance of extracting features from echoes related to reflections has been also studied. Figure 10 shows the RMSE obtained by the different predictors, depending on the number of selected features. The different curves show the performance depending on the combination of frequencies considered, in the case of only using features from the wrap-around echo, and in the case of including features from the reflected echo. The combination of frequencies and the classifier used in each case are those that provide the best results in terms of RMSE. It can be seen that the results improve (less RMSE) as the number of evaluated frequencies is increased. It happened especially until 158, 350 and 548 kHz frequencies were evaluated. The results slightly improve when the number of frequencies studied is increased, so there would be no need to make the system more complex to reduce a few millimeters the final rate.

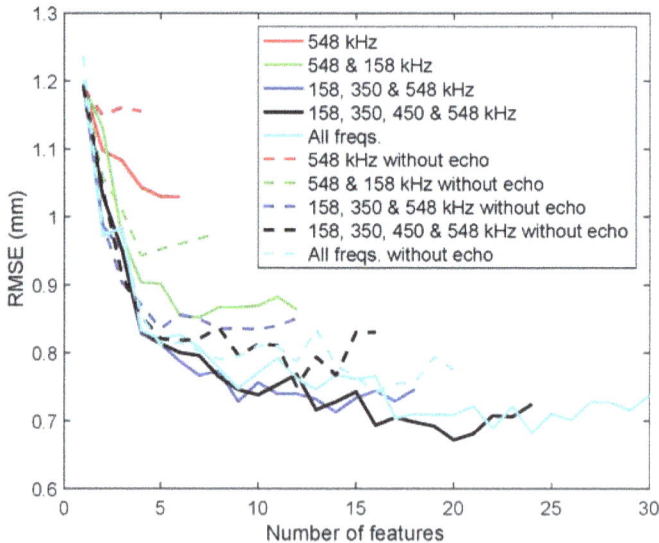

**Figure 10.** RMSE obtained with neural networks predictor depending on the number of features selected at different frequencies.

In terms of the reflected echo features, if they are not used (dashed lines) the tendency of the graphs is similar to the previous one. In general, the results are not significantly different from those using echo features, especially when all frequencies are used. Because of that, the features extracted from the echo signals are not essential to evaluate the problem and it is possible to get a good result in case the return of the signal cannot be evaluated.

With regard to the used predictors, Figure 11 shows the RMSE depending on the number of features employed and the number of neurons configured in the MLP (1, 2, 3, 4 and 5). Although the results are good using just one neuron, they can be improved by using two. However, if the number of neurons is further increased (3–5), the result is not much better than before.

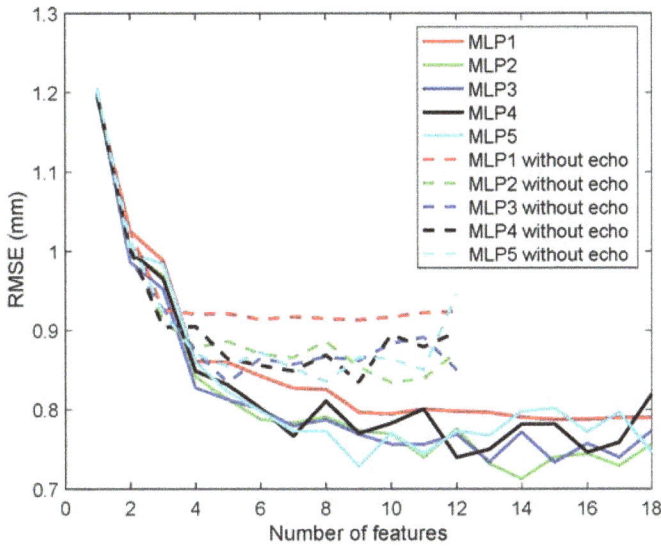

**Figure 11.** RMSE obtained depending on the number of selected features considering different number of neurons in the MLPs.

In all the cases above it is clear that as the number of features is higher, the results improve. It happens especially in the range between 1 and 5 features, where the error falls considerably in Figures 10 and 11.

Now the features selected at different frequencies from the proposed ones are going to be studied. With this purpose the manuscript will focus on the case of mixing eight features from three frequencies: 158, 350 and 548 kHz (dark blue line). The results do not improve substantially when more frequencies are added or when the number of features is increased. In Tables 2 and 3 the ratios of selection of the features are shown. The first one includes the case of considering the reflected echo features and the second the case of not considering them. The RMSE associated to these two cases are 0.79 mm and 0.87 mm, with and without reflected echo features, respectively.

**Table 2.** Ratios of selection of the features, including echo features.

| Feature | Frequency (kHz) | | |
|---|---|---|---|
| | 158 | 350 | 548 |
| Maximum Amplitude | 0% | 92% | 0% |
| Phase Delay | 0% | 0% | 0% |
| Average Energy | 100% | 100% | 100% |
| Group Delay | 0% | 0% | 58% |
| Maximum Amplitude of Echo | 8% | 50% | 0% |
| Average Energy of Echos | 92% | 100% | 100% |

**Table 3.** Ratios of selection of the features without echo features.

| Feature | Frequency (kHz) | | |
|---|---|---|---|
| | 158 | 350 | 548 |
| Maximum Amplitude | 100% | 100% | 92% |
| Phase Delay | 0% | 0% | 0% |
| Average Energy | 100% | 100% | 100% |
| Group Delay | 8% | 100% | 100% |

The results show that the most important feature is the average energy of the received pulse of the signal, since it is selected in all the frequencies considered. It confirms that the energy is the most affected parameter when the signal travels through a defect. This reasoning applies for both the wrap around echo and the reflected echo.

Other features which work properly are the maximum amplitude and the group delay, but mainly when the features extracted from the reflected echo are not taken into consideration. It means that they are good features, but not as much as the energy from the reflected echo. Furthermore, features like phase delay are not relevant for the study, because they are not selected to get a better result. In fact, this feature has 0% of selection in all cases of study. This is caused by the problems with the uncertainty in measuring this parameter.

These results demonstrate again that the reflected echo features are not essential for the problem. Using them the model fits just 0.1 mm better with the target.

## 5.2. Defect Sizing in a Real Pipeline

Now the results obtained during the experimental trials in a pipe mock-up are presented. Data has been collected using MRUT PMX scanner on a pipe, as mentioned before. The inspection was performed moving the sensor axially (sending the waves circumferentially) on a pipe which includes three flat defects with different depths. An image of the pipe is shown in Figure 12, while its specifications are shown in Table 4.

**Figure 12.** Image of the real pipe.

**Table 4.** Specifications of the real pipe.

| Parameter | Value |
| --- | --- |
| Material | Structural Steel S355NH |
| Thickness (mm) | 9.27 |
| Depth of Defect 1 (mm) | 6.18 |
| Depth of Defect 2 (mm) | 4.63 |
| Depth of Defect 3 (mm) | 1.85 |

The inspections were carried out at different frequencies. The distortions introduced by the defect over the analyzed signals are shown in Figure 13. Axial scans sending the waves circumferentially have been carried out every millimeter of the pipeline, and the main wave packages of the wrap-around echos have been stored. Different behaviors are depicted depending on the depth of the defects and the excited frequency, $f$ = 158 kHz (Figure 13a) and $f$ = 548 kHz (Figure 13b).

The graphs included in Figure 13 show that the defects change the amplitude and the phase of the received signals. Furthermore, these distortions are different depending on the size of the defects and the excited frequency $f$.

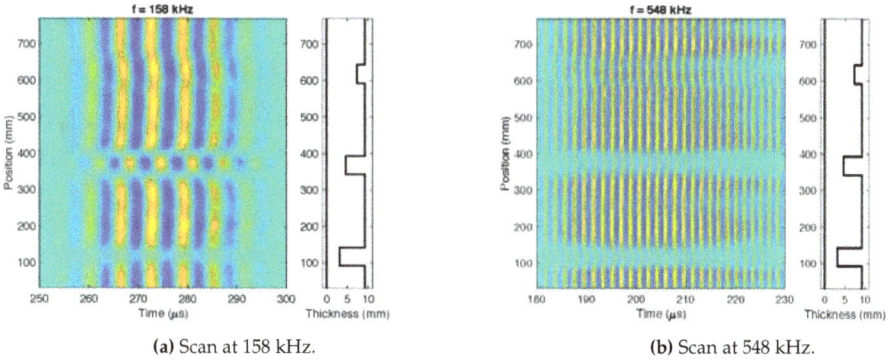

**(a)** Scan at 158 kHz.

**(b)** Scan at 548 kHz.

**Figure 13.** Axial scan of the real pipeline at two frequencies.

The profile of the real pipe will be modeled using the features explained above. In this experiment, the predictor trained was used along with the synthetic database and it was also tested with the signals sensed in the real laboratory trials. Therefore, it has not been necessary to apply the cross validation technique, as the training and test subsets were clearly defined. Three cases were developed: considering 4 features at 158 kHz, 4 features at 548 kHz and mixing all of them. The 2 features from the reflected echo (maximum amplitude of the reflected echo and average energy of the reflected echo) have not been considered because the return from the signal was sometimes overlapped with the excitation pulse, so it was not posible to extract any information from it.

In Figure 14 and Table 5 the results of the described experiments are shown. The model that better fits with the pipeline is the predictor which considers eight features and the two frequencies. It is the one which better distinguishes between defect and non-defect areas, since the estimation of the defect depth is very close to zero in non-defect areas. In fact, the RMSE is the best of the three predictors (1.48 mm). When mixing the information from the two frequencies the results are quite improved. With the other models the estimation was not so good, especially when there were no defects in the profile.

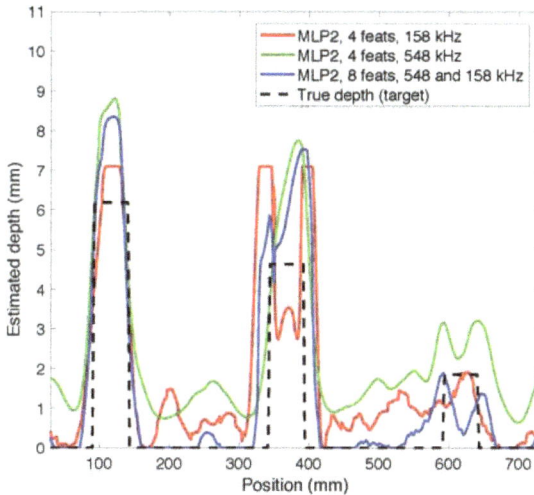

**Figure 14.** Estimation of the real pipeline model.

**Table 5.** Results of real database.

| Predictor | Number of feats | Frequency (kHz) | RMSE (mm) |
|---|---|---|---|
| | 4 | 158 | 1.84 |
| MLP 2 neurons | 4 | 548 | 1.80 |
| | 8 | 158, 548 | 1.48 |

## 6. Discussion

In this work the defect detection and sizing in steel pipelines have been studied. With this purpose, Lamb waves have been generated with EMAT-based techniques. These have been analyzed with SSP methods, in order to try to model the pipeline. After the experiments described above have been carried out, the following conclusions have been drawn:

- The shape of the defect causes differences in the received signal. It is not feasible to obtain an analytical solution for all the cases.
- The extracted features are useful for the pipeline sizing problem, since the results are good in terms of RMSE. The average energy and the maximum amplitude from the signals are particularly relevant for the study.
- It is important to excite the waves at several frequencies because the behavior and velocity of the Lamb modes is totally different depending on this parameter.
- Related to the predictors, a large number of neurons in the MLPs is not required. When it is increased above two or three neurons, the results do not improve significantly.

In the future it would be possible to increase the amount and variety of defects in real pipelines. Besides, more features could be applied, such as some from the signal in successive wrap-arounds.

## 7. Conclusions

Pipeline inspection problem can be approached in many different ways. Lamb wave generation through EMAT actuators proves to be a very effective and useful one. However, the amount of information provided by the wrap-around signals needs to be processed by advanced techniques, such as smart sound processing algorithms. Thanks to them, it is feasible to get good estimation results of the pipeline defects, in both real and simulated signals. In the manuscript it has been demonstrated the importance of applying a multi-frequency study for defect sizing problem, the relevance of some features from the signals (e.g., energy and amplitude) and the absence of the need to greatly increase the complexity of the classifiers to get a good estimation in the problem at hand.

**Acknowledgments:** This work has been funded by Innerspec Technologies Europe S.L through the "Chair of modeling and processing of ultrasonic signals" (CATEDRA2007-001), and by the Spanish Ministry of Economy and Competitiveness/FEDER under Project TEC2015-67387-C4-4-R.

**Author Contributions:** J.G., R.G. and M.R. designed and performed the experiments, as well as analyzed the results; J.G. and R.G. wrote the paper; A.R., J.A.J. and V.G. provided the hardware equipment and the real pipeline measurements.

**Conflicts of Interest:** The authors declare no conflict of interest.

## References

1. Blitz, J.; Simpson, G. *Ultrasonic Methods of Non-Destructive Testing*; Springer Science & Business Media: New York, NY, USA, 1995; Volume 2.
2. Silk, M.G. *Ultrasonic Transducers for Nondestructive Testing*; CRC Press: Boca Ratón, FL, USA, 1984.
3. Vanaei, H.; Eslami, A.; Egbewande, A. A review on pipeline corrosion, in-line inspection (ILI), and corrosion growth rate models. *Int. J. Press. Vessels Piping* **2017**, *149*, 43–54.
4. Green, R.E. Non-contact ultrasonic techniques. *Ultrasonics* **2004**, *42*, 9–16.

5. Zhai, G.; Jiang, T.; Kang, L. Analysis of multiple wavelengths of Lamb waves generated by meander-line coil EMATs. *Ultrasonics* **2014**, *54*, 632–636.

6. Park, J.H.; Kim, D.K.; Kim, H.j.; Song, S.J.; Cho, S.H. Development of EMA transducer for inspection of pipelines. *J. Mech. Sci. Technol.* **2017**, *31*, 5209–5218.

7. Salzburger, H.J.; Niese, F.; Dobmann, G. EMAT pipe inspection with guided waves. *Welding World* **2012**, *56*, 35–43.

8. Andruschak, N.; Saletes, I.; Filleter, T.; Sinclair, A. An NDT guided wave technique for the identification of corrosion defects at support locations. *NDT E Int.* **2015**, *75*, 72–79.

9. Clough, M.; Fleming, M.; Dixon, S. Circumferential guided wave EMAT system for pipeline screening using shear horizontal ultrasound. *NDT E Int.* **2017**, *86*, 20–27.

10. Wang, S.; Huang, S.; Zhao, W.; Wei, Z. 3D modeling of circumferential SH guided waves in pipeline for axial cracking detection in ILI tools. *Ultrasonics* **2015**, *56*, 325–331.

11. Singh, R. *Pipeline Integrity: Management and Risk Evaluation*; Gulf Professional Publishing: Houston, TX, USA, 2017.

12. Demma, A. The Interaction of Guided Waves With Discontinuities in Structures. Ph.D. Thesis, University of London, London, UK, 2003.

13. Cobb, A.C.; Fisher, J.L. Flaw depth sizing using guided waves. In *AIP Conference Proceedings*; AIP Publishing: Melville, NY, USA, 2016; Volume 1706, p. 030013.

14. Nurmalia; Nakamura, N.; Ogi, H.; Hirao, M. EMAT pipe inspection technique using higher mode torsional guided wave T (0, 2). *NDT E Int.* **2017**, *87*, 78–84.

15. Hirao, M.; Ogi, H. An SH-wave EMAT technique for gas pipeline inspection. *NDT E Int.* **1999**, *32*, 127–132.

16. Layouni, M.; Hamdi, M.S.; Tahar, S. Detection and sizing of metal-loss defects in oil and gas pipelines using pattern-adapted wavelets and machine learning. *Appl. Soft Comput.* **2017**, *52*, 247–261.

17. Mohamed, A.; Hamdi, M.S.; Tahar, S. A machine learning approach for big data in oil and gas pipelines. In Proceedings of the 2015 3rd International Conference on Future Internet of Things and Cloud (FiCloud), Rome, Italy, 24–26 August 2015; pp. 585–590.

18. Mohamed, A.; Hamdi, M.S.; Tahar, S. An adaptive neuro-fuzzy inference system-based approach for oil and gas pipeline defect depth estimation. In Proceedings of the SAI Intelligent Systems Conference (IntelliSys), London, UK, 10–11 November 2015; pp. 35–42.

19. Luangvilai, K. *Attenuation of Ultrasonic Lamb Waves With Applications To Material Characterization and Condition Monitoring*; Georgia Institute of Technology: Atlanta, GA, USA, 2007.

20. Su, Z.; Ye, L. *Identification of Damage Using Lamb Waves: From Fundamentals To Applications*; Springer Science & Business Media: New York, NY, USA, 2009; Volume 48.

21. Garcia, V.; Boyero, C.; Jimenez, J.A. Corrosion detection under pipe supports using EMAT Medium Range Guided Waves. In Proceedings of the 19th World Conference on Non-Destructive Testing, Munich, Germany, 13–17 June 2016; pp. 1–9.

22. Gil-Pita, R.; García-Gomez, J.; Bautista-Durán, M.; Combarro, E.; Cocana-Fernandez, A. Evolved frequency log-energy coefficients for voice activity detection in hearing aids. In Proceedings of the 2017 IEEE International Conference on Fuzzy Systems (FUZZ-IEEE), Naples, Italy, 9–12 July 2017; pp. 1–6.

23. Weisz, L. Pattern Recognition Statistical Structural And Neural Approaches. *Pattern Recogn.* **2016**, *1*, 2.

24. Rosenblatt, F. The perceptron: A probabilistic model for information storage and organization in the brain. *Psychol. Rev.* **1958**, *65*, 386.

25. Hagan, M.T.; Menhaj, M.B. Training feedforward networks with the Marquardt algorithm. *IEEE Trans. Neural Netw.* **1994**, *5*, 989–993.

26. Refaeilzadeh, P.; Tang, L.; Liu, H. Cross-validation. In *Encyclopedia of Database Systems*; Springer: New York, NY, USA, 2009; pp. 532–538.

*sensors*

MDPI

*Article*

# A Collaborative Approach for Surface Inspection Using Aerial Robots and Computer Vision

Martin Molina *, Pedro Frau and Dario Maravall

Department of Artificial Intelligence, Technical University of Madrid, Boadilla del Monte, Madrid 28660, Spain; pedro.frauamar@gmail.com (P.F.); dmaravall@fi.upm.es (D.M.)
* Correspondence: martin.molina@upm.es; Tel: +34-910-672-912

Received: 26 December 2017; Accepted: 15 March 2018; Published: 17 March 2018

**Abstract:** Aerial robots with cameras on board can be used in surface inspection to observe areas that are difficult to reach by other means. In this type of problem, it is desirable for aerial robots to have a high degree of autonomy. A way to provide more autonomy would be to use computer vision techniques to automatically detect anomalies on the surface. However, the performance of automated visual recognition methods is limited in uncontrolled environments, so that in practice it is not possible to perform a fully automatic inspection. This paper presents a solution for visual inspection that increases the degree of autonomy of aerial robots following a semi-automatic approach. The solution is based on human-robot collaboration in which the operator delegates tasks to the drone for exploration and visual recognition and the drone requests assistance in the presence of uncertainty. We validate this proposal with the development of an experimental robotic system using the software framework Aerostack. The paper describes technical challenges that we had to solve to develop such a system and the impact on this solution on the degree of autonomy to detect anomalies on the surface.

**Keywords:** Automated visual inspection; human-robot collaboration; surface inspection; aerial robotics

## 1. Introduction

The maintenance of certain infrastructures requires periodical inspections of surfaces (e.g., the surface of a dam, the facade of a building, an indoor wall, etc.) to find defects (e.g., holes, fissures, mould, spots, humidity, etc.) as symptoms of potential problems due to, for example, structural imperfections. Aerial robots can help human operators to inspect this type of surfaces. For example, operators can manually tele-operate aerial robots equipped with cameras to observe areas that are difficult to reach by other means.

In order to be more effective in this task, it is desirable for aerial robots to have a higher degree of autonomy. With the current state of technology, robots could perform automatically certain tasks related to surface inspection, but not with complete autonomy. Therefore, it may be more realistic to follow an approach based on a human-robot collaboration in which the operator delegates certain routine tasks to the robot, but the robot asks for assistance in the presence of uncertainty. In this domain, different techniques can be considered to provide autonomy to robots (e.g., path planning, computer vision, obstacle avoidance, visual alignment, coordination of multi-robot inspection, etc.).

According to this, this paper presents a solution for visual inspection that increases the degree of autonomy of aerial robots following semi-automatic approach. In this work, computer vision techniques are used and extended to detect surface anomalies in a collaborative scheme. We validate our proposal with the development of an experimental robotic system. For the development of such a system, an existing software framework for aerial robotics called Aerostack was used, adding new components for visual recognition and user-system interaction.

The remainder of the paper is organized as follows. First, the paper describes the related work and presents how we consider the problem of surface inspection based on a collaborative work between human and robot. Then, the paper describes our approach for automated visual recognition of defects, using an existing method that we extended to be used with operator assistance. Next, the paper describes the software framework for aerial robotics Aerostack that we used to construct the robotic system, describing how we extended this framework with specific components for surface inspection. Finally, the paper describes a set of experiments that we performed to evaluate our solution. The experiments showed a good behavior for user-system interaction and visual recognition in the context of a collaborative surface inspection. As supplementary material, the paper provides the dataset of images from surfaces that we constructed and used to evaluate the quality of the visual recognition method.

## 2. Related Work

In general, the problem of automated visual inspection is a well-known specialized subfield of computer vision (for a general revision of methods and systems in this type of problem see for example references [1–3]). In automated visual inspection, different techniques in pattern recognition and image processing can be applied. This process usually includes image segmentation (e.g., by means of histogram-based thresholding or by means of edge-based filtering techniques plus heuristic post-processing) and the use of classifiers (e.g., statistical methods, kernel-based methods, neural networks, etc.).

For example, computer vision techniques have been used in problems such as the following: atmospheric corrosion detection on petroleum plant equipment exposed to marine atmosphere [4], crack detection (which is a hard defect to detect in civil structures) [5] using Gabor filters invariant to rotation, and the inspection of surfaces of vessels [6] for rust detection using a magnetic robot crawler with a classifier ensemble (called PICARD) that combines different classification techniques (SVM, Bayesian classifier and random forest).

Our work deals with surface inspection by means of aerial robots. This type of robots have been recently used for different inspection problems, such as the inspection of buildings or large civil structures [7,8] (using techniques image reconstruction and monitoring), the inspection of power lines [9] (with solutions for the communication and collaboration of robot teams), inspection of marine vessels [10] (using visual-based recognition techniques for robot motion estimation and self-localization near the surfaces), fire detection [11] (using an heterogeneous fleet of aerial vehicles), or inspection of penstocks and tunnel-like environments [12] (proposing a self-localization method for this type of environment).

State-of-the-art technology in automated visual inspection has achieved high performance, but not the level of perfection required to perform inspection tasks in full autonomy. Therefore, we follow an approach based on using semi-autonomous aerial vehicles assisted by human operators. Our work can be considered an example on a novel paradigm in robotics systems with growing interest, namely, human-robot collaborative interaction [13]. This emerging approach is an answer to the trend to design more robust robotic systems by completing robot autonomy with online human intervention. This poses a huge challenge to robot system designers, since their robots must be able to operate as true partners with humans, which requires the implementation of the robot's cognitive, communicative and perceptual capabilities that lie at the frontiers of the current state of the art of advanced robots.

The specific subfield addressed in this paper (vision-based human-robot collaborative inspection) is a novel research area about which we have found only few proposals in other domain problems. For example, Suzuki et al. (2000) [14] present an experimental prototype to cooperate with multiple teleoperated ground robots for inspection, and Kimura and Ikeuchi (1999) [15] use computer vision to recognize human motion while human and robot cooperate to assemble components.

## 3. Collaborative Surface Inspection with Aerial Robots

Surface inspection is a task that is usually done as part of the periodical maintenance work of certain infrastructures to detect symptoms of potential problems. In general, we consider that the goal of surface inspection is to explore a spatial area of a given surface in order to detect the presence of anomalies and classify them into prefixed categories. A simple example of this problem is to find imperfections such as fissures or holes in the surface of a wall. Other imperfections include spots, mould or humidity.

A simple method to carry out this task with aerial robots is by teleoperation where the human operator manually guides a robot equipped with a camera. One of the immediate advantages of this approach is that it is possible to observe, with lower costs, areas that are difficult to reach by other means (e.g., extreme points on the surface of a dam or the surfaces of the pillars of a large bridge).

In our work, we assume that there is an adaptive inspection in the sense that the environment conditions may affect the way the inspection is carried out. This means, for example, that a certain area may be difficult to be observed clearly during the inspection (due to bad illumination, excessive distance or another reason). Thus, it is not possible to determine in advance a fixed trajectory to be done to inspect the surface. Instead, it may be necessary for operators to manually teleoperate the drone to access new areas of interest during inspection and adjust the position or orientation (or turn on lights, etc.) to obtain better images of potential anomalies.

In contrast to a fully manual teleoperation, it is possible to conceive a more supported method following a collaborative approach, in which the operator delegates certain routine tasks to the robot, but the robot asks for assistance in the presence of uncertainty. In this case, aerial robots perform certain tasks autonomously with less intervention of the human operator. This may contribute to perform safer operation (e.g., robots may be able to avoid obstacles that the operator does not perceive clearly from a distance such as overhead cables, vegetation, etc.), more accurate inspections (e.g., robots may generate automatically accurate trajectories to cover a surface to inspect), or more productivity by reducing the cognitive load of operators (e.g., robots can detect automatically the presence of certain surface anomalies).

Figure 1 summarizes this type of human-robot interaction in which the operator can adopt two different roles corresponding to two interaction modes: supervised mode and assisted mode. The following sections describe in more detail how we understand such modes.

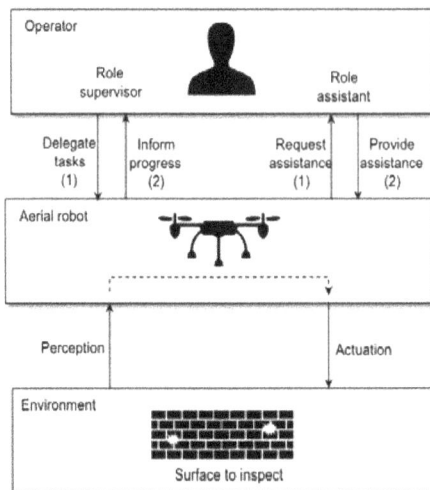

**Figure 1.** Human-robot interaction in surface inspection.

## 3.1. Supervised Mode

The supervised mode is related to the notion of supervisory control [16] in which a human operator is intermittently acting on the robot to delegate tasks. The robot closes an autonomous control loop through effectors to the environment. This concept has been used to design flexible interaction models, for example, for military mission planning of UAVs [17], swarming networks [18] or remote surveillance system [19].

To clarify the interaction modes, messages can be divided into categories according to the theory of speech acts [20–22]: assertive messages, which are sent to give certain information to the receiver (for example, the robot informs the operator the completion of a task), and directive messages, which cause the receiver to take a particular action. Within directive messages, it is possible to distinguish between two categories: action directives (requests for action) and information requests.

In the supervised mode, the operator sends action directives to the robot in order to delegate mission tasks. The operator may ask the aerial robot to perform an inspection mission, specifying the area to cover and the exploration strategy. In this case, the relation between operator and robot follows a hierarchical authority (as supervisor-subordinate schema) in which the operator delegates a set of tasks.

During the development of the mission, the operator observes the robot behaviour and the robot sends assertive messages to inform about the mission execution progress (e.g., completed task or finished mission). These messages are useful for the operator to verify that the mission is being developed as expected. The operator can interrupt the mission under certain circumstances (for example, to avoid wrong behaviours).

In this interaction mode, the robot shows autonomy to adapt to a dynamic environment while tries to reach its goal [23]. But the robot also shows autonomy to accept or reject the proposed actions according to characteristics of the environment and its own goals (e.g., safety goals) [24].

## 3.2. Assisted Mode

In the assisted mode, the human operator works as a resource for the robot, providing additional information. The robot may ask the operator questions as it works, to obtain assistance with perception and cognition. This allows the human to compensate for limitations of autonomy. This is related to the idea of collaborative control in which human and robot work together [25]. The human and the robot dialogue to exchange information, to ask questions, and to resolve differences. This interaction scheme is a kind of mixed initiative approach [26]. Both, the operator and the robot, may take the initiative of the conversation during the dialogue.

This interaction mode is required because robots have partial knowledge and are not completely self-sufficient. In this case, the robot works like the field technician (i.e., it is skilled, but may need help) and the operator is like the expert (i.e., she or he can provide assistance when needed) as it is considered in human collaborative control [27]. In the design of these systems, it is important to minimize the interaction time (i.e., the expected amount of time that a human must interact with the robot) [28].

In the particular case of inspection missions, robots may have recognition abilities for certain defects but, sometimes, certain defects are difficult to classify automatically. In addition, unexpected changes of the environment (e.g., shadows, wind, etc.) may require attention from the operator to decide the appropriate response.

The interaction mode for assistance starts, for example, when the robot is not able to recognize the category (or it is uncertain about a selected category) of a detected defect in the surface. In this case, it sends an information request to ask the operator for the category of the detected defect. The operator answers the category or rejects the detection. In addition, the operator may help the robot proposing motion actions to have better views of the surface.

This interaction mode may also start when the robot is not able to complete a requested task because there is a problem in the environment such as: low visibility, low battery, lost position, high vibrations, impassable barrier, or unstable ground. This includes also the time out event that happens when it is not able to complete the task in the expected time due to unknown reasons. The operator helps the robot saying how to respond to these events.

In general, it is also possible to consider that the robot may delegate certain specialized tasks to other robots. For example, the robot can transfer part of the mission to other robot because it does not have enough battery charge, or it can delegate a certain specialized task that require specialized actuators (e.g., use a special device to mark the detected defect on the wall).

## 4. On-Board Camera for Surface Inspection

In this work, we consider that the inspection task is performed with the help of a drone with a camera on board. This is a type of technology for which there are different types of available solutions in the market. For example, there are commercial aerial platforms, equipped with cameras, with characteristics that may allow inspection of surfaces. In some of the experiments described in this paper, we used Parrot AR.Drone 2.0 (Parrot, Paris, France)—a platform with an on-board front camera (Figure 2)—a flight time of about 35 min (power edition) and Wi-Fi communication with a maximum range of approximately 50 m. Table 1 shows examples of other general-purpose commercial platforms with different features that could be selected, for example, to obtain better images or to use bigger communication range.

**Figure 2.** Commercial drones with on-board cameras: Parrot AR.Drone 2.0 (Parrot, Paris, France) (**left**) and DJI Inspire 1 (DJI, Shenzhen, China) (**right**).

**Table 1.** Examples of commercial drones with on-board cameras.

| Platform | Camera | | | Flight Time | Range | |
|---|---|---|---|---|---|---|
| | Resolution | Frame Rate | Field of View | | Wi-Fi | Transmitter |
| Parrot AR.Drone 2.0 | HD 1280 × 720 px | 30 fps | 93° | 35 min | 0 m | |
| Parrot Bebop 2 | FHD 1920 × 1080 px | 30 fps | 180° | 25 min | 100 m | 2.2 Km |
| DJI Mavic Pro | UHD 4096 × 2160 px | 24 fps | 79° | 27 min | 80 m | 7 Km |
| DJI Phantom 3 Pro | UHD 4096 × 2160 px | 30 fps | 94° | 23 min | | 5 Km |
| DJI Inspire 1 Pro | UHD 4096 × 2160 px | 30 fps | 72° | 15 min | | 5 Km |
| Yuneec Typhoon H | UHD 4096 × 2160 px | 30 fps | 94° | 25 min | 400 m | |

Other robotic systems for inspection can be used with more specific designs, combining advanced platforms (e.g., DJI Inspire 1 Pro or Yuneec Tornado H920, DJI, Shenzhen, China) with better quality cameras (e.g., Zenmuse X5, DJI, Shenzhen, China; CGO4, Yuneec, Jiangsu, China; etc.) and mechanical camera stabilizers (gimbals).

*On-Board Camera*

An on-board camera oriented to the front of the aerial vehicle can be used to inspect vertical and flat surfaces. The specific features of the camera will depend on the desired performance of the inspection task. For example, the resolution of the camera to be used will depend on the minimum size of the anomalies to be detected. Cameras in commercial drones normally use a CMOS image sensor with a resolution that may range for example from high definition HD with 1280 × 720 pixels (AR.Drone 2.0, Parrot, Paris, France) to ultra-high definition UHD with 4096 × 2160 pixels (Phantom 3 Pro, DJI, Shenzhen, China or Yuneec Typhoon H, Yuneec, Jiangsu, China).

The minimum size of the anomalies to be detected (represented here by horizontal length $l_h$ and vertical length $l_v$) can be estimated using the resolution $R_h \times R_v$ (horizontal × vertical) and the following two equations:

$$l_h = \frac{r}{R_h} 2d \tan(\alpha) \tag{1}$$

$$l_v = \frac{r}{R_v} 2d \frac{\tan(\beta)}{\cos(\omega)} \tag{2}$$

where $d$ is the distance from the camera lens to the surface to be inspected, $\alpha = HFOV/2$, $\beta = VFOV/2$ (*HFOV* and *VFOV* are respectively the horizontal and vertical field of views), $\omega$ is the vertical inclination of the surface, and $r$ is a parameter that represents the minimum number of pixels needed to recognize a defect (in practice, it is possible to follow the general rule of thumb $r = 10$ or $r = 20$). The surface to be inspected may present certain vertical inclination (Figure 3). Considering that the front camera is fixed, if the value of inclination $\omega$ increases, then pixels represent more distance in the surface and, therefore, the minimum vertical length of anomalies to recognize increases, as it is established by Equation (2).

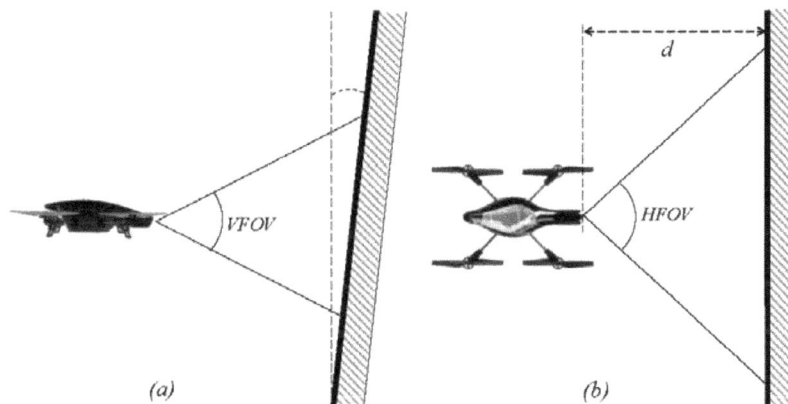

**Figure 3.** Location of the air vehicle with respect to the surface. The figure shows the vertical field of view VFOV of the camera (**a**), the horizontal field of view HFOV (**b**), vertical surface inclination $\omega$ and distance of the surface $d$.

When inspecting surfaces, it is useful to be able to fly at a short distance from the surface in order to detect smaller anomalies. This distance must be established taking into account safety requirements (based on the stability of the aerial platform and the environment conditions). For example, in our experiments, we used distance values of 50 cm < $d$ < 150 cm.

In order to scan the complete surface, the vehicle can develop automatically a flight trajectory. This is an important advantage compared to manual inspection, since more precise trajectories can be automatically developed in large surfaces, preventing gaps from remaining uninspected zones due to drifts and other errors caused by the human operation.

The specific waypoints that define such trajectories must be planned before the flight (for example, using the software tool Aerostack), considering the field of view of the camera. In the case of a horizontal inspection, the movement of the drone develops a sequence of adjacent horizontal scans, changing the altitude at the end of each scan. The difference in altitude between two adjacent horizontal scans (value *h* in Figure 4) depends on the vertical field of view and a small overlap between two adjacent scans (represented with parameter *g* that is manually adjusted for each particular case). This overlap is recommended to avoid gaps and it is also useful when edges of the photo are too distorted. To estimate the value of *h*, the following equation can be used:

$$h = \left[ d \left( \frac{\sin(\beta)}{\cos(\omega + \beta)} + \frac{\sin(\beta)}{\cos(\omega - \beta)} \right) - g \right] \cos(\omega) \tag{3}$$

Other considerations about the on-board camera can be taken into account for an adequate inspection performance. For example, the camera captures images that may be blurred due to vehicle movement This can be reduced with the help of mechanical gimbal stabilization (e.g., as it used in DJI Phantom 3 Pro) or digital imaging stabilization (Parrot Bebop 2), although the digital solution loses resolution. Camera optics may also generate distorted images that affect the inspection process. For example, AR. Drone 2.0 uses a wide-angle lens that produces a certain distortion that affects how our algorithm spatially locates the position of anomalies. To reduce this problem, it is possible to increase the overlap between adjacent scans (which increases the required number of scans). It may be also possible to use algorithms to automatically correct this distortion or to use lens cameras that do not produce distortion (e. g., the DJI Phantom Pro camera).

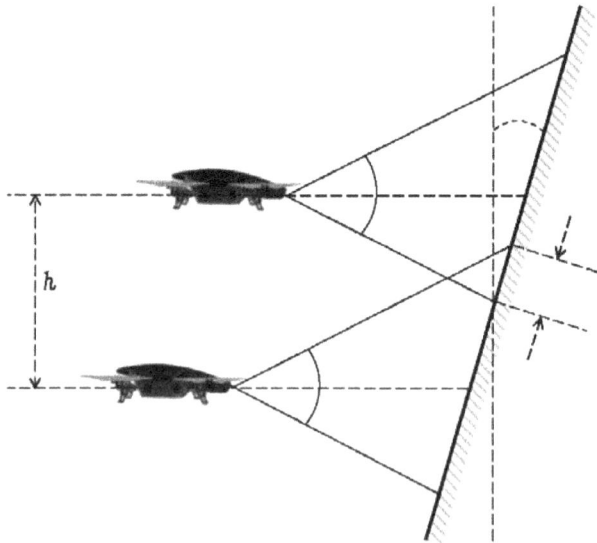

**Figure 4.** Vertical distance *h* between consecutive horizontal scans.

## 5. The Visual Recognition Method

### 5.1. The Method for Defect Recognition

In order to detect defects in still images of surfaces, we selected and implemented one of the existing methods in the literature. This method was based on the approach of frequency histogram of connected elements (FHCE) [29], which is a generalization of the conventional image histogram aimed at detecting and segmenting textured surfaces. This approach was selected for our problem due to its efficiency for image segmentation as it has been proved in practical problems such as wooden pallets [30] and road segmentation for autonomous car driving [31].

This approach uses the concept of neighbourhood. For a given pixel $(i, j)$ of an image, its neighborhood is formed by a set of pixels which distances to $(i, j)$ are not greater than two integer values $r, s$ and is defined as $\varphi_{(i,j)}^{r,s}$. A connected element is the neighborhood selected such as the intensity $I$ of a pixel $(k, l)$ is a subset of a given grayscale range $[T - \varepsilon, T + \varepsilon]$:

$$C_{(i,j)}(T) = \varphi_{(i,j)}^{r,s} : I(k,l) \subset [T - \varepsilon, T + \varepsilon], \forall (k,l) \in \varphi_{(i,j)}^{r,s}$$

Given the previous definitions, $H(T)$ is defined as the sum of all the connected elements for each pixel of an image on different grey levels $T$ where $T$ is greater than 0 and inferior than the maximum intensity minus one:

$$H(T) = C_{(i,j)}(T) 0 \leq T \leq I_{max} - 1 \tag{4}$$

Algorithm 1 summarizes the main steps of the version the FHCE method implemented to recognize anomalies on the surface. The first line of the algorithm is the identification of the interval of grey values $[k_1, k_2]$ of an anomaly, which is determined by the analysis of the frequency histogram. This analysis can be done dynamically for each image or, in certain domains with similar defects, this could be done in advance and use the same values of $k_1$ and $k_2$ for all the images.

The next lines in the algorithm (lines 2–4) assign to each pixel a new distinctive colour $s$ (e.g., white colour) if its grey value belongs to this interval. As a result, an image is obtained with the flaws separated from the background (Figure 5).

---

**Algorithm 1.** Defect recognition.

---

**Input:** digital image $I$ (a still picture)
1.    $[k_1, k_2] \leftarrow$ interval of grey values for anomalies (based on histogram analysis)
2.    **for each** pixel $(i, j)$ **in** input image $I$
3.        $g_{ij} \leftarrow$ grey value of pixel $(i, j)$
4.        **if** $(g_{ij} \in [k_1, k_2])$ **then**
5.            substitute the value of pixel $(i, j)$ by the distinctive colour $s$
6.    $R_1 \leftarrow$ rectangles obtained as contour of images of colour $s$
7.    $R_2 \leftarrow$ rectangles $r_k \in R_1$ such as $area(r_k) \in [k_3, k_4]$
8.    $D \leftarrow$ classified rectangles $r_k \in R_2$ according to shape conditions
9.    **return**$(D)$

---

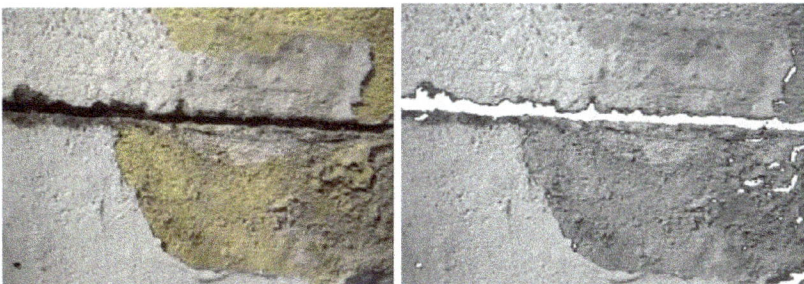

**Figure 5.** Example of original versus resulting image after flaw separation.

Then, the algorithm applies a heuristic approach to identify categories of imperfections. The last lines of the algorithm (lines 5–7) obtain the contour of the images of distinctive colour $s$ as a rectangle and selects rectangles whose area $a \in [k_3, k_4]$. This constraint assumes that only imperfections of certain size are considered and, therefore, the wall must be observed from an appropriate distance.

The method finally classifies the selected rectangles into categories according to their shape. For example, if $l_1$ and $l_2$ are the length of the sides in pixels, a rectangle with square shape, which satisfies that $|l_1 - l_2| < 30$ pixels, can be considered as a hole. Fissures can be characterized as a linear flaw, which is why the method searches for rectangles along a certain linear direction. As a result, the method obtains delimited areas for both flaw types (see Figure 6).

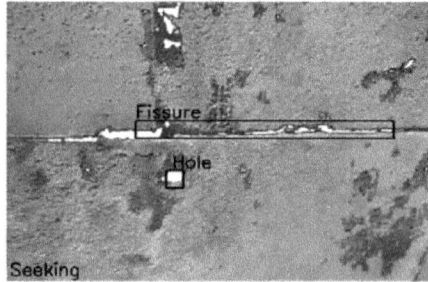

**Figure 6.** Example of image classification.

*5.2. Defect Recognition with Operator Assistance*

This section describes our solution to extend the previous method in order to be used in the collaborative approach presented in this paper. In general, the information recorded by a robot camera is received as a stream of still pictures (frames) at a certain frequency. For example, the aerial vehicle AR Drone 2.0 uses a front camera with a frame rate of 15 or 30 frames per second. The practical experience using robot cameras shows that the visual recognition usually presents gaps in the classified categories of the video stream, with some intermediate frames where the categories are misclassified.

In our approach, the presence of these gaps is used as a way to estimate a credibility value associated to the recognized category. This credibility represents the robot's confidence in the recognition, and it may be a solution to decide when the robot may need operator assistance. We assume that a classification with a high credibility value corresponds to a sequence of classifications in the video stream with narrow gaps. On the other hand, a classification with low credibility presents wider gaps.

To implement this, our method uses a sliding window technique to analyse the video stream. The method uses a window size of $n$ frames from the video stream. At each step, the content of this window is a frame sequence $F = \{f_i\}$, an ordered set of frames $i = 1, \ldots, n$ where $|F| = n$ is the cardinality of this set.

The uncertainty associated with an anomaly $a_j$ is represented with the credibility $Cr(a_j) \in [0, 1]$. To obtain this value, the relative frequency of the detection of anomaly $a_j$ in the set of images $F$, is computed using the following equation:

$$Cr(a_j) = \frac{n_j}{n} \tag{5}$$

where:

$$n_j = \sum_{i=1}^{n} \delta(f_i, a_j)$$

$$\delta(f_i, a_j) = \begin{cases} 1 \ if \ anomaly \ a_j \ is \ detected \ in \ frame \ f_i \\ 0 \ otherwise \end{cases}$$

The credibility is used to decide when the robot needs to ask operator assistance. For this purpose, our method uses an interval defined by two parameters $[k_5, k_6]$ that establishes the uncertainty range when the robot requires assistance. Thus, the robot asks for operator assistance to confirm the recognition of anomaly $a_j$ when the credibility satisfies $Cr(a_j) \in [k_5, k_6]$. The robot rejects (or confirms) automatically the recognition of anomaly $a$ when $Cr(a_j) < k_5$ (or when $Cr(a_j) > k_6$) without asking for operator assistance.

Algorithm 2 summarizes this process. This algorithm uses the procedure described in Algorithm 1 to detect defects in still pictures (line 3). The recognized defects in the frames are collected to compute finally their credibility (line 8) using Equation (4). Each anomaly is automatically confirmed if its credibility is greater than $k_6$ (line 10) or, if the credibility is greater than $k_5$, a confirmation is requested to the operator (line 11). Otherwise the anomaly is discarded.

The values of the interval $[k_5, k_6]$ need to be calibrated to have an adequate performance of the method. In general, if we increase $k_6$, then the number of false positives decreases, and if we decrease $k_5$, the number of false negatives decreases. However, very low values of $k_5$ or very high values of $k_6$ increase significantly the number of requests for operator assistance. For example, with the extreme values $k_5 = 0.0$ and $k_6 = 1.0$, all the detected anomalies must be confirmed by the operator. As Section 6 describes (experimental evaluation), we use parameter values such as $k_5 = 0.2$ and $k_6 = 0.5$ with a good performance in our domain. But our approach is general to accept other values according to the needs of certain applications or other visual recognition methods.

---

**Algorithm 2.** Defect recognition in a frame sequence with operator assistance.

---

**Input:** $F \leftarrow \{f_i\}$ ordered set of the last $n$ frames of a video stream
1.  $A = \varphi, R = \varphi, n(a_j) = 0$
2.  **for** $(i = 1, \ldots, n)$
3.      $D \leftarrow \{a_j, a_j \in \textbf{defect\_recognition}(f_i)\}$  // *Algorithm 1*
4.      **if** $(D \neq \varphi)$ **then**
5.          $A = A \cup D$
6.              **for** $(j = 1, \ldots, |D|)$   $n(a_j) = n(a_j) + 1$
7.  **for** $(j = 1, \ldots, |A|)$
8.      $Cr(a_j) = n_j / n$   // *Equation (5)*
9.      **if** $(Cr(a_j) > k_6)$ **then** $R = R \cup \{a_j\}$
10.     **else if** $(Cr(a_j) \geq k_5)$ **then**
11.         **if** (**operator\_confirm\_anomaly**$(a_j)$) **then** $R = R \cup \{a_j\}$
12. **return**$(R)$

---

## 6. The Aerostack Framework

To implement the collaborative approach for surface inspection described above in a robotic system, it is necessary to build a complex control architecture that integrates both reactive low-level components (e.g., for perception and motion control) and deliberative high-level components (e.g., for operator interaction). For this development, we have used Aerostack (http://www.aerostack.org) [32,33].

Aerostack is an open-source software framework that helps developers design and build the control architecture of aerial robotic systems, integrating multiple heterogeneous computational solutions (e.g., computer vision algorithms, motion controllers, self-localization and mapping methods, planning algorithms, etc.). Aerostack is useful for building autonomous aerial systems in complex and dynamic environments and it is also a useful research tool for aerial robotics to test new algorithms and architectures.

Figure 7 shows the main functional components of the architecture. This design follows the hybrid reactive/deliberative paradigm, that is, an architecture that integrates both a deliberative and reactive approaches [34]. The design includes several layers: physical, reactive, executive, deliberative and social. The three layers reactive-executive-deliberative are based on the popular hybrid design known as the three-layer architecture [35]. The architecture includes also a social layer with communication abilities, as it is proposed in multiagent systems and other architectures with social coordination (e.g., [36]).

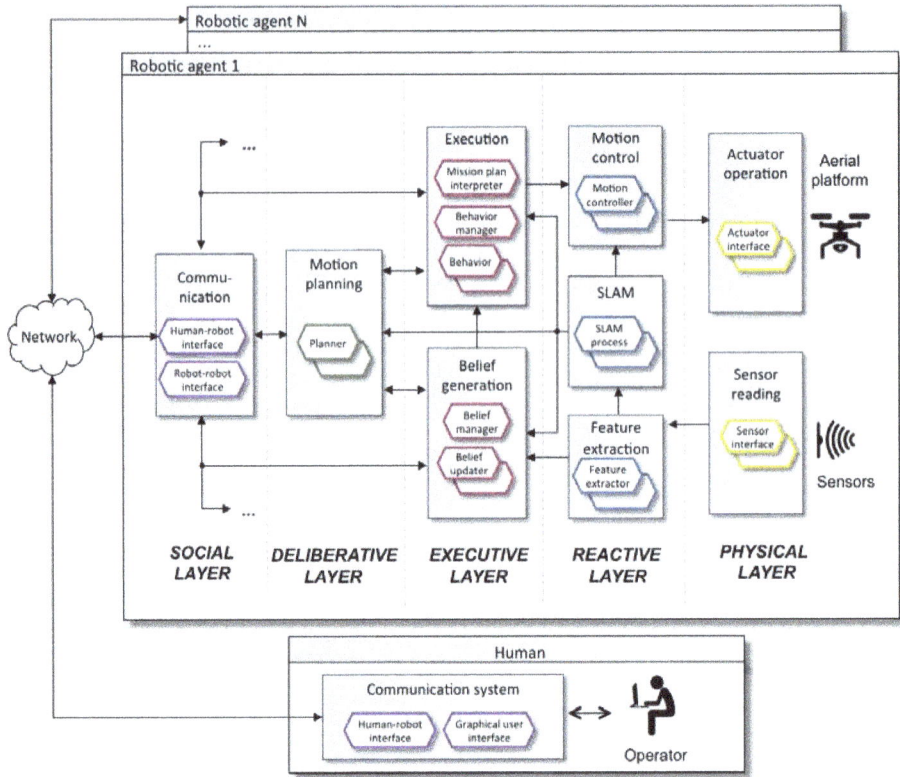

**Figure 7.** The multi-layered architecture of Aerostack.

Aerostack (version 2.0) follows a behaviour-based approach which has been traditionally used in robotics [37–39]. A behaviour provides abstraction, hiding low level technical details and complexity, and it is appropriate to simplify how to specify complex missions. Aerostack provides a library of behaviours including basic motion behaviours (take off, land, keep hovering, keep moving, go to point, rotate, etc.) that can be used for surface inspection besides others (follow object image, self-localize by visual markers, wait, etc.). This is an open library that can be extended with additional behaviours to provide additional capabilities. A new behaviour can be included in the library by adding new processes. Aerostack uses ROS (Robot Operating System, http://www.ros.org). A new process can be created by programming a new ROS node in C++. The node may be integrated in the library by using inter-process communication mechanisms provided by ROS (e.g., publish/subscribe topics or service calls).

The operator can specify a mission plan in Aerostack using three alternative options: TML language, Python language or with behaviour trees. For example, the TML language [40,41] uses a task-based approach with hierarchies of tasks in which groups of behaviours are activated. Aerostack coordinates the execution of multiple behaviours avoiding conflicts when incompatible behaviours are requested to be active at the same time.

Aerostack provides a graphical user interface (Figure 8) with which the operator can delegate tasks to the robot and the robot sends information to the operator to inform about its progress (this interaction mode corresponds to what we have called in this paper supervised mode). For example, the operator can use this interface to start the execution of a particular mission using the control panel. Once the mission has been initiated, the operator can use the interface to observe dynamic information (e.g., position coordinates and orientation), the location of the robot in an environment map with the existing obstacles, the camera image, and other information (parameter values, active behaviours, etc.). In addition, the graphical user interface includes buttons to help the operator to interrupt the mission execution (e.g., emergency land or abort mission).

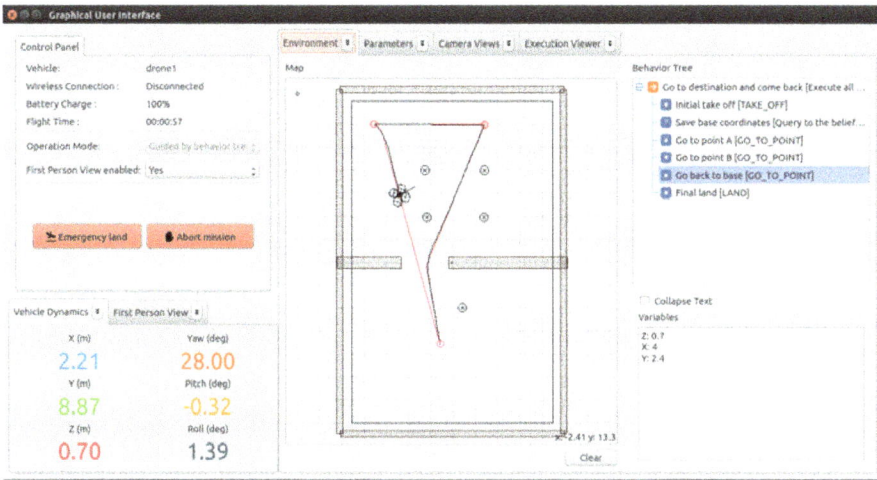

**Figure 8.** Example of screen presented by the graphical user interface of Aerostack to execute and supervise a mission plan.

The operator can also use the keyboard to tele-operate the robot by executing certain basic motions (Figure 9), such as take-off, land, move forwards, turn clockwise, and so forth. These actions activate the corresponding behaviours to perform the motions.

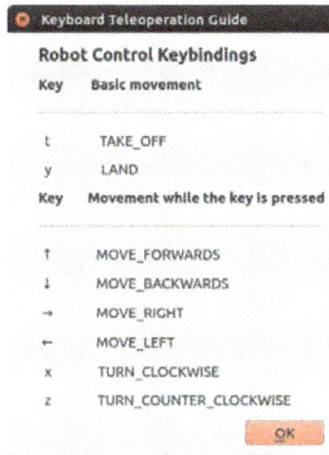

**Figure 9.** Available commands for tele-operation using the keyboard.

## 7. Extensions for Surface Inspection

Aerostack provides a number of useful solutions for our surface inspection problem, but there are specific functionalities that are not covered. In particular, Aerostack does not provide two capabilities: (1) recognize automatically surface imperfections and (2) support the assisted mode in human-robot interaction, as it is required in our collaboration approach. This section describes how we extended Aerostack with additional components to cover such capabilities.

Basically, the extension for surface inspection was done by adding new behaviours. In addition, it was necessary to modify an existing Aerostack component (the behaviour coordinator) to facilitate the support for assisted mode. Part of these changes are integrated in Aerostack (www.aerostack.org) to be available as open-source in the software framework.

Figure 10 shows new processes that were developed (marked with a red dotted line) and how they were integrated with other Aerostack processes. For example, the behaviour called *recognize surface imperfections* controls the execution of a specialized process called *surface inspector* that uses the computer vision technique presented in this paper to recognize automatically surface imperfections. This process receives an input image from the front camera together with the estimated pose of the robot and, as a result, generates the following information when an imperfection is detected: (1) the characteristics of the detected anomaly (category and position) that is stored as a symbolic belief and (2) the camera image with added marks indicating where the recognized defects are located.

Other processes specialized in human-robot communication were developed to support the assisted mode. Figure 10 shows a process that supports the behaviour called *request operator confirm recognition*. This behaviour controls the interaction with the operator in order to confirm or reject the recognized imperfection. The following section describes in more detail the specific techniques used to support the assisted mode.

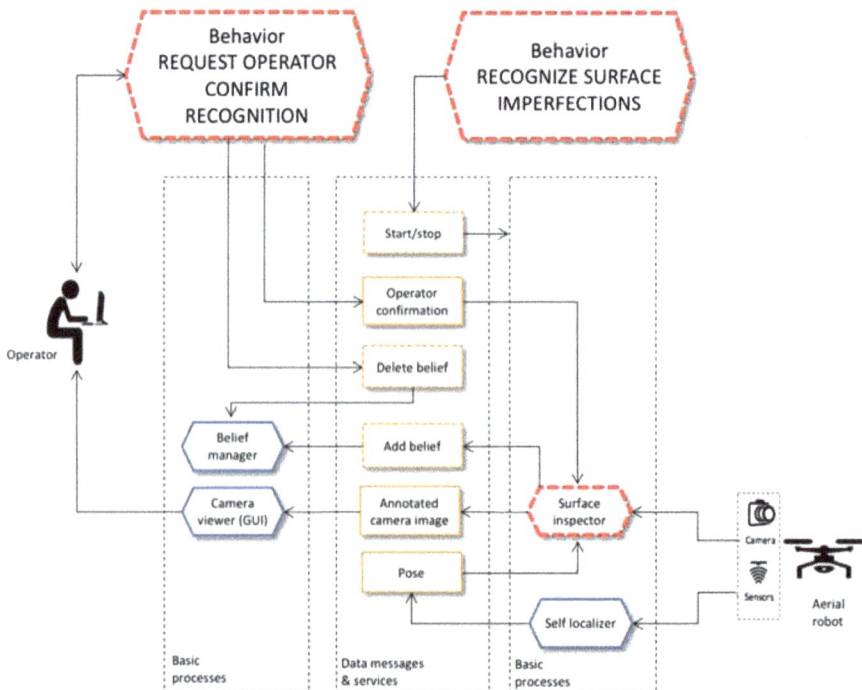

**Figure 10.** Example processes developed to support surface inspection (marked with a red dotted line).

*Assistance Mode in Human-Robot Interaction*

As described above, the collaborative scheme for surface inspection uses two interaction modes between operator and robot: supervised and assisted. Aerostack only supports the first mode with a graphical user interface. Therefore, it was necessary to extend Aerostack to support the second mode. In principle, this extension could be done by adding specialized behaviours for user interaction that request information to the operator through specialized windows. However, as explained below, this solution is not enough, because it is also necessary to extend the way Aerostack coordinates concurrent behaviours to change correctly between the two different interaction modes (supervised and assisted).

To request assistance to the operator, some windows were implemented to present images and texts and wait for the operator answer. For example, Figure 11 shows windows to request confirmation of recognized anomalies. This is implemented with the behaviour *request operator confirm recognition*. The operator can observe the camera image with the camera viewer where the detected imperfections are presented. A pop-up window informs the operator that the mission has been stopped and the type of recognized figure.

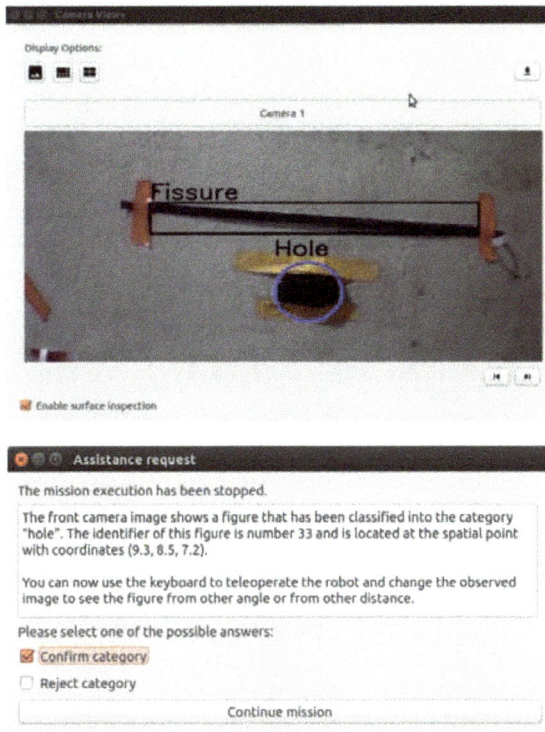

**Figure 11.** Example of windows to request confirmation of a recognized figure.

If more than one defects are detected, the image highlights in blue colour the figure for which the system is requesting confirmation. The robot keeps a visual tracking of the selected defect until the operator selects one of two options (confirm or reject category) and continue the mission. Meanwhile, the other defects are highlighted in black. The robot memorizes the answer provided by the operator to avoid repeating questions about the same defect. The robot memorizes both the position of the defect (world coordinates) and the operator answer (confirmation or rejection).

Figure 12 shows another type of window to interact with the operator implemented with the behaviour *request operator assistance*. This is a general window that presents different options to the operator and it is useful to use other interaction patterns [42]. As the figure shows, this window can be used by the robot to know how respond in the presence of low light.

As mentioned above, it was also necessary to add another functionality to Aerostack to support the assisted mode. Aerostack provides languages to the operator to specify mission plans (e.g., TML language). During the mission execution, Aerostack interprets the plan written by the operator in these languages to perform the complete mission. This interpretation automatically activates groups of concurrent behaviours according to what is specified in the plan.

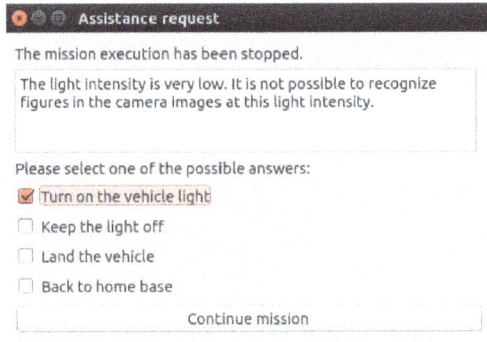

**Figure 12.** Example of window to request assistance.

The problem is that the assisted mode requires to interrupt the execution of the mission and continue later, after the operator gives an answer. While the mission is paused, the user must be able to execute other behaviours manually, to observe the surface from other angles or distances. This requires stopping part of the active behaviours corresponding to the mission execution, let the operator activate manually other behaviours (e.g., by teleoperation) and finally restore the stopped behaviours to continue the mission.

As a solution for this, Aerostack was extended at the level of behaviour coordination, with a mechanism that informs about the set of active behaviours that are incompatible with a given behaviour. This solution was implemented with a ROS service called *consult incompatible behaviours*. With the help of this, it is possible to pause the execution of a mission to request assistance as it is illustrated by Algorithm 3.

Let us suppose that when a question is presented to the operator using the corresponding window, we want the robot to stop its movement and keep hovering while waiting for the operator's response. In general, instead of the behaviour keep hovering, we could use any other behaviour (e.g., align visually at a certain distance in front of the recognized object). Algorithm 3 uses the term *waiting behaviour* to identify this behaviour in a general way.

The first step of the Algorithm 3 is to consult the incompatible behaviours with the waiting behaviour. For illustration purposes, let us consider the incompatible behaviour is go to point (3.0, 5.5, 2.5). Then, the waiting behaviour is activated (keep hovering), which stops the movement. This automatically deactivates the incompatible behaviour (go to point). Then, the window that requests information to the operator is displayed, showing alternative options to select $\{o_1, o_2, \ldots, o_n\}$. Then, there is a loop to wait until the operator selects one of the options (or until timeout). If the operator selects an option (line 9), the result of this selection is written in the belief memory to be used later during the mission execution. Once this has been done (line 11), the initial behaviours are activated again. In our example, this means that the behaviour go to point (3.0, 5.5, 2.5) is restored to continue the mission.

---

**Algorithm 3.** Assistance request.

| | |
|---|---|
| 1. | $B \leftarrow$ consult incompatible behaviours with the waiting behaviour $b$ |
| 2. | activate the waiting behaviour $b$ |
| 3. | display window with options $O = \{o_1, o_2, \ldots, o_n\}$ |
| 4. | $t \leftarrow 0$ |
| 5. | **repeat** |
| 6. | $\quad$ sleep($t_s$ seconds) |
| 7. | $\quad$ $t \leftarrow t + t_s$ |
| 8. | **until** (operator has selected option $o_i$) or ($t >$ time-out) |
| 9. | **if** (operator has selected option $o_i$) **then** |
| 10. | $\quad$ write the result of option $o_i$ in the belief memory |
| 11. | **for each** $b_j$ **in** $B$ |
| 12. | $\quad$ activate behaviour $b_j$ |

---

With this solution, the operation can change consistently between two methods (guided by a mission plan or guided by teleoperation) to support the collaborative scheme described in this paper. This solution is general because it is possible be used by any behaviour that needs to interact with the operator.

## 8. Experimental Evaluation

In order to evaluate our solution, two separate types of tests were applied: (1) tests to evaluate the visual recognition method, and (2) tests to validate the semi-autonomous operation. The following sections describe the details and results of these tests.

### 8.1. Testing the Visual Recognition Method

The quality of our method for defect recognition in video streams was evaluated with a dataset of frame sequences. For this purpose, we built a dataset with 650 frames organized in 13 sequences of 50 frames (Figure 13). This dataset is provided as supplementary material of this paper.

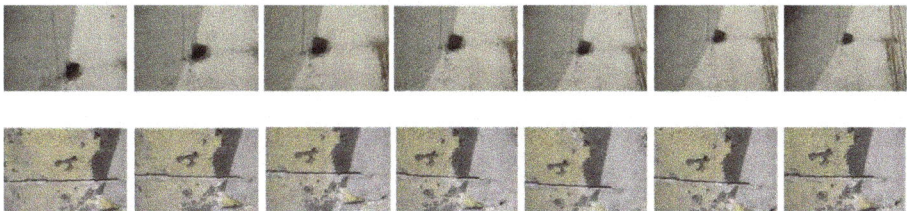

**Figure 13.** Sample of the dataset of surface imperfections used for the evaluation. The figure presents part of two frame sequences of the video stream take by the camera.

The dataset includes 6 sequences for holes (3 positive and 3 negative) and 7 sequences for fissures (5 positive and 2 negative). For each positive example, the dataset provides its category (hole or fissure), its size defined by a rectangle covering the figure, and its location (coordinates of the centre of the rectangle in pixels). Figure 14 shows how this information is presented graphically.

The dataset was separated in two parts: (1) a design dataset to estimate the values of parameters $\{k_1, \ldots k_6, n\}$, and (2) an evaluation dataset to evaluate the quality of the resulting model. The design dataset includes 7 frame sequences and the evaluation dataset 6 frame sequences.

As evaluation metrics, we used precision and recall (and their combination in F-measure). In addition, we also estimated the percentage of times that the recognition method needs operator assistance. The way these values were computed is the following. For precision and recall, true positives are positive examples of the dataset that satisfy $Cr(a_j) \geq k_5$, false negatives are positive examples of the dataset that satisfy $Cr(a_j) < k_5$, and false positives are negative examples that satisfy $Cr(a_j) \geq k_6$. To evaluate the level of operator assistance required by the recognition method, we defined a metric called autonomy degree. To obtain the value of this metric, we added the number of examples that satisfy the expression $[Cr(a_j) < k_5] \lor [Cr(a_j) > k_6]$ (i.e., the cases when the robot does not require operator assistance) and divided this value by the total number of the examples.

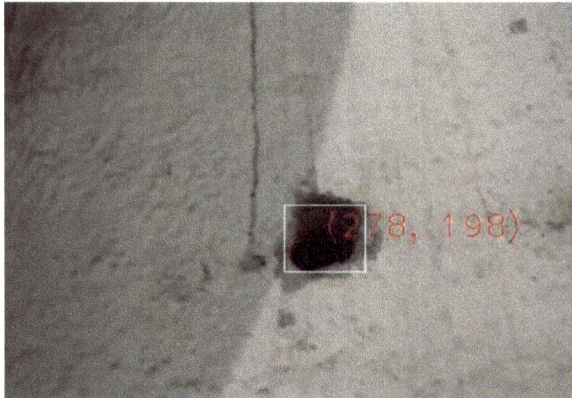

**Figure 14.** Visualization of data associated to an image of the evaluation dataset.

The design dataset was used to manually calibrate the parameters in order to obtain good values for the evaluation metrics. The parameter values that we found were $\{k_1 = 0, k_2 = 90, k_3 = 120, k_4 = 11,000, k_5 = 0.2, k_6 = 0.5, n = 10\}$. Table 2 shows the results obtained for the evaluation metrics with the design dataset. We obtained the maximum value for F-measure (1.0) and a value for autonomy degree of 0.96, meaning that the 4% of the examples require operator assistance.

**Table 2.** Evaluation results of the recognition methods.

| Evaluation Metric | Design Dataset | Evaluation Dataset | |
|---|---|---|---|
| | Method with HC | Method without HC | Method with HC |
| Precision | 1.00 | 0.93 | 1.00 |
| Recall | 1.00 | 0.81 | 0.97 |
| F-measure | 1.00 | 0.87 | 0.98 |
| Autonomy degree | 0.96 | 1.00 | 0.63 |

Then, the evaluation dataset was used to analyse the performance of our method using a with-and-without comparison. Two methods were compared:

- *Method without human collaboration*, which corresponds to the FHCE defect recognition method in still pictures described in Section 5.1.
- *Method with human collaboration*, which corresponds to our solution with operator assistance, described in Section 5.2.

The evaluation shows that our solution with human collaboration obtains a high recognition performance (F-measure: 0.98) compared to the other method (F-measure: 0.87) at the cost of reducing the degree of autonomy (0.63) that is, having a partial operator assistance in 27% of the examples. These results are useful to estimate the amount of improvement of the recognition method with the collaborative approach presented in this paper with semi-autonomous operation and partial operator assistance.

## 8.2. Testing the Semi-Autonomous Operation

In addition to the visual recognition tests, other tests were performed to evaluate the capacity of the robot for interaction and autonomous operation. For this purpose, we designed several flight experiments using both simulated flights and real flights. The experiments were designed to validate the correct change between interaction modes (supervised and assisted modes), and the ability to fly efficiently to cover the area to inspect of a surface.

For example, in one of the experiments performed by simulation, a surface to inspect with two walls in a corner was defined. Figure 15 shows some of the resulting trajectories generated by our system for the same mission plan in different situations. Figure 15a shows a trajectory to cover the two surfaces of the corner, without obstacles and without interruptions. Figure 15b shows the execution of the same mission, but with a column in one of the walls as an obstacle. In this case, our system adapts correctly the execution of the mission to avoid the obstacle. Figure 15c shows the execution of the same mission, but with two interruptions. The robot stops at these points and asks for assistance to the operator. The operator moves manually (by tele-operation) the robot, answers the question and returns the control to the interpreter of the mission plan to continue the mission execution.

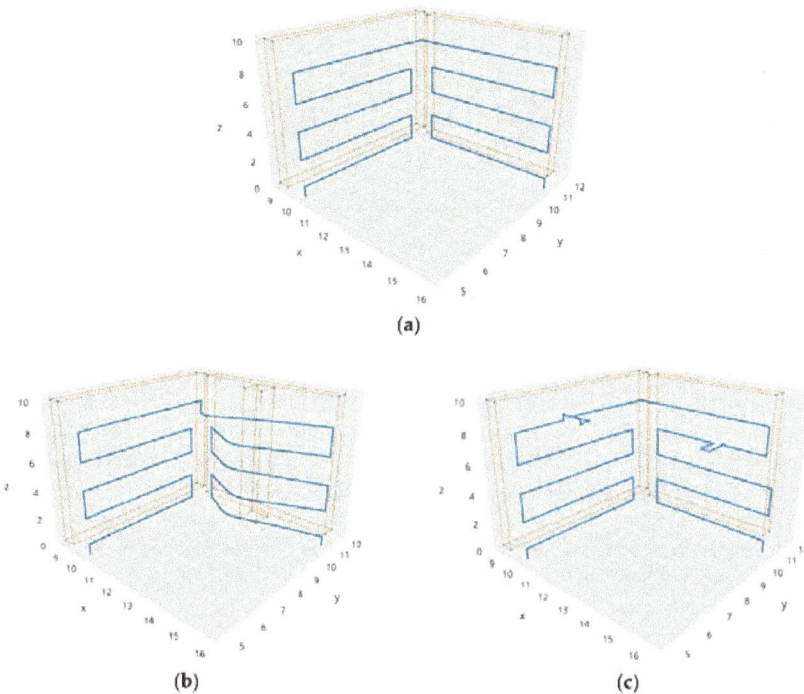

(a)

(b)

(c)

**Figure 15.** Example trajectories generated by our system (simulated flights). The three figures show the execution of the same mission plan in three different situations. Figure (**a**) shows the trajectory to cover the surfaces defined in the mission plan, figure (**b**) shows how the same plan is adapted when there is an obstacle, figure (**c**) shows two points where the robot stops and asks for assistance to the operator. In the figure, orange lines represent walls and blue lines represent trajectories followed by the aerial vehicle.

This experiment proves that the Aerostack language is expressive enough to formulate the movements required for surface inspection and that the robot is able to fly autonomously, correctly following the plan and efficiently describing a trajectory to cover the surface of the walls.

This experiment also verified that the robot was able to interact correctly with the operator to request assistance. Table 3 shows partially the sequence of the behaviour activation. In this experiment, the robot starts the execution guided by a mission plan, that is, the execution is controlled by the plan interpreter that activates behaviours according to what is written in the mission plan. The table shows that the first behaviour to be active is take off. Then, the next behaviours are keep hovering and go to a point with coordinates (9.9, 10.2, 0.7).

**Table 3.** Partial sequence of behaviour activation.

| Order | Operation Method | Behaviour |
|---|---|---|
| 01 | Guided by mission plan | TAKE_OFF |
| 02 | Guided by mission plan | KEEP_HOVERING |
| 03 | Guided by mission plan | GO_TO_POINT coordinates: [9.9, 10.2, 0.7] |
| ... | ... | ... |
| 11 | Guided by mission plan | GO_TO_POINT coordinates: [9.9, 10.8, 8.4] |
| 12 | Guided by mission plan | REQUEST_OPERATOR_CONFIRM_RECOGNITION |
| 13 | Guided by mission plan | KEEP_HOVERING |
| 14 | Guided by teleoperation | KEEP_MOVING direction: UP |
| 15 | Guided by teleoperation | KEEP_HOVERING |
| 16 | Guided by teleoperation | KEEP_MOVING direction: BACKWARDS |
| ... | ... | ... |
| 20 | Guided by teleoperation | KEEP_HOVERING |
| 21 | Guided by mission plan | GO_TO_POINT coordinates: [9.9, 10.8, 8.4] |
| 22 | Guided by mission plan | KEEP_HOVERING |
| 23 | Guided by mission plan | ROTATE angle: 90 |
| ... | ... | ... |

After some steps, in step 11, the active behaviour is go to a point with coordinates (9.9, 10.8, 8.4). While the robot is going to this point, the robot recognizes a figure on the wall and automatically activates a behaviour to request assistance to the operator to confirm the recognition (step 12). This behaviour displays a confirmation window to the operator (similar to the window presented in Figure 11) and activates the behaviour keep hovering.

The operator now manually moves the robot by teleoperation (activating other motion behaviours) to observe the surface from other angles or distances. In this case, the operator tele-operates the robot by using the keyboard and activates behaviours such as move up and move backwards to observe the figure from other positions and distances. Finally, the operator confirms the recognition (by clicking the corresponding buttons in the confirmation window) and the behaviour that requests assistance restores the behaviour go to a point with coordinates (9.9, 10.2, 0.7) and returns the control to the plan interpreter to continue the mission.

This type of experiment was repeated with other defects in different positions. The experiments proved that the operation changed consistently between two operation methods (guided by a mission plan or guided by teleoperation) to support correctly the collaborative scheme described in this paper.

In addition to simulated flights, we also performed tests with real flights. The main goal of these tests was to verify that the robot was able to execute efficiently an inspection plan, doing intermediate stops to have closer views of the wall. We used the platform Parrot AR Drone 2.0 and indoor walls with several imperfections (Figure 16). Figure 17 shows the trajectory developed by the robot in one of the experiments. The robot develops autonomously a horizontal exploration starting from the take-off point (2.0, 2.0, 0.0) and landing at the same point after the exploration. There are two points where the robot zooms in and out to have closer views of the wall: point with coordinates (2.0, 3.0, 1.0) and point with coordinates (6.0, 3.0, 1.0).

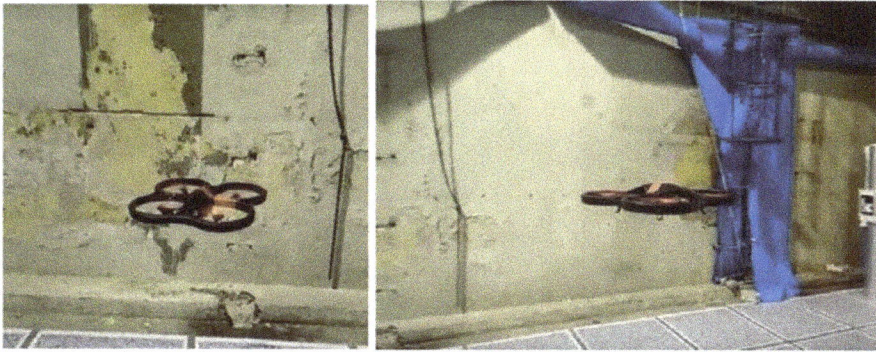

**Figure 16.** The aerial robot (AR Drone 2.0) during wall inspection.

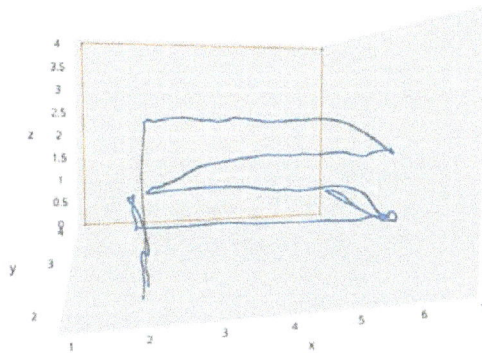

**Figure 17.** Example exploration trajectory followed by the robot in a real flight. In the figure, the orange line represents the wall and the blue line represents the trajectory followed by the aerial vehicle.

In this experiment, the position of the quadcopter was determined by visual odometry. Although this positioning method is not very precise, it was used in the experiments to simplify the practical execution of flight tests. The generated trajectory in this case shows certain irregularities (Figure 17) compared to the ideal trajectories showed by simulations (Figure 15), but this can be improved by using other positioning methods (e.g., with visual markers, lidar, etc.), which is outside the scope of the work presented in this paper. In the experiment, we obtained satisfactory results about efficiency (the whole mission was completed in 1 min and 19 s) and the correct sequence of steps.

## 9. Conclusions

This paper has presented the results of our work on surface inspection using aerial robots with cameras on board. The results show that it is possible to increase the degree of robot autonomy by using an automated visual recognition method together with a human-robot collaboration approach.

In this work, an existing visual recognition method (frequency histogram of connected elements) was extended to be able to operate with partial human assistance. This extension is based on a credibility-based solution that analyses the video stream from the camera to decide when to request assistance to the operator. The evaluation of the assisted method shows a high recognition performance (F-measure: 0.98) compared to a method without assistance (F-measure: 0.87) at the cost of receiving partial operator assistance (in 27% of the cases).

We built a robotic system to validate the type of operation proposed in this paper. The experiments showed that the control changed correctly between different interaction modes (supervised and assisted) and different operation methods (guided by mission plan or by teleoperation) to support the collaborative scheme.

The software framework Aerostack (www.aerostack.org) was useful to develop our robotic system. This framework provided an architectural pattern that guided the design of our system together with a set of reusable software components to support autonomous behaviour (e.g., motion control, obstacle avoidance, mission plan interpreter, etc.). But it was also necessary to design, implement and integrate new components to support the collaborative approach in surface inspection presented in this paper.

It is also possible to identify potential improvements that could be achieved by further research. For example, our work paid attention to visual recognition and assistance, but the inspection problem includes other autonomous behaviours that can be considered to achieve more autonomy. For example, it would be possible to use elaborated path planning methods to generate automatically the best trajectory to inspect a given surface.

Our visual recognition method has an acceptable performance, according to the goals of a collaborative work, with two types of anomalies (holes and fissures) and vertical surfaces using the frontal camera. Further research can improve the recognition method to increase its performance in order gain more autonomy with less operator assistance and extend its capacity for other types of imperfections (mould, humidity, spots, etc.) together with other camera orientations to cover other types of surfaces.

**Supplementary Materials:** The evaluation dataset and a video illustrating the work presented in this paper are available online at http://www.mdpi.com/1424-8220/18/3/893/s1.

**Acknowledgments:** This research work has been partially supported by the Spanish Ministry of Economy and Competitiveness through the project VA4UAV (Visual autonomy for UAV in Dynamic Environments), reference DPI2014-60139-R. This research work has been carried out in the research group Computer Vision and Aerial Robotics (CVAR) (www.visionaerialrobotics.org). The authors would like to thank the following members of CVAR for their help in software programming, testing, and useful comments: Alberto Camporredondo, Rafael Artiñano, Daniel Rabasco, Guillermo Echegoyen, and Hriday Bable.

**Author Contributions:** Martin Molina supervised the research, analyzed the related work, designed the collaborative extension of the visual recognition method, analyzed the related work, designed its integration in Aerostack, conceived the evaluation procedure, analyzed results, organized the paper, and wrote the paper. Pedro Frau adapted and implemented the visual recognition method, implemented a drone-operator collaborative behaviour, acquired images for the dataset, designed and performed experiments, analyzed results, and wrote part of the paper. Dario Maravall supervised the research, analyzed the related work and wrote part of the paper.

**Conflicts of Interest:** The authors declare no conflict of interest.

## References

1. Newman, R.S.; Jain, A.K. A survey of automated visual inspection. *Comput. Vis. Image Underst.* **1995**, *61*, 231–262. [CrossRef]

2. Neogi, N.; Mohanta, D.K.; Dutta, P.K. Review of vision-based steel surface inspection systems. *EURASIP J. Image Video Process.* **2014**, *2014*, 50. [CrossRef]

3. Xie, X. A review of recent advances in surface defect detection using texture analysis techniques. *ELCVIA Electron. Lett. Comput. Vis. Image Anal.* **2008**, *7*. [CrossRef]

4. Medeiros, F.N.; Ramalho, G.L.; Bento, M.P.; Medeiros, L.C. On the evaluation of texture and colour features for nondestructive corrosion detection. *EURASIP J. Adv. Signal Process.* **2010**, *2010*, 817473. [CrossRef]

5. Medina, R.; Llamas, J.; Zalama, E. Crack detection in concrete tunnels using a Gabor Filter invariant to rotation. *Sensors* **2017**, *17*, 1670. [CrossRef] [PubMed]

6. Maglietta, R.; Milella, A.; Caccia, M.; Bruzzone, G. A vision-based system for robotic inspection of marine vessels. *Signal Image Video Process.* **2017**, *12*, 471–478. [CrossRef]

7. Kaamin, M.; Idris, N.A.; Bukari, S.M.; Ali, Z.; Samion, N.; Ahmad, M.A. Visual Inspection of Historical Buildings Using Micro UAV. In *MATEC Web of Conferences*; EDP Sciences: Paris, France, 2017; Volume 103, p. 07003.

8.  Hallermann, N.; Morgenthal, G.; Rodehorst, V. Unmanned Aerial Systems (UAS)–Case Studies of Vision Based Monitoring of Ageing Structures. In Proceedings of the International Symposium Non-Destructive Testing in Civil Engineering, Berlin, Germany, 15–17 September 2015; pp. 15–17.

9.  Deng, C.; Wang, S.; Huang, Z.; Tan, Z.; Liu, J. Unmanned aerial vehicles for power line inspection: A cooperative way in platforms and communications. *J. Commun.* **2014**, *9*, 687–692. [CrossRef]

10. Ortiz, A.; Bonnin-Pascual, F.; Garcia-Fidalgo, E. Vessel inspection: A micro-aerial vehicle-based approach. *J. Intell. Robot. Syst.* **2014**, *76*, 151–167. [CrossRef]

11. Merino, L.; Caballero, F.; Ollero, A. Cooperative fire detection using unmanned aerial vehicles. In Proceedings of the ICRA 2005, Barcelona, Spain, 18–22 April 2005.

12. Özaslan, T.; Shen, S.; Mulgaonkar, Y.; Michael, N.; Kumar, V. Inspection of penstocks and featureless tunnel-like environments using micro UAVs. In *Field and Service Robotics*; Springer: Cham, Switzerland, 2015; pp. 123–136.

13. Goodrich, M.A.; Schultz, A.C. Human-Robot Interaction: A Survey. *Found. Trends Hum. Comput. Interact.* **2007**, *1*, 23–275. [CrossRef]

14. Suzuki, T.; Sekine, T.; Fujii, T.; Asama, H.; Endo, I. Cooperative formation among mobile robot teleoperation in inspection task. In Proceedings of the 2000 IEEE Conference on Decision and Control, Sydney, Australia, 12–15 December 2000.

15. Kimura, H.; Ikeuchi, T. Task-model based Human-Robot Cooperation using vision. In Proceedings of the IEEE/RSJ International Conference on Intelligent Robots and Systems 1999—IROS 99, Kyonju, Korea, 17–21 October 1999.

16. Sheridan, T.B. *Telerobotics, Automation, and Human Supervisory Control*; MIT Press: Cambridge, MA, USA, 1992.

17. Miller, C.A.; Parasuraman, R. Designing for flexible interaction between humans and automation: Delegation interfaces for supervisory control. *Hum. Factors J. Hum. Factors Ergon. Soc.* **2007**, *49*, 57–75. [CrossRef] [PubMed]

18. Cummings, M. Human supervisory control of swarming networks. In Proceedings of the 2nd Annual Swarming: Autonomous Intelligent Networked Systems Conference, Arlington, VA, USA, June 2004; pp. 1–9.

19. Lee, J.S.; Zhou, M.C.; Hsu, P.L. An application of Petri nets to supervisory control for human-computer interactive systems. *IEEE Trans. Ind. Electron.* **2005**, *52*, 1220–1226. [CrossRef]

20. Searle, J.R. *Speech Acts*; Cambridge University Press: Cambridge, UK, 1969.

21. Searle, J.R. *Language Mind and Knowledge Minnesota*; Studies in the Philosophy of Science Chapter A Taxonomy of Illocutionary Acts; University of Minnesota Press: Minneapolis, MN, USA, 1975.

22. Austin, J.L. *How to Do Things with Words*; Oxford University Press: Oxford, UK, 1975.

23. Murphy, R. *Introduction to AI robotics*; MIT Press: Cambridge, MA, USA, 2000.

24. Castelfranchi, C. Guarantees for autonomy in cognitive agent architecture. In *Intelligent Agents*; Springer: Berlin/Heidelberg, Germany, 1994; pp. 56–70.

25. Fong, T.; Thorpe, C.; Baur, C. Multi-robot remote driving with collaborative control. *IEEE Trans. Ind. Electron.* **2003**, *50*, 699–704. [CrossRef]

26. Horvitz, E. Principles of mixed-initiative user interfaces. In Proceedings of the SIGCHI Conference on Human Factors in Computing Systems, Pittsburgh, PA, USA, 15–20 May 1999; ACM: New York, NY, USA, 1999; pp. 159–166.

27. Fong, T.; Thorpe, C.; Baur, C. Robot as partner: Vehicle teleoperation with collaborative control. In *Multi-Robot Systems: From Swarms to Intelligent Automata*; Springer: Dordrecht, The Netherlands, 2002; pp. 195–202.

28. Crandall, J.W.; Goodrich, M.A.; Olsen, D.R.; Nielsen, C.W. Validating human-robot interaction schemes in multitasking environments. *IEEE Trans. Syst. Man Cybern. Part A Syst. Hum.* **2005**, *35*, 438–449. [CrossRef]

29. Maravall, D.; Patricio, M.A. Image Segmentation and Pattern Recognition. A novel concept: The histogram of connected elements. In *Pattern Recognition and String Matching*; Cheng, X., Ed.; Kluwer Academic Publishers: Dordrecht, The Netherlands, 2003; pp. 411–463.

30. Maravall, D.; Patricio, M.A. A novel generalization of the grayscale histogram and its application to the automated visual measurement and inspection of wooden pallets. *Image Vis. Comput.* **2007**, *25*, 805–816.

31. Maravall, D.; Patricio, M.A. Segmentation of Traffic images for automatic car driving. In *Computer—Aided Systems Theory. EUROCAST 2003*; Moreno, R., Pichler, F., Eds.; Lectures Notes on Computer Science; Springer: Berlin/Heidelberg, Germany, 2004; Volume 2809, pp. 314–325.

32. Sanchez-Lopez, J.L.; Suarez-Fernandez, R.A.; Bavle, H.; Sampedro, C.; Molina, M.; Pestana, J.; Campoy, P. AEROSTACK: An Architecture and Open-Source Software Framework for Aerial Robotics. In Proceedings of the 2016 International Conference on Unmanned Aircraft Systems (ICUAS), Arlington, VA, USA, 7–10 June 2016.

33. Sanchez-Lopez, J.L.; Molina, M.; Bavle, H.; Sampedro, C.; Suarez-Fernandez, R.A.; Campoy, P. A Multi-Layered Component-Based Approach for the Development of Aerial Robotic Systems: The Aerostack Framework. *J. Intell. Robot. Syst.* **2017**, *88*, 683–709. [CrossRef]

34. Arkin, R.C.; Riseman, E.M.; Hansen, A. AuRA: An Architecture for Vision-Based Robot Navigation. In Proceedings of the DARPA Image Understanding Work-shop, Los Angeles, CA, USA, 23 February 1987; pp. 417–431.

35. Gat, E. On Three-Layer Architectures. In *Artificial Intelligence and Mobile Robots*; Kortenkamp, D., Bonnasso, R.P., Murphy, R., Eds.; AAAI MIT Press: Cambridge, MA, USA, 1998.

36. Duffy, B.R. The Social Robot. Ph.D. Thesis, Department of Computer Science, University College Dublin, Dublin, Ireland, 2001.

37. Brooks, R.A. A Robust Layer Control System for a Mobile Robot. *IEEE J. Robot. Automat.* **1986**, *2*, 14–23. [CrossRef]

38. Mataric, M. Interaction and intelligent behavior. Ph.D. Thesis, MIT, Cambridge, MA, USA, 1994.

39. Michaud, F.; Nicolescu, M. Behaviour-based systems. In *Springer Handbook of Robotics*; Springer International Publishing: Basel, Switzerland, 2016; pp. 307–328.

40. Molina, M.; Suarez-Fernandez, R.A.; Sampedro, C.; Sanchez-Lopez, J.L.; Campoy, P. TML: A language to specify aerial robotic missions for the framework Aerostack. *Int. J. Intell. Comput. Cybern.* **2017**, *10*, 491–512. [CrossRef]

41. Molina, M.; Diaz-Moreno, A.; Palacios, D.; Suarez-Fernandez, R.A.; Sanchez-Lopez, J.L.; Sampedro, C.; Bavle, H.; Campoy, P. Specifying Complex Missions for Aerial Robotics in Dynamic Environments. In Proceedings of the International Micro Air Vehicle Conference and Competition, IMAV 2016, Beijing, China, 17–21 October 2016.

42. Molina, M.; Frau, P.; Maravall, D.; Sanchez Lopez, J.L.; Bavle, H.; Campoy, P. Human-Robot Cooperation in Surface Inspection Aerial Missions. In Proceedings of the 2017 International Micro Air Vehicle Conference and Flight Competition (IMAV), Toulouse, France, 18–21 September 2017.

![sensors logo] *sensors*

MDPI

*Article*

# Dual-Polarization Ku-Band Compact Spaceborne Antenna Based on Dual-Reflectarray Optics [†]

**Carolina Tienda [1,\*], Jose A. Encinar [2], Mariano Barba [2] and Manuel Arrebola [3]**

[1]  Airbus Defence and Space, Gunnels Wood Road, Stevenage SG1 2AS, UK
[2]  Information, Processing and Telecommunications Center, Universidad Politécnica de Madrid (UPM), 28040 Madrid, Spain; jose.encinar@upm.es (J.A.E.); mbarba@etc.upm.es (M.B.)
[3]  Department of Electrical Engineering, University of Oviedo, 33203 Gijón, Spain; arrebola@uniovi.es
\*  Correspondence: Carolina.TIENDA-HERRERO@airbus.com; Tel.: +44-143-877-3087
†  The authors dedicate this work to the memory of their beloved friend and colleague, the late Mariano Barba Gea, who actively supported the team with his brilliant professional skills.

Received: 10 January 2018; Accepted: 31 March 2018; Published: 5 April 2018

**Abstract:** This article demonstrated an accurate analysis technique for dual-reflectarray antennas that take into account the angle of incidence of the impinging electric field on the main reflectarray cells. The reflected field on the sub and the main reflectarray surfaces is computed using Method of Moments in the spectral domain and assuming local periodicity. The sub-reflectarray is divided into groups of elements and the field radiated by each group is used to compute the incident and reflected field on the main reflectarray cells. A 50-cm demonstrator in Ku-band that provides European coverage has been designed, manufactured and tested to validate the analysis technique. The measured radiation patterns match the simulations and they fulfill the coverage requirements, achieving a cross-polar discrimination better than 25 dB in the frequency range: 12.975–14.25 GHz.

**Keywords:** dual-reflectarray; contoured beam; Ku-band

## 1. Introduction

Synthetic Aperture Radar [1] and long distance communications are two of the most popular and demanding applications for spaceborne antennas that require contoured beams [2,3], multibeam or beam scanning. Reflectarray antennas are a very competitive solution for these applications compared to conventional reflector and phased array antennas. The manufacturing photo-etching technology on multiple layer printed circuit boards, is a very controlled and well-known process. The bandwidth limitations of the printed elements and the differential spatial phase delay [4,5] can be overcome with several techniques like: multi-layer designs [6–8] multi-resonant elements, aperture-coupled patches with delay lines [9], or faceted configurations [10].

Dual-reflectarray antennas present some advantages respect to the well-known single offset ones [11], like the reduction of the volume for antenna geometries with a large focal distance. Another advantage is the possibility of using two reflectarray surfaces to increase the functionality of the antenna, or improve its electric characteristics. In the last years, a reflectarray as sub reflector has been combined with large main parabolic reflectors, the high gain characteristics of the reflector are combined with the degree of freedom of a reflectarray subreflector. A sub-reflectarray can improve the behavior of the antenna, either correcting the surface errors of the deployable surface [12], or providing beam scanning within certain angular range [12,13]. When using electronic controlled elements like MEMs [14–16], PIN diodes [17,18], varactor diodes [19] or liquid crystal elements [20–22] for frequencies higher than 100 GHz.

Antennas implemented with two reflectarrays [13,23,24], one as sub reflector and one as main reflector, provide a high number of degrees of freedom. In a dual-reflectarray antenna, the phase control in both surfaces can be used to improve the performance of the antenna in several ways, depending on the application.

Some of the advantages of using dual offset reflectarray antennas are: They have a high degree of compactness. They are useful to design multibeam antennas with a stable performance in all beams. By implementing the principle of bifocal antennas [25], the sub reflectarray can be used to change from linear to circular polarization or to re-configure the beam, while using the main reflectarray for additional functionalities, or improvements of the antenna performance. Some of the applications of this kind of antennas are Synthetic Aperture Radar [26], radiometric remote-sensing and multibeam antennas for mobile communications.

Reference [24] reported an analysis technique for dual-reflectarray antennas. The reflected coefficient of the elements in both reflectarrays were computed by Spectral-Domain Method of Moments, assuming local periodicity. The reflection coefficient is a matrix that relates the incident and reflected fields of every cell of the main or the sub reflectarrays.

The accurate analysis of a dual-reflectarray antenna is complicated and requires a high computational effort. For every single pair combination of elements of the sub and the main reflectarray, a reflection coefficient matrix must be computed. If $NMR$ and $NSR$ are the number of cells in the main and sub reflectarrays respectively, $NMR \times NSR$ would be the number of reflection coefficient matrices that must be computed for an accurate analysis of the antenna.

The incident electric field on every cell of the main reflectarray was approximated by a single plane wave. In the fix incident angle approximation, the reflected coefficient matrices of the main reflectarray are computed $NMR$ times instead of $NMR \times NSR$ times as it is done in the full wave analysis. This approximation reduces significantly the computation time. In this approximation, the reflected field on every cell of the main reflectarray is computed only once, assuming the incident angle as the one coming from the center of the sub reflectarray, as the superposition of all the field contributions from every cell of the sub reflectarray. There are some antenna geometries where the reflected field on every cell on the main reflectarray is practically not affected by the incident angle from different sub reflectarray cells. In that cases, the approximation of the fix incident angle is valid, like the compact dual-reflectarray antenna in Ku-band to produce a pencil beam presented in [27]. In that antenna, the elements in the main reflectarray were chosen to be less sensitive to the incident angles, the period was very small ($0.4\lambda \times 0.4\lambda$ at 15 GHz) and even when having incident angles up to $50°$ in some elements at the border of the main reflectarray, a good agreement between measurements and simulations was achieved. However, in other antenna configurations, the approximation of the fix angle used in the main reflectarray may not be accurate enough. When using sensitive elements to the incident angle, large antennas or contoured beam antennas. In this case, for Direct Broadcast Satellite (DBS) applications [2,3,28], phase errors may produce non acceptable distortions on the contoured radiation pattern.

A general analysis technique is presented in this paper to provide an accurate model for dual-reflectarray antennas. The incident angles of every element of the sub and main reflectarrays are considered in the analysis, improving significantly the accuracy in the prediction of the radiation pattern. The analysis technique presented in [24] is a particular and simplified case of the general analysis technique presented in this article. Particularly it corresponds to the approximate and less accurate case and cannot be used for a good prediction of contoured beams and crosspolar component of the radiation patterns, which is shown in this paper. A dual-reflectarray technique for contoured beam applications has been implemented. A demonstrator to provide DBS European coverage has been designed, manufactured and tested. The size of the demonstrator was limited to 50 cm to simplify the manufacturing process. The antenna geometry was selected to limit the phase variation on both reflectarray surfaces to $360°$ and avoid the need of multi-layer phase-delay elements. This paper demonstrates that a dual-reflectarray antenna can produce a contoured beam, emulating a dual offset parabolic reflector of a large $F/D$ ($F/D = 4$). The antenna was manufactured and tested at

Universidad Politécnica de Madrid, achieving good agreement between simulations and measurements in the whole designed frequency band (12.975–14.250 GHz).

## 2. Analysis Technique for Dual-Reflectarray Antennas

The proposed analysis technique is used for a dual-reflectarray antenna that consists of a flat sub-reflectarray (SRA) illuminated with a feed horn and a flat main reflectarray (MRA). Both reflectarray surfaces are based on phasing elements of any type implemented with one or more layers of printed patches arranged in a regular lattice. The technique consists of four steps: first, the computation of the incident field on the SRA coming from the primary feed horn. Second, the analysis of the SRA using SD-MoM, assuming local periodicity, and taking into account the angle of incidence of the electric field coming from the feed-horn. Third, the SRA is divided into groups of elements (sub-arrays), and the elements of the MRA are analyzed considering the incident angle and the field contribution coming from each of the SRA groups, using the same technique as for the SRA: SD-MoM and local periodicity. The last step consists on the computation of the radiation pattern, through a 2D-FFT of the reflected field on the MRA. This is a very general approach since it allows adjusting the level of accuracy in the analysis considering a higher or lower number of groups in the SRA. The election of the number of groups is done depending on the antenna geometry demand, based on the maximum angles of incidence and the periodicity. The reflection coefficients of every cell in the MRA $(m, n)$, due to the plane wave coming from a cell of the SRA $(p, q)$, are giving by a matrix $R_{(p,q)}(m, n)$. The coefficients of this matrix depend on which defines the maximum angles of incidence on the MRA, on the period and type of cells used in the MRA. The CPU time computation of the $R_{(p,q)}(m, n)$ matrix is very high since all combinations of cells $(m, n)$-$(p, q)$ must be calculated.

For some antenna geometries where the MRA is in the Fresnel zone of the SRA, the incident angles of the field contributions from different cells are very similar. These incident angles can be approximated as done in [24] where $R_{(p,q)}(m, n) \approx R(m, n)$, being $R(m, n)$ the reflection matrix corresponding to the incident field coming from the center of the SRA. For those antenna geometries with the cells of the MRA in the near field of the SRA, although the contributions from neighbor cells are still similar, the variation in the incidence angle of the field contributions from separated cells can be important. In this case the SRA is divided in groups of elements. The field contribution on every cell of the MRA is computed for each group of elements of the SRA and the incidence angle will be different for the contribution of each group impinges. This angle is considered for the computation of the reflection matrix of each pair SRA group and MRA cell.

The reflected coefficients on every cell of the MRA, are computed as many times as the number of groups in which the SRA is divided. The size of the groups is selected in such a way, that the far field conditions on every cell of the MRA are fulfilled for every group of the SRA.

All the field contributions of the elements belonging to the same group are added up before computing its corresponding reflected coefficient matrix. The incident angle is calculated assuming that each of those contributions starts in the center of its group. Equations (1) and (2) show the incident and reflected field on every $(m,n)$ cell of the MRA due to one group of cells of the SRA:

$$E_{inc(g)}^{X/Y}(m, n) = \sum_{r=1}^{Ngx}\sum_{s=1}^{Ngy} E_{inc(r,s)}^{X/Y}(m, n), \tag{1}$$

being $Ngx$ and $Ngy$ the number of elements for each group $(g)$ in the $x$ and $y$ axis respectively. $E_{inc(g)}^{X/Y}(m, n)$ is the electric field impinging on the cell $(m,n)$ of the MRA from the group of cells $(g)$ of the SRA, according to Figure 1, for $X$ or $Y$ polarization. The tangential reflected field at each cell $(m, n)$ of the MRA is:

$$E_{ref}^{X/Y}(m, n) = \sum_{g=1}^{NG} R_g(m, n) \cdot E_{inc(g)}^{X/Y}(m, n), \tag{2}$$

where:

$$R_{(g)}(m,n) = \begin{pmatrix} \rho_{(g)\,xx}^{m,n} & \rho_{(g)\,yx}^{m,n} \\ \rho_{(g)\,xy}^{m,n} & \rho_{(g)\,yy}^{m,n} \end{pmatrix}. \tag{3}$$

$R_{(g)}(m,n)$ is the reflection matrix for the cell $(m,n)$ of the MRA due to the group $g$ of the SRA and NG is the number of groups. The components $\rho_{(g)\,xx}^{m,n}$ and $\rho_{(g)\,yx}^{m,n}$ are the direct and cross polarization reflection coefficients for an incident wave polarized in $X_{SR}$ direction (without component in $Y_{SR}$). On the other hand, $\rho_{(g)\,yy}^{m,n}$ and $\rho_{(g)\,xy}^{m,n}$ are the reflection coefficients for incident electric with $Y_{SR}$ polarization.

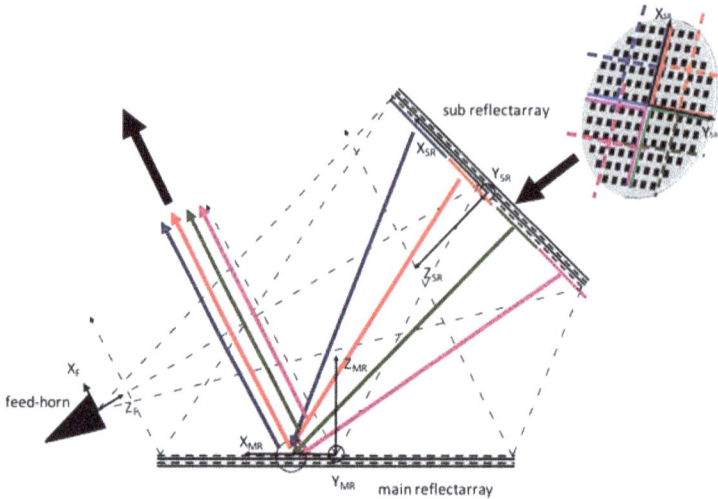

**Figure 1.** Scheme of the dual-reflectarray designed antenna with the sub-reflectarray divided in groups of elements.

The discussed analysis technique is a general approach; the two extreme cases correspond to a full-wave analysis and the fixed single angle analysis. In the first one, each group of elements contains only one element and in the second one, there is only one group of elements including all cells of the SRA. In the full-wave analysis, the incident angle of the electric field coming from each cell of the SRA impinging on each cell of the MRA, is taken into account. In the fix-angle case, there is only one angle of incidence to consider from all the field contributions from the SRA elements as it was considered in [24].

Finally, the 2D Inverse Discrete Transform computes the co-polar and cross-polar components of the radiation patterns by using an 2D-FFT based algorithm.

## 3. Design of a Contoured Beam Dual-Reflectarray Antenna

A demonstrator with limited size has been manufactured and tested to provide a contoured beam for European coverage. The antenna geometry main parameters are summarized in Table 1, they are similar to the antenna previously reported in [27]. The antenna geometry, including the relative positions of the feed, the SRA and the MRA remain equal to the antenna presented in [27]. The phase distribution on the SRA is also the same and emulates a hyperboloid to produce an equivalent parabolic reflector system with $F/D = 4$. In this design the main difference lies on the phase distribution of the MRA that has been synthesized to provide a contoured beam in order to fulfill the requirements of a DBS mission.

**Table 1.** Geometrical parameters of the dual-reflectarray antenna.

| Main Reflectarray (MRA) | |
|---|---|
| Main-Reflectarray size | 492 mm × 444 mm |
| Rows × Columns | 41 × 37 elements |
| Period | 12 mm × 12 mm |
| **Sub-Reflectarray (SRA)** | |
| Center of the coordinate system | (−217, 0, 370) mm |
| Direction cosines matrix (relates MRA and SRA Coordinate Systems) | $\begin{bmatrix} 0.715 & 0 & 0.698 \\ 0 & -1 & 0 \\ 0.698 & 0 & -0.715 \end{bmatrix}$ |
| Sub-Reflectarray size | 380 mm × 380 mm |
| Rows × Columns | 38 × 38 elements |
| **Feed-Horn** | |
| Coordinates of phase-center | (193.73, 0, 635.54) mm |
| Intersection of feed axis and SRA | (15, 0, 0) mm |
| Coordinates of phase-center | (193.73, 0, 635.54) mm |

A dual-linear polarized corrugated horn illuminates the SRA with the electric field polarized in X- and Y-axis of the feed coordinate system. The antenna is accommodated with the Y-axis of the MRA pointing to the North of the Earth, the V-polarized field is parallel to the Y-axis of the MRA. Using a dual circular horn, the antenna could operate in dual circular polarization since both linear polarizations are provided to the coverage. In the antenna geometry presented, the SRA is in the near field of the horn. A near field model of the feed-horn [29–32] has been used to make an accurate analysis, based on the spherical mode wave expansion of the pattern. The feed illuminates the SRA with a taper of −13.48 dB at the edges and the MRA is illuminated with −16.24 dB. The phase distribution on the SRA is computed using ray tracing and the phase shift in each cell is calculated as a function of the difference between the cell to the feed and the virtual focus of the antenna [27], as shown in Figure 1.

### 3.1. Phase Synthesis

The phase synthesized provide an European coverage from a geostationary satellite located at 10° E longitude and 0° latitude. A minimum required gain of 25 dBi including beam pointing error in the enlarge region is defined. The phase distribution on the MRA is synthesized to provide the coverage defined by the blue line in Figure 2.

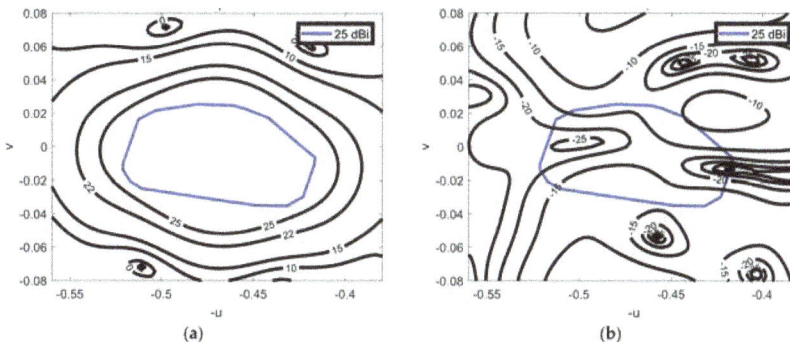

**Figure 2.** Synthesized pattern for vertical polarization at 13.75 GHz superimposed to the coverage mask: (**a**) co-polar component; (**b**) cross-polar component.

The synthesis of the phase distribution on the MRA to produce a contoured beam is done using the technique based on the intersection approach algorithm reported in [32]. In the SRA, the phase-shift is calculated as a function of the difference between two distances: the distance between the virtual focus of the equivalent parabolic system and every cell of the SRA, and the distance from the feed-horn and every cell of the SRA. Figure 2 shows the superposition of the required mask for the European coverage, and the computed radiation pattern using the phase-shift distribution that fulfills the requirements. The radiation pattern fulfills the 25 dBi gain requirements, but it does not follow the gain contours of the mask accurately due to the limited size of the demonstrator.

### 3.2. Design of Main- and Sub-Reflectarrays

After computing the illumination and the phase distribution on both surfaces, next step consists on the design of both SRA and MRA. The dimensions of the patches are computed assuming local periodicity and the iterative process based on SD-MoM reported in [6]. The design has been carried out for dual linear polarization at 13.75 GHz. Table 2 shows the sandwich configuration where the same thickness and electric properties are considered for both reflectarray surfaces. The same sandwich structure with two layers of printed patches in a rectangular lattice is used for the SRA and MRA. The period of the regular layout for the MRA is 12 mm × 12 mm, it is arranged in 41 × 37 elements. For the SRA, the period is 10 mm × 10 mm arranged in a grid layout of 38 × 38 elements.

**Table 2.** Lay-up of the sandwich and radio-electric characteristics of the materials.

| Dielectric/Metallization | $\varepsilon_r$ | Tag $\delta$ | Thickness (mm) |
|---|---|---|---|
| Copper (Printed Patch Layer 2) | - | - | 0.05 |
| (SRA) Cuclad 217LX/(MRA)Diclad880B | 2.17 | 0.0009 | 1.524 |
| (SRA/MRA)Cuclad6250 (Thermoplastic bonding film) | 2.32 | 0.0013 | 0.038 |
| Copper (Printed Patch Layer 1) | - | - | 0.05 |
| (SRA) Cuclad 217LX/(MRA)Diclad880B | 2.17 | 0.0009 | 1.524 |
| Copper (Ground Plane) | - | - | 0.05 |

Figure 3 shows the phase-shift as a function of the patch size at the frequency design (13.75 GHz) for 0°, 30° and 50° incident angle. The curves show a very linear variation over 400° on patches with a lateral size between 4 mm and 9 mm on the first layer, for up to 50° incident angles for both linear polarizations.

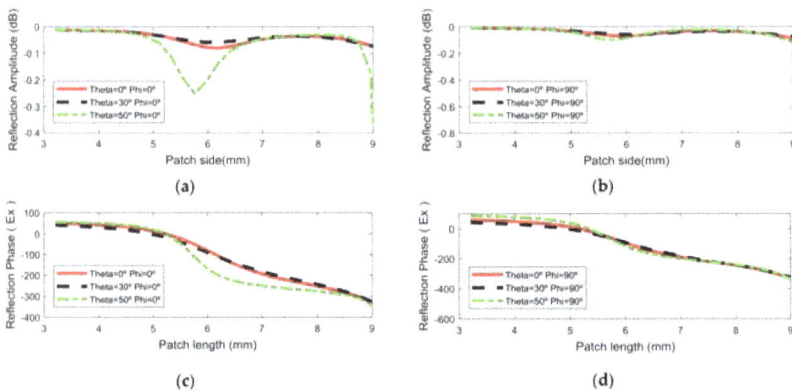

**Figure 3.** Reflection coefficient vs. The patch length in the first layer for the cell used in the MRA for: (**a**) Amplitude and; (**c**) Phase of EX field component at φ = 0° (Horizontal polarization) and; (**b**) Amplitude and (**d**) Phase of EY field component, φ = 90° (Vertical polarization) at 13.75 GHz.

The average losses for the patch sandwich are lower than 0.1 dB. In Figure 3c, there are phase variations of up to 80° in the MRA and 30° in the SRA for 50° incidence angles. For 9 mm patch lengths and 50° incident angles, the losses increase drastically due to the resonance, these patches are closer to the edge of the reflectarray where the illumination is 10 dB lower than in the center, their effect is negligible.

## 4. Comparison between Simulations and Experimental Results

For the validation of the process, the designed dual-reflectarray antenna has been analyzed, manufactured and then tested in an anechoic chamber. The analysis has been carried out with different degrees of accuracy to demonstrate the efficiency of the analysis technique. In the first analysis, groups of one element in the SRA are considered, what it has been previously mentioned as full-wave analysis.

Then, the approximation of the fix incident angle is used in the analysis where only one group of elements in the SRA is considered. Then the antenna is analyzed when dividing the SRA in a progressive higher number of groups, increasing the accuracy in the radiation pattern prediction, until getting closer to the results obtained with the full wave analysis, ensuring the convergence of the analysis method. To compute the optimum number of groups in the SRA that makes the results to converge to the full-wave case analysis, with a reasonable computational time. The maximum group size in the SRA is calculated to fulfill the far field conditions with every cell of the MRA. Considering the minimum distance between the sub and the main reflectarray, 236 mm, the diameter of one group in the sub must be less than 50 mm to fulfill far-field conditions and between 50 mm and 140 mm to be in the Fresnel zone. With a period of 10 mm in the SRA, groups of 5 × 5 elements would be in the far field of the cells in the MRA. The SRA has 38 × 38 elements, when making groups of 13 × 13 elements, 9 groups are considered in the analysis, when making groups of 8 × 8 elements, 25 groups are considered in the analysis. Still, with 25 groups, the elements of the MRA closer to the SRA are in the Fresnel Zone, but most of the elements in the MRA are in the far field of the 25 groups of the SRA.

Figure 4 shows the copolar component of the radiation pattern for several cases: for the full-wave analysis (blue line), when the SRA is divided into one (black line), nine (red line), 25 (green line) groups, for both linear polarizations. When the SRA is divided into nine groups, a good agreement in the main beam is achieved, when it is divided in 25 groups, the green line converges to the full-wave case. For vertical polarization, the fix-angle approach is good since the reflectarray elements are not very sensitive to the incidence angle for 9 mm patch dimensions. Performing an analysis on an antenna, like the one presented in the paper requires a computational time of 100 h in the case of the full-wave analysis and 3 min using the approximated method, using the same computer set-up.

As shown in Figure 5, the prototype of the antenna has been measured in the Compact Antenna Test Range (CATR) of Universidad Politécnica de Madrid in an anechoic chamber. In this measurement system, the probe is placed at the focus of a parabolic reflector so that, the reflector transforms the spherical wave coming from the probe to a plane wave that illuminates the Antenna Under Test (AUT). This system permits to measure the antenna in far field conditions in a smaller room that a conventional spherical range system. The AUT is placed in the quite zone of the test range which is the volume where the plane wave conditions are fulfilled. The AUT can be rotated along two axes (roll and azimuth) so the entire sphere can be measured by the probe, while the RX antenna can be rotated along its axis so the co and crosspolar patterns can be measured.

The measurements and the simulated radiation pattern fulfill the requirements for the three measured frequencies (12.975, 13.75 and 14.25 GHz). Good agreement between simulations (25 groups in the SRA) and measurements at the central frequency is achieved. The small misalignment between the layers of the printed patches on the MRA produces small mismatching between simulations and measurements. Those misalignments produced up to 30° phase errors in some areas of the reflectarray where the incident angles are high. When considering theses phase errors in the analysis of the antenna and making a new comparison with the measurements, much better agreement was achieved as shown in Figure 6c,d.

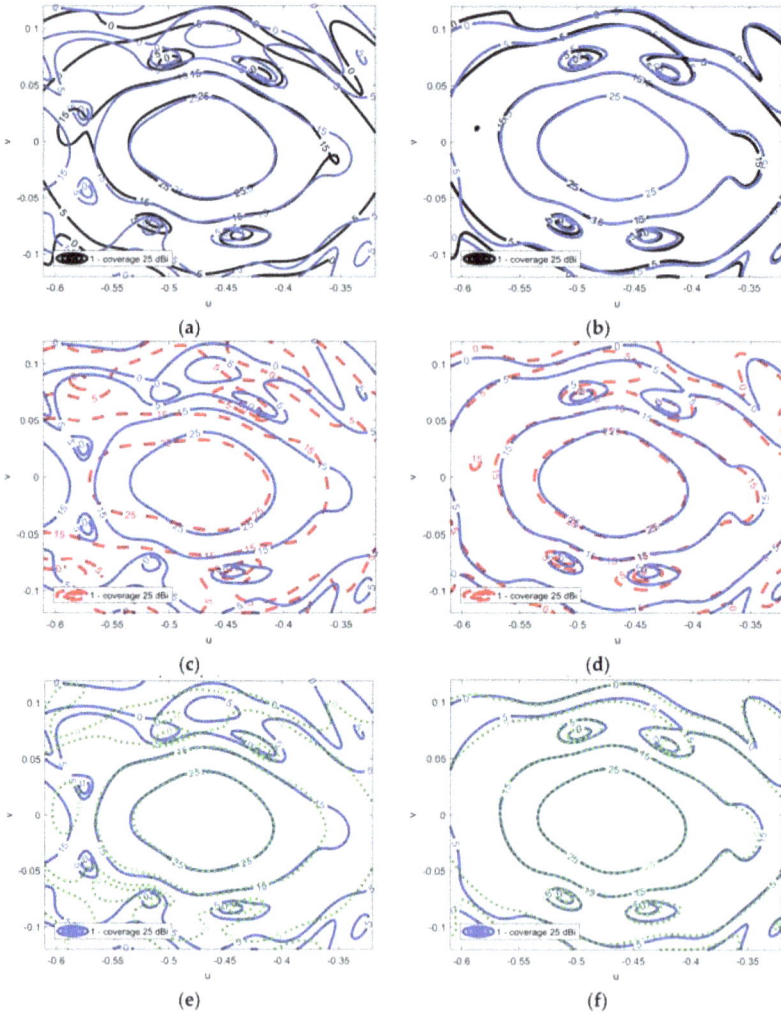

(a)

(b)

(c)

(d)

(e)

(f)

**Figure 4. (a)** Contour plots of the copolar components of the radiation pattern full wave analysis (solid blue line (-)) and approximate case (dashed black line (–)) vertical polarization; **(b)** for horizontal polarization. **(c)** Contour plots of the copolar components of the radiation pattern simulated for the full wave analysis (solid blue line (-)) and the more general approach dividing the SRA in 9 (dashed dot red line (-)) for vertical polarization and **(d)** for horizontal polarization. **(e)** Contour plots of the copolar components of the radiation pattern simulated for the full wave analysis (solid blue line (-)) and the more general approach dividing the SRA in 25 (dashed green line (:)) for vertical polarization and **(f)** for horizontal polarization.

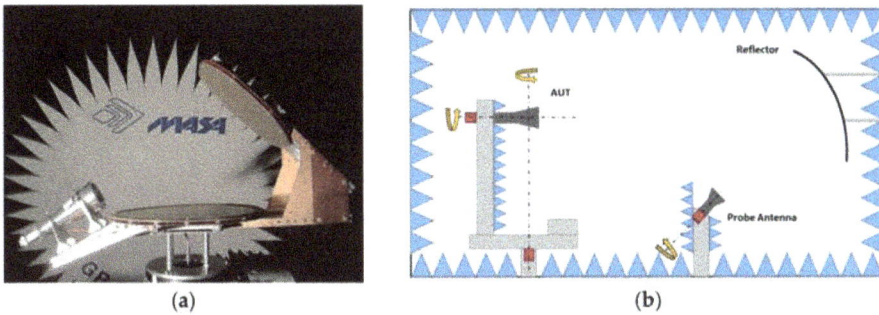

**Figure 5.** (**a**) Dual-reflectarray antenna for contoured beam in the compact antenna test rage (anechoic chamber) (**b**) Sketch of the CATR measurement system.

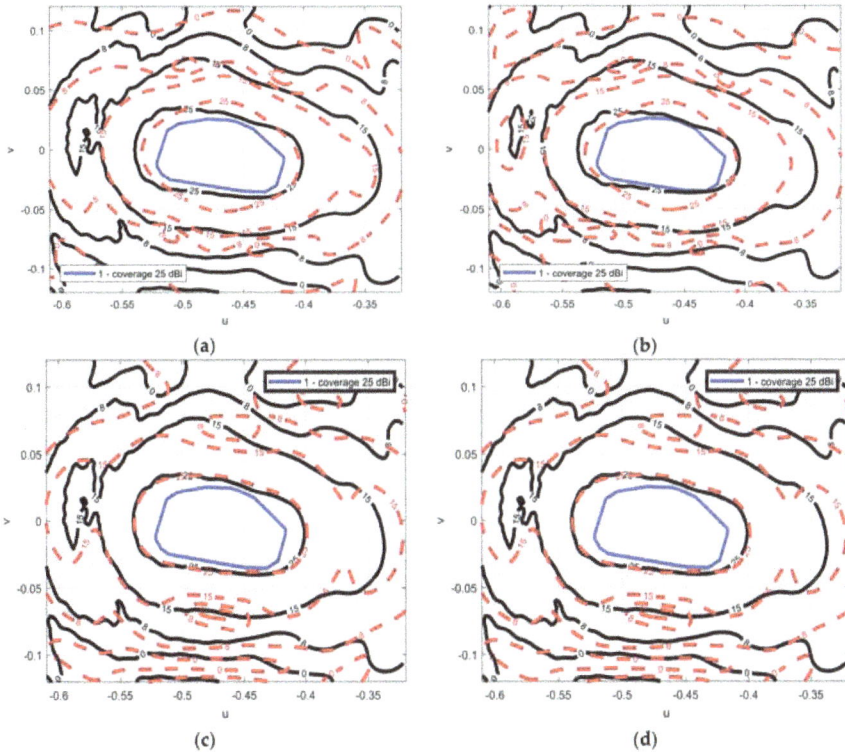

**Figure 6.** (**a**) Contour plots of the simulated copolar components of the radiation pattern (dashed red line (–)), measurements (solid black line (-)) and the mask (thicker solid blue line) for vertical polarization at 13.75 GHz; (**b**) for horizontal polarization at 13.75 GHz; (**c**) The simulations are repeated accounting for the phase errors (red line) and compared with measurements for vertical polarization; (**d**) for horizontal polarization.

The antenna was measured in the frequency range (12.975–14.25 GHz), as Figure 7 points out, a very stable performance in the required frequency band was achieved. The cross polar discrimination is larger than 25 dB, Figure 8 shows the crosspolar component of the radiation pattern for both linear polarizations at 13.75 GHz.

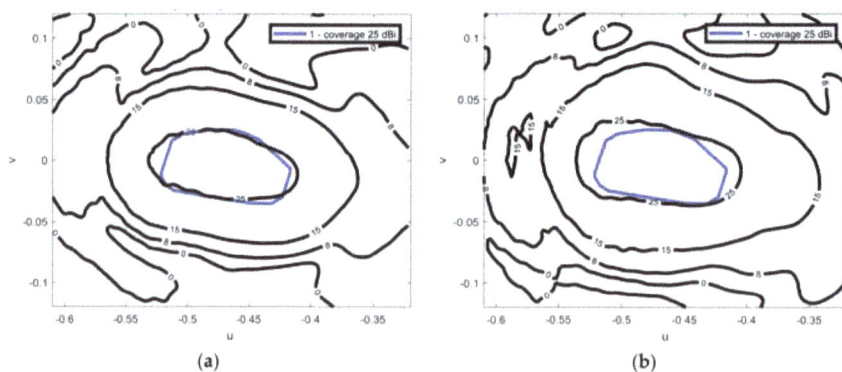

(a)                                    (b)

**Figure 7.** Measured contoured patterns in dBi (solid black line) superimposed to the coverage mask (thicker solid blue line) for vertical polarizations: (**a**) at 12.975 GHz; (**b**) at 14.25 GHz.

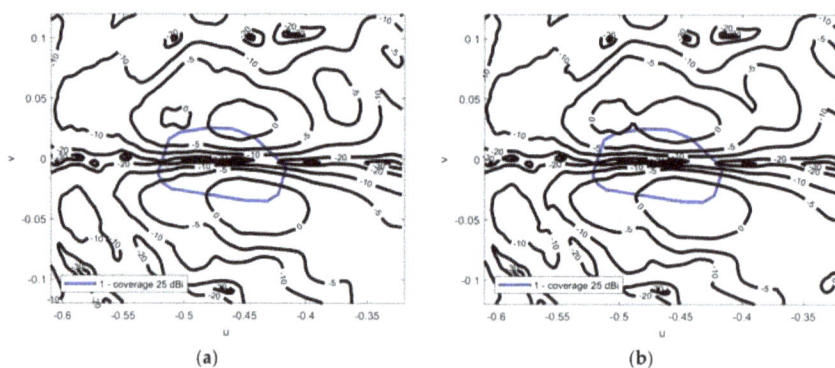

(a)                                    (b)

**Figure 8.** (**a**) Measured contoured plots of the cross-polar components of the radiation pattern (solid black line) superimposed to the coverage mask (thicker solid blue line) for vertical polarization at 13.75 GHz; (**b**) for horizontal polarization at 13.75 GHz.

## 5. Conclusions

This article shows the validation of a general and accurate analysis technique for dual-reflectarray antennas. When computing the incident and reflected field on every cell of the MRA with this technique, the subreflectarray (SRA) is divided in groups of elements. The contribution of each group is considered for every element of the main reflectarray (MRA) that is analyzed. An increase on the number of groups derives in an increase of the computation time but also an improvement of the accuracy. A limited size demonstrator was designed, manufactured and tested to validate the analysis technique. The MRA size is 450 mm × 500 mm and it provides a contoured beam for DBS European coverage with dual linear polarization. The cross-polarization discrimination is higher than 25 dB and remains stable within the frequency range: from 12.975 GHz to 14.25 GHz. Although, the comparison between the simulations and the measurements shows a good agreement, there are slight differences in the shape of the patterns due to manufacturing tolerance errors.

This paper demonstrated the capability of dual-reflectarray antennas to provide contoured beams with low levels of cross-polarization in Ku-band. A similar configuration could be used for multi-spot antennas in Ka-band, since the simultaneous phase adjustment in two surfaces can be used for multi focal designs.

**Acknowledgments:** This work was supported in part by the European Space Agency (ESA) under contract ESTEC/AO/1-7064/12/NL/MH (RAIPAD); by the Ministerio de Economía y Competitividad, under projects TEC2016-75103-C2-1-R (MYRADA) and TEC2017-86619-R (ARTEINE); and by the Regional Government of Madrid under project SPACERADAR-CM (P2013/ICE-3000).

**Author Contributions:** C.T., J.A.E. and M.A. conceived the antenna; C.T. developed the analysis tool and designed the antenna; C.T. and M.A. defined the antenna optics and carried out the pattern synthesis; M.B. manufactured the antenna and performed the experiments; C.T. and J.A.E. analyzed the data; C.T. wrote the paper.

**Conflicts of Interest:** The authors declare no conflict of interest.

## References

1. Di Maria, A.; Limbach, M.; Horn, R.; Reigber, A. Reflectarray membrane study for deployable SAR antenna. In Proceedings of the 3rd European Conference on Antennas and Propagation—EuCAP 2009, Berlin, Germany, 23–27 March 2009.
2. Pozar, D.M.; Targonski, S.D.; Pokuls, R. A shaped-beam microstrip patch reflectarray. *IEEE Trans. Antennas Propag.* **1999**, *47*, 1167–1173. [CrossRef]
3. Encinar, J.A. Design of two-layer printed reflectarrays using patches of variable size. *IEEE Trans. Antennas Propag.* **2001**, *49*, 1403–1410. [CrossRef]
4. Zhou, M.; Sorensen, S.B.; Kim, O.S.; Jorgensen, E.; Meincke, P.; Breinbjerg, O. Direct Optimization of Printed Reflectarrays for Contoured Beam Satellite Antenna Applications. *IEEE Trans. Antennas Propag.* **2013**, *61*, 1995–2004. [CrossRef]
5. Huang, J. Bandwidth study of microstrip reflectarray and a novel phased reflectarray concept. In Proceedings of the AP-S. Digest Antennas and Propagation Society International Symposium, Newport Beach, CA, USA, 18–23 June 1995; pp. 582–585.
6. Pozar, D.M. Bandwidth of reflectarrays. *Electron. Lett.* **2003**, *39*, 1490–1491. [CrossRef]
7. Nayeri, P.; Yang, F.; Elsherbeni, A.Z. Broadband Reflectarray Antennas Using Double-Layer Subwavelength Patch Elements. *IEEE Antennas Wirel. Propag. Lett.* **2010**, *9*, 1139–1142. [CrossRef]
8. Chaharmir, M.R.; Shaker, J.; Cuhaci, M.; Ittipiboon, A. A broadband reflectarray antenna with double square rings. *Microw. Opt. Technol. Lett.* **2006**, *48*, 1317–1320. [CrossRef]
9. Carrasco, E.; Encinar, J.A.; Barba, M. Bandwidth improvement in large reflectarrays by using true–time delay. *IEEE Trans. Antennas Propag.* **2008**, *56*, 2496–2503. [CrossRef]
10. Legay, H.; Bresciani, D.; Labiole, E.; Chiniard, R.; Girard, E.; Caille, G.; Calas, D.; Gillard, R.; Toso, R. A 1.3 m facetted reflectarray in Ku band. In Proceedings of the 15th International Symposium on Antenna Technology and Applied Electromagnetics (ANTEM), Toulouse, France, 25–28 June 2012; pp. 1–4.
11. Khayatian, B.; Rahmat-Samii, Y.; Huang, J. Radiation characteristics of reflectarray antennas: Methodology and applications to dual configurations. In Proceeding of the First European Conference on Antennas and Propagation, Nice, France, 6–10 November 2006.
12. Menudier, C.; Koleck, T. Sub-Reflectarrays Performances for Reconfigurable Coverages. *IEEE Trans. Antennas Propag.* **2012**, *60*, 3476–3481. [CrossRef]
13. Wenfei, H.; Arrebola, M.; Cahill, R.; Encinar, J.A.; Fusco, V.; Gamble, H.S.; Alvarez, H.Y.; Las-Heras, F. 94 GHz dual-reflector antenna with reflectarray sub-reflector. *IEEE Trans. Antennas Propag.* **2009**, *57*, 3043–3050. [CrossRef]
14. Sorrentino, R. MEMS-based reconfigurable reflectarrays. In Proceedings of the Second European Conference on Antennas and Propagation, Edinburgh, UK, 11–16 November 2007.
15. Perruisseau-Carrier, J.; Skriverviky, A.K. Monolithic MEMS-based reflectarray cell digitally reconfigurable over a 360-phase range. *IEEE Antennas Wirel. Propag. Lett.* **2008**, *7*, 138–141.
16. Bayraktar, O.; Civi, O.A.; Akin, T. Beam Switching Reflectarray Monolithically Integrated with RF MEMS Switches. *IEEE Trans. Antennas Propag.* **2012**, *60*, 854–862. [CrossRef]

*Sensors* **2018**, *18*, 1100

17. Yang, H.; Yang, F.; Cao, X.; Xu, S.; Gao, J.; Chen, X.; Li, M.; Li, T. A 1600-Element Dual-Frequency Electronically Reconfigurable Reflectarray at X/Ku-Band. *IEEE Trans. Antennas Propag.* **2009**, *57*, 364–372. [CrossRef]

18. Riel, M.; Laurin, J.J. Design of an electronically beam scanning reflectarray using aperture–coupled elements. *IEEE Trans. Antennas Propag.* **2007**, *55*, 1260–1266. [CrossRef]

19. Hum, S.V.; Okoniewski, M.; Davies, R.J. Realizing an electronically tunable reflectarray using varactor diode–tuned elements. *IEEE Microw. Wirel. Compon. Lett.* **2005**, *15*, 422–424. [CrossRef]

20. Mössinger, A.; Marin, R.; Mueller, S.; Freese, J.; Jakoby, R. Electronically reconfigurable reflectarrays with nematic liquid crystals. *Electron. Lett.* **2006**, *42*, 899–900. [CrossRef]

21. Bildik, S.; Dieter, S.; Fritzsch, C.; Menzel, W.; Jakoby, R. Reconfigurable Folded Reflectarray Antenna Based Upon Liquid Crystal Technology. *IEEE Trans. Antennas Propag.* **2015**, *63*, 122–132. [CrossRef]

22. Perez-Palomino, G.; Baine, P.; Dickie, R.; Bain, M.; Encinar, J.A.; Cahill, R.; Barba, M.; Toso, G. Design and experimental validation of liquid crystal-based reconfigurable reflectarray elements with improved bandwidth in F-Band. *IEEE Trans. Antennas Propag.* **2013**, *61*, 1704–1713. [CrossRef]

23. Xu, S.; Rahmat-Sami, Y.; Imbriale, W.A. Subreflectarrays for reflector surface distortion compensation. *IEEE Trans. Antennas Propag.* **2009**, *57*, 364–372. [CrossRef]

24. Tienda, C.; Arrebola, M.; Encinar, J.A.; Toso, G. Analysis of a dual-reflectarray antenna. *IET Microw. Antennas Propag.* **2011**, *5*, 1636–1645. [CrossRef]

25. Menzel, W.; Al-Tikriti, M.; Leberer, R. A 76 GHz multiple-beam planar reflector antenna. In Proceedings of the 32nd European Microwave Conference, Milano, Italy, 23–26 September 2002; pp. 977–980.

26. Braun, H.M.; Knobloch, P.E. SAR on small satellites-shown on the SAR-Lupe example. In Proceedings of the International Radar Symposium, Cologne, Germany, 5–7 September 2007.

27. Tienda, C.; Encinar, J.A.; Arrebola, M.; Barba, M.; Carrasco, E. Design, manufacturing and test of a dual-reflectarray antenna with improved bandwidth and reduced cross-polarization. *IEEE Trans. Antennas Propag.* **2012**, *61*, 1180–1190. [CrossRef]

28. Encinar, J.A.; Arrebola, M.; de la Fuente, L.F.; Toso, G. A transmit–receive reflectarray antenna for direct broadcast satellite applications. *IEEE Trans. Antennas Propag.* **2011**, *59*, 3255–3264. [CrossRef]

29. Hansen, J.E. *Spherical Near-Field Antenna Measurements*; IET Electromagnetic Waves Series 26; IET: Stevenage, UK, 1988.

30. Jorgensen, E.; Lumholt, M.; Meincke, P.; Zhou, M.; Sorensen, S.B.; Borries, O.; Cappellin, C.; Frandsen, P.E. New Modelling capabilities in commercial software for high-gain antennas. In Proceedings of the 6th European Conference on Antennas and Propagation (EUCAP), Prague, Czech Republic, 26–30 March 2012.

31. Pontoppidan, K. *Technical Description of GRASP8 (TICRA)*; TICRA: Copenhagen, Denmark, 2003.

32. Bucci, O.; Franceschetti, G.; Mazzarella, G.; Panariello, G. Intersection approach to array pattern synthesis. *IEE Proc.* **1990**, *137*, 349–357. [CrossRef]

MDPI

St. Alban-Anlage 66

4052 Basel, Switzerland

Tel. +41 61 683 77 34

Fax +41 61 302 89 18

http://www.mdpi.com

*Sensors* Editorial Office

E-mail: sensors@mdpi.com

http://www.mdpi.com/journal/sensors

www.ingramcontent.com/pod-product-compliance
Lightning Source LLC
Chambersburg PA
CBHW051707210326
41597CB00032B/5403